Neutronics of Advanced Nuclear Systems

Yican Wu

Neutronics of Advanced Nuclear Systems

Springer

Yican Wu
Institute of Nuclear Energy Safety
Technology (INEST)
Chinese Academy of Sciences
Hefei, Anhui, China

ISBN 978-981-13-6519-5 ISBN 978-981-13-6520-1 (eBook)
https://doi.org/10.1007/978-981-13-6520-1

Library of Congress Control Number: 2019931510

This Springer imprint is published by the registered company Springer Nature Singapore Pte Ltd.
The registered company address is: 152 Beach Road, #21-01/04 Gateway East, Singapore 189721, Singapore

Foreword I

Energy is essential for the survival and the quality of human life. Nuclear energy is expected to be efficient and with low carbon content, opening an important role in the energy choices. However, three severe nuclear accidents (Three Miles Island, Chernobyl, and Fukushima) have significantly slowed down the pace of nuclear energy development in the world. Furthermore, the current nuclear energy raises problems of radioactive waste disposal and uranium shortage.

Therefore, the development of new concepts of a nuclear energy with much higher levels of safety represents a major progress. Advanced nuclear systems including advanced fission, fusion and hybrid systems have been proposed in view of their high sustainability, safety and reliability, economic efficiency as well as proliferation resistance. Advanced nuclear energy is attracting worldwide increasing attention.

The discovery of the neutron has opened the door to the practical use of nuclear energy. The neutron is a key factor in the development of advanced nuclear systems, triggering the fission reactions and as the main energy carrier of fusion reactions.

With the help of breeding reactions of a neutron and nuclide Li-7, of fertile nuclides U-238 or Th-232, nuclear tritium or fissile nuclides fuels could be bred, solving the problems of the limited resources of nuclear fuels. Transmutation is also an important approach to reduce the nuclear waste, since minor actinides and long-lived fission products are transformed into short-lived or stable nuclides. Furthermore, with the help of a better control of neutrons, it would be easier to decrease radiation risks such as material irradiation damage, radioactive waste, occupational radiation exposure, radiation related environmental impact. The development of advanced nuclear systems strongly depends on the technology of generation, control, utilization, and protection of neutrons, calling for more research on the neutronics of the advanced nuclear systems.

The author, Prof. Yican Wu has devoted more than 30 years to research in advanced nuclear system and he is a world-recognized leader in the field of neutronics. He and his team have performed systematic researches in the advanced nuclear systems, including innovative neutronics simulation methods and codes, the neutron source and related technologies, the advanced nuclear system concepts and the R&D technologies as described in this book. Accurate modeling and calculations with multi-processes direct coupling have been developed and implemented in the unique neutronics simulation code SuperMC, overcoming the main challenges in complex nuclear system design & safety analysis. This has been accepted as a very important tool of this field. The High Intensity D–T Neutron Generator HINEG has achieved the highest level neutron yield in the world among the accelerator-based fusion neutron sources. This contribution represents a highlight in the field of nuclear science and technology in the recent years and it will promote the further development and validation of advanced nuclear systems. Based on neutron physics and technology, the safety of the fusion systems have been systematically illustrated. Several concepts of China Lead-based Reactors (CLEAR) have been proposed, with special reference to the Chinese mini-reactor HeDianBao. The implementation of such innovative mini-reactor with inherent safety and high efficiency in fuel utilization may bring remarkable changes in the future use of nuclear energy. Furthermore, those technologies have been extendedly applied in order to develop the radiotherapy system KylinRay, which provides cutting-edge solutions for accurate radiotherapy and new hopes for cancer patients. These creative achievements have represented a major contribution to the progress in recent years in the field of neutronics and of advanced nuclear system.

I am very impressed by the comprehensive and clear book of Prof. Wu. I believe it will effectively promote the research of advanced nuclear energy and I recommend this as a reference in the field.

Geneva, Switzerland

Carlo Rubbia
Nobel Laureate, Physics 1984
Former Director General of CERN

Foreword II

Nuclear energy plays an important role in securing environmental and climate-friendly energy supply. Currently, there are more than 450 nuclear power plants in operation around the world, producing 11% of the world's electricity, and nuclear energy is clearly recognized as a key technology in the fighting against climate change. However, four major challenges are faced for long-term development of nuclear energy: meeting increasingly stringent safety requirements, adhering to the criteria of sustainable development, improvement of the economic competitiveness, and non-proliferation. To continue exploitation of nuclear energy as well as to overcome the key issues and challenges that current traditional nuclear system faces, the development of advanced nuclear systems is necessary and very urgent.

The advanced nuclear systems usually refer either to the Gen IV systems, accelerator driven subcritical systems or to fusion systems. In recent years more and more countries have put very serious efforts in the research on advanced nuclear systems. Understanding of neutronics is a key issue in development of advanced nuclear systems. The advanced nuclear systems have complicated characteristics, such as complex structures and material compositions, strong anisotropy and wide energy range of the neutron interactions with matter. External neutron source coupled with subcritical reactor brings new challenges to neutronics, including lack of the neutron interaction cross sections in high-energy range, difficulties in accurate and efficient simulation of neutronics, lack of experimental facilities with specified neutron spectrum, etc. New theories, methods, cross section data and technologies dealing with neutronics must be developed for advanced nuclear systems. Along with emerging of advanced nuclear systems, there is an increasing demand for a comprehensive text books focused on the neutronics of such systems. As one of the researchers pioneering development of advanced nuclear systems, and ADS in particular since 1990's, I'm very pleased to see that Prof. Yican Wu who has been dedicated to neutronics related research for whole his career, wrote this book. It is really the first comprehensive book on the neutronics of advanced nuclear systems.

Prof. Wu has been investigating advanced nuclear systems and its related neutronics all along for more than 30 years, He and his team for Frontier Development of Science (FDS) at the Institute of Nuclear Energy Safety Technology, Chinese Academy of Sciences have already become one of the world-leading and creative forces in the field of advanced nuclear energy research. I am impressed by his pioneering and remarkable work, including the neutronics simulation methods and code development, neutronics experiment technologies, reactor concepts and design, which are well described in this book. Some of the subjects presented in this book are particularly exceptional:

a) On methodology, the unique CAD-based neutronics modelling and simulation methods of the entire nuclear system have been developed offering a very high accuracy. The superior code SuperMC has been developed with capabilities of a very high accuracy high efficiency and used as intelligent tool for the nuclear researchers and engineers.
b) On neutron physics experimental technique, a novel multilayer neutron target has been developed, a real technical innovation firstly appeared in this field, with the significantly improved neutron yield under the same ion beam parameters. Consequently, the neutron yield of D-T neutron source HINEG reached the level of 6.4×10^{12} n/s, so that the HINEG has achieved the highest neutron yield in the world among accelerator-based fusion neutron sources in operation. HINEG is a unique neutron physics testing facility coupled with an experimental station by developing creative multi-type modules and inverse optimization technology for the first time in this field. This has unprecedentedly overcome the challenges on accurate generation of various neutron spectra for advanced nuclear systems. These special features make it a cutting-edge facility for the research of advanced nuclear systems.
c) On the reactor concept development, the innovative Advanced External Neutron Source Driven Nuclear Energy System (CLEAR-A) has been proposed, and it shows very appealing features in terms of inherent safety, environment friendliness, and economics. Being based on practical technologies, it provides an ideal new path for the application of advanced nuclear systems in the near future.

Overall, those contributions represented highlights in recent years in this field. I believe in this book you can find more creative perspectives that inspire your work.

It should be noted and highly appreciated that Prof. Wu is describing the entire area of neutronics of advanced nuclear systems in a holistic way. He guides the readers from basic theory to engineering applications. This book is a timely contribution appearing at a time that the importance of advanced nuclear systems is becoming increasingly recognized, and provide an important impetus for the innovation of nuclear systems. I strongly recommend this book to anyone who is interested in neutronics and advanced nuclear energy research. Thanks for the efforts by Prof. Wu and this book.

Stockholm, Sweden

Waclaw Gudowski
Professor, Royal Institute of Technology—KTH

Foreword III

Nuclear power makes an important contribution to the reduction of greenhouse gas emissions while playing an increasing role in supplying much-needed energy for global economic development. Nuclear power produces virtually no greenhouse gas emissions or air pollutants during their operation and only very low emissions over its full life cycle. The advantages of nuclear power in terms of climate change are an important reason why many countries intend to introduce nuclear power or expand their existing programmes in the coming decades. All countries have the right to use nuclear technology for peaceful purposes, as well as the responsibility to do so safely and securely.

The accident at the Fukushima Daiichi nuclear power plant in March 2011 raised concerns again about nuclear power safety and reminded the world that nuclear safety could never be taken for granted even though we know that the highest standards of safety can be maintained and the challenges regarding nuclear safety and security can be overcome through innovation so that the commercial use of nuclear power can be made more sustainable and more cost-effective. Advanced nuclear energy systems, including the Gen IV systems, hybrid systems, and fusion systems, have been proposed and widely studied in many countries.

Neutronics is the basis for the development of advanced nuclear systems. The complicated characteristics of advanced nuclear systems pose new challenges to existing neutronic theories, methods, and technologies.

Professor Yican Wu's research focus has been on the development of advanced nuclear systems and neutronics for over 30 years. He and his FDS team have made world-impact contributions to the study of neutronics and its role in developing advanced nuclear systems, including the accurate neutronics simulation methods and codes, neutron control technologies and a series of experimental facilities, and new concepts of advanced nuclear systems.

The innovative CAD-based whole-process neutronics modeling and simulation methods, for example, have been developed and implemented in the unique comprehensive neutronics code SuperMC, the first code of this kind, in addition to the virtual power plant Virtual4DS, which provides for the first time a fast and comprehensive simulation platform for the safety evaluation, environmental impact assessment and nuclear emergency simulation of power plants. It has also been applied to the safety evaluation of advanced nuclear systems, such as ITER and CLEAR.

On the neutron control technologies and experimental facilities, some interesting technologies have been proposed and developed, such as, the novel neutron generating target with multi-layer compositions, in-core neutron spectrum detection technology based on ex-core measured data and multiple sets of response matrix. And it is encouraging to see that the large-scale D–T neutron source experimental facility HINEG has been developed on the basis of such technologies. In spite of its daunting challenges, it has been uniquely coupled with a neutron control station, which can be flexibly reassembled to generate specified neutron spectrum and make it a new international precedent.

In terms of advanced nuclear systems, a series of innovative and interesting concepts have been proposed, such as a series of fusion and hybrid reactors FDS-I/II/III/MFX/GDT, and a series of China Lead-based Reactors CLEAR-A/M. Especially, Professor Wu has proposed an innovative idea to use plutonium in fusion-driven subcritical hybrid systems to improve the efficiency and safety of transmutation, which charts a new path for the disposal of high radioactive nuclear waste permanently.

Nuclear energy has benefited from Professor Wu's research achievements and such benefits have been extended in related fields. For example, the above technologies have been used in the systematic safety assessment of the fusion system and the radical viewpoint of "fusion being safe" has been corrected.

In this book, Professor Wu summarizes and shares his work on neutronics in advanced nuclear systems in detail. It is the first systematic book specializing in this field internationally. It significantly updates the existing knowledge and will work as a strong driving force for the development of advanced nuclear systems. I highly recommend this book as a valuable reference for researchers working on advanced nuclear systems.

Kowloon Tong, Hong Kong

Way Kuo
President and University
Distinguished Professor
City University of Hong Kong

Preface

In 1938, it is found that nuclear fission can be triggered by the bombardment of uranium with neutrons, and this work laid the physical foundation for the application of nuclear energy and opened the era of the development and utilization of nuclear energy. After nearly 80 years of development, nuclear energy has become an important part of the world's energy supply and is considered to be a highly efficient low-carbon form of energy. In the development of nuclear energy, nuclear safety is a common concern of people around the world. In addition, the problems of disposal of nuclear waste and the limitation of fuel resources have become increasingly prominent. To face the challenges of public acceptability, economy, sustainability, etc., the R&D of advanced nuclear systems has become a major trend in the development of nuclear energy, although pressurized water reactors are the primary reactors used in the nuclear power industry. Advanced nuclear systems mainly include fourth-generation nuclear systems, fusion systems, and hybrid nuclear systems, which are characterized by high safety, high economic efficiency, high sustainability, and the effective prevention of nuclear proliferation.

The neutron is pivotal in generating nuclear energy and releasing radiation, triggering the fission reactions, and as the main energy carrier of fusion reactions. The behavior of neutrons has a significant impact on the safety and economic efficiency of nuclear systems. The neutronics of nuclear energy focuses on the motions of neutrons and interactions of neutrons with materials, and is a comprehensive subject including multiple complex physical processes. Compared with traditional nuclear systems, advanced nuclear systems have complicated characteristics, such as complex geometries and multiple special nuclides, the interaction of neutrons with matter over wide energy range, the coupling of external sources, and subcritical reactors/blankets. These characteristics exist in the whole-life design and safety operations of advanced nuclear systems and bring new challenges to neutronics research. Nevertheless, there is still a lack of a dedicated publication that would provide a systematic introduction to the neutronics of advanced nuclear systems.

The author and his team (known as the FDS Team) have been engaged in neutronics-related research of advanced nuclear systems for 30 years. This book was written over the years and is an extension of the author's earlier book "Fusion Neutronics". This work is also based on research experiences in advanced fission systems and hybrid nuclear systems performed by the author and his team over the past few decades. The book consists of three parts with 15 chapters in total, and covers aspects of neutronics theories and methodologies, neutronics design principles, and experimental technologies for advanced nuclear systems.

The first part begins with the integrated neutron transport theory, then describes the processes of steady-state neutron transport, neutron kinetics, transmutation, radiation dosimetry and biologic safety, material neutron irradiation damage, etc., and finally focuses on comprehensive neutronics simulation of neutron and multiple neutron-related physical effects for the entire space and life cycle of nuclear systems.

The second part is dedicated to neutronics characteristics, design principles, and objectives of advanced fission systems, fusion systems, and hybrid nuclear systems, then focuses on the introduction to the methods of neutronics design, using China Lead-based Reactors (CLEAR), International Thermonuclear Experimental Reactor (ITER), and the fusion-driven subcritical system as examples to illustrate the application of the corresponding theories and methodologies in the engineering practice of advanced nuclear systems.

The last part introduces technologies and typical facilities used when conducting neutronics experiments of advanced nuclear systems, summarizes the methods and typical examples of neutronics experiments, and finally introduces the practice and discussion of the author and his team on neutronics experiments of advanced nuclear systems.

The author wishes to take this opportunity to express sincere thanks to Prof. Carlo Rubbia, Prof. Waclaw Gudowski, Prof. Way Kuo, Prof. Zehua Li, Prof. Hangang Liu, Prof. Yongqian Shi, Prof. Albert Gu, etc., for their painstaking review of the manuscript and for providing many valuable suggestions in the process of writing this book. In addition, the author also wishes to express deep gratitude to the members of the FDS Team, for their long-term joint efforts and collaboration support in related research work and in the writing of this book.

In light of the wide-ranging contents and continuous development of related theories, methods, and technologies, deficiencies are inevitable. We sincerely invite readers to provide valuable comments and suggestions concerning this book. Finally, the author hopes that the publication of "Neutronics of Advanced Nuclear Systems" can provide a reference for researchers and scholars who are interested in nuclear energy, and provide a modest contribution to the promotion of original innovation in nuclear energy science and technology and to the sustainable and healthy development of nuclear energy.

Hefei, China
December 2018

Prof. Yican Wu
Director General
Institute of Nuclear Energy Safety Technology
Chinese Academy of Sciences

Contents

Chapter 1
Introduction

Energy is indispensable for social and economic development and is vital for improving human welfare and quality of life. Fossil fuels, such as coal, oil, and natural gas, have been major sources of energy over the past two centuries. However, fossil fuel reserves are limited and nonrenewable, so they will have difficulty meeting society's energy demands. Furthermore, the combustion of fossil fuels leads to environmental pollution. Nuclear energy is considered to be a clean energy source that could replace fossil energy on a large scale and has become an important component of global energy.

Nuclear systems include fission systems, fusion systems, and hybrid nuclear systems. As of October 30, 2018, there were 454 commercial nuclear power plants in operation worldwide, accounting for 11% of total global power generation [1]. All of these plants are traditional fission systems, which face problems in terms of radioactive leakage, nuclear waste disposal, and limited fissile fuel resources. To solve these problems, the development of advanced nuclear systems is imperative [2].

The neutron, whose behavior determines the safety and economic efficiency of nuclear systems, is pivotal in generating nuclear energy and releasing radiation. The cornerstone of nuclear systems R&D is neutronics research which particularly focuses on the generation, measurement, control, and utilization of neutrons, as well as radiation protection. This chapter briefly introduces the development of nuclear systems, typical examples of advanced nuclear systems, as well as the characteristics and content relating to neutronics research on advanced nuclear systems.

1.1 The Development of Nuclear Systems

Nuclear systems generate energy by controlled nuclear reactions. In 1942, the graphite reactor at the University of Chicago achieved the first controlled self-sustaining nuclear chain reaction, which signified the beginning of the peaceful use of nuclear energy.

Y. Wu, *Neutronics of Advanced Nuclear Systems*,
https://doi.org/10.1007/978-981-13-6520-1_1

1.1.1 Basic Principles of Nuclear Energy Generation

According to Einstein's mass–energy equation, the mass lost in a nuclear reaction can be converted into energy, as is shown in Eq. (1.1):

$$\Delta E = \Delta mc^2 \tag{1.1}$$

where Δm is the mass that is lost in the reaction, and c is the speed of light.

Nuclear energy includes fission energy, fusion energy, and nuclear decay energy. The nuclear systems introduced in this book refer to systems that utilize fission energy and fusion energy, while nuclear decay systems are not included.

The energy released during a nuclear fission reaction is called fission energy. Nuclides such as ^{233}U, ^{235}U, and ^{239}Pu can be induced to fission with a high probability, especially with low-energy neutrons. They are therefore often referred to as fissile nuclides. Nuclides such as ^{232}Th, ^{238}U, and ^{240}Pu are not fissionable by low-energy neutrons, but can be converted into fissile nuclides by the absorption of neutrons. They are therefore often referred to as fertile nuclides. A typical nuclear fission reaction is shown in Eq. (1.2):

$$^{235}\mathrm{U} + \mathrm{n} \rightarrow \left(^{236}\mathrm{U}\right)^* \rightarrow {}^{A_1}_{Z_1}\mathrm{X} + {}^{A_2}_{Z_2}\mathrm{Y} + \nu\mathrm{n} + Q \tag{1.2}$$

where n denotes a neutron; ${}^{A_1}_{Z1}\mathrm{X}$ and ${}^{A_2}_{Z2}\mathrm{Y}$ are two different medium-mass nuclei called fission products; ν is the average number of neutrons released per fission reaction, which is approximately 2–3; and Q is the energy released per fission reaction, which is approximately 200 MeV.

It is estimated that the energy released by the fission of 1 kg of ^{235}U is approximately equal to the energy released by the combustion of 2,800 tons of standard coal. The neutrons produced by fission reactions are referred to as fission neutrons in this book. The fission neutrons will further initiate new fission reactions as shown in Fig. 1.1, which is known as the fission chain reaction. This process is the basis for the sustained operation of fission systems.

On the other hand, the energy released during fusion reactions is called fusion energy. Some representative fusion reactions are given below:

$$\mathrm{D} + \mathrm{D} \rightarrow {}^3\mathrm{He}\,(0.82\,\mathrm{MeV}) + \mathrm{n}\,(2.45\,\mathrm{MeV}) \tag{1.3}$$

$$\mathrm{D} + \mathrm{D} \rightarrow \mathrm{T}\,(1.01\,\mathrm{MeV}) + \mathrm{p}\,(3.02\,\mathrm{MeV}) \tag{1.4}$$

$$\mathrm{D} + \mathrm{T} \rightarrow {}^4\mathrm{He}\,(3.52\,\mathrm{MeV}) + \mathrm{n}\,(14.06\,\mathrm{MeV}) \tag{1.5}$$

$$\mathrm{D} + {}^3\mathrm{He} \rightarrow {}^4\mathrm{He}\,(3.67\,\mathrm{MeV}) + \mathrm{p}\,(14.67\,\mathrm{MeV}) \tag{1.6}$$

where D denotes deuterium, T denotes tritium, and p denotes a proton.

Fig. 1.1 Schematic diagram of the fission chain reaction of ^{235}U

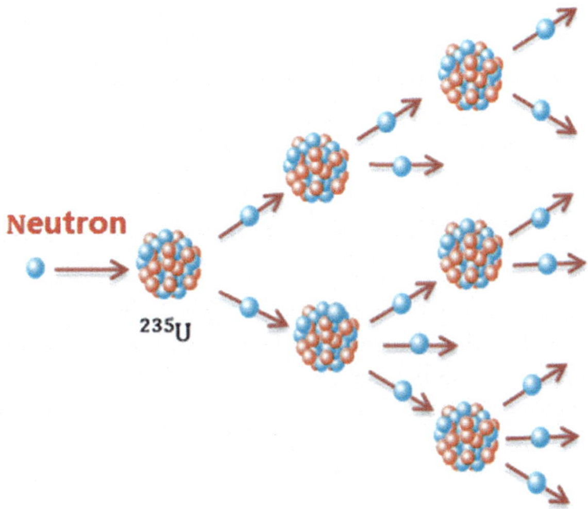

The neutrons produced by fusion reactions are called fusion neutrons. Among the reactions above, 75–80% of the fusion energy is carried by the produced neutrons or protons. Reactions (1.3) and (1.5) produce neutrons with energies of 2.45 MeV and 14.06 MeV, respectively, and reactions (1.4) and (1.6) produce protons with energies of 3.02 MeV and 14.67 MeV, respectively.

The relationship between the cross section of the fusion reaction and the energy of the deuterium particle is shown in Fig. 1.2 (data are taken from the ENDF/B-VII.1 database). It can be observed that the cross section of the D-T reaction is the highest under the achievable plasma temperature. Therefore, the D-T reaction is considered as the first choice for a fusion system in spite of its issues such as limited tritium resources and shielding problems caused by high-energy neutrons. In this book, the

Fig. 1.2 Cross sections of various fusion reactions with variations in the deuterium energy (the curve for D-D represents a sum over the cross sections of all reaction branches)

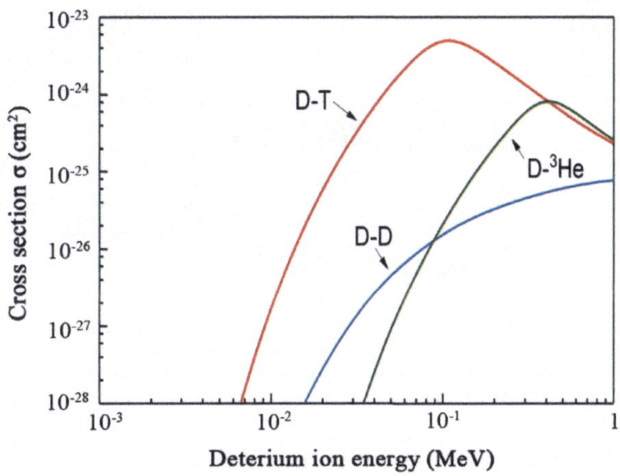

terms "fusion reactions" and "fusion neutrons" refer to the D-T reaction and the generated 14.06 MeV neutrons, respectively. It is estimated that the energy released by the fusion reaction of 1 kg of D and 1.5 kg of T is equal to that by the combustion of 29,000 tons of standard coal.

1.1.2 The Development of Nuclear Systems

The development of nuclear systems is shown in Fig. 1.3. The traditional nuclear systems in this book include the first-generation early prototype reactors, the second-generation commercial power reactors, and the third-generation advanced light water reactors. Nuclear systems of the fourth and next generations, which are called advanced nuclear systems in this book, have the features of high sustainability, high safety and reliability, high economic efficiency, and effective prevention of nuclear proliferation [3].

The first-generation nuclear systems are the early prototype reactors that demonstrated the feasibility of electricity generation by nuclear reactors. These systems include a series of prototype nuclear power plants established in various countries in the 1950s, such as the fast-neutron breeder reactor Fermi 1, the Magnox reactor with graphite as the moderator and gas as the coolant, and the Shippingport Atomic Power Station.

Second-generation nuclear systems are commercial power reactors that have been in operation since the 1960s and proved the economic feasibility of nuclear energy. The main types include the pressurized water reactor (PWR), boiling water reactor

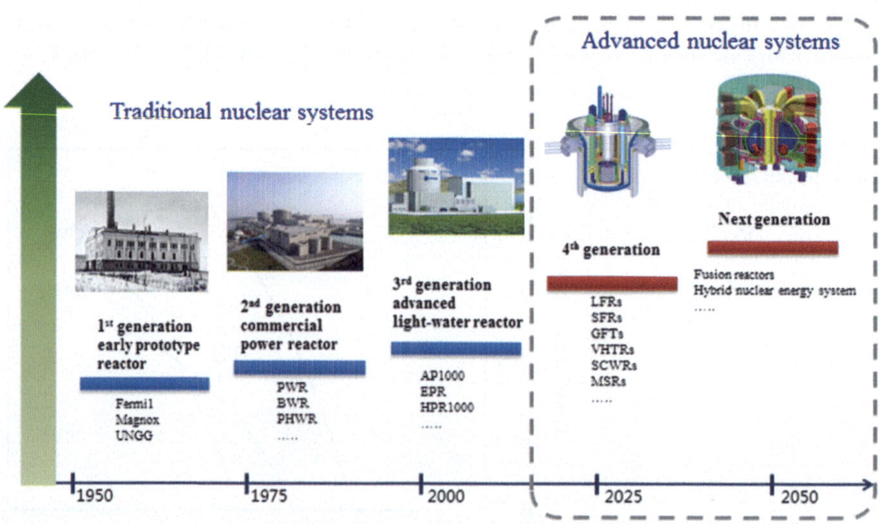

Fig. 1.3 Development of nuclear systems

(BWR), and heavy water reactor (HWR). However, the accidents at Chernobyl, Three Mile Island, and Fukushima have resulted in public skepticism about the safety of nuclear power.

Third-generation nuclear systems refer to safer systems, based on lessons learned from the nuclear accidents mentioned above. Currently, the third-generation nuclear systems include the Advanced Passive Pressurized Water Reactor 1000 MW (AP1000), the European Pressurized Water Reactor (EPR) in France, the VVER Pressurized Water Reactor in Russia, the Hualong Pressurized Water Reactor 1000 MW (HPR-1000), the China Advanced Passive Pressurized Water Reactor 1400 MW (CAP1400) developed by China, etc.

Fourth-generation nuclear systems are advanced fission systems with significant innovations in reactor concept and the fuel cycle [4]. Their main objectives are to achieve high sustainability, high safety and reliability, high economic efficiency, and effective prevention of nuclear proliferation. In terms of sustainability, they aim to achieve sustainable energy production, to promote efficient utilization of fuels, to minimize nuclear wastes generation and to reduce the long-term management burden of nuclear wastes. In terms of safety and reliability, the objective is to have a high reliability and excellent safety performance, as well as an extremely low probability and extent of core damage, and to eliminate the need for off-site emergency response. In terms of economic efficiency, these systems should have a life cycle cost advantage over other energy sources with a comparable level of financial risk. In terms of nuclear proliferation prevention and physical protection, fourth-generation nuclear systems are a very unattractive route for the diversion or theft of weapon-usable materials and can provide increased physical protection against acts of terrorism. Fourth-generation nuclear systems mainly include six types of reactors recommended by the Generation IV International Forum (GIF) [5, 6], namely lead-based fast reactor (LFR), very-high-temperature reactor (VHTR), sodium-cooled fast reactor (SFR), supercritical water-cooled reactor (SCWR), molten salt reactor (MSR), and gas-cooled fast reactor (GFR). The technology roadmap of the fourth-generation nuclear systems released by GIF in 2014 is shown in Fig. 1.4. It can be seen that as of now, most of the fourth-generation nuclear systems have completed the feasibility verification and come into the stage of performance verification. The LFR could be the first system to finish the performance verification and enter into the stage of industrial demonstration.

Advanced nuclear systems also include fusion systems, hybrid nuclear systems, etc. Among them, fusion systems have the advantages of inexhaustible resources, relatively few long-lived and high-level wastes, which would correspond to a low level of radioactive hazards, and good inherent safety. Hybrid nuclear systems include the fusion-driven subcritical system (FDS) [6] and the accelerator-driven subcritical system (ADS) [7]. These systems generate power by a combination of an accelerator or fusion reactor with a subcritical reactor/blanket, having inherent safety and excellent performance in nuclear waste transmutation and nuclear fuel breeding.

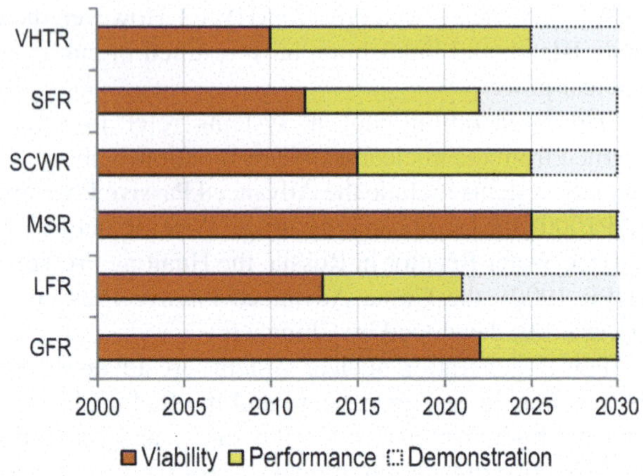

Fig. 1.4 The technology roadmap of the fourth-generation nuclear systems [5]

1.2 Brief Introduction to Advanced Nuclear Systems

This section provides a brief introduction to advanced fission systems, fusion systems, and hybrid nuclear systems.

1.2.1 Advanced Fission Systems

1.2.1.1 Lead-Based Fast Reactor (LFR)

LFRs are reactors using Pb or a Pb alloy as the coolant (hereinafter referred to as lead-based reactor) [5]. The coolant (lead-based materials) has relatively low neutron moderation and capture cross sections, a high thermal capacity, a low melting point, a high boiling point, and no exothermic reaction with water or air. Thus, LFRs have good performance in neutron economy, thermal safety, and chemical safety, and are characterized by their inherent safety, fine sustainability, proliferation resistance, and high economic efficiency [8–10].

(1) Inherent safety. First, the operation will excel in safety and reliability. Molten lead has the advantage of allowing for the operation of the primary system at a low (atmospheric) pressure. A low dose to the operators can also be predicted, due to its low vapor pressure and high capability of trapping fission products and high shielding of gamma radiation. In the case of accidental air ingress, particularly during refueling, any produced lead oxide can be reduced by the injection of hydrogen into lead and then reactor operation can resume safely. The moderate temperature difference between the core inlet and outlet temperatures reduces the thermal stress during transients, and the relatively low core outlet temperature minimizes the creep effects in the steels. Second, the system

has a low likelihood and degree of core damage. It is possible to design fuel assemblies with fuel pins spaced further apart, which allows for a large coolant fraction as in the case of the water reactor. This results in a moderate pressure loss through the core of approximately 1 bar, in spite of the high density of lead, with an associated improved heat removal by natural circulation and the possibility of an innovative reactor layout such as installing the primary pumps in the hot collector to improve several aspects affecting safety. Lead allows for a high level of natural circulation of the coolant, which results in less stringent requirements for the timing of operations and simplification of the control and protection system. In the case of leakage of the reactor vessel, the free level of the coolant can be designed to maintain a level that ensures the coolant circulation throughout, ensuring safe heat removal from the core. Any leaked lead would solidify without significant chemical reactions affecting the operation or performance of the surrounding equipment or structures. Third, there is no need for off-site emergency response. With high-density lead as a coolant, fuel dispersion dominates over fuel compaction, making the occurrence of complex sequences leading to re-criticality less likely. In fact, lead, with its higher density than oxide fuel and its natural convection flow, makes it difficult to lead to fuel aggregation with subsequent formation of a secondary critical mass in the event of postulated fuel failure.

(2) Fine sustainability: Lead is abundant and hence available, even in the case of deployment of a large number of reactors. Because lead is a coolant with very low neutron absorption and moderation, it is possible to maintain a fast neutron flux even with a large amount of coolant in the core. This allows for an efficient utilization of excess neutrons. Reactor designs can readily achieve a breeding ratio of above 1, a long core life, and a high fuel burnup. Also, a fast neutron flux significantly reduces waste generation, and Pu recycling in a closed cycle is the condition recognized by GEN IV for waste minimization. The capability of the LFR systems to safely burn recycled long-lived minor actinides within the fuel will add to the attractiveness of the LFR.

(3) Proliferation resistance: The use of MOX fuel containing minor actinide (MA) in LFRs increases proliferation resistance. The neutron spectrum can be easily adjusted, and a long period reactivity equilibrium can be achieved, which is beneficial for the prevention of nuclear proliferation. Additionally, the use of a coolant that is chemically compatible with air and water and a system operation at an ambient pressure enhances the physical protection.

(4) High economic efficiency: The high energy utilization caused by a high coolant outlet temperature leads to the good economic efficiency of LFRs.

However, LFRs have some problems that remain to be unsolved, such as corrosion of structural materials by liquid lead-based materials, measurement and control technology of oxygen, in-service maintenance in nontransparent media, and earthquake and shock resistance in environments with a high-density coolant.

Currently, the development plans of LFRs have been launched in many countries and several demonstration projects are being implemented worldwide. The

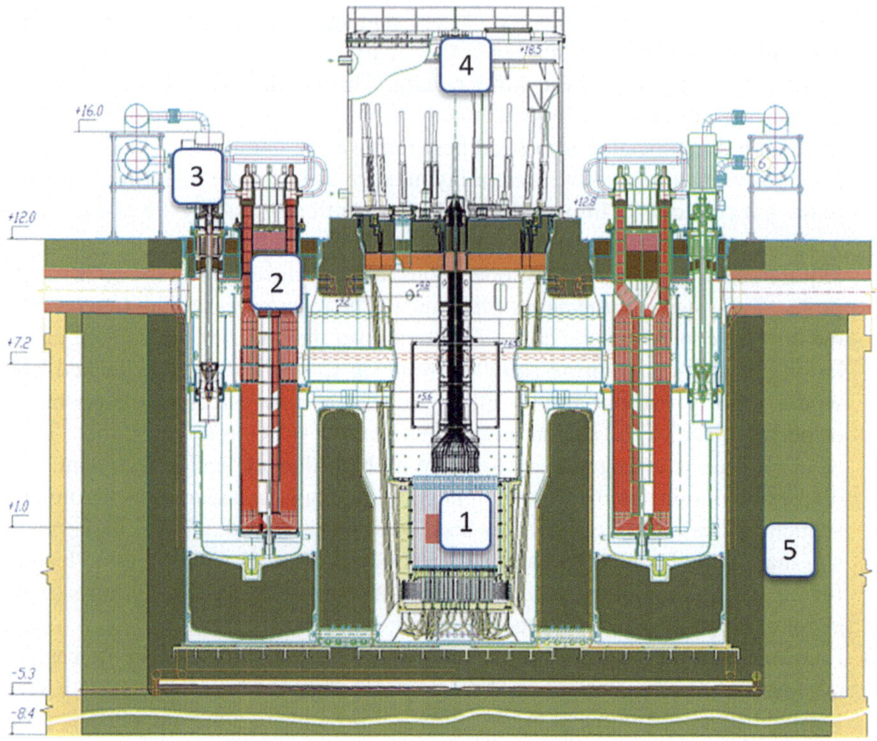

Fig. 1.5 Schematic diagram of the BREST-300 [8]

application of LFRs for commercial utilization is now actively promoted in Russia, and the lead–bismuth-cooled reactor SVBR-100 (Lead-Bismuth Fast Reactor of l00MWe) and lead-cooled fast reactor BREST300 (Lead-Cooled Fast-Neutron Reactor of 300MWe) project have been proposed. Figure 1.5 shows the schematic diagram of the BREST-300. In the European Union, under the support of EU FP5, FP6, FP7 framework, a huge project with detailed plans for LFRs was launched with more than 20 institutions involved. Among them, the design and configuration of the Advanced Lead Fast Reactor European Demonstrator (ALFRED) was implemented in Italy. Research on the Small Secure Transportable Autonomous Reactor (SSTAR) was conducted in the United States. In Korea, design and related researches on thermal hydraulics and materials of PEACER (Proliferation-resistant, Environment-friendly, Accident-tolerant, Continual, and Economical Reactor) and URANUS (Ubiquitous, Robust, Accident-forgiving, Nonproliferating and Ultra-lasting Sustainer) have been conducted. The plan of separation and transmutation of high-level nuclear wastes has also been performed in Japan, and the design of a lead–bismuth reactor with an industrial scale has been completed.

In China, LFR research began in the mid-1980s with the China Lead-based Reactor (CLEAR) proposed by INEST, CAS (the FDS Team). The FDS Team has undertaken several projects, including the International Atomic Energy Agency (IAEA) and International Energy Agency (IEA) cooperation projects, the International Thermonuclear Experimental Reactor (ITER) related domestic and international projects, the National High-tech R&D Program of China ("863 Program"), the National Basic Research Program of China ("973 Program"), the National Magnetic Confinement Fusion Science Program of China, the Strategic Priority Research Program of CAS, etc. Continuous fundamental research and key technical research have been conducted. Over 30 years of development, China has achieved great progress and several breakthroughs in key technologies of LFRs design and research. The design theory and method have been developed, and CLEAR series LFRs have been designed. Breakthroughs on lead-based coolant, professional components and equipment, structural materials and nuclear fuels, reactor operation, control, etc., have been achieved. A series of engineering verification platforms have been built, including the comprehensive liquid lead alloy loop, experimental facilities for LFR neutron physics, integrated experimental facilities for LFR engineering technology, virtual lead-based nuclear reactor, etc.

Based on the compact configuration of the LFR, the FDS Team proposed and designed the China Lead-based Mini-Reactor (CLEAR-M) for energy production. It features excellent safety, small modular and compact configuration, long operation time, and high reliability. The system can generate powers over a broad range, from kilowatts to megawatts, and is flexible for a variety of needs. It can be widely used in power supply for island and offshore platforms, ship power, space power, distributed power supply in desert and remote areas, and cogeneration in independent industrial parks.

Additionally, under the support of the National Basic Research Program of China ("973 Program"), the Chinese Academy of Sciences conducted preliminary design research for lead–bismuth fast reactors. Recently, LFR technology has also aroused interests in other related institutes which are conducting related studies.

1.2.1.2 Very-High-Temperature Reactor (VHTR)

VHTRs are graphite-moderated and helium-cooled reactors with a thermal neutron spectrum [5, 11, 12]. The core outlet temperatures are between 700 and 950 °C and could reach 1000 °C in the future. The most common type is a pebble-bed reactor or prismatic reactor. In addition to the inherent safety, VHTRs have the advantages of a high output temperature, high power generation efficiency, and good economic efficiency. However, VHTRs also present challenges in materials, thermal hydraulics, and power conversion.

In the GIF technology roadmap released in 2014, the modular high-temperature gas-cooled reactor was defined as the first stage and current research objective of the VHTR, and a reactor with an outlet temperature of 1000 °C or above was defined as the second stage and the long-term development goal. The modular high-temperature gas-cooled reactor was developed on the basis of the carbon dioxide-cooled reactor Magnox and the Advanced Gas-cooled Reactor (AGR) built in the United Kingdom, which uses helium as the coolant and high-temperature resistant all-ceramic materials in the core structure and fuel cladding material. The coated particle fuel is dispersed in the graphite matrix to better encapsulate the fission products.

The high-temperature gas-cooled reactors that have been operated and retired include the Dragon in the United Kingdom, the AVR (Arbeitsgemeinschaft Versuchsreaktor) and the THTR-300 (Thorium High-Temperature Nuclear Reactor) in Germany, and the Peach Bottom II in the United States. The high-temperature gas-cooled reactors that are in operation include Japan's HTR (High Temperature Engineering Test Reactor) and China's HTR-10 (10 MW High Temperature gas-cooled Reactor-test module). At present, a High Temperature Reactor Pebble Module (HTR-PM) is under construction in Shandong, China. A typical VHTR system is shown in Fig. 1.6.

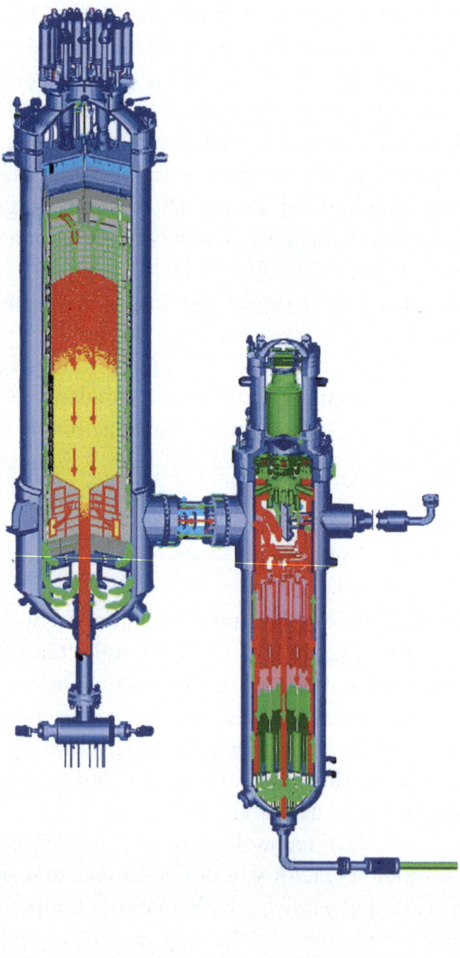

Fig. 1.6 Schematic diagram of the HTR-PM [13]

1.2.1.3 Sodium-Cooled Fast Reactor (SFR)

SFRs are reactors that use liquid sodium as the coolant with a core outlet temperature from 500 to 550 °C [5, 14]. In SFRs, fission reactions are initiated by fast neutrons, which allows for achieving nuclear fuel breeding and nuclear waste transmutation and can solve the problems of nuclear fuel shortage and long-lived waste disposal. Due to the preferable thermophysical properties of sodium coolants, SFRs have a higher power density and lower coolant volume fraction, and thus a high economic efficiency. However, sodium is chemically active and reacts violently when in direct contact with air or water. Leakage and combustion of sodium are very serious safety issues in SFRs, so the basic safety criterion for SFR design is to effectively prevent sodium leakage and alleviate the consequences.

Research on SFRs has been conducted in such countries as France, Russia, Japan, the United Kingdom, Korea, and India. The related experimental reactors and commercial demonstration reactors that have been built include Phoenix in France, the BN series in Russia (BN-350, BN-600, BN-800, etc.), JOYO and Monju in Japan, PFR (Prototype Fast Reactor) in the United Kingdom, FBTR (Fast Breeder Test Reactor) in India, etc. Other SFRs under construction include ASTRID (Advanced Sodium Technological Reactor for Industrial Demonstration) in France, BN-1200 in Russia, PFBR (Prototype Fast Breeder Reactor) in India, PGSFR (Prototype Gen-IV Sodium-cooled Fast Reactor) in Korea, etc. The schematic diagram of the PGSFR is shown in Fig. 1.7. In China, the China Experimental Fast Reactor (CEFR) has been built, and the China Demo Fast Reactor CFR-600 with 600 MW is under design and construction.

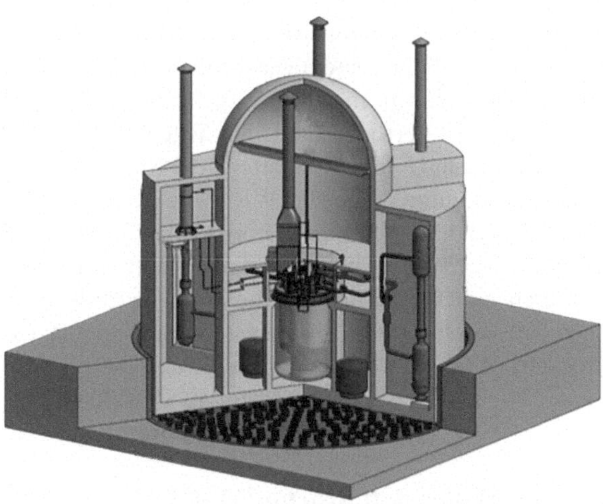

Fig. 1.7 Schematic diagram of PGSFR [15]

1.2.1.4 Supercritical Water-Cooled Reactor (SCWR)

SCWRs are water reactors operating above the thermodynamic critical point of water with a high temperature, high pressure, and core outlet temperature higher than 500 °C [5]. Depending on the core design, the reactor core may have a thermal or a fast-neutron spectrum, and light water or heavy water could be used as the moderator.

SCWR has many advantages. For example, the increase in the enthalpy of supercritical water in the core from inlet to outlet is greater, the amount of water required to cool the core is much less than that of a boiling water reactor or pressurized water reactor, and the thermal efficiency is at least 30% higher than that of current light water reactors. Moreover, in an SCWR, compared with pressurized water reactor, cooling pumps and steam generators are not needed, and compared with boiling water reactors, steam separators and dryers are not needed. Additionally, the containment system and the turbine system can be designed to be smaller. These general features offer potential lower capital costs for the given electric power of the plant and can improve the fuel utilization, providing a significant economic advantage compared with current water-cooled reactors. However, the development of SCWRs still has many technical challenges. For example, non-uniformities of the local power and the coolant mass flow rate in the core may cause hot spots due to the larger enthalpy rise of the coolant, the higher coolant temperature results in higher fuel cladding temperatures, and the large density variation within the core could lead to instability and subsequently large neutronic variation and a high fuel cladding temperature.

In the 1950s, the idea of using supercritical water as the reactor coolant was proposed by scientists in the United States and the Soviet Union, and exploratory research has been conducted. But the research was suspended due to the limitation of science and technology at that time. In the 1990s, the preliminary conceptual design of an SCWR with a thermal neutron spectrum JSCWR (Japanese Supercritical Water-Cooled Reactor) was proposed by Japanese scientists, and then research on SCWRs began in the European Union, Canada, China, and other countries. In 2000, the High Performance Light Water Reactor (HPLWR) project was launched in the European Union for supercritical water reactor research. Figure 1.8 shows the schematic diagram of the HPLWR. Based on the CANDU reactor, research on a pressure tube-type SCWR was performed in Canada by the Atomic Energy of Canada Limited (AECL). In China, Shanghai Jiaotong University, the China Nuclear Power Research and Design Institute, the China Institute of Atomic Energy, etc., jointly conducted research on SCWRs. They received support from the National Basic Research Program of China ("973 Program"), and then proposed the concept of a Chinese Supercritical Water-cooled Reactor (CSR1000).

1.2.1.5 Molten Salt Reactor (MSR)

MSRs use molten salt as the coolant. The core outlet temperature can reach 700 °C or higher [5, 17–19]. MSRs can be designed as either a thermal neutron reactor or a fast-neutron rector for different application purposes. According to the fuel type, it

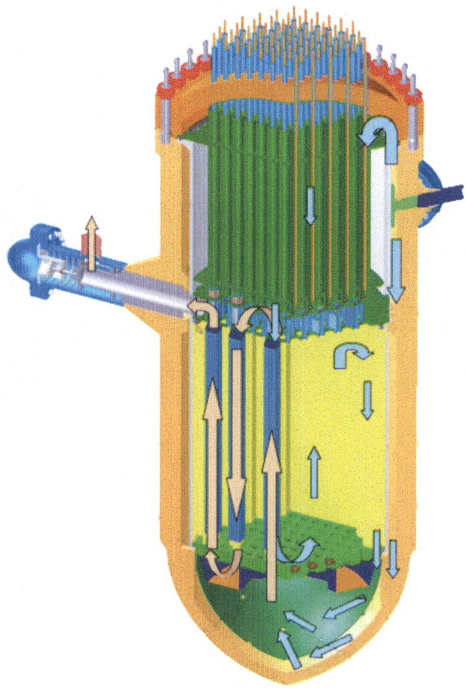

Fig. 1.8 Schematic diagram of the HPLWR [16]

could be classified as either a liquid or solid fuel MSR. The liquid fuel MSR is taken as an example in this book. Figure 1.9 shows the schematic diagram of a typical liquid fuel MSR.

The main advantages of a liquid fuel MSR include the following: (1) High inherent safety. For example, dissolution of fuel in liquid fluoride salt could avoid a core melting accident caused by coolant loss; a lower operating pressure reduces the probability of a loss of coolant (LOCA) accident due to the rapid solidification of the liquid molten salt at ambient temperatures even in the case of a LOCA accident. (2) Flexible fuel circulation. The reactor is convenient for various forms of fuel circulation. (3) Effective prevention of nuclear proliferation. Nuclear fuel such as uranium, plutonium, and thorium can be directly used without special treatment, reducing the use of enriched uranium. However, liquid fuel MSRs still have the following problems to be solved, including (1) problems in MSR technology which is different from other solid fuel reactors because of the flow characteristics of fuel salts; (2) the necessity of mature reactor design, safety analysis methods, and safety assessment specifications; (3) the feasibility of online posttreatment technology of fuel salt that requires further experimental validation; (4) the solubility of lanthanide and actinide nuclides; (5) the compatibility of molten salt with structural materials and graphite after irradiation; and (6) metal segregation and the tritium control problem.

The study of MSRs originated in the United States in the late 1940s. From 1950 to 1976, the ARE (Aircraft Reactor Experiment) and the MSRE (Molten Salt Reac-

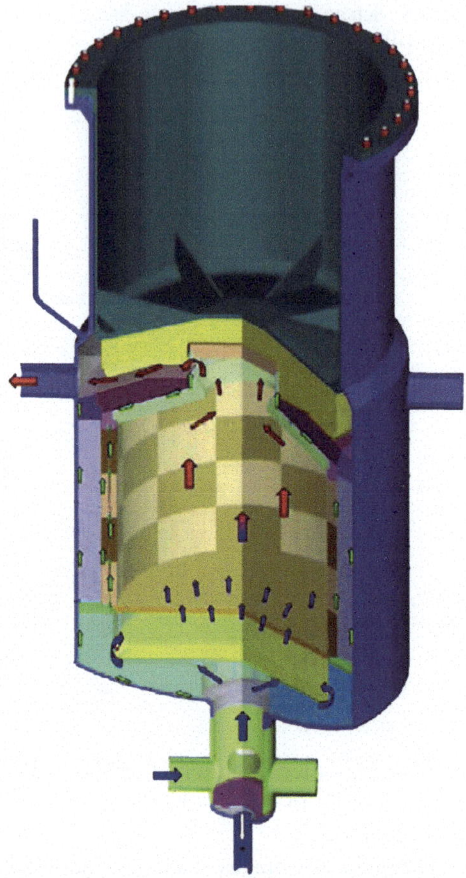

Fig. 1.9 Schematic diagram of the MOSART [20]

tor Experiment) were established and operated, and a 1000 W MSBR (Molten Salt Breeder Reactor) was designed. Since then, a large amount of MSRs research has been conducted throughout the world, including MSFR (Molten Salt Fast Reactor) in France, MOSART (Molten Salt Advanced Reactor Transmute) in Russia, Fuji-MSR (Molten Salt Reactor) in Japan, etc. Figure 1.9 shows the schematic diagram of the MOSART. The Chinese Academy of Sciences launched the Strategic Priority Research Program "Future Advanced Nuclear Fission Energy—Thorium-based Molten Salt Reactor (TMSR)" for the implementation of MSR research in 2011.

1.2.1.6 Gas-Cooled Fast Reactor (GFR)

GFRs are helium-cooled fast-spectrum reactors with a core outlet temperature of 850 °C [5, 21]. The GFR with a high sustainability can perform nuclear fuel breeding and long-lived actinides transmutation. In addition, as a high-temperature reactor, it has a high thermal efficiency and could be used in industrial applications such as hydrogen production. However, there are some technological challenges associated with the use of a gas coolant. Its low thermal inertia leads to a rapid heat up of the core followed by a loss of forced cooling. Since the power density is high, the HTR-type "conduction cooldown" will not work for the removal of decay heat, and other solutions must be considered. Also, the density of the gas coolant is too low to achieve sufficient natural circulation to cool the core, and the power requirements for the blower are important at low pressure. Finally, additional consideration needs to be given to the effects of the fast-neutron dose on the reactor pressure vessel in the absence of core moderation.

In the early time, GFRs studies have been conducted or are being conducted in such countries as Germany, the United States, Russia, the United Kingdom, and Japan. In 2010, the ALLEGRO (ALLEGRO—implementing Advanced Nuclear in Central Europe) research project was launched by the Czech Republic, Slovakia, and Hungary, and the V4G4 Centre of Excellence was established in company with Poland in 2013. ALLEGRO is expected to begin its construction in 2020 and to begin operating in 2025. Figure 1.10 shows the schematic diagram of the ALLEGRO.

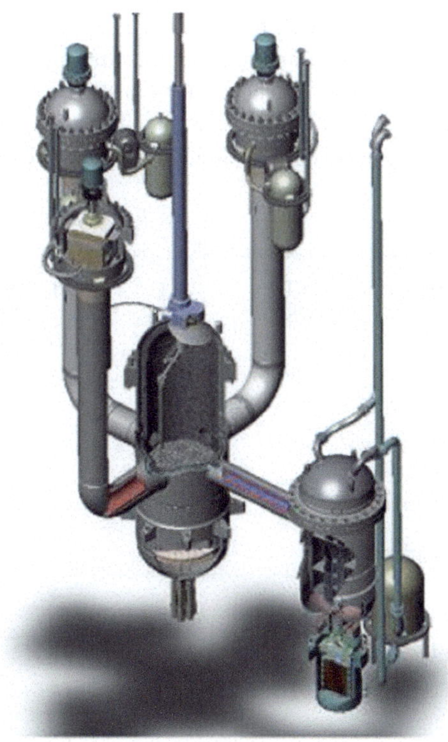

Fig. 1.10 Schematic diagram of the ALLEGRO [22]

1.2.2 Fusion Systems

Fusion systems utilize energy released by controlled nuclear fusion reaction, which can be realized by MCF (magnetic confinement fusion), ICF (inertial confinement fusion), beam–target fusion [23], beam–beam fusion [23], inertial electrostatic confinement [23], muon-catalyzed fusion [23], pyroelectric fusion [23], etc. Among them, MCF and ICF are currently the prominent research areas.

1. Magnetic confinement fusion

MCF uses a magnetic field to confine the plasma to achieve a fusion reaction in a quasi-equilibrium steady state. There are two forms of magnetic field configuration, one is the so-called closed or toroidal configuration facilities represented by tokamak, stellarator, and reversed field pinch, and the other one is linear open-ended facilities, represented by magnetic mirror and z-pinch. Currently, the most representative facilities are the Tokamak Fusion Test Reactor (TFTR) in the United States [24], the Joint European Tokamak (JET) in the European Union [25], the Japan Torus (JT-60) and JT-60SA (updated facility of JT-60) in Japan [26], the superconducting tokamak TORE SUPRA in France [27], the Experimental Advanced Superconducting Tokamak (EAST) in China [28], and the Korea Superconducting Tokamak Advanced Research (KSTAR) in Korea [29]. Table 1.1 lists the main parameters of typical tokamak experimental facilities throughout the world.

Based on the successful construction and operation of a large number of tokamak devices, the International Thermonuclear Experimental Reactor (ITER) [29] is being jointly built in Cadarache, France by China, the European Union, India, Japan,

Table 1.1 Main parameters of typical experimental tokamak facilities throughout the world [29]

Parameter	TFTR (USA)	JET (EU)	TORE SUPRA (France)	EAST (China)	KSTAR (Korea)	JT-60SA (Japan, EU)
Major radius/m	2.40	2.96	2.25	1.75	1.8	3.16
Minor radius/m	0.80	0.96	0.70	0.43	0.5	1.02
Plasma current/MA	3.0	6.0	2.0	0.5	2.0	5.5
Magnetic field/T	6.0	4.0	4.5	5.0	3.5	2.7
Elongation	1	1.6	1	1.6–2	2.0	1.93
Fusion power/MW	10.5	16.1	–	–	–	–
Operation date/year	1982	1983	1988	2006	2008	2019 (plan)

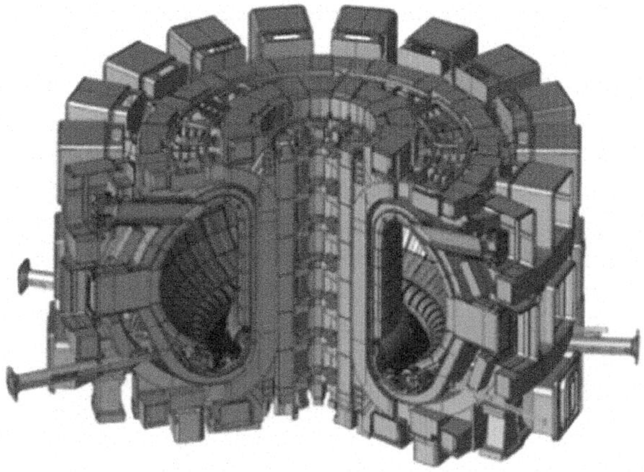

Fig. 1.11 Schematic diagram of the ITER host

South Korea, Russia, and the United States. It is scheduled to begin low-power operation in 2025. The ITER host is shown in Fig. 1.11. Based on the ITER program, the development of fusion energy for commercial application has been launched in many countries. For example, a variety of concepts [23, 30] have been proposed in China, including engineering experimental reactors (such as the Multi-Functional eXperiment Reactor, FDS-MFX, China Fusion Engineering Test Reactor, CFETR), industrial demonstration reactor (such as the Chinese DEMOnstration reactor, C-DEMO) and fusion prototype power plant (such as the Fusion Power Reactor FDS-II, the High-Temperature Fusion Reactor FDS-III and the Spherical Tokamak-based Reactor FDS-ST).

2. Inertial confinement fusion

Inertial confinement fusion (ICF) uses the inertia of hot and highly compressed fuels to keep them together for nanoseconds to achieve fusion (hence the term inertial confinement). ICF facilities achieve fusion reactions mainly through the following four processes:

(1) The fuel pellet surface is heated by an external power source, typically produced by high-intensity lasers, X-rays or ions.
(2) This heating causes severe ablation and an explosion of the outer layers of the pellet, leading to an inward momentum input by Newton's Third Law (similar to a rocket).
(3) The DT fuel in the core is compressed to approximately 1000 times its original density.
(4) D-T reactions are ignited by shock waves, the injection of a petawatt laser beam, or a relativistic electron beam, and a fusion burn wave occurs.

ICF was first proposed by the LLNL in the early 1970s. Appearing to be a practical approach to fusion power generation, the field flourished. Throughout the 1980s and 1990s, many experiments were conducted to study the complex interaction of high-intensity laser and plasma. These experiments led to the design of new machines with a larger size, which aimed at achieving ignition. In recent years, the development of high-energy and high-power laser facilities has promoted research into laser-driven ICFs. The most representative facilities [23] include the National Ignition Facility (NIF) in the United States, the Laser Mégajoule facility (LMJ) in France, SGI/SGII/SGIII laser facilities in China, etc. To confirm the feasibility of laser fusion power reactors, researchers in Europe, Japan, and the United States proposed laser fusion research programs (HiPER, KOYO-F, and LIFE, respectively) and performed a series of fusion power reactor conceptual designs.

1.2.3 Hybrid Nuclear Systems

Hybrid nuclear systems include fusion-driven subcritical systems (FDS), accelerator-driven subcritical systems (ADS), etc. In these systems, neutrons with a high flux and high intensity are generated from accelerators or fusion devices, and then drive a continuous chain reaction with fissile materials in a subcritical reactor/blanket [23]. On the one hand, the fission reactor operates in the subcritical mode and will be automatically shut down in the event of an accident, avoiding the occurrence of a supercritical accident, which can occur in traditional fission systems. On the other hand, under the drive of the high-energy external neutron source, it has stronger fuel breeding and waste transmutation abilities.

1. Fusion-driven subcritical systems

In recent years, the scientific feasibility of fusion reactors has been verified. However, there are still a series of important physical and technical problems that need to be solved, and there is still a long way to go before achieving commercial applications of fusion energy. Currently, based on the application development of fusion and fission nuclear energy, fusion-driven subcritical systems have a good technical foundation and could be used for the disposal of nuclear wastes, breeding of nuclear fuel, production of tritium, energy supply, etc. It can promote the development of fusion energy and solve the problems brought about by the development of fission energy. It is regarded as a promising early application of fusion energy. The fusion-driven subcritical system is currently still in the conceptual design stage.

Fusion-driven subcritical systems mainly include a breeding system, a transmutation system, and an energy generation system. Preliminary research of the breeding reactor has been conducted at LLNL since 1954. Fifteen years later, researchers at LLNL proposed the fission suppressed blanket. The main concept is to increase the neutron flux by adding a moderator and breeder, reduce the average neutron energy, and increase the capture to fission ratio of the breeding material in the blanket. In 1972, the concept of a fast-fission blanket for breeding neutrons and producing energy, which uses high-energy fusion neutrons to induce fission reactions for neutron breeding and energy generation in the blanket, was proposed by the researchers at LLNL. In the early phase of the study, the fission suppressed blanket played a dominant role, and then the fast-fission blanket gradually became the prominent research area. Transmutation system research began in 1973, and the transmutation of fission products in the fusion system blanket was studied by researchers at Pacific Northwest National Laboratory (PNNL), later studying the transmutation of the actinide nuclides. Afterward, a series of research works were also performed in the United States and Turkey on fusion-driven subcritical transmutation system. In 1978, the concept of the fusion-driven subcritical energy generation system was first proposed by the Princeton Plasma Physics Laboratory (PPPL). Then, research in this area was conducted at the University of Tokyo, Kyushu University, Osaka University, Westinghouse, and in Turkey.

Fusion–fission hybrid studies began in the 1980s in China [23]. In the early stages, the studies were supported by the National High-tech R&D Program ("863 Program"). The related research institutes include the FDS Team, the China Academy of Engineering Physics (CAEP), the Southwestern Institute of Physics (SWIP), and several universities and colleges. Since China joined the ITER project in 2006, national support for fusion-related research has been greatly strengthened and has played a vital role in promoting research in hybrid reactors. The systematic physical design in terms of electricity generation, breeding, and transmutation for the fusion-driven subcritical system was conducted by the FDS Team, the concept of FDS-I for the transmutation and incineration of nuclear wastes was proposed, and research on key materials and critical loop experimental technologies was performed. Figure 1.12 shows the schematic diagram of FDS-I. In addition, a conceptual design of the hybrid reactors for fissile fuel breeding and nuclear waste transmutation based on conventional tokamaks has been presented by SWIP, and a hybrid reactor concept driven by Z-pinch has been proposed by CAEP.

2. Accelerator-driven subcritical systems

The concept of ADS was proposed in the 1950s, but due to the limitations of accelerator technology, the related research did not begin to flourish until the 1990s. Nobel Prize winner Professor Carlo Rubbia proposed the concept of "Energy Amplifier EA" in the 1990s. EA, which could be used to generate energy and transmute nuclear

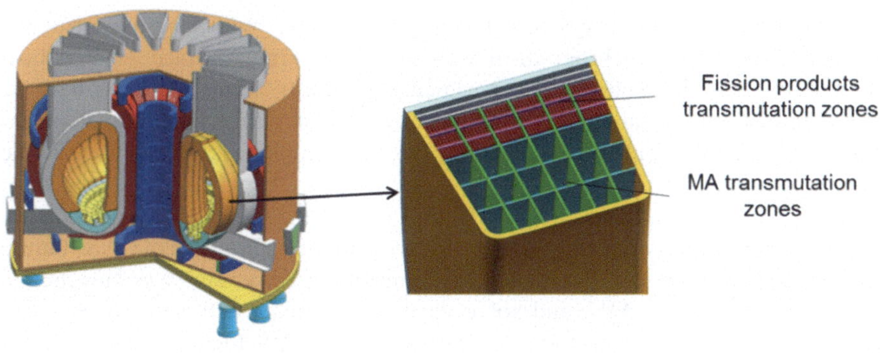

Fig. 1.12 Schematic diagram of the FDS-I

wastes, is a subcritical system based on the U-Th cycle and driven by a high-intensity proton accelerator. After the concept of EA was proposed, ADS aroused extensive research interest in several countries.

The medium- and long-term development roadmaps for ADS have been developed in China, the EU, Japan, Russia, the United States, etc., and now the key technology research has gradually shifted to the construction of devices for principle research of ADS with integrated systems [23]. A group of 16-member advisors led by Carlo Rubbia was established by the European Union to draw up the EU-ROTRANS program, supporting more than 40 universities and institutes participation under the EU FP6 framework. The original Preliminary Design Studies of an Experimental Accelerator Driven System (PDS-XADS) program was extended to form: (1) Conceptual design of a 50–100 MWt principle demonstration device (Experimental Transmutation Accelerator Driven System, XT-ADS), and (2) conceptual design of the European Facility for Industrial Transmutation (EFIT) of a transmutation reactor with hundreds of MW (including lead target) driven by a 16 MW accelerator. An accelerator-driven lead–bismuth-cooled fast-neutron subcritical nuclear system was scheduled to be built by SCK CEN (Belgium), under the support of the MYRRHA (Multi-purpose hYbrid Research Reactor for High-tech Applications) program.

In 1999, the Accelerator Transmutation of Waste (ATW) program was launched by the United States, and the Advanced Accelerator Applications (AAA) program was implemented in 2001 as the imperative part of the advanced nuclear fuel cycle (AFCI). The Research and Development on Nuclear Partitioning and Transmutation Technology program called OMEGA was launched in Japan for the disposal of nuclear wastes in 1988, regarding ADS as the best option for transmutation of MA. Figure 1.13 shows the schematic diagram of OMEGA. In the 1990s, ADS research was conducted by the Institute for Theoretical and Experimental Physics (ITEP, Russia) cooperating with the Los Alamos National Laboratory (LANL, America). In 1998, the ADS research program was launched by the Federal Agency on Atomic Energy (FAEA, Russia), which involved experiments, theoretical and computer software development, optimization design of the ADS experimental simulation device, the R&D of a 1 GeV/30 mA proton linear accelerator, research on advanced nuclear fuel cycles, etc.

In the 1990s, ADS concept research began in China. With the support of the National Basic Research Program of China ("973 Program") since 1999, a fast-thermal coupled ADS subcritical experimental platform "Venus 1#" was built by the China Institute of Atomic Energy. In the meantime, research and development of superconducting accelerator technology has been emphatically supported by the Chinese Academy of Sciences, and the major project "ADS Pre-Study" has been arranged. To support ADS research, in 2011, faced with the strategic needs of Chinese nuclear energy development, the Strategic Priority Research Program of CAS "Future Advanced Nuclear Fission Energy—ADS Transmutation System" was launched to verify the principle of ADS and solve the key technical challenges in the ADS accelerator, spallation target, and reactor systems. The FDS Team is responsible for the design of reactor systems and the detailed design of the LFR CLEAR-I with a critical and subcritical dual mode [32]. In 2015, the China Initiative Accelerator

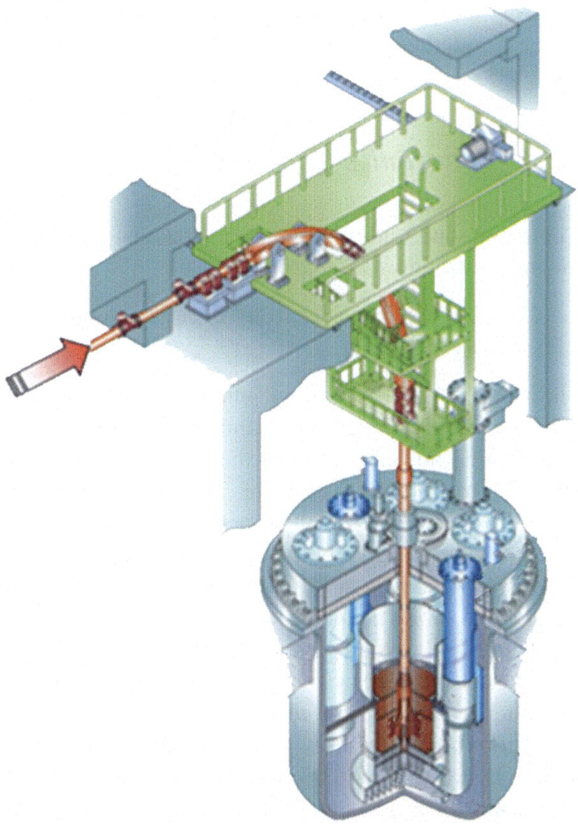

Fig. 1.13 Schematic diagram of the accelerator-driven subcritical system OMEGA [31]

Driven System (CIADS), a major national science and technology infrastructure program, was approved by the National Development and Reform Commission to build an ADS research facility in Huizhou City, Guangdong Province.

1.3 Neutronics Research Overview of Advanced Nuclear Systems

Compared with traditional nuclear systems, to achieve sustainability, safety and reliability, economic efficiency, and prevent nuclear proliferation, the advanced nuclear systems have become more complicated in terms of research, design, and operation. Specifically, the strong anisotropic transport of neutrons, which brings new challenges to neutronics research, results from the high intensity of neutrons, the complex spectrum and angular distribution of neutrons, complicated material composition, large space span with complex geometry, complicated multi-physics fields, etc. This section first introduces the neutronics characteristics in advanced nuclear systems, and then the research contents of advanced nuclear energy neutronics.

1.3.1 Neutronics Characteristics

Compared with traditional nuclear systems, advanced nuclear systems have the following neutronics characteristics:

1. Complex neutron spectrum structure, angular distribution, and high neutron intensity

The neutrons of a traditional nuclear system are derived from a fission reaction, with energies below 10 MeV and weak anisotropic angular distribution. However, in advanced nuclear systems, there is the coupling of an external neutron source and fission reactions in reactor. Neutrons may be derived from fission reactions, fusion reactions, and spallation reactions, with energies up to hundreds of MeV, and the spectrum may be fission, fusion spectra, and spallation spectra. Moreover, the energy spectrum at different locations in advanced nuclear systems also varies. Figure 1.14 shows the neutron energy spectrum of the first wall, blanket, and vacuum vessel in ITER, which is more complicated than that of traditional nuclear systems. In advanced nuclear systems, due to the high neutron energy, the angular distribution after scattering is more complicated and the anisotropy is stronger. In addition, neutrons with high energy and intensity result in more serious radiation damage to materials, for example, the radiation damage of the first wall (FW) material in the fusion reactor is 150 dpa (units of displacements per atom) or even higher, requiring better anti-irradiation performance for materials. In advanced nuclear systems, due to the higher neutron intensity, materials with strong absorption are generally used for sufficient shielding. Compared with traditional nuclear systems, neutron flux attenuation is stronger, resulting in a more complex deep-penetration problem.

2. Complicated material composition

There are multiple special nuclides in advanced nuclear systems, such as lead, lithium, and fluorine, which have a strong influence on the nuclear systems design and operation. Also, the cross section data of reactions between some nuclides and high-energy neutrons are missing or have poor accuracy, requiring corresponding neutronics experiments for cross section measurement and validation. At the same time, materials in advanced nuclear systems have more phase states, a larger variation of temperature and density, and a continuous distribution. For example, the density of liquid water coolant in supercritical water-cooled reactors changes continuously, which could not be accurately described by the traditional spatial discrete homogenization method. In liquid MSRs, fuel will move with the flowing coolant, and the density of fuel nuclides and delayed neutron precursors will change.

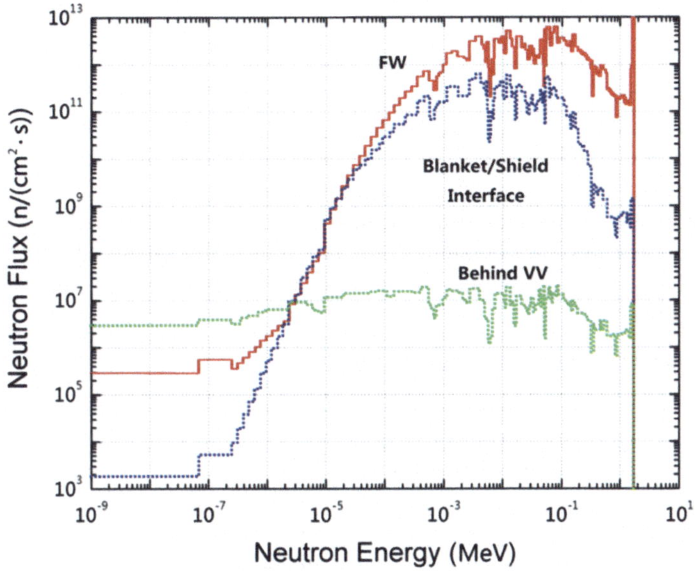

Fig. 1.14 Neutron spectrum at different positions in a fusion reactor

3. Large spatial span and complex geometry

Compared with traditional nuclear systems, advanced nuclear systems have a larger spatial span. A fusion system has a typical building structure with the size of several hundred meters in three dimensions, but there are also millimeter-scale holes and slits. In addition, the geometry of advanced nuclear systems is more complicated. For example, in a fusion reactor, the tokamak contains thousands of parts that can be described by complex surfaces such as sweeping surfaces and B-spline surfaces. In the very-high-temperature reactor, fuel pebbles are randomly distributed in the core without a fixed arrangement. These characteristics bring difficulties to the neutronics simulation of advanced nuclear systems and also bring new challenges to the neutronics design and experiments of advanced nuclear systems.

4. Complicated multi-physics fields and extreme conditions

There may be multi-physics fields, such as a radiation field, temperature field, flow field, and electromagnetic field, in advanced nuclear systems. These physics fields are tightly coupled with one another, and some of them are under extreme conditions, making neutron behaviors in advanced nuclear systems more complex. For example, the temperature may change from −269 °C to over 1000 °C, which results in a large variation in the reaction cross sections, the temperature field, the radiation field, and other fields, which will have an influence on the structural materials, leading to deformation. These phenomena will directly affect the neutron transport process and bring new challenges to neutronics experiments for advanced nuclear systems.

1.3.2 Research Content

The neutronics of advanced nuclear systems, focused on the generation, control, measurement, utilization, and protection of neutrons, have become a basic discipline in advanced nuclear energy research after decades of development. According to the research content, this book is divided into three parts on neutronics research of advanced nuclear systems: neutronics theories and methodologies for advanced nuclear systems (Chaps. 2–8); neutronics design of advanced nuclear systems (Chaps. 9–11); and neutronics experiments of advanced nuclear systems (Chaps. 12–15).

1. Neutronics theories and methodologies

Neutronics theories describe the movement of neutrons in a medium and the nuclear reaction process, i.e., the neutron transport process [33], which is the basis and the core of neutronics research. Neutron transport includes neutron motion, nuclear transmutation, energy deposition, etc. Among them, neutron motion studies the motion discipline of neutrons in the medium, including steady-state neutron transport and time-dependent neutron transport. Time-dependent neutron transport can be divided into short-time (milliseconds to seconds), medium-time (hours to days), and long-time (month to 1 year) neutron transport processes. In the short-time neutron transport process, expected or unexpected changes in the nuclear system affect the neutron transport, leading to a rapid change in the neutron flux in the system, which is the content of neutron kinetics. In medium-time and long-time neutron transport processes, the changes in nuclides (including the change in combustible toxicants and the change in nuclear fuel) due to transmutation of nuclides (including depletion, nuclear waste transmutation, nuclear fuel breeding, and material activation) have an effect on neutron transport. Energy deposition studies the process of conversion from kinetic energy and energy released by a nuclear reaction into heat during neutron transport in the medium, which is the basis of nuclear power generation and is also the main factor that causes human radiation dose hazard and radiation damage to devices. Moreover, the theories and disciplines introduced above are based on nuclear data, which are the research foundation for nuclear science and technology as well as nuclear system designs.

In the neutronics design of advanced nuclear systems, neutron transport is highly anisotropic, and the processes of neutron transport are tightly coupled and affect one another. The neutronics simulation involves all the processes of neutron transport, nuclear transmutation, and energy deposition as well as the coupling of neutronics processes with other physical processes. It reveals the physical and engineering characteristics of fusion systems on the entire scale within the lifetime, which are essential for ensuring that the designs of fusion systems fulfill all necessary requirements in terms of safety, reliability, sustainability, and economic performance.

2. Neutronics design

In the neutronics design of advanced nuclear systems, the complex structure and harsh service environment bring great challenges, especially in the core physics design, shielding design, and radioactive wastes management. With the design objectives of safety, economic efficiency performance, and environmental friendliness for advanced nuclear systems, the neutronics design must cover the entire life cycle of the design, operation, and decommissioning of advanced nuclear systems.

(1) Neutronics design of advanced fission systems. Neutronics design of advanced fission systems includes the core physics and radiation protection designs. The core physics design focuses on fuel selection and fuel management in the reactor. The core physics design optimization, the design of reactivity control and fuel management, etc., are conducted by coupling and iteration between neutronics calculations (such as the critical calculation, reactivity feedback analysis, power distribution analysis, fuel depletion analysis, transient response analysis) and structural mechanics and thermal-hydraulic analysis, to ensure the realization of high sustainability, high safety and reliability, high economic efficiency, etc. In radiation protection design, the radiation source distribution is first analyzed, and then the radiation safety analysis of the occupational exposure dose, radionuclide migration, and environmental impact assessment are performed. Based on this, the shielding material and shielding layout are optimized. Additionally, radioactive wastes need to be assessed and managed to ensure the safety and environmental friendliness of the nuclear system.
(2) Neutronics design of fusion systems. The neutronics design of a fusion system is divided into two parts: the blanket neutronics design, and the machine and building neutronics design. For the neutronics design of the blanket, the layout of the blanket structure and material selection are optimized to achieve the tritium self-sustaining and high-efficiency energy conversion as well as to reduce the nuclear thermal deposition, radiation damage, and radioactivity of the outer layer of the blanket as much as possible, ensuring the safe operation of the fusion systems. The neutronics design of the tokamak and the building is mainly to optimize the shielding scheme and to conduct a safety analysis and environmental impact assessment to ensure the safety of personnel, equipment, and the environment during the operation of fusion systems.
(3) Neutronics design of hybrid nuclear systems. The neutronics design of hybrid nuclear systems includes the optimization of neutron parameters of the external neutron source, a subcritical reactor/blanket neutronics design, and a radiation shielding design. In the optimization of neutron parameters of the external neutron source, the influence of parameters, such as the intensity and shape of the neutron source, and the neutron spectrum on the system fuel loading, discharge burnup, burnup reactivity loss, power peak factor and material irradiation damage, etc., is analyzed first. Then, together with the neutron source design constraints, the optimal neutronics parameters of the external neutron source can be selected. In the subcritical reactor/blanket neutronics design, the best possible performance of nuclear fuel breeding and nuclear waste transmutation are obtained by optimizing the subcritical reactor/blanket structure layout and material selection, and the safe operation of the hybrid nuclear system is ensured through the design of the reactivity control scheme. The shielding

design of fusion-driven subcritical systems is essentially the same as that of the traditional fusion machine and building. Compared with traditional fission reactors, the shielding design of accelerator-driven subcritical systems needs to be focused on the complicated radiation field produced by the coupling of the accelerator, heavy metal target, and the subcritical reactor to ensure the safety of workers and equipment during operation.

3. Neutronics experiments

Neutronics experiments are an important part of neutronics research of advanced nuclear systems, as well as the application and validation of theories. Based on neutron source, experimental facilities as well as measurement technologies of neutron flux, neutron energy spectrum, and others, neutronics experiments are in essence the experimental research on testing nuclear data, validating codes and methods, validating nuclear design and critical equipment, etc. Neutronics experiments in early times were mainly focused on traditional fission systems, and a large number of experimental data were accumulated. Compared with traditional nuclear systems, neutronics experiments of advanced nuclear systems should comprehensively consider the neutronics features of advanced nuclear systems in the aspects of experimental facilities, methods, and contents.

In terms of experimental facilities, neutronics experiments of advanced nuclear systems are mainly conducted on the accelerator neutron source device and the experimental reactor. In terms of experimental methods, in order to obtain neutron field with a specific energy, angle, spatial, and time distribution, it is necessary to conduct experiments on the precise control of neutrons. To accurately detect the flux and energy spectrum, it is necessary to conduct research on corresponding detection equipments and experimental methods.

In terms of experimental contents, according to the features of various nuclear systems, neutronics experiments of advanced fission systems mainly focus on the determination of the reactor's critical state, the measurement of power, reactivity, kinetic parameters, and other experimental research related to reactor operation safety. Due to the high neutron energy and strong source, neutronics experiments of fusion systems should attach more importance to the activation of materials, tritium breeding, materials irradiation damage, shielding, and other experiments that affect the stable and safe operation of fusion systems. With consideration of the previous experiments, neutronics experiments of hybrid nuclear systems are more concerned with such experiments as subcriticality measurements and transmutation measurements.

References

1. Agneta R (2018) World nuclear performance report 2018
2. Li SN (1997) Advanced nuclear energy systems without long-lived high-level waste (HLW). At Energyence Technol 06:79–88
3. Roberto JB, Rubia TDDL (2007) Basic research needs for advanced nuclear energy systems. Jom J Minerals Metals Mater Soc 59(4):16–19

4. Schultz K, Marder J, Rath B (2007) Advanced nuclear power systems. Adv Mater Process
5. GENIV International Forum (2014) Technology roadmap update for generation IV nuclear energy systems
6. Wu YC, Team FDS (2016) Design and R&D progress of China lead-based fast reactor for ADS research facility. Engineering 2(1):124–131
7. Wu YC, Qiu LJ (2000) A fusion neutron source driven sub-critical clear nuclear energy system: a way of early application of fusion energy technology. Nucl Technol 08:519–525
8. Alemberti A, Smirnov V, Smith CF et al (2014) Overview of lead-cooled fast reactor activities. Prog Nucl Energy 77:300–307
9. Zhan WL, Xu HS (2012) Advanced fission energy program-ADS transmutation system. Bull Chin Acad Sci 27(3):375–381
10. Hu DP, Yuan HQ (1995) Reactor physics characteristics of lead cooled fast reactor—a new type fast neutron reactor. Nucl Power Eng 03:194–198
11. Sabharwall P, Bragg-Sitton SM, Stoots C (2013) Challenges in the development of high temperature reactors. Energy Convers Manag 74:574–581
12. Fu XM, Wang J (2006) Summary of the development of high temperature gas cooled reactors in China. Mod Electr Power 05:70–75
13. Sun X, Chen Z, Sun J et al (2018) CFD investigation of bypass flow in HTR-PM. Nucl Eng Des 329:147–155
14. He JR, Guo ZR (2013) Development status of sodium cooled fast reactor. Dongfang Electr Rev 03:36–43
15. Ahn SJ, Park GY, Lee KL et al (2018) The sensitivity analysis for IHTS and SG due to the Large-scale Sodium-Water Reaction event in PGSFR. Ann Nucl Energy 118:26–34
16. Kebwaro JM, Zhao Y, He C (2015) Design and optimization of HPLWR high pressure turbine gamma ray shield. Nucl Eng Des 284:293–299
17. Jiang MH, Xu HJ, Dai ZM (2012) Advanced fission energy program-TMSR nuclear energy system. Bull Chin Acad Sci 03:366–374
18. Cai XZ, Dai ZM, Xu HJ (2016) Thorium molten salt reactor nuclear energy system. Physics 45(9)
19. Qiu SZ, Zhang DL, Su GH et al (2009) Research on inherent safety and relative key issues of a molten salt reactor. At Energy Sci Technol 43(s1):64–75
20. Seo SB, Shin Y, Bang IC (2018) Numerical analysis on spatial universality of similarity technique inside molten salt reactor system. Int J Heat Mass Transf 116:569–580
21. Rooijen WFG (2009) Gas-cooled fast reactor: a historical overview and future outlook. Sci Technol Nucl Install 965757
22. Stainsby R, Peers K, Mitchell C et al (2011) Gas cooled fast reactor research in Europe. Nucl Eng Des 241(9):3481–3489
23. Wu YC (2017) Fusion neutronics. Springer Nature Singapore Pte, Ltd
24. Strachan JD, Bitter M, Ramsey AT et al (1987) High-temperature plasmas in a Tokamak fusion test reactor. Phys Rev Lett 58(10):1004
25. Rebut PH, Bickerton RJ, Keen BE (1985) The joint european torus: installation, first results and prospects. Nucl Fusion 25(9):1011
26. Tsuji S, Ushigusa K, Ikeda Y et al (1990) Observation of the limiter h mode in the JT-60 Tokamak with lower-hybrid current drive. Phys Rev Lett 64(9)
27. Van D, Supra ET (1993) One minute pulse operation in the tore supra Tokamak. Nucl Fusion 33(1):137
28. Wan YX, Li JG, Weng PD et al (2006) First engineering commissioning of EAST Tokamak. Plasma Sci Technol 8(3):253–254
29. Yuan BS, Jiang SF, Lu ZH (2011) Tokamak device engineering foundation. Atomic Energy Publishing House, Beijing

30. Wu YC, Wang HY, KE Y et al (2006) Status of development and design of magnetic confinement fusion reactors and ITER test blanket modules. Nucl Phys Rev 23(2):89–95
31. Sasa T, Yang JA, Oigawa H (2005) Shielding analysis at the upper section of the accelerator-driven system. Radiat Prot Dosim 116(1–4 pt 2):256–258
32. Wu YC (2018) Development of high intensity D-T fusion neutron generator HINEG. Int J Energy Res 42(2):68–72
33. Wu YC (2018) Multi-functional neutronics calculation methodology and program for nuclear design and radiation safety evaluation. Fusion Sci Technol 74(4):321–329

Part I
Neutronics Theories and Methodologies

Chapter 2
Steady-State Neutron Transport Theory and Simulation

Neutron transport theory is the research basis for neutronics and focuses on the description of neutron motion in media and the corresponding laws. There are two types of methods for neutron transport calculation: the Monte Carlo method (also called the probabilistic method or the stochastic method) and the deterministic method. The Monte Carlo method is a numerical method based on probability and statistical theories, and can explicitly describe the characteristics of randomly moving particles and the process of physical experiments. In contrast, in the deterministic method, a group of mathematical–physical equations are first built up to explain the physical characteristics of the target system. Then, by discretizing the variables in these equations, including direction, energy, space, and time, an approximate solution can be obtained with numerical calculation. There are complicated features for advanced nuclear systems, such as the complex neutron spectrum and angular distribution, complicated material composition, large spatial span, complex geometry, etc. The Monte Carlo method uses the continuous-energy cross section, and can be used to address any complex geometry, with prominent advantages for neutron transport simulations for advanced nuclear systems. However, some challenges, such as the slow convergence rate and difficulty in addressing problems of deep penetration, still exist. The deterministic method is faster, but falls short in addressing advanced nuclear systems with complex geometries, strong anisotropy of neutron scattering, and complicated energy spectrums. In recent years, the method of characteristics (MOC) and the discrete ordinates method with unstructured meshes have been developed with improved geometry processing abilities. However, problems, such as the ray effects and the high cost of large-scale systems, still need to be solved. The Monte Carlo–deterministic coupling method, which combines the advantages of both the Monte Carlo and deterministic methods, is one of the most efficient and accurate methods for solving transport problems in advanced nuclear systems.

This chapter first introduces the neutron transport theory, followed by the Monte Carlo computational method, deterministic computational method, Monte Carlo–deterministic coupling method, and software for the steady-state neutron transport problems of advanced nuclear systems.

Y. Wu, *Neutronics of Advanced Nuclear Systems*,
https://doi.org/10.1007/978-981-13-6520-1_2

2.1 Neutron Transport Theory

Transport theory has a history of more than 100 years. In 1872, L. Boltzmann derived the conservation expression for the transport of microscopic molecules in a medium. Similar conservation equations for neutrons, photons, and various other particles have been derived, and these relations are known as the particle transport equations or Boltzmann equations [1–3]. The neutron transport equation is the basic equation used to study the neutron transport process in a medium. It is essential to understanding neutron transport theory and performing reactor physics analysis. In this section, the interaction of neutrons with matter, relevant basic physical quantities and the foundation of neutron transport theory will be introduced.

2.1.1 Interaction of Neutrons with Matter

Interactions of neutrons with matter mainly occur between the neutrons and the nuclei of the matter, and can be divided into three fundamentally different mechanisms: potential scattering, compound nucleus formation, and direct interaction [1, 2].

The simplest type of nuclear reaction that occurs in a nuclear system is potential scattering, in which a neutron scatters off a nuclear potential without ever penetrating the nucleus itself. Potential scattering is a type of elastic scattering that can occur for neutrons of any energy. In this scattering mode, both the total kinetic energy and the total momentum of the system (neutron and target nucleus) are conserved.

In the formation of a compound nucleus, an incident neutron is absorbed by a target nucleus to form a compound nucleus in an excited state, the energy (above the ground state) of which is equal to the sum of the kinetic energies of the incident neutron and the target nucleus, and their binding energy, as shown in Fig. 2.1. The excited compound nucleus may decay or disintegrate through various pathways, such as resonance elastic scattering, resonance inelastic scattering or fission, etc.

Neutrons of comparatively high energy can also interact with nuclei via direct interaction. In this process, an incident neutron collides directly with the nucleons inside a nucleus. As a result, nucleons can be knocked out of the nucleus, whereas the incident neutron may be retained by the target nucleus, such as (n, p) reaction. After absorbing the kinetic energy of the incident neutron, the target nucleus is excited to some higher energy state. Eventually, this excited nucleus will de-excite by releasing gamma rays.

Depending on the results of the reactions, the interactions between neutrons and nuclei can be classified into two types: neutron scattering and neutron absorption. In neutron scattering, a neutron is released after the reaction, and the difference in energy between the incident and outgoing neutrons becomes an internal or kinetic energy of the target nucleus. Neutron scattering (elastic and inelastic scattering) is the main reaction process for neutron moderation. In neutron absorption, the incident neutron is absorbed by the target nucleus, and particles such as photons and protons

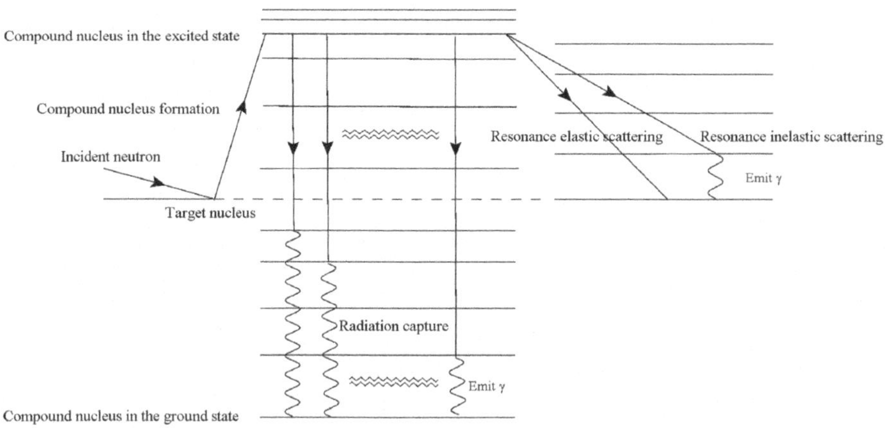

Fig. 2.1 Formation and decay of a compound nucleus

are released thereafter. Such reactions include (n, γ), (n, f), (n, p), (n, d), (n, t), (n, ^{3}He), and (n, α).

The relevant basic physical quantities include the reaction cross section, the differential cross section, and the reaction rate.

There are two types of reaction cross sections: microscopic cross section and macroscopic cross section. The microscopic cross section, for simplification, can be visualized as the effective target area of a single target nucleus for an incident particle. The larger the effective area is, the greater the reaction probability is. By contrast, the macroscopic cross section represents the effective target area of all of the nuclei contained in a volume of material. For example, consider a plate of area A and thickness x, as shown in Fig. 2.2, for which the density of nuclei in the plate is N (1/m^3). An incident neutron beam with intensity I_0 (1/m^2 s) travels in the direction perpendicular to the plate. Then, the number of times P that the incident neutrons interact with the nuclei in the plate per unit time is proportional to I_0, N, A, and x, the equation is as follows:

$$P = \sigma I_0 N A x \tag{2.1}$$

where the proportionality coefficient σ is the microscopic cross section, in units of m^2 (in practical applications, the unit is expressed in barn "b", where 1b = 10^{-28} m^2 or 10^{-24} cm^2). The relationship between the macroscopic cross section (Σ, in units of m^{-1}) and the microscopic cross section σ is

$$\Sigma = N\sigma \tag{2.2}$$

After the collision of a neutron with a nucleus, the probability of the neutron emission is related to the exit angle of the emitted neutron, which is often described by the differential cross section, denoted by $\mathrm{d}\sigma/\mathrm{d}\boldsymbol{\Omega}$, with units of b/sr. In cases where

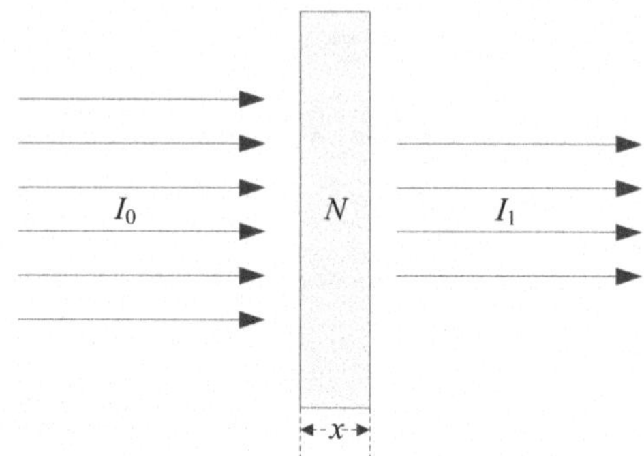

Fig. 2.2 Neutron beam passing through a plate

the probability of neutron emission is related to both the exit angle and the energy, the double-differential cross section, denoted by $d^2\sigma/(d\boldsymbol{\Omega}dE)$ and expressed in units of $b/(sr \cdot MeV)$, is used to describe the emission probability. The reaction rate $(1/m^3\ s)$ is the total number of times that neutrons interact with nuclei in a unit volume per unit time and can be expressed as

$$R = nv\Sigma \tag{2.3}$$

where n is the neutron density, which describes the number of neutrons per unit volume ($1/m^3$), and v is the velocity of neutron movement (m/s). Σ can be explained as the reciprocal of the neutron mean free path, and nv is the total travel length in unit volume and unit time.

The interaction cross section of a neutron with a nucleus strongly depends on the incident neutron's energy. For an incident neutron with energy between 1 eV and 1 MeV, the cross section for neutron interaction with a heavy nucleus or medium-mass nucleus may oscillate with the neutron energy. Many large peaks, which are called resonance peaks, appear on the curves of σ versus energy of the neutron, and this phenomenon is called resonance. During the interaction process, the resonance peaks is broaden as the temperature of matter increases, which is called Doppler broadening.

2.1.2 Foundation of Neutron Transport Theory

Seven reasonable assumptions are made to simplify the transport problems when deriving the neutron transport equation [2, 3]: (1) neutrons are considered as classical points; (2) neutrons are without charge and thus travel in straight lines between point collisions; (3) neutron–neutron interactions are neglected; (4) collisions are instantaneous; (5) the material properties are assumed to be isotropic; (6) the nuclear properties and compositions of the materials under consideration are assumed to be known and time-independent unless explicitly stated otherwise; (7) only the expected or mean value of the neutron density distribution is considered.

Under these assumptions, the main research aims of neutron transport theory are to calculate the neutron distribution in the medium and develop corresponding models and methods. To describe the neutron distribution in a medium, it is necessary to obtain the neutron distribution with space coordinates $\boldsymbol{r}(x, y, z)$, energy E, direction of its velocity $\boldsymbol{\Omega}(\theta, \varphi)$, and time t. This neutron distribution is usually described by the neutron angular density $n(\boldsymbol{r}, E, \boldsymbol{\Omega}, t)$, neutron angular flux $\phi(\boldsymbol{r}, E, \boldsymbol{\Omega}, t)$, and neutron current $J(\boldsymbol{r}, E, t)$.

The neutron angular density $n(\boldsymbol{r}, E, \boldsymbol{\Omega}, t)$ is defined as the probable (or expected) number of neutrons at position $\boldsymbol{r}$ with direction $\boldsymbol{\Omega}$ and energy E at time t per unit volume per unit solid angle per unit energy. The commonly used unit of the neutron angular density is $1/(\mathrm{cm}^3\ \mathrm{sr})$.

The neutron angular flux $\phi(\boldsymbol{r}, E, \boldsymbol{\Omega}, t)$ is defined by the product of the neutron angular density $n(\boldsymbol{r}, E, \boldsymbol{\Omega}, t)$ and neutron velocity v. The unit $1/(\mathrm{cm}^2\ \mathrm{sr\ s})$ is typically used in practical applications. It can be expressed as the number of neutrons at position $\boldsymbol{r}$ with energy E traveling through the unit surface area perpendicular to direction $\boldsymbol{\Omega}$ at time t, as shown in Fig. 2.3. $\phi(\boldsymbol{r}, E, \boldsymbol{\Omega}, t)\mathrm{d}V\mathrm{d}\boldsymbol{\Omega}\mathrm{d}E\mathrm{d}t$ describes the total path length traveled during $\mathrm{d}t$ by all the particles in the incremental phase space volume $\mathrm{d}V\mathrm{d}\boldsymbol{\Omega}\mathrm{d}E$.

The neutron flux $\phi(\boldsymbol{r}, E, t)$ is obtained by integrating the neutron angular flux over all directions, as shown in Eq. (2.4). The unit then becomes $1/(\mathrm{cm}^2\ \mathrm{s})$. The neutron fluence is then obtained by integrating the neutron flux over time, and the unit is $1/\mathrm{cm}^2$. The neutron flux equals the neutron fluence rate according to the following definition.

$$\phi(\boldsymbol{r}, E, t) = vn(\boldsymbol{r}, E, t) = \int_{4\pi} \phi(\boldsymbol{r}, E, \boldsymbol{\Omega}, t)\mathrm{d}\boldsymbol{\Omega} = \int_{4\pi} vn(\boldsymbol{r}, E, \boldsymbol{\Omega}, t)\mathrm{d}\boldsymbol{\Omega} \tag{2.4}$$

The neutron current $\boldsymbol{J}(\boldsymbol{r}, E, t)$ in units of $1/(\mathrm{cm}^2 \cdot \mathrm{s})$ is obtained by integrating the quantity $\boldsymbol{\Omega}\phi(\boldsymbol{r}, E, \boldsymbol{\Omega}, t)$ over all directions.

$$\boldsymbol{J}(\boldsymbol{r}, E, t) = \int_{4\pi} \boldsymbol{\Omega}\,\phi(\boldsymbol{r}, E, \boldsymbol{\Omega}, t)\mathrm{d}\boldsymbol{\Omega} \tag{2.5}$$

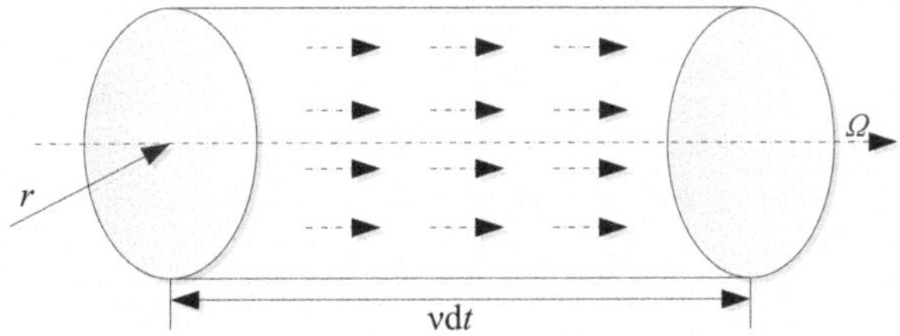

Fig. 2.3 Differential neutron beam

2.1.2.1 Neutron Transport Equation

The balance of the neutron number gain or loss, is a basic principle in the study of neutron transport [2, 3], which means that the rate of change of the neutron angular density with time is equal to the neutron production rate minus the leakage and removal rate in a given microelement ($\mathrm{d}V\mathrm{d}E\mathrm{d}\boldsymbol{\Omega}$), as shown in Fig. 2.4, where $\boldsymbol{\Omega}$ perpendicular to $\mathrm{d}A$. In a steady-state system, the rate of change of the neutron angular density with time, $\partial n/\partial t$, is zero.

$$\frac{\partial n}{\partial t} = \frac{1}{v}\frac{\partial \phi}{\partial t} = \text{Production rate (Q)} - \text{Leakage rate (L)} - \text{Removal rate (R)} \tag{2.6}$$

1. Production rate

There are three sources of neutrons in $\mathrm{d}V\mathrm{d}E\mathrm{d}\boldsymbol{\Omega}$: (1) the scattering source (Q_s), which refers to the neutrons produced after incident neutrons scatter from energy E' and direction $\boldsymbol{\Omega}'$ to energy E and direction $\boldsymbol{\Omega}$; (2) the fission source (Q_f); and (3) the independent external source (S), such as a spontaneous fission source, a natural radioactive source and others. Thus, the production rate is

$$\begin{aligned} Q &= Q_\mathrm{s} + Q_\mathrm{f} + S \\ &= \mathrm{d}V\mathrm{d}E\mathrm{d}\boldsymbol{\Omega}\left[\int_0^\infty \mathrm{d}E' \int_{\boldsymbol{\Omega}'} \Sigma_\mathrm{s}(\boldsymbol{r}, E') f(\boldsymbol{r}, E' \to E, \boldsymbol{\Omega}' \to \boldsymbol{\Omega}) \times \phi(\boldsymbol{r}, E', \boldsymbol{\Omega}', t)\mathrm{d}\boldsymbol{\Omega}' \right. \\ &\left. + \frac{\chi(E)}{4\pi}\int_0^\infty \mathrm{d}E' \int_{\boldsymbol{\Omega}'} \upsilon\Sigma_\mathrm{f}(\boldsymbol{r}, E')\phi(\boldsymbol{r}, E', \boldsymbol{\Omega}', t)\mathrm{d}\boldsymbol{\Omega}' + S(\boldsymbol{r}, E, \boldsymbol{\Omega}, t)\right] \end{aligned} \tag{2.7}$$

where $f(\boldsymbol{r}, E' \to E, \boldsymbol{\Omega}' \to \boldsymbol{\Omega})$ is the scattering function, which is defined as the probability of a neutron being scattered from $(E', \boldsymbol{\Omega}')$ to $(E, \boldsymbol{\Omega})$; Σ_s is the macroscopic scattering cross section; Σ_f is macroscopic fission cross section; $\chi(E)$ is the fission spectrum; and υ is the mean number of the fission neutrons released per fission. It

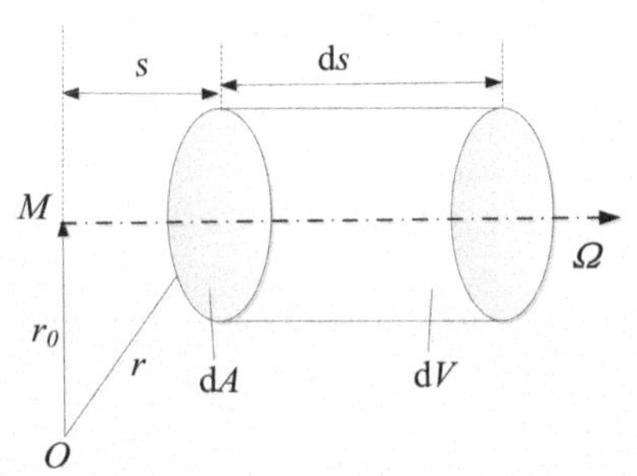

Fig. 2.4 Neutron in microelement $\mathrm{d}V$

should be noted that the time-delaying effects of delayed neutrons are neglected in Eq. (2.7).

2. Leakage rate

The neutron leakage rate is defined as the difference between the numbers of neutrons exiting and entering the elemental volume ($\mathrm{d}V$) per unit time. As shown in Fig. 2.4, the number of neutrons entering the volume $\mathrm{d}V$ from the surface $\mathrm{d}A$ at position $\boldsymbol{r}_0 + s\boldsymbol{\Omega}$ with energy from E to $E + \mathrm{d}E$ with a direction interval $\mathrm{d}\boldsymbol{\Omega}$ at $\boldsymbol{\Omega}$ per unit time is $\phi(\boldsymbol{r}_0 + s\boldsymbol{\Omega}, E, \boldsymbol{\Omega}, t)\mathrm{d}A\mathrm{d}E\mathrm{d}\boldsymbol{\Omega}$. The number of neutrons streaming out from another surface at position $\boldsymbol{r}_0 + (s + \mathrm{d}s)\boldsymbol{\Omega}$ of volume $\mathrm{d}V$ with energy from E to $E + \mathrm{d}E$ with a direction interval $\mathrm{d}\boldsymbol{\Omega}$ at $\boldsymbol{\Omega}$ per unit time is $\phi(\boldsymbol{r}_0 + (s + \mathrm{d}s)\boldsymbol{\Omega}, E, \boldsymbol{\Omega}, t)\mathrm{d}A\mathrm{d}E\mathrm{d}\boldsymbol{\Omega}$. Thus, the number of neutrons leaking out from element $\mathrm{d}V$ per unit time can be expressed as in the following Eq. (2.8), where the condition of $\mathrm{d}s \to 0$ should be applied.

$$\begin{aligned} L &= \phi(\boldsymbol{r}_0 + (s + \mathrm{d}s)\boldsymbol{\Omega}, E, \boldsymbol{\Omega}, t)\mathrm{d}A\mathrm{d}E\mathrm{d}\boldsymbol{\Omega} - \phi(\boldsymbol{r}_0 + s\boldsymbol{\Omega}, E, \boldsymbol{\Omega}, t)\mathrm{d}A\mathrm{d}E\mathrm{d}\boldsymbol{\Omega} \\ &= \frac{\mathrm{d}\phi(\boldsymbol{r}_0 + s\boldsymbol{\Omega}, E, \boldsymbol{\Omega}, t)}{\mathrm{d}s}\mathrm{d}V\mathrm{d}E\mathrm{d}\boldsymbol{\Omega} = \boldsymbol{\Omega}\cdot\nabla\phi\mathrm{d}V\mathrm{d}E\mathrm{d}\boldsymbol{\Omega} \end{aligned} \tag{2.8}$$

3. Removal rate

Neutrons can be removed from $\mathrm{d}V\mathrm{d}E\mathrm{d}\boldsymbol{\Omega}$ in two ways: (1) neutrons can be absorbed in $\mathrm{d}V$, and (2) neutrons can be scattered out from energy E and direction $\boldsymbol{\Omega}$ in $\mathrm{d}V$. Therefore, the neutron number removed from $\mathrm{d}V\mathrm{d}E\mathrm{d}\boldsymbol{\Omega}$ per unit time is

$$R = (\Sigma_\mathrm{s} + \Sigma_\mathrm{a})\phi(\boldsymbol{r}, E, \boldsymbol{\Omega}, t)\mathrm{d}V\mathrm{d}E\mathrm{d}\boldsymbol{\Omega} = \Sigma_\mathrm{t}\phi(\boldsymbol{r}, E, \boldsymbol{\Omega}, t)\mathrm{d}V\mathrm{d}E\mathrm{d}\boldsymbol{\Omega} \tag{2.9}$$

where Σ_s is the macroscopic scattering cross section, Σ_a is the macroscopic absorption cross section, and Σ_t is the total macroscopic cross section.

Considering unit volumes, unit energy, and unit solid angle and substituting Eqs. (2.7), (2.8), and (2.9) into Eq. (2.6), the conservation equation of neutrons in phase space ($\boldsymbol{r} \times E \times \boldsymbol{\Omega}$) at any instant t is derived as shown in Eq. (2.10), which is a classical time-dependent differential–integral neutron transport equation. The steady-state neutron transport equation can be obtained by setting $\partial\phi/\partial t$ to zero.

$$\begin{aligned} &\frac{1}{v}\frac{\partial\phi}{\partial t} + \boldsymbol{\Omega}\cdot\nabla\phi + \Sigma_\mathrm{t}(\boldsymbol{r}, E)\phi \\ &\quad = \int_0^\infty \mathrm{d}E' \int_{\boldsymbol{\Omega}'} \Sigma_\mathrm{s}(\boldsymbol{r}, E')f(\boldsymbol{r}, E' \to E, \boldsymbol{\Omega}' \to \boldsymbol{\Omega})\phi(\boldsymbol{r}, E', \boldsymbol{\Omega}', t)\mathrm{d}\boldsymbol{\Omega}' \\ &\quad + \frac{\chi(E)}{4\pi}\int_0^\infty \mathrm{d}E' \int_{\boldsymbol{\Omega}'} v\Sigma_\mathrm{f}(\boldsymbol{r}, E')\phi(\boldsymbol{r}, E', \boldsymbol{\Omega}', t)\mathrm{d}\boldsymbol{\Omega}' + S(\boldsymbol{r}, E, \boldsymbol{\Omega}, t) \end{aligned} \tag{2.10}$$

The time-dependent integral neutron transport equation can be obtained by integrating Eq. (2.10) along the characteristic curve [2, 3].

$$\phi(\boldsymbol{r}, E, \boldsymbol{\Omega}, t) = \int_{0}^{\infty} \exp\left[-\int_{0}^{l} \Sigma_{\mathrm{t}}(\boldsymbol{r} - l'\boldsymbol{\Omega}, E)\mathrm{d}l'\right] Q\left(\boldsymbol{r} - l\boldsymbol{\Omega}, E, \boldsymbol{\Omega}, t - \frac{l}{v}\right)\mathrm{d}l \tag{2.11}$$

where Q is the neutron source term, including the external neutron source, scattering source, and fission source. For the advanced fission system, the source term contains a scattering source and a fission source, but it does not contain an external source. For the fusion system, the source term contains a scattering source and a neutron source generated from the fusion reaction, but it does not contain any fission source (because there is no fissile or fertile material). For the hybrid nuclear system, fertile materials in the breeding or transmutation blanket introduce a fission source term in addition to the scattering source and external source (fusion neutron source or acceleration neutron source).

4. Boundary and initial conditions

To solve the neutron transport equation, boundary and initial conditions are necessary. Commonly used boundary conditions are as follows:

(1) The neutron angular flux must be a finite, nonnegative real number in regions where the equation applies.
(2) The neutron angular flux is continuous across the boundary between two different media if there is no source at the interface.
(3) The outer surface boundary conditions (i.e., explicit boundary conditions) can be divided into two types. (i) If no neutrons enter from the external regions, or if a neutron cannot return once it leaves the surface, then the surface is called a free surface, for which the incident neutron angular flux is zero. A convex surface that meets the vacuum is one example. (ii) The second type is a source surface with a given incident flux. Usually, the incident flux is treated as an imaginary surface source, and thus it can be treated as a free surface.
(4) Implicit boundary conditions can be categorized into three types according to the relationship between the incident and the outgoing neutron fluxes. (i) The incident neutron flux in one direction is equal to the outgoing neutron flux in the reflected direction on the reflective boundary that appears on the symmetry plane of the symmetry system. (ii) The incident flux on the albedo boundary is equal to a known isotropic albedo, $\alpha(E)$, times the outgoing flux on the same boundary in the direction corresponding to the spectral reflection. The special case of $\alpha = 1.0$ is referred to as a reflective boundary condition since all outgoing particles are "reflected" back. (iii) All particles passing out of volume V through a surface element return with an isotropic distribution on the white boundary.

To solve the time-dependent neutron transport equation, the initial condition (i.e., the angular neutron flux distribution in the phase space at the initial moment) must be given.

$$\phi(\boldsymbol{r}, E, \boldsymbol{\Omega}, t)|_{t=0} = \phi_0(\boldsymbol{r}, E, \boldsymbol{\Omega}) \tag{2.12}$$

The neutron transport equation contains many independent variables, such as the space, the coordinates, the energy, the neutron motion direction, and the time. In practice, the complexity of the geometry and structure, the heterogeneity of the spatial distribution of the materials, and the dependence of cross sections on the neutron energy for various materials (fissile and non-fissile) should all be considered. Thus, it is very difficult to obtain an exact analytical solution for the energy-dependent neutron transport equation, even in the steady-state case. Therefore, approximate numerical methods must be adopted to solve the transport equation. The neutron transport exhibits weak anisotropy in traditional nuclear systems, and the neutron transport equation can be simplified as a diffusion equation, which plays an important role in traditional nuclear system physics analysis. The neutron transport exhibits strong anisotropy in advanced nuclear systems, and it cannot satisfy the diffusion approximation assumption.

When performing deep-penetration simulations for radiation shielding, sensitivity and uncertainty analysis, and neutron kinetics analysis, among others, the adjoint neutron flux is often used to characterize the contribution of neutrons with position $\boldsymbol{r}$, time t, energy E, and direction $\boldsymbol{\Omega}$ to the target response. The adjoint neutron flux is also known as the neutron importance function.

The adjoint neutron transport equation, which is also known as the neutron importance conservation equation, can be derived according to the mathematical conjugation of the neutron transport equation [2, 3].

$$\begin{aligned}
&-\frac{1}{v}\frac{\partial\phi^*}{\partial t}-\boldsymbol{\Omega}\cdot\nabla\phi^*+\Sigma_{\mathrm{t}}\phi^* \\
&=\int_0^\infty \mathrm{d}E'\int_{\boldsymbol{\Omega}'}\Sigma_{\mathrm{s}}(\boldsymbol{r},E)f(\boldsymbol{r},E\to E',\boldsymbol{\Omega}\to\boldsymbol{\Omega}')\phi^*(\boldsymbol{r},E',\boldsymbol{\Omega}',t)\mathrm{d}\boldsymbol{\Omega}' \\
&+\frac{v\Sigma_{\mathrm{f}}(\boldsymbol{r},E)}{4\pi}\int_0^\infty \mathrm{d}E'\int_{\boldsymbol{\Omega}'}\chi(E')\phi^*(\boldsymbol{r},E',\boldsymbol{\Omega}',t)\mathrm{d}\boldsymbol{\Omega}'+S^*
\end{aligned} \tag{2.13}$$

By solving Eq. (2.13), the adjoint neutron flux can be obtained. In Monte Carlo neutron transport simulation, the adjoint neutron flux can be obtained by the iterated fission probability method.

Because the neutrons leaving the surface of the system no longer contribute to the system, the boundary condition at boundary Γ of the region is

$$\phi^*(\boldsymbol{r},E,\boldsymbol{\Omega},t)=0,\quad \text{if}\,(\boldsymbol{\Omega}\cdot n)>0\ \text{ and }\ \boldsymbol{r}\in\Gamma \tag{2.14}$$

where n is the normal vector of boundary Γ.

The physical meaning of the time-independent adjoint function can be understood by considering a steady-state system with an arbitrary steady source $S(\boldsymbol{r}, E, \boldsymbol{\Omega})$. S^* characterizes a certain response. For instance, in a fission reaction, $S^* = \Sigma_{\mathrm{f}}$, and

$\phi^*(\boldsymbol{r}, E, \boldsymbol{\Omega})$ represents the contribution of a neutron with position $\boldsymbol{r}$, energy E, and direction $\boldsymbol{\Omega}$ to a certain response.

For time-independent multiplying media without an external neutron source, $S^* = 0$, and $\phi^*(\boldsymbol{r}, E, \boldsymbol{\Omega})$ denotes the response of a neutron with position $\boldsymbol{r}$, energy E, and direction $\boldsymbol{\Omega}$ to nuclear power.

The neutron transport equation is converted to the steady-state neutron transport equation without consideration of the time term. The following sections of this chapter will introduce the solution of the steady-state neutron transport equation.

2.1.2.2 Fixed-Source and Eigenvalue Problems

The neutron transport equation can be expressed in operator notation as follows:

$$\boldsymbol{M}\phi = \boldsymbol{F}\phi + \boldsymbol{S} \tag{2.15}$$

where $\phi = \phi(r, E, \boldsymbol{\Omega})$, $\boldsymbol{M}$ is the transport operator, and $\boldsymbol{F}$ is the fission operator.

$$\begin{aligned}\boldsymbol{M}\phi = {} & \boldsymbol{\Omega}\cdot\nabla\phi + \Sigma_t(\boldsymbol{r}, E)\phi \\ & - \int_0^\infty \mathrm{d}E' \int_{\boldsymbol{\Omega}'} \Sigma_s(\boldsymbol{r}, E') f(\boldsymbol{r}, E' \to E, \boldsymbol{\Omega}' \to \boldsymbol{\Omega}) \phi(\boldsymbol{r}, E', \boldsymbol{\Omega}') \mathrm{d}\boldsymbol{\Omega}'\end{aligned} \tag{2.16}$$

$$\boldsymbol{F}\phi = \frac{\chi(E)}{4\pi} \int_0^\infty \mathrm{d}E' \int_{\boldsymbol{\Omega}'} \upsilon\Sigma_f(\boldsymbol{r}, E') \phi(\boldsymbol{r}, E', \boldsymbol{\Omega}') \mathrm{d}\boldsymbol{\Omega}' \tag{2.17}$$

Solutions to the neutron transport equation can be divided into fixed-source problems and eigenvalue problems according to whether an external neutron source exists. If neutrons from a time-independent external source are supplied, the subcritical system will eventually reach an equilibrium state characterized by a time-independent flux distribution in which the production rate of external and fission neutrons is in equilibrium with the absorption and leakage. However, if the system is critical or supercritical, no such equilibrium can exist in the presence of an external source; thus, the neutron flux distribution will be an increasing function of time. This is the so-called fixed-source problem. Eigenvalue problems need to be solved (i.e., via criticality calculation) to determine whether a system is subcritical or supercritical and to obtain its effective multiplication factor k_{eff}.

A key step in fixed-source problems is obtaining the neutron flux. For a system without a multiplying medium, such as a fusion system, the fission neutron source Q_f is zero. Then, the fixed-source problem is expressed by using the following inhomogeneous equation

$$\boldsymbol{M}\phi = \boldsymbol{S} \tag{2.18}$$

For systems with a multiplying medium, such as a hybrid nuclear system, the fixed-source problem is expressed by the following inhomogeneous equation:

$$\boldsymbol{M}\phi = \boldsymbol{F}\phi + \boldsymbol{S} \tag{2.19}$$

Eigenvalue problems can be categorized into k-eigenvalue problems and α-eigenvalue problems. Taking the k-eigenvalue problem as an example, it is necessary to solve for the eigenvalue and the eigenfunction of the homogeneous Eq. (2.20). For instance, the effective multiplication factor k_{eff} and the neutron flux of the system must be determined for fission systems or hybrid nuclear systems.

$$\boldsymbol{M}\phi = \frac{1}{k_{\text{eff}}}\boldsymbol{F}\phi \tag{2.20}$$

In a hybrid nuclear system with an external neutron source, the effective multiplication factor k_{eff} is not sufficient to characterize the breeding property of the system. Therefore, the source multiplication factor k_{s} is introduced to demonstrate the breeding property of the system with k_{eff}. Thus, k_{s} characterizes the breeding property of the system by considering the external neutron source.

$$k_{\text{s}} = \frac{\boldsymbol{F}\phi}{\boldsymbol{M}\phi} = \frac{\boldsymbol{F}\phi}{\boldsymbol{F}\phi + \boldsymbol{S}} \tag{2.21}$$

2.2 Monte Carlo Computational Methods

The basic idea underlying the Monte Carlo method is as follows: When the solution to a problem is the probability of an event or the mathematical expectation of a random variable, the occurrence frequency of the event or the arithmetical mean of several specific observations of the random variable can be obtained through numerical experiments. Thus, a solution to that problem is obtained. The advantages of Monte Carlo methods include the realistic descriptions of the characteristics of objects with stochastic natures, the ability to simulate physical experiments, few limitations on geometric conditions, and parallel computing adaptability. The disadvantages of Monte Carlo methods include the slow convergence rate and statistical uncertainty [2, 3].

The following section is composed of three parts. The first part discusses the basic principles of Monte Carlo neutron transport calculations. The second part introduces the physical quantity estimation methods, and the last part presents the main acceleration methods.

2.2.1 Basic Principles

When the Monte Carlo method is applied to neutron transport simulations, the random motion history of a single neutron in a certain geometry is simulated first, and then, sufficient random experimental values (or sampled values) are obtained by tracking a large number of neutron histories. Finally, an estimator of the numerical characteristics of a random variable is obtained as the solution to the problem. Three processes are involved in solving the neutron transport problem: (1) source sampling from its probability distribution; (2) tracking of the neutron locations, energies and directions; and (3) contribution recording and result analysis. The simulation process is shown in Fig. 2.5.

By tracking a large number of neutron histories and recording the contribution of each neutron, physical quantities, such as the neutron flux, dose rate, energy deposition, reaction rates, eigenvalues, reactivity coefficient, and kinetic parameter, can be obtained by statistical methods, and the statistical errors on these quantities can be determined. The statistical method is described in the following section, and examples for the estimation of the neutron flux, eigenvalue k_{eff}, reactivity coefficient, and kinetic parameter are presented.

2.2.2 Estimation Methods for Physical Quantities

2.2.2.1 Estimation of Neutron Flux

Neutron flux calculations are essential in neutron transport simulations. There are three types of flux: point flux, surface flux, and volume flux. The methods used to estimate point flux include the pointing probability method and the reciprocity method. The estimation methods for surface flux include the analytical estimator, the weighting method, the point flux substitution method, and the volume flux substitution method. Finally, those used to estimate volume flux include the analytical estimator, the track length estimator, the collision estimator, the absorption estimator, and the point flux substitution method. In this section, volume flux estimation is taken as an example.

The volume flux in a given cell (volume V) is obtained by

$$\phi(V) = \frac{1}{V}\int_V \phi(\boldsymbol{r}, E, \boldsymbol{\Omega})\mathrm{d}V\mathrm{d}E\mathrm{d}\boldsymbol{\Omega} \tag{2.22}$$

where the angular flux $\phi(\boldsymbol{r}, E, \boldsymbol{\Omega})$ can be expressed as the sum of the contributions of each scattering.

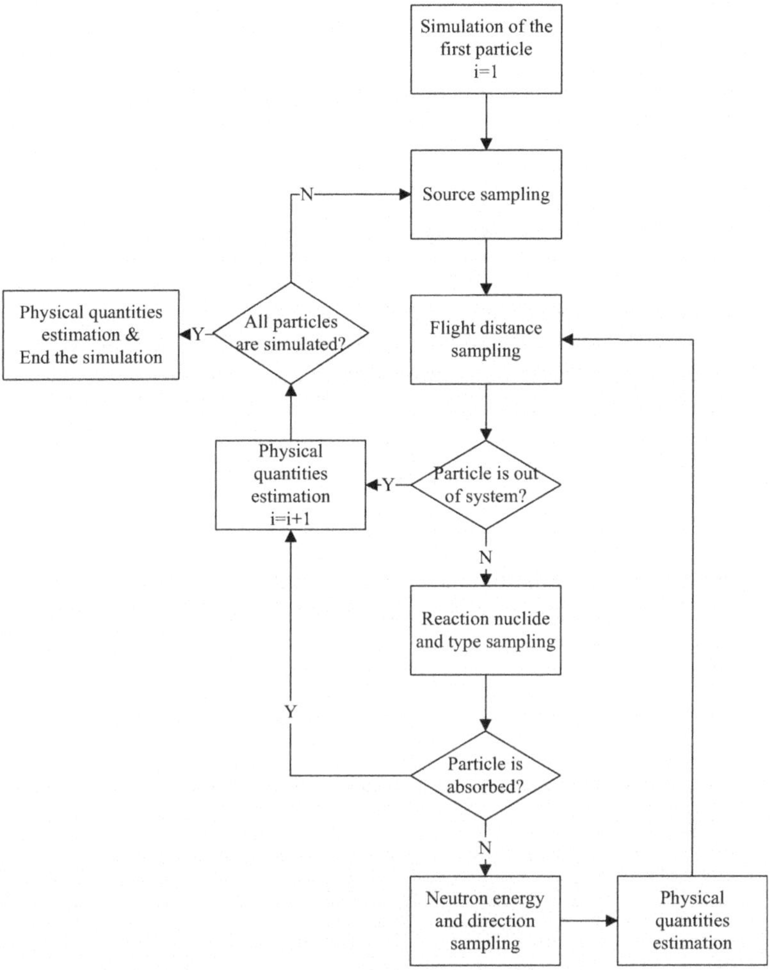

Fig. 2.5 Schematic diagram of Monte Carlo particle transport simulation

$$\phi(\boldsymbol{r}, E, \boldsymbol{\Omega}) = \frac{1}{N}\sum_{i=1}^{N}\phi_i(\boldsymbol{r}, E, \boldsymbol{\Omega}) = \frac{1}{N}\sum_{i=1}^{N}\sum_{n=1}^{M}\phi_{i,n}(\boldsymbol{r}, E, \boldsymbol{\Omega}) \tag{2.23}$$

where $\phi_i(\boldsymbol{r}, E, \boldsymbol{\Omega})$ represents the track length of the ith neutron at the phase space $\mathrm{d}V\mathrm{d}E\mathrm{d}\boldsymbol{\Omega}$. $\phi_{i,n}(\boldsymbol{r}, E, \boldsymbol{\Omega})\mathrm{d}V\mathrm{d}E\mathrm{d}\boldsymbol{\Omega}$ is the average track length of the ith neutron between the nth and $(n + 1)$th scatterings at microelement $\mathrm{d}V\mathrm{d}E\mathrm{d}\boldsymbol{\Omega}$.

The following is an example of volume flux estimation using the track length estimator. Assuming that the track length from the nth to the $(n + 1)$th scattering of the ith neutron is s, the range of volume element V is $s_1 \sim s_2$, and the flux contribution of the nth scattering is

$$\phi_{i,n}(V) = \begin{cases} w_n(s - s_1)/V & s_1 \le s \le s_2 \\ w_n(s_2 - s_1)/V & s > s_2 \\ 0 & s < s_1 \end{cases} \tag{2.24}$$

where w_n is the neutron weight after the nth scattering.

Equations (2.23) and (2.24) can be used to calculate the contribution of a single neutron to the flux ϕ_i. Then, the neutron flux in a certain cell can be obtained by Monte Carlo simulation of N neutrons.

$$\bar{\phi} = \frac{1}{N}\sum_{i=1}^{N}\phi_i \tag{2.25}$$

In the Monte Carlo calculation, the estimated value of the variance is required, in addition to the estimated flux value for a certain cell. The estimated variance (S^2) is

$$S^2 = \frac{1}{N-1}\sum_{i=1}^{N}\left(\phi_i - \bar{\phi}\right)^2 \tag{2.26}$$

where S is the standard deviation of the N neutrons' contributions. Thus, the statistical error of $\bar{\phi}$ is

$$S_{\bar{\phi}} = S\frac{1}{\sqrt{N}} \tag{2.27}$$

2.2.2.2 Eigenvalue Estimation

Theoretically, the effective multiplication factor k_{eff} is defined as the ratio of the neutron counts of two adjacent generations. From neutron transport theory, this quantity can be expressed as

$$k_{\text{eff}} = \lim_{n\to\infty} k^{(n+1)} = \lim_{n\to\infty} \frac{\int_{4\pi}\int_0^{E_{\max}}\int_V \upsilon(\boldsymbol{r}, E')\Sigma_{\text{f}}(\boldsymbol{r}, E')\phi^{(n+1)}(\boldsymbol{r}, E', \boldsymbol{\Omega}')\,\mathrm{d}\boldsymbol{r}\mathrm{d}E'\mathrm{d}\boldsymbol{\Omega}'}{\int_{4\pi}\int_0^{E_{\max}}\int_V \upsilon(\boldsymbol{r}, E')\Sigma_{\text{f}}(\boldsymbol{r}, E')\phi^{(n)}(\boldsymbol{r}, E', \boldsymbol{\Omega}')\,\mathrm{d}\boldsymbol{r}\mathrm{d}E'\mathrm{d}\boldsymbol{\Omega}'} \tag{2.28}$$

where

$$\phi^{(n+1)}(\boldsymbol{r}, E, \boldsymbol{\Omega}) = \int_0^{\infty} e^{-\int_0^{l} \Sigma_t(\boldsymbol{r}-l'\boldsymbol{\Omega}, E)\mathrm{d}l'}$$

$$\left[\frac{1}{k^{(n)}} \int_0^{E_{\max}} \int_{4\pi} \frac{\chi(r, E' \to E)}{4\pi} \upsilon(\boldsymbol{r}', E') \Sigma_f(\boldsymbol{r}', E') \phi^{(n)}(\boldsymbol{r}', E', \boldsymbol{\Omega}') \mathrm{d}\boldsymbol{\Omega}' \mathrm{d}E' \right.$$

$$\left. + \int_0^{E_{\max}} \int_{4\pi} \Sigma_s(\boldsymbol{r}'; E', \boldsymbol{\Omega}' \to E, \boldsymbol{\Omega}) \phi^{(n)}(\boldsymbol{r}', E', \boldsymbol{\Omega}') \mathrm{d}\boldsymbol{\Omega}' \mathrm{d}E' \right] \mathrm{d}l \quad (2.29)$$

The exponential iteration method is widely used to solve k-eigenvalue problems in Monte Carlo eigenvalue calculations, which calculates the effective multiplication factor k_{eff}. The neutrons and their respective secondary neutrons are all simulated generation by generation. During the simulation of one generation, the neutrons generated from the fission reaction are stored as the source neutrons for the next generation. Because the spatial distribution of the fission source for the initial generations (i.e., the non-active generations) is not convergent, and the statistical error is large, the k_{eff} of one cycle begins to accumulate from the active generation. The collision estimator, absorption estimator, track length estimator, and their combinations are often used to estimate k_{eff} [4, 5].

The reactivity ρ is usually used to represent the global property of fission systems and hybrid nuclear systems, and it is defined as

$$\rho = \frac{k_{\mathrm{eff}} - 1}{k_{\mathrm{eff}}} \quad (2.30)$$

2.2.2.3 Reactivity Coefficient Estimation

The reactivity coefficient, a very important physical quantity in fission systems and hybrid nuclear systems, is closely related to system safety. The reactivity coefficient describes the reactivity change rate with certain parameters, such as temperature, density, size, etc., including the temperature coefficient (coolant temperature coefficient, fuel temperature coefficient, etc.), density coefficient (coolant density coefficient, fuel density coefficient, etc.), expansion coefficient (axial, radial, etc.), cavitation coefficient, etc. The definition of the reactivity coefficient is presented in Eq. (2.31).

$$\alpha_i = \frac{\partial \rho}{\partial x_i} \quad (2.31)$$

where α_i is the reactivity coefficient of the ith parameter x_i of the system and ρ is the reactivity. To calculate the reactivity coefficient, the parameter x_i can be slightly changed by Δx_i, and two calculations are performed to obtain the reactivity ρ and ρ' corresponding to x_i and $x_i + \Delta x_i$, as well as the corresponding eigenvalues k_{eff1} and k_{eff2}. Then, the reactivity coefficient can be calculated according to the following equation.

$$\alpha_i = \frac{\partial \rho}{\partial x_i} = \frac{\rho' - \rho}{\Delta x_i} = \frac{1}{\Delta x_i} \frac{k_{\mathrm{eff2}} - k_{\mathrm{eff1}}}{k_{\mathrm{eff1}} k_{\mathrm{eff2}}} \tag{2.32}$$

When the variation Δx_i is small, the reactivity change will be very small, even less than the calculation accuracy of the eigenvalue k_{eff}. Thus, the reactivity coefficient will be difficult to estimate, and more computational resources are required to improve the calculation accuracy. Under this condition, the perturbation theory can be used to calculate the reactivity coefficient.

According to Eq. (2.20), assuming the operator $P = M - F/k_{\mathrm{eff}}$, the transport equation before perturbation is

$$P\phi = 0 \tag{2.33}$$

The corresponding adjoint equation is

$$P^*\phi^* = 0 \tag{2.34}$$

Assuming that some perturbations are introduced into the system, such as the changes of temperature, density, size, etc., the operator P will become $P' = P + \delta P$, and the perturbed neutron transport equation becomes

$$P'\phi' = P\phi' + \delta P\phi' = 0 \tag{2.35}$$

in which $\phi' = \phi + \delta\phi$. The inner product of the solution of the unperturbed adjoint equation ϕ^* and the perturbed transport equation is

$$\langle \phi^*, P\phi' \rangle + \langle \phi^*, \delta P\phi' \rangle = 0 \tag{2.36}$$

Because P and P^* are adjoint, assuming that the boundary condition is not perturbed, then

$$\langle \phi', P^*\phi^* \rangle + \langle \phi^*, \delta P\phi' \rangle = 0 \tag{2.37}$$

According to Eq. (2.34), $P^*\phi^* = 0$, and then

$$\langle \phi^*, \delta P\phi' \rangle = 0 \tag{2.38}$$

Equation (2.38) is established regardless of the magnitude of δP. However, the perturbation neutron flux ϕ' is difficult to obtain, so some approximation needs to be introduced for simplification. Assuming that the perturbation is small, there is no significant change in flux, and thus the second-order amount of flux $\delta\phi$ can be eliminated. Then, $\phi' = \phi + \delta\phi \approx \phi$, and the basic equation of the perturbation theory can be obtained from Eq. (2.38).

$$\langle \phi^*, \delta P\phi \rangle = 0 \tag{2.39}$$

This is the perturbation equation, in which the perturbation operator δP includes perturbation of various parameters and the corresponding changes of $k_{\rm eff}$. In the Monte Carlo neutron transport simulation, by the iterated fission probability method, the adjoint neutron flux ϕ^* can be calculated, and the neutron flux ϕ before perturbation can be obtained. According to the perturbation of temperature, density, size, and shape of the nuclear system, the corresponding change of k can be calculated, and then the reactivity coefficient can be calculated from Eq. (2.32).

2.2.2.4 Kinetic Parameter Estimation

The kinetic parameters, including the effective delayed neutron fractions ($\beta_{\rm eff}$), prompt neutron lifetime (l), neutron generation time (Λ), reactor period (T), and α-eigenvalue, are the important parameters in the safety analysis of fission systems and hybrid nuclear systems. The effective delayed neutron fraction is the ratio of the number of fission reactions caused by delayed neutrons to the total number of fission reactions, as shown in Eq. (2.40), characterizing the value of delayed neutrons. Prompt neutron lifetime describes the average time from a generation to absorption or removal from the system for prompt neutrons. Neutron generation time refers to the average time of neutron production between adjacent generations, namely, the time from the production of a neutron in the previous generation of fission to the production of a neutron in this generation of fission. Reactor period is the time required for the neutron density to change e times exponentially. α-eigenvalue describes the change of the neutron flux with time. There are close relationships between these kinetic parameters, and based on two kinetic parameters and the reactivity ρ, other kinetic parameters can be calculated. The estimation method for effective delayed neutron fractions and prompt neutron lifetime will be introduced in this section.

The effective delayed neutron fraction is defined as [6]

$$\beta_{\rm eff} = \frac{\int \phi^*(r, E', \Omega')\chi_{\rm d}(E')\upsilon_{\rm d}(E)\Sigma_{\rm f}(r, E)\phi(r, E, \Omega)dEd\Omega dE'd\Omega'dr}{\int \phi^*(r, E', \Omega')\chi(E')\upsilon(E)\Sigma_{\rm f}(r, E)\phi(r, E, \Omega)dEd\Omega dE'd\Omega'dr} \tag{2.40}$$

In Monte Carlo transport calculation, the adjoint neutron angle flux $\phi^*(r, E', \Omega')$ can be obtained by the iterated fission probability method. The neutron angular flux $\phi(r, E, \Omega)$ can be directly obtained, and the fission spectrum $\chi(E')$, the delayed neutron fission spectrum $\chi_{\mathrm{d}}(E')$, the fission neutron yield $\upsilon(E)$, the delayed neutron yield $\upsilon_{\mathrm{d}}(E)$, and the fission reaction cross section $\Sigma_{\mathrm{f}}(r, E)$ can be obtained from the nuclear data. Then, the delayed neutron fractions can be calculated by Eq. (2.40).

The prompt neutron lifetime is defined as [6]

$$l = \left\langle \phi^*, \frac{1}{v}\phi \right\rangle \tag{2.41}$$

where v is the neutron speed. In Monte Carlo transport calculation, the adjoint neutron flux ϕ^* can be obtained by the iterated fission probability method, $\frac{1}{v}\phi$ can be obtained directly, and then the prompt neutron lifetime can be obtained by Eq. (2.41).

With effective delayed neutron fractions and prompt neutron lifetime, other kinetic parameters can be obtained by Eqs. (2.42) through (2.44) [6].

$$\Lambda = \frac{l + \sum_i \beta_i/\lambda_i}{k_{eff}} \tag{2.42}$$

$$T = \frac{l + \sum_i \beta_i/\lambda_i}{k_{eff} - 1} \tag{2.43}$$

$$\alpha = \frac{\rho - \beta_{eff}}{l(1 - \rho)} \tag{2.44}$$

where β_i and λ_i are the fractions and the decay constant of the ith delayed neutron precursor nucleus, respectively.

2.2.3 Acceleration Methods

Considering the slow convergence of the Monte Carlo method, a number of methods have been proposed to accelerate Monte Carlo transport calculations. These methods include variance reduction techniques, source convergence acceleration methods, efficient geometry processing methods, and parallel computing methods.

2.2.3.1 Variance Reduction Techniques

During a transport simulation, the statistical error is generally required to be less than a certain value. For example, in a point detector tally, the general requirement of the statistical error should be less than 5%. Because it is proportional to $N^{-1/2}$, the statistical error can be reduced by increasing the number of simulated particles or using variance reduction techniques. In the variance reduction techniques, the stochastic probability model that makes the variance as low as possible is selected to reduce the statistical error while ensures that the calculation result is unbiased.

Variance reduction techniques are divided into two varieties: basic variance reduction techniques and adaptive variance reduction techniques.

1. Basic variance reduction techniques

Basic variance reduction techniques are divided into truncation methods, population control methods, modified sampling methods, and partially deterministic methods, among others.

Truncation methods are the simplest variance reduction methods. These methods accelerate calculations by truncating parts of the phase space that do not contribute significantly to the solution. Typical cutoff methods include the energy cutoff and time cutoff methods. In the energy cutoff method, particles with energy less than the cutoff value are killed to reduce the simulation time cost associated with each particle. In contrast, in the time cutoff method, particles with transport simulation times exceeding the cutoff value are killed to reduce the simulation time of each particle.

In population control methods, particle splitting and Russian roulette are used to control the numbers of sampled particles in various regions of the phase space. In important regions, many sampled particles with low weights are tracked, whereas in unimportant regions, few sampled particles with high weights are tracked. Weight adjustment is performed to ensure that the problem solution remains unbiased. Typical population control methods include geometry/energy/time split, Russian roulette, weight cut, and weight window. The weight window is a combination of the split and Russian roulette methods. The basic idea is to define the upper and lower bounds of the weight window in space, energy, and time. If the particle weight is less than the lower bound of the weight window, Russian roulette occurs; then, either the particle is killed or the particle weight is increased to be within the range of the weight window. If the particle weight exceeds the weight window upper limit, the particle is split up so that the weights of all split particles lie within the weight window. If the particle weight is within the weight window, the particle will continue to be transported without any special treatment. Using reasonable settings for the weight window parameters can effectively reduce the variance in the Monte Carlo transport calculation results.

In modified sampling methods, the sampling processes are altered to increase the contribution of each particle to tallies. For any random event, it is possible to sample from any arbitrary distribution rather than the physical probability as long as the particle weights are adjusted to compensate. Thus, in modified sampling methods, sampling is performed from distributions that send particles in desired directions or into other desired regions of the phase space, such as time or energy, or that change the location or type of collisions. Typical modified sampling methods include exponential transformations, implicit capture, forced collisions, and source biases.

Partially, deterministic methods are the most complicated class of variance reduction techniques. They avoid the normal random walk process by using deterministic-like techniques, such as next-event estimators, or by controlling the random number sequence. Typical partially deterministic methods include point detectors and correlated sampling.

2. Adaptive variance reduction techniques

The basic variance reduction techniques usually require the user to set parameters manually; thus, they are dependent on the user's knowledge and experience. Therefore, adaptive variance reduction techniques, that can automatically generate variance parameters and adapt to different problems, must be developed. According to the tally targets, adaptive variance reduction techniques can be divided into local adaptive variance reduction techniques and global adaptive variance reduction techniques. The local adaptive variance reduction techniques include Consistent Adjoint Driven Importance Sampling (CADIS) [7], which was developed by Oak Ridge National Laboratory (ORNL), as well as the automatic adaptive mesh generation method and CADIS based on weight window smoothing [8], both of which were developed by the Institute of Nuclear Energy Safety Technology (INEST), Chinese Academy of Sciences · FDS Team. The global adaptive variance reduction techniques include Forward Weight-CADIS (FW-CADIS) [9] (developed by ORNL), the Method of Automatic Generation of Importance by Calculation (MAGIC) [10] (developed by the Culham Centre for Fusion Energy, CCFE), the Global Weight Window Generator (GWWG) [11] (developed by the FDS Team), and Global Weight Window Generator & Uniform Fission Site (GWWG & UFS, developed by the FDS Team) [12]. The following sections describe adaptive variance reduction techniques, including the automatic adaptive mesh generation method, CADIS method based on weight window smoothing, GWWG method and GWWG & USF method.

(1) Automatic adaptive mesh generation method

In the automatic adaptive mesh generation method, the mesh can be generated automatically according to the neutron mean free path during the transport process, thereby eliminating the dependence of the weight window parameters on the user's experience. Additionally, proper numbers of collisions are guaranteed in every weight window mesh to ensure the validity of the statistical results. A flowchart of the automatic adaptive mesh generation method is shown in Fig. 2.6. First, according to the geometry characteristics of the target area and radiation source distribution, the interested mesh region is determined, and the parameters of the coarse mesh boundary are obtained. Then, according to the radiation source energy, the spatial scale of the problem and the geometric complexity, the neutron mean free path in the current region is obtained by pre-calculation. Finally, the optimal mesh boundary parameters are automatically generated after refining the mesh boundary according to the neutron mean free path.

(2) CADIS method based on weight window smoothing

The CADIS method based on weight window smoothing is a local variance reduction technique that involves deterministic precomputation. In this method, the computer-aided design (CAD) model is first converted into a deterministic calculation model. Then, the adjoint calculation is performed by using the deterministic method to obtain the adjoint flux. During the calculation of the lower bound of the weight window,

Fig. 2.6 Flowchart of the adaptive mesh generation method

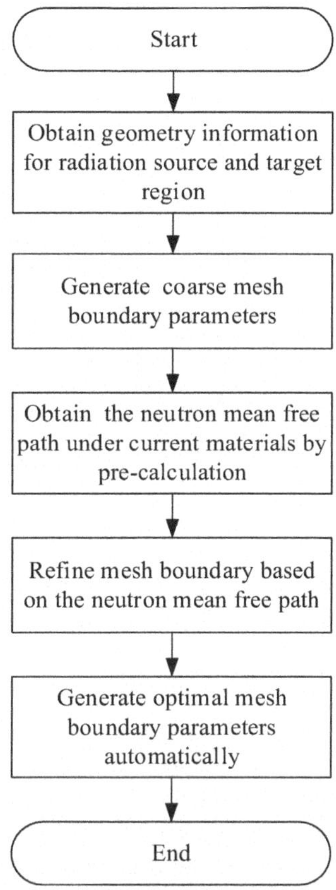

the particle is split excessively in the region with drastically changing adjoint flux, causing the computational efficiency to decrease. Therefore, the smoothing factor ρ ($0 < \rho \leq 1$) is introduced, and the lower bound of the weight window W_L' is calculated as follows:

$$W_L' = C\left(\frac{W_L}{W_{Lr}}\right)^{\rho} \tag{2.45}$$

where W_L is the lower bound of the weight window in the CADIS method, W_{Lr} is the lower bound of the weight window at the reference point, and C is a normalization constant. The lower bound of the weight window calculated in the CADIS method will be decreased significantly in the region where the material shielding performance is strong as shown in Fig. 2.7. By setting a proper value of ρ, the tendency of the rapid weights change can be avoided, the excessive splitting of the particles through these regions can be reduced, and the computational efficiency can be increased while maintaining the computational accuracy.

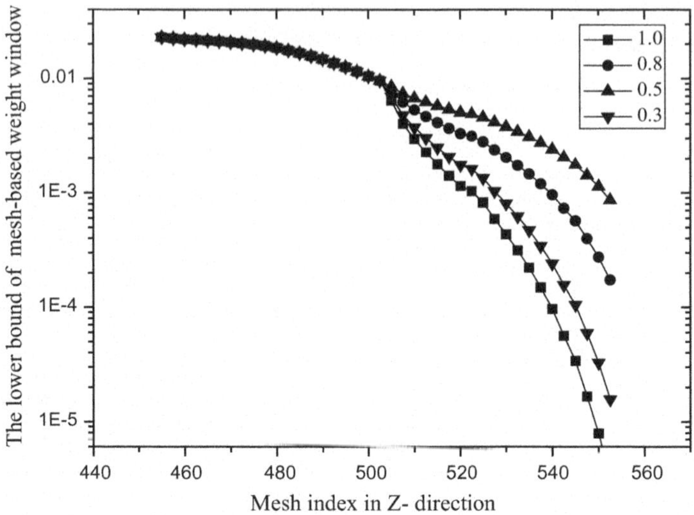

Fig. 2.7 Lower bound smoothing of the mesh weight window in the z-direction

(3) GWWG method

The GWWG method, which utilizes the particle density nonuniformity and historical mesh importance, is a physical feature prediction-based efficient particle transport simulation method. This method constructs the particle density uniformity function to represent the mesh importance and accumulates historical contributions iteratively to optimize the importance distribution. The mesh importance is defined as

$$I_i = \frac{C_i}{W_i} \tag{2.46}$$

where I_i is the importance of mesh P_i, C_i represents the contributions of particles entering mesh P_i to the particle density uniformity, and W_i is the total weight of the particles entering mesh P_i. The mesh importance is related to the contributions to the uniformity of the global particle density distribution. With the generated weight window, the flattening of the density distribution of particles can be effectively realized, and the globally uniform statistical error can be obtained. To enhance the efficiency of iterative weight window optimization, the contribution to the weight window from the current iteration step and the total particle weight values in previous steps are multiplied by the discount factor q and then delivered to the next iteration step. After simulating step k, the importance of mesh P_i is

$$I_{ik} = \frac{\sum_{j=1}^{k} q^{k-j} c_j}{\sum_{j=1}^{k} q^{k-j} w_j} \tag{2.47}$$

where c_j is the cumulated contribution in mesh P_i in step j of the simulation, and w_j is the total weight of the particles entering mesh P_i in step j of the simulation. After obtaining the importance of all meshes, the lower bound of the weight window of the entire space can be calculated. Then, the convergent neutron flux of the entire space can be obtained by neutron transport calculation based on the lower bound of the weight window.

The GWWG method was benchmarked with the International Thermonuclear Experimental Reactor (ITER) C-lite model, which is one of the series of basic models released by ITER IO, and the neutron flux of the entire space was calculated by setting the number of simulated particles to 2×10^9. This model has a large spatial scale. Thus, because of the significant deep-penetration problem, the flux distribution of the entire space cannot be effectively calculated with the traditional Monte Carlo method. Based on the GWWG method, the improvement in the Figure of Merit (FOM, also computational efficiency) of the neutron flux in the entire space is 637.4. Figure 2.8 shows the results of the C-lite model.

(4) GWWG and UFS coupling method

The flux/power distribution of the fission system or hybrid nuclear system is calculated by the exponential iterative method. The source in each generation is the fission neutrons sampled in the previous generation simulation, distributed in the entire model and related to the power distribution. In the calculation, the inhomogeneity power distribution in the reactor will result in the inhomogeneity fission neutron source distribution, which affects the convergence speed of the calculation. The UFS method biases the produced probability of the fission source based on the inhomogeneity distribution of the fission source, which can obtain an approximately uniform full-core source distribution. However, this method will generate source particles with weights that are too low, and it is difficult to speed up the transport calculation in the low-power region of the reactor. The Weight Window can speed up the transport calculation in the low-power region of reactor, but it may cause a large number of splits when the source particle is generated, thus reducing the computational efficiency, and cannot solve the problem of inhomogeneity fission neutron source distribution. To solve the above problems, the FDS Team has developed the GWWG & UFS coupling method, which uses the UFS method for source bias, uses the GWWG method to generate the global weight window to indirectly bias the particle transport, and guides particle transport as uniformly as possible in the entire reactor core. In this process, the weight of the source particle generated by the UFS method and the global weight window generated by the GWWG method are balanced, so the problem of a large split or truncation for the source particle will not occur, and the flux/power distribution calculation of the system will be accelerated. The coupling of the GWWG and UFS methods essentially achieves a consistent bias between source and transport, combining the advantages of the two methods.

The GWWG and UFS coupling method was benchmarked with the BEAVRS model, which was released by the Massachusetts Institute of Technology (MIT). The neutron flux of the entire space was calculated by setting the number of simulated activation cycles to 800 and the number of simulated particles of each cycle to 2 ×

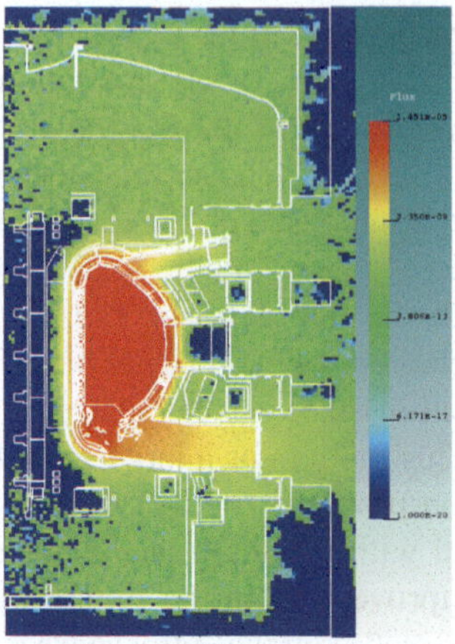

(a) Neutron flux distribution without GWWG

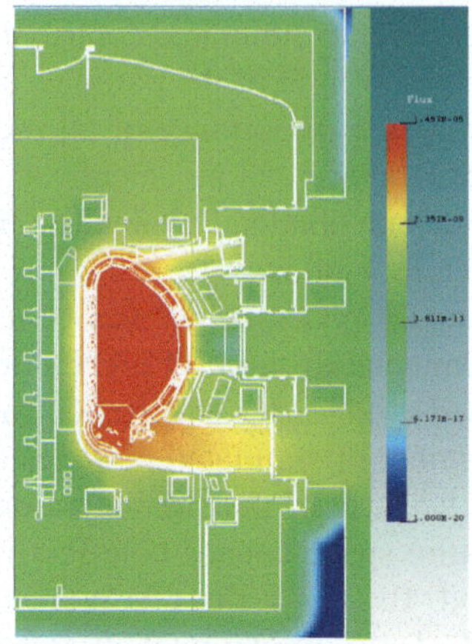

(b) Neutron flux distribution with GWWG

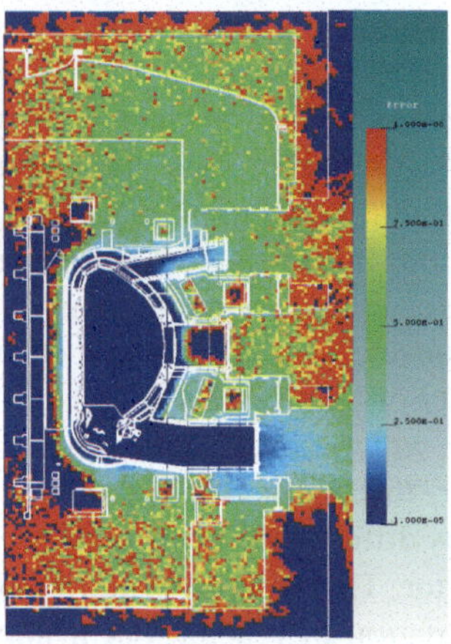

(c) Standard error distribution without GWWG

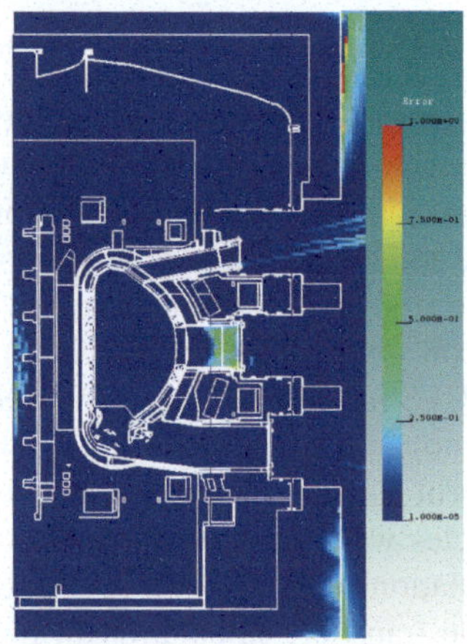

(d) Standard error distribution with GWWG

Fig. 2.8 Comparison of the calculation results of the ITER C-lite model

10^5. The results show that the calculation acceleration rate of the thermal neutron flux distribution is 19.3 times, while it is 12.6 times for the fast neutron flux distribution. Figures 2.9 and 2.10 show the statistical error distribution of a thermal neutron and fast neutron.

2.2.3.2 Source Convergence Acceleration Method

In the Monte Carlo eigenvalue calculation, the fission source distribution of the system must converge within a prescribed tolerance to achieve reliable estimates for k_{eff} and the power. To ensure the convergence of the fission source, a large number

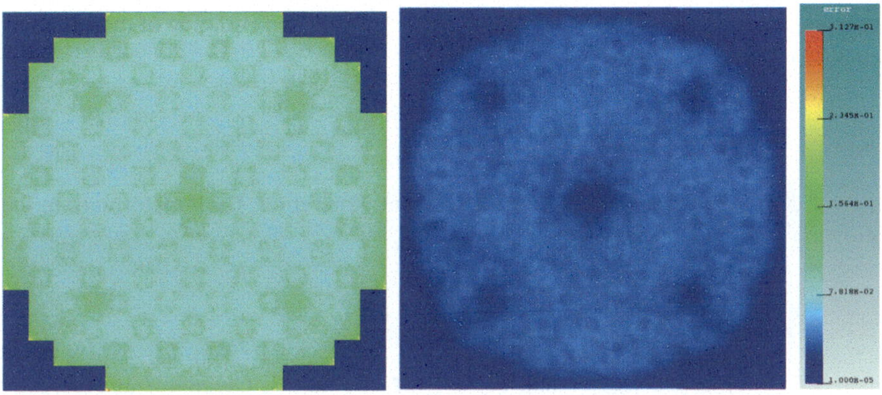

(a) Analog simulation (b) GWWG&UFS simulation

Fig. 2.9 Relative error maps of the thermal group of BEAVRS

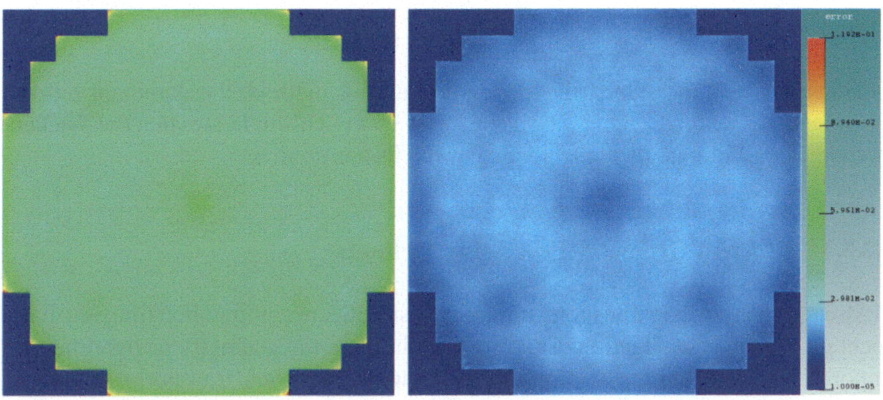

(a) Analog simulation (b) GWWG&UFS simulation

Fig. 2.10 Relative error maps of the fast group of BEAVRS

of non-active generations must be simulated, and the Shannon entropy is employed to evaluate the convergence of the fission source distribution. Simulating non-active generation consumes a large amount of unnecessary time, highlighting the need for fission source convergence acceleration methods.

In Monte Carlo eigenvalue calculations, many methods can be used to accelerate the convergence of the fission source, including the Wieland method, the finite difference method, the fission matrix method, and the zero-variance method. Among them, the Wieland method is most mature with a good acceleration effect. The basic idea underlying the Wieland method is introduced below.

In the Wieland method, the dominance ratio of the system is reduced by subtracting a fixed fission source from both sides of the transport Eq. (2.20), and fast convergence of the fission source distribution is thus achieved. The transformation of the transport equation in the Wieland method is

$$(\boldsymbol{M} - \frac{1}{k_e}\boldsymbol{F})\phi = (\frac{1}{k_{\text{eff}}} - \frac{1}{k_e})\boldsymbol{F}\phi, \quad k_e > k_{\text{eff}} \tag{2.48}$$

where $\boldsymbol{F}\phi/k_e$ is the fission source of the current generation. By transformation, the following equation can be obtained:

$$\phi^{(n+1)} = \frac{1}{k^{(n)}}\boldsymbol{F}_1\phi^{(n)} \tag{2.49}$$

where $k^{(n)} = (1/k_{\text{eff}} - 1/k_e)^{-1}$ and $\boldsymbol{F}_1 = \boldsymbol{F}(\boldsymbol{M} - \boldsymbol{F}/k_e)^{-1}$. Given $k^{(0)}$ and $\phi^{(0)}$, the final k_{eff} and ϕ can be obtained by iteration.

According to Eq. (2.49), in the Wieland method, the dominance of the system is

$$\rho^W = \frac{k_e - k_0}{k_e - k_1} \cdot \rho \tag{2.50}$$

However, with the standard exponential iterative method, the dominant ratio of the system is $\rho = k_1/k_0$. $\rho^W < \rho$ because $k_e > k_0 > k_1$. This indicates that the Wieland method can accelerate the convergence of the fission source.

2.2.3.3 Efficient Geometry Processing Methods

In Monte Carlo particle transport simulations, significant computation time is usually spent on geometry navigation. Therefore, optimizing the geometry navigation is an important research aiming to improve the efficiency of particle transport simulations. In particle transport simulations, both the particle location and step length need to be calculated for every step of the random walk, resulting in low-efficiency calculations. In the traditional method, all geometry objects should be traversed to identify the geometry object in which the particle is located, and then the shortest distance to the geometry boundary along the particle movement direction is determined. The time complexity of this process grows linearly as the number of geometry objects increases. To accelerate geometry navigation and improve the efficiency of particle transport calculations, efficient probabilistic transport computational methods based on the prediction of position characteristics have been developed. These methods include the neighbor list method based on CAD preprocessing, geometry selection method based on bounding boxes and optimized grid segmentation method based on the cost estimation function [13].

1. Neighbor list methods based on CAD preprocessing

A neighbor list is constructed for each solid geometry in a complex model by analyzing the topological relationships between the geometry and its neighbor in the CAD model. The neighbor list includes the geometry that the particle may enter after it

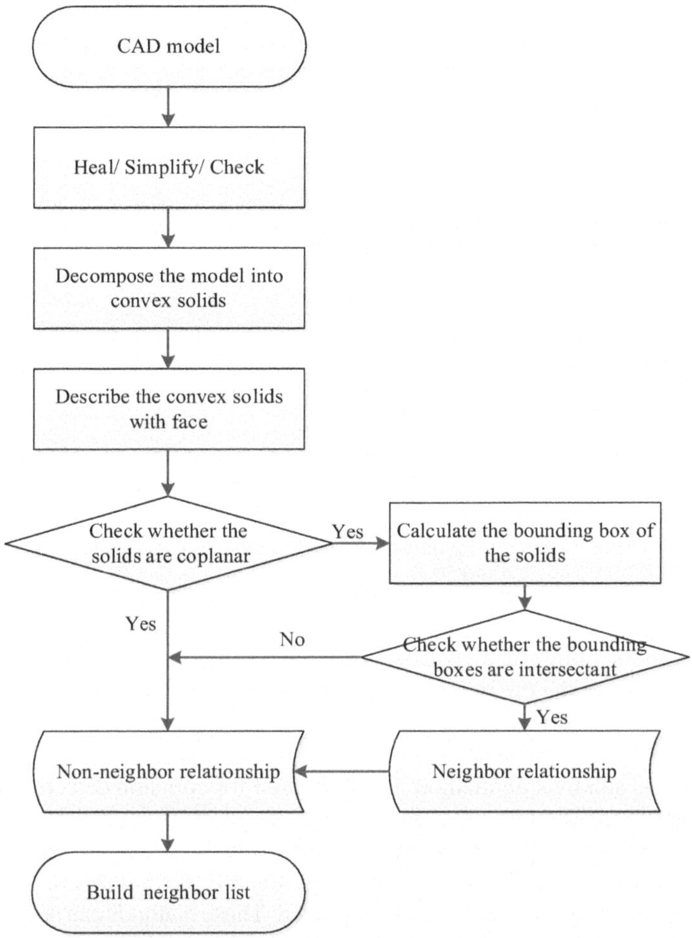

Fig. 2.11 The process of constructing a neighbor list based on CAD preprocessing

leaves the current geometry. When the geometry is located based on the neighbor list, only the neighbor geometry, not the entire model, is traversed, which reduces the average time for the geometry search and improves the efficiency of transport calculations. Figure 2.11 shows the process of constructing a neighbor list based on CAD preprocessing.

2. Geometry selection method based on bounding boxes

The bounding box method is widely used to accelerate ray tracing and collision detection in computer graphics. Based on the CAD model, the bounding boxes of all geometries are constructed. When the geometry steps are calculated, the intersection of the ray along the particle motion and the bounding boxes is used to quickly

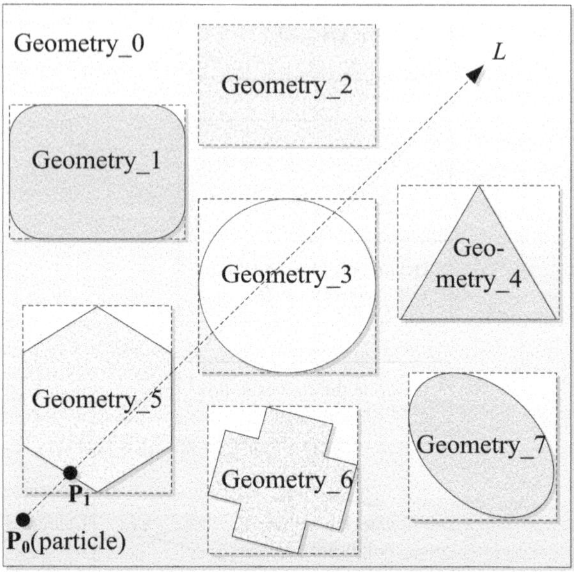

Fig. 2.12 Geometry selection method based on bounding boxes

determine whether the particle track intersects the geometry in the bounding box. This process can improve the efficiency of geometry processing, as shown in Fig. 2.12.

3. Optimal spatial subdivision method based on the cost estimation function

The spatial grid subdivision method is widely used for collision detection in virtual reality simulations. The basic idea is to divide the geometry model into a series of virtual space grids according to certain rules. In collision detection, only objects in the certain spatial grid need to be detected; consequently, the number of objects to be detected is reduced, and the efficiency is improved. These methods can be divided into two types: the uniform spatial grid subdivision method and nonuniform spatial grid subdivision method. The advantage of the uniform spatial grid subdivision method is that the subdivision algorithm is simple, and all of the grids are exactly the same; thus, the grid location with time complexity O(l) can be realized according to the particle position. The biggest drawback of this method is that it does not consider the specific distribution of geometry objects in the model, resulting in large differences in geometry object numbers in the model. The advantage of the nonuniform spatial grid subdivision method is that it can consider the spatial distribution of geometry objects in the model, and all of the geometry objects are divided into different grids. The disadvantage is that the subdivision algorithm for spatial grid subdivision is relatively complex, and traversing the tree data structure of the model is time-consuming. Additionally, this method cannot locate the grid with time complexity O(1).

The optimal spatial subdivision method based on the cost estimation function can overcome the shortcomings of the traditional spatial grid subdivision method. The cost estimation function is

$$F(C) = \frac{\sum_{J=1 \sim N_C} M_{J,C}}{\sum_{J=1 \sim N_C} N_{J,C}} \tag{2.51}$$

where C represents the current subdivision axis, $J(1 \sim N_C)$ is the mesh index, N_C stands for the total number of grids subdivided along axis C, and $M_{J,C}$ is the number of geometry objects in the Jth grid ($N_{J,C} = 1$ when $M_{J,C} > 0$; $N_{J,C} = 0$ when $M_{J,C} = 0$). $F(C)$ is the cost estimation function of the spatial subdivision along the coordinate axis C. $F(C)$ is used to express the average number of geometry objects in each non-empty grid. As $F(C)$ decreases, the geometry navigation cost decreases, and the efficiency increases.

When the grid is divided, the minimum bounding box size along each axis is selected as the grid width. Then, the cost estimation function value for division along each axis is calculated, and the grid corresponding to the smallest cost estimation function value is selected as the optimal subdivision scheme. The above operations are repeated for each of the divided grids until all obtained grid cost estimation function values are less than the prescribed optimal values. Figure 2.13 shows the basic process of optimal spatial subdivision.

Geometry localization based on the optimal spatial subdivision model is briefly described below. First, according to the position of the particle and the size of the grid, the grid containing this particle can be quickly found. Then, the geometry objects contained in the grid are traversed to determine in which geometry object the particle is. Thus, the particle is in a geometry object or stays in the parent geometry object. In this method, it is necessary only to obtain the geometry information contained in the grid intersecting the particle trajectory. If a grid does not intersect the particle

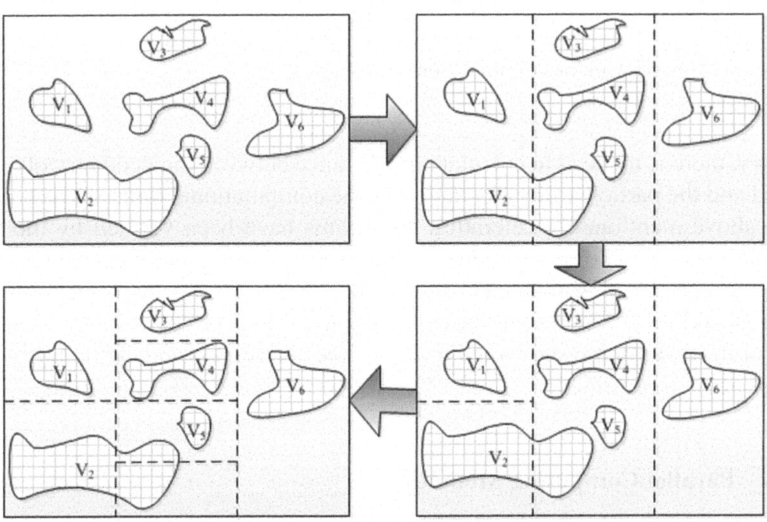

Fig. 2.13 Optimal spatial subdivision based on the cost function

Table 2.1 Calculation time comparisons for the ITER neutronics benchmark model

Test cases	Calculation time (min)	Acceleration rate
Traditional geometry processing method	130.60	–
Coupled neighbor list and bounding box method	62.91	2.08
Optimal spatial subdivision method	14.10	9.26
Coupled neighbor list, bounding box, and optimal spatial subdivision method	13.05	10.0

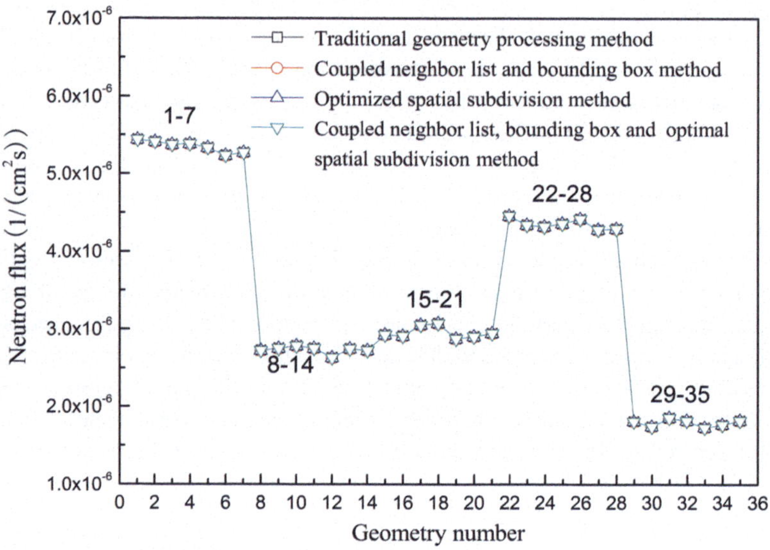

Fig. 2.14 ITER benchmark model simulation results

trajectory, there is no need to calculate the distance between the geometry objects in this grid and the particle trajectory, and thus the computational time can be reduced.

The above mentioned acceleration algorithms have been verified by the ITER benchmark model. The neutron flux of the divertor component was calculated by setting the number of simulated particles to 10^6. The above three methods were compared, and Table 2.1 presents the comparison results. The neutron flux calculation results of the divertor are shown in Fig. 2.14. The correctness and efficiency of these methods are demonstrated.

2.2.3.4 Parallel Computing Method

In Monte Carlo particle transport simulations with high fidelity, the computational speed and memory often are great challenges. The parallel computing method is an effective way to overcome these challenges. Because it performs independent simulations of each particle, the Monte Carlo method has natural parallel characteristics. There are three types of typical parallel computing methods: parallel on particles, region decomposition, and data decomposition.

Parallel on particles is the basic parallel method for Monte Carlo particle transport simulations and can be divided into two varieties: parallel computing for fixed-source problems and parallel computing for eigenvalue problems. The main idea underlying parallel computing for fixed-source problems is to equally distribute particles to each process for independent simulation. After each process is complete, the simulation results are merged into the main process to obtain the final results. In parallel computing for eigenvalue problems, the fission source particles are sampled and stored in each process during the simulation of a certain generation of particles. After this generation is simulated, the fission source particles are redistributed between the processes based on the neighbor list algorithm or bidirectional traversal algorithm for next-generation calculation.

When using the Monte Carlo method for high-fidelity calculations, especially in burnup calculations, the memory required may exceed the physical memory of a single computer node. To solve this problem, region decomposition and data decomposition algorithms can store memory data in the memory of different nodes. The main idea of a region decomposition algorithm is that the entire model is divided into several regions, which are assigned to each process. Each process only stores the geometry, material, and tally data for that region. When the particles travel across boundaries from one region to another, the particles will be passed to the new region by data communication between processes and then simulated by the corresponding process. Region decomposition algorithms can effectively solve the problem of memory, but other problems, such as load imbalance, still need to be solved. The main idea underlying data decomposition algorithm is that the tally cells are uniformly distributed to each process and then numbered. During the simulation, the particles that move through the tally cell generate tally information, which will be sent to the process for storage according to the process number corresponding to the tally mesh. Compared with the region decomposition algorithm, the data decomposition algorithm has higher parallel efficiency.

2.3 Deterministic Computational Methods

In deterministic methods, the continuous variables in transport equations are replaced by a set of discrete values. Thus, a matrix equation can be obtained and then solved using the matrix-calculation method. The energy is usually discrete with the multi-energy-group approximation. Angle discretization can be performed by direct discretization or function expansion. Space discretization is achieved based on the MOC and the finite difference, the finite element, the nodal, and other methods. For time-dependent problems, the direct discrete method is also used for time variables. In deterministic transport methods, several strategies for the discretization of the angle variable, such as the discrete ordinates method, the spherical harmonics method, the collision probability method and the transmission probability method, can be used. Regarding the discretization of spatial variables, the neutron transport equation can be solved by the MOC, the finite difference method, the finite element method, the nodal method, and other methods.

Compared with the Monte Carlo method, the numerical calculation process of the deterministic method is simpler and has a faster convergence rate. However, the deterministic method has poor adaptability to complex geometries, and its computational time increases significantly with the dimension of the problem.

In this section, some deterministic methods, including the discrete ordinates method, the spherical harmonics method, the MOC, and some typical deterministic acceleration methods, are introduced. The numerical process of the discrete ordinates method is relatively simple, but it has a "ray effects" problem and poor adaptability to complex geometries. The geometric description of MOC is rather flexible, but its computational time increases significantly with the complexity of the problem. The spherical harmonics method does not have ray effects, but for complex multidimensional geometry, it is far more complicated to obtain the solution.

2.3.1 The Discrete Ordinates Method

The discrete ordinates method (also known as the S_N method) was first implemented by B. Carlson to solve the neutron transport equation and has become one of the most effective methods for studying neutron and photon transport problems in recent years [3].

The basic idea underlying the discrete ordinates method is that the angular distribution of the neutron flux is evaluated in a number of discrete directions instead of using spherical harmonics in phase space ($\boldsymbol{r} \times E \times \boldsymbol{\Omega}$). Specifically, for the angular neutron flux $\phi(\boldsymbol{r}, E, \boldsymbol{\Omega})$, the direction $\boldsymbol{\Omega}$ is discretized into a sequence of discrete points ($\boldsymbol{\Omega}_1, \boldsymbol{\Omega}_2, \ldots, \boldsymbol{\Omega}_N$) or subdomains ($\boldsymbol{\Delta\Omega}_i, i = 1, \ldots, N$). Then, after obtaining $\phi(\boldsymbol{r}, E, \boldsymbol{\Omega}_i)$ values for these discrete points or subdomains, the flux $\phi(\boldsymbol{r}, E, \boldsymbol{\Omega})$ is determined by approximately weighting these values. The solution of the transport equation with the discrete ordinates method can be divided into three steps.

The first step is to choose a reasonable and accurate quadrature set $\{w_m, \boldsymbol{\Omega}_m\}$, in which w_m is the quadrature weight coefficient in direction $\boldsymbol{\Omega}_m$. In a one-dimensional (1D) problem, only one variable is needed to represent direction $\boldsymbol{\Omega}$, whereas in two-dimensional (2D) or 3D problems, two variables are required. The choice of the quadrature sets influences the computational accuracy and efficiency of the discrete ordinates method. No "best" quadrature set is suitable for all problems. The related literature [3] can be referred to understand the principle of selecting quadrature sets in detail. Taking the fully symmetric quadrature sets as an example, Fig. 2.15 depicts the distribution of quadrature points on a 1/8 spherical surface when N equals 2, 4, 6, and 8. N represents the discrete number of angular directions in one ordinate. Different numbers on the spherical surface represent different types of quadrature points, and

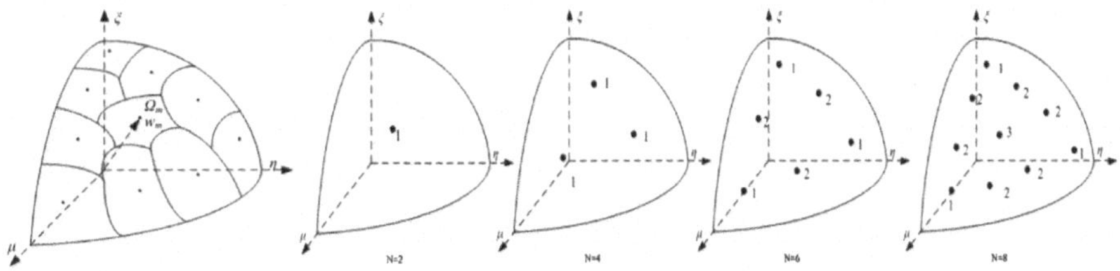

Fig. 2.15 Quadrature points on a 1/8 spherical surface when $N = 2, 4, 6, 8$

the maximum value is equal to the number of quadrature weight coefficients w_m to be determined.

The second step is to establish the difference equation. After determining the quadrature sets, the direction variables are then discretized. Next, the multigroup transport Eq. (2.52) is integrated over $\Delta\boldsymbol{\Omega}_m$, which is near each selected discrete direction $\boldsymbol{\Omega}_m$ (μ_m, η_m, ξ_m). Ultimately, the discrete ordinates Eq. (2.56), in which the energy group symbol g is omitted for brevity, can be obtained using Eqs. (2.53)–(2.55).

$$\boldsymbol{\Omega} \cdot \nabla\phi_g(\boldsymbol{r}, \boldsymbol{\Omega}) + \Sigma_{t,g}\phi_g(\boldsymbol{r}) = Q_g(\boldsymbol{r}, \boldsymbol{\Omega}) \tag{2.52}$$

$$\int\limits_{\Delta\boldsymbol{\Omega}_m} \boldsymbol{\Omega} \cdot \nabla\phi(\boldsymbol{r}, \boldsymbol{\Omega})\mathrm{d}\boldsymbol{\Omega} = w_m[\boldsymbol{\Omega} \cdot \nabla\phi(\boldsymbol{r}, \boldsymbol{\Omega})]_m \tag{2.53}$$

$$\int\limits_{\Delta\boldsymbol{\Omega}_m} \phi(\boldsymbol{r}, \boldsymbol{\Omega})\mathrm{d}\boldsymbol{\Omega} = w_m\phi_m(\boldsymbol{r}) \tag{2.54}$$

$$\int\limits_{\Delta\boldsymbol{\Omega}_m} Q(\boldsymbol{r}, \boldsymbol{\Omega})\mathrm{d}\boldsymbol{\Omega} = w_m Q_m(\boldsymbol{r}) \tag{2.55}$$

$$[\boldsymbol{\Omega} \cdot \nabla\phi(\boldsymbol{r}, \boldsymbol{\Omega})]_m + \Sigma_t\phi_m(\boldsymbol{r}) = Q_m(\boldsymbol{r}) \tag{2.56}$$

Then, the discretization of the spatial variables is performed (the finite difference method is generally used). In a 3D spatial geometry, the difference equation for a discrete space mesh (i, j, k) is

$$\begin{aligned} &w_m\mu_m\left(A_{i+1/2}\phi_{i+1/2,j,k,m} - A_{i-1/2}\phi_{i-1/2,j,k,m}\right) \\ &\quad + w_m\eta_m B_{ik}\left(\phi_{i,j+1/2,k,m} - \phi_{i,j-1/2,k,m}\right) \\ &\quad + w_m\xi_m C_{i,j}\left(\phi_{i,j,k+1/2,m} - \phi_{i,j,k-1/2,m}\right) \\ &\quad + \left(A_{i+1/2} - A_{i-1/2}\right)\left(a_{m+1/2}\phi_{i,j,k,m+1/2} - a_{m-1/2}\phi_{i,j,k,m-1/2}\right) \\ &\quad + w_m\Sigma_{t,i,j,k}\phi_{i,j,k,m}V_{i,j,k} = w_m Q_{i,j,k,m}V_{i,j,k} \end{aligned} \tag{2.57}$$

Table 2.2 Area and volume elements in commonly used coordinate systems

Ordinates		$A_{i\pm1/2}$	$B_{i,k}$	$C_{i,j}$	$V_{i,j,k}$
Cartesian	x, y, z	$\Delta y_j \Delta z_k$	$\Delta x_i \Delta z_k$	$\Delta x_i \Delta y_j$	$\Delta x_i \Delta y_i \Delta z_k$
Cylindrical	r, φ, z	$2\pi r_{i\pm1/2} \Delta\varphi_j \Delta z_k$	$\Delta r_i \Delta z_k$	$R_i \Delta\varphi_j$	$R_i \Delta\varphi_j \Delta z_k$

Note (a) $R_i = R_i = r_{i+1/2}^2 - r_{i-1/2}^2$, $\Delta\varphi$ is normalized by 2π

where Q is the neutron source term; μ_m, η_m, and ξ_m constitute the direction cosine of the discrete direction Ω; and $a_{m\pm1/2}$ is the coefficient to be determined, which satisfies the condition $a_{m+1/2} - a_{m-1/2} = -w_m\mu_m$. Coefficients A, B, C, and V have different expressions in different coordinate systems, as shown in Table 2.2.

The third step is to solve the difference Eq. (2.57) with the inner and outer iteration method.

The numerical process of the discrete ordinates method is relatively simple. When the iterative method is used to solve the equation, the neutron source term is already known, and the equations in each discrete direction are independent of each other. The numerical processes for solving the equations in each direction are similar, and thus they are easy to code. However, this method requires a large amount of storage and long computational time when applied to complex 3D problems.

The "ray effects" problem exists in the discrete ordinates method, especially for media with relatively small values of the scattering ratio Σ_s/Σ_t. Unrelated to the error of the numerical process, this problem is caused by approximating the original continuous direction of the transport equation with a finite number of discrete directions. The most direct method to eliminate the ray effect is to increase the number of discrete directions. However, increasing the number of discrete directions increases the computational time, which is not considered to be effective. Another more effective method is to convert the discrete ordinate equation into an equivalent form of the spherical harmonics approximation equation.

In general, the structured mesh is used in the discrete ordinates method to discretize the spatial variable $\boldsymbol{r}$. However, for irregular cells in complex models, a regular structured mesh for space discretization may result in large errors. Additionally, for a fusion system with complicated geometry, solving the problem with a structured mesh S_N is difficult. In recent years, the unstructured mesh S_N method has been developed to discretize irregular geometries by using tetrahedral meshes. This method has a stronger ability to deal with complex geometry models, especially in fusion reactors, but long computational time and large computational resources are required.

2.3.2 The Method of Characteristics

The MOC, which combines the advantages of both the discrete ordinates method and the collision probability method, was proposed by Askew [14]. In this method, the differential neutron transport equation is used. This differential equation is then

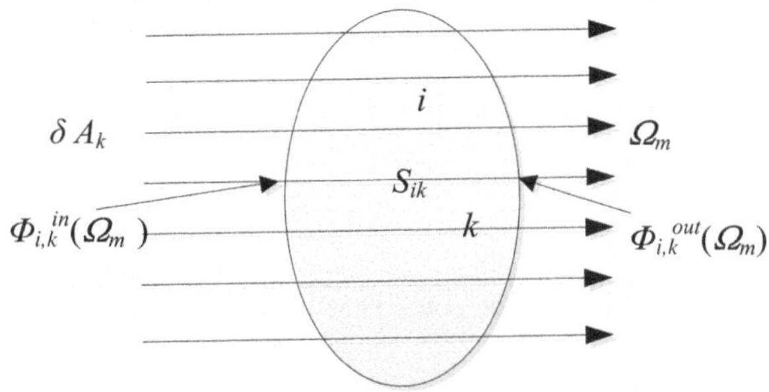

Fig. 2.16 Characteristic line

solved by integrating along the characteristics of the differential operator that correspond to the tracking lines.

The MOC can be divided into two types: the MOC with the flat source approximation and the MOC with the linear source approximation. Furthermore, several types of sub-methods have been developed. The step MOC developed from the flat source approximation MOC is the most intuitive, stable, and widely used. This MOC requires the system to be divided into several subregions, and the total cross section is assumed to be constant. The neutron source is also assumed to be constant over a certain characteristic line segment k.

Similar to the discrete ordinates method, the first step in the MOC is to choose a reasonable, accurate quadrature group $\{w_m, \boldsymbol{\Omega}_m\}$ for discretizing the direction variable. Then, in each discrete direction $\boldsymbol{\Omega}_m$, a large number of parallel rays pass through all regions, as shown in Fig. 2.16. The neutron flux is finally calculated by integrating the differential neutron transport equations along these rays. In Fig. 2.16, $S_{i,k}$ is the kth characteristic line in region i. δA_k is the projection area of region i along the direction of neutron motion. $\phi_{i,k}^{\text{in}}(\boldsymbol{\Omega}_m)$ is the incoming neutron flux, and $\phi_{i,k}^{\text{out}}(\boldsymbol{\Omega}_m)$ is the outgoing neutron flux.

The multi-group differential neutron transport equation in a steady state is

$$\boldsymbol{\Omega} \cdot \nabla \phi_g(\boldsymbol{r}, \boldsymbol{\Omega}) + \Sigma_{\text{t},g} \phi_g(\boldsymbol{r}, \boldsymbol{\Omega}) = Q_g(\boldsymbol{r}, \boldsymbol{\Omega}) \tag{2.58}$$

According to the assumptions of the MOC, in region i, both the total cross section $\Sigma_{\text{t},i}$ and neutron source term $Q_i(\boldsymbol{\Omega})$ are constant. $\phi_{i,k}(s, \boldsymbol{\Omega})$ is the neutron flux along the characteristic line k in region i. Thus, the equation in direction $\boldsymbol{\Omega}_m$ and energy group g is

$$\frac{d\phi_{i,k}(s, \boldsymbol{\Omega}_m)}{ds} + \Sigma_{\text{t},i} \phi_{i,k}(s, \boldsymbol{\Omega}_m) = Q_{i,k}(\boldsymbol{\Omega}_m) \tag{2.59}$$

Equation (2.59) is a first-order ordinary differential equation, and the general solution is shown in Eq. (2.60), in which the energy index g is omitted for simplicity.

$$\phi_{i,k}(s, \boldsymbol{\Omega}_m) = C \cdot \exp(-\Sigma_{\text{t},i} s) + \frac{Q_{i,k}(\boldsymbol{\Omega}_m)}{\Sigma_{\text{t},i}} \tag{2.60}$$

where C is a constant. The constant C, the neutron flux along the characteristic line $\phi_{i,k}(s, \boldsymbol{\Omega}_m)$ and the outgoing neutron flux $\phi_{i,k}^{\text{out}}(\boldsymbol{\Omega}_m)$ can be obtained as in Eqs. (2.61)–(2.63), respectively, according to the incoming neutron flux $\phi_{i,k}^{\text{in}}(\boldsymbol{\Omega}_m)$.

$$C = \phi_{i,k}^{\text{in}}(\boldsymbol{\Omega}_m) - \frac{Q_{i,k}(\boldsymbol{\Omega}_m)}{\Sigma_{\text{t},i}} \tag{2.61}$$

$$\phi_{i,k}(s, \boldsymbol{\Omega}_m) = \phi_{i,k}^{\text{in}}(\boldsymbol{\Omega}_m)\exp\left(-\Sigma_{\text{t},i}s\right) + \frac{Q_{i,k}(\boldsymbol{\Omega}_m)}{\Sigma_{\text{t},i}}\left[1 - \exp\left(-\Sigma_{\text{t},i}s\right)\right] \tag{2.62}$$

$$\phi_{i,k}^{\text{out}}(\boldsymbol{\Omega}_m) = \phi_{i,k}^{\text{in}}(\boldsymbol{\Omega}_m)\exp\left(-\Sigma_{\text{t},i}s_{i,k}\right) + \frac{Q_{i,k}(\boldsymbol{\Omega}_m)}{\Sigma_{\text{t},i}}\left[1 - \exp\left(-\Sigma_{\text{t},i}s_{i,k}\right)\right] \tag{2.63}$$

The average neutron angular flux along the characteristic line can be obtained by integrating the above equation along the characteristic line.

$$\begin{aligned}
\overline{\phi}_{i,k}(\boldsymbol{\Omega}_m) \cdot s_{i,k} &= \int_0^{s_{i,k}} \phi_{i,k}(s, \boldsymbol{\Omega}_m)\text{d}s \\
&= \frac{Q_{i,k}(\boldsymbol{\Omega}_m)}{\Sigma_{\text{t},i}}s_{i,k} + \frac{\phi_{i,k}^{\text{in}}(\boldsymbol{\Omega}_m)}{\Sigma_{\text{t},i}} - \frac{\phi_{i,k}^{\text{out}}(\boldsymbol{\Omega}_m)}{\Sigma_{\text{t},i}} \\
&= \frac{Q_{i,k}(\boldsymbol{\Omega}_m)}{\Sigma_{\text{t},i}}s_{i,k} + \frac{\Delta\phi_{i,k}(\boldsymbol{\Omega}_m)}{\Sigma_{\text{t},i}}
\end{aligned} \tag{2.64}$$

The average neutron angular flux in region i is volume weighted, as shown in Eq. (2.65).

$$\overline{\phi}_i(\boldsymbol{\Omega}_m) = \frac{\sum_k \overline{\phi}_{i,k}(\boldsymbol{\Omega}_m)s_{i,k}\delta A_k}{V_i} \approx \frac{\sum_k \overline{\phi}_{i,k}(\boldsymbol{\Omega}_m)s_{i,k}\delta A_k}{\sum_k s_{i,k}\delta A_k} \tag{2.65}$$

After calculating the neutron angular flux in a certain direction $\boldsymbol{\Omega}_m$, energy group g and region i, the neutron flux is obtained by angular integration of the neutron angular flux over all directions using the following numerical integral formula, where M is the number of discrete directions

$$\overline{\phi}_i = \int_{4\pi} \overline{\phi}_i(\boldsymbol{\Omega})d\boldsymbol{\Omega} \approx \sum_{m=1}^{M} \omega_m \overline{\phi}_i(\boldsymbol{\Omega}_m) \tag{2.66}$$

The main advantage of the MOC compared to the other deterministic transport methods is its flexible geometric description. Therefore, more attention has been paid to the MOC in recent years. In the MOC, to achieve high computational accuracy, a large number of parallel lines must pass through each region, which requires longer computational time and much larger memory storage. Thus, the 3D MOC transport calculations consume substantial amounts of computer resources, and the 3D MOC is very costly for advanced nuclear systems because of their complex geometry.

2.3.3 The Spherical Harmonics Method

The spherical harmonics method (also known as the P_N method) was the first method developed for neutron transport calculations. Some functions with the direction variable $\boldsymbol{\Omega}$, such as the neutron angular flux $\phi(\boldsymbol{r}, E, \boldsymbol{\Omega})$ in the original transport equation, are expanded by the spherical harmonic function $Y_n^m(\theta, \varphi)$ [15]. Then, the transport equation can be transformed into a group of differential equations. Each coefficient of the expansion series is determined by solving these differential equations. The neutron angular flux is expanded in terms of spherical harmonics functions as follows:

$$\phi(\boldsymbol{r}, E, \boldsymbol{\Omega}) = \sum_{n=0}^{\infty} \frac{2n+1}{4\pi} \sum_{m=-n}^{n} a_{n,m} \phi_{n,m}(\boldsymbol{r}, E) Y_n^m(\theta, \varphi) \tag{2.67}$$

where $a_{n,m}$ and $Y_n^m(\theta, \varphi)$ are listed

$$a_{n,m} = \frac{(n-|m|)!}{(n+|m|)!} \frac{2}{(1+\delta_{0m})} \tag{2.68}$$

$$Y_n^m(\theta, \varphi) = \begin{cases} \mathrm{P}_n^{|m|}(\cos\theta)\sin|m|\varphi, & m = -1, \ldots, -n \\ \mathrm{P}_n^m(\cos\theta)\cos m\varphi, & m = 0, 1, 2, \ldots, n \end{cases} \tag{2.69}$$

In Eq. (2.69), δ_{0m} is the Kronecker symbol. $\delta_{0m} = 1$ when $m = 0$, and $\delta_{0m} = 0$ when $m \neq 0$. $\phi_{n,m}(\boldsymbol{r}, E)$ is a coefficient to be determined and can be expressed as

$$\phi_{n,m}(\boldsymbol{r}, E) = \int_{\boldsymbol{\Omega}} \phi(\boldsymbol{r}, E, \boldsymbol{\Omega}) Y_n^m(\theta, \varphi) d\boldsymbol{\Omega} \tag{2.70}$$

During calculations, we usually take the expansion of the former $N + 1$ item, that is, n ranges from 0 to the order N, where N is the expansion order. The neutron angular flux is

$$\phi(\boldsymbol{r}, E, \boldsymbol{\Omega}) = \sum_{n=0}^{N} \frac{2n+1}{4\pi} \sum_{m=-n}^{n} a_{n,m} \phi_{n,m}(\boldsymbol{r}, E) Y_n^m(\theta, \varphi) \tag{2.71}$$

Similarly, the scattering cross section is expanded using Legendre polynomials as follows:

$$\Sigma_s\left(\boldsymbol{r}; E' \rightarrow E, \mu_0\right) = \sum_{n=0}^{N} \frac{2n+1}{2} \Sigma_{sn}\left(\boldsymbol{r}; E' \rightarrow E\right) P_n(\mu_0) \tag{2.72}$$

where $\mu_0 = \boldsymbol{\Omega}' \cdot \boldsymbol{\Omega}$. $\Sigma_{sn}(\boldsymbol{r}; E' \rightarrow E)$ is the Legendre expansion coefficient, which is expressed by

$$\Sigma_{sn}\left(\boldsymbol{r}; E' \rightarrow E\right) = \int_{-1}^{+1} \Sigma_s\left(\boldsymbol{r}; E' \rightarrow E, \mu_0\right) P_n(\mu_0) \mathrm{d}\mu_0 \tag{2.73}$$

The scattering source is

$$Q_s(\boldsymbol{r}, E, \boldsymbol{\Omega}) = \int_0^{\infty} \mathrm{d}E' \int_{\boldsymbol{\Omega}'} \Sigma_s\left(\boldsymbol{r}; E' \rightarrow E, \mu_0\right) \phi\left(\boldsymbol{r}, E', \boldsymbol{\Omega}'\right) \mathrm{d}\boldsymbol{\Omega}' \tag{2.74}$$

By substituting Eqs. (2.71) and (2.72) into Eq. (2.74) and using the additive formula and orthogonal relation of Legendre polynomials, the scattering source can be derived as follows

$$Q_s(\boldsymbol{r}, E, \boldsymbol{\Omega}) = \int_0^{\infty} \mathrm{d}E' \sum_{n=0}^{N} \Sigma_{sn}\left(\boldsymbol{r}; E' \rightarrow E\right) \sum_{m=-n}^{n} \frac{2n+1}{4\pi} a_{n,m} \phi_{n,m}\left(\boldsymbol{r}, E'\right) Y_n^m(\theta, \varphi) \tag{2.75}$$

Substituting Eqs. (2.71) and (2.75) into the neutron transport equation and using the additive formula and orthogonal relation of spherical harmonic functions, a set of closely coupled equations can be obtained. Take a 1D Cartesian coordinate system and a single energy group as an example. $N + 1$ equations are written as follows:

$$\frac{n+1}{2n+1} \frac{d\phi_{n+1}(z)}{dz} + \frac{n}{2n+1} \frac{d\phi_{n-1}(z)}{dz} + \Sigma_n \phi_n(z) = S(z) \delta_{0n}, \quad n = 0, 1, 2, \ldots, N \tag{2.76}$$

where $\Sigma_n = \Sigma_t - \Sigma_{sn}$, $S(z)$ is the neutron source term. The undetermined coefficients of the neutron flux can be found after applying the appropriate boundary conditions.

For the angular flux expanded with spherical harmonics, the direction variable is continuous. This expansion has a characteristic of rotational invariance at any angle. Thus, the spherical harmonics method does not have ray effects. Moreover, the variables can be treated with the finite element method, and the spherical harmonics method is very effective for unstructured meshes of irregular geometry. However, for complex multidimensional geometry, it is far more complicated to obtain the solution in the spherical harmonics method. Both the equations and the processing of the numerical solution are different for different expansion orders. Therefore, coding for equations with different orders is difficult and eventually leads to substantial inconvenience under practical engineering calculations. These shortcomings hinder the application of the high-order spherical harmonics method.

In the simplified spherical harmonics method (SP_N), by replacing the 1D spatial differential operators of spherical harmonic equations with 3D spatial differential operators, the SP_N equations in a 3D Cartesian coordinate system can be obtained. The simplified spherical harmonics approximation reduces the number of coupled differential equations without resulting in substantial loss of computational accuracy.

2.3.4 Acceleration Methods

The neutron transport equations are transformed into linear equations after numerical discretization and solved by matrix-calculation methods. Because the order of the matrix is so high, an iterative method is required. The convergence rate of the commonly used source iteration method is very slow, and sometimes, the solution does not converge. Therefore, it is time-consuming to solve the neutron transport equation for practical problems. Adopting the new iterative strategy, speeding up the iterative convergence rate, and adopting parallel computing are necessary to accelerate the calculation. A variety of acceleration methods have been developed, such as the Discrete Nodal Transport Method (DNTM) [16], the coarse mesh rebalance method [3], the diffusion synthetic acceleration method, the transport synthetic acceleration method, and the Krylov subspace method [17]. This section briefly introduces several acceleration methods.

1. DNTM

The discrete ordinates method discretizes continuous direction variables with a finite number of discrete directions. To ensure accuracy, the spatial mesh cannot be too large and thus, the traditional fine mesh finite difference method is used. However, the computational cost of the fine mesh finite difference in the discrete ordinates method is very high. Discretizing the spatial mesh with the nodal method can greatly improve the computational efficiency. DNTM combines the ideas of the nodal Green's function method and the traditional discrete ordinates method, and thus the neutron transport equation can be solved effectively in a Cartesian coordinate system. However, in actual engineering problems, curve coordinate systems (i.e., cylindrical and spherical systems) are often used, and the neutron scattering is generally anisotropic. Problems arise from the DNTM method due to the scattering anisotropy and the angular redistribution. DNTM-related theory and a solution method in the curvilinear coordinate system have been developed by the FDS Team. This method can solve the above problems and perform more efficient simulation for engineering problems while consuming less resources. The following is an example of the DNTM method in a 1D curvilinear coordinate system.

In a 1D spherical coordinate system, $\mu = \boldsymbol{\Omega} \cdot \mathbf{e}_r$, and

$$\frac{\mu}{r^2}\frac{\partial(r^2\phi)}{\partial r}+\frac{1}{r}\frac{\partial\left[(1-\mu^2)\phi\right]}{\partial\mu}+\Sigma_t(r)\phi(r,\mu)=q(r,\mu) \tag{2.77}$$

In a 1D cylindrical coordinate system, $\mu=\boldsymbol{\Omega}\cdot\mathbf{e}_r$, $\eta=\boldsymbol{\Omega}\cdot\mathbf{e}_\theta$, $\xi=\boldsymbol{\Omega}\cdot\mathbf{e}_z$; then, we have

$$\frac{\mu}{r}\frac{\partial(r\phi)}{\partial r}-\frac{1}{r}\frac{\partial(\eta\phi)}{\partial\chi}+\Sigma_t(r)\phi(r,\xi,\chi)=q(r,\xi,\chi) \tag{2.78}$$

The second terms on the left-hand sides of Eqs. (2.77) and (2.78) are the so-called angular redistribution terms (ARTs) and appear only in curvilinear geometries. Suppose that there are several nodes along the radial direction r and that the width of the ith node is Δr_i. Equations (2.77) and (2.78) are multiplied by r^j ($j=1$ for cylindrical geometry; $j=2$ for spherical geometry), and the variable transformation $r=r_i+a_i r'$ ($r'\in[-1,1]$, $a_i=\Delta r_i/2$) is performed. When Eqs. (2.77) and (2.78) are rearranged (i.e., by moving the ART from the left side to the right side) and combined, the following equation is obtained (where r' is still represented by r for convenience).

$$\mu\frac{\partial}{\partial r}F(r,\boldsymbol{\Omega})+\Sigma_i F(r,\boldsymbol{\Omega})=a_i R(r,\boldsymbol{\Omega}) \tag{2.79}$$

where the dimensionless (local) variable r is written in terms of the node half-width a_i. r_i is the radius of the ith node midpoint, and Σ_i, $F(r,\boldsymbol{\Omega})$ and $R(r,\boldsymbol{\Omega})$ are, respectively, represented by

$$\boldsymbol{\Omega}=\begin{cases}\mu, & \text{for spherical geometry}\\ (\xi,\chi), & \text{for cylindrical geometry}\end{cases} \tag{2.80}$$

$$\Sigma_i=a_i\Sigma_t \tag{2.81}$$

$$F(r,\boldsymbol{\Omega})=(r_i+a_ir)^j\phi(r_i+a_ir,\boldsymbol{\Omega}) \tag{2.82}$$

$$R(r,\boldsymbol{\Omega})=\begin{cases}(r_i+a_ir)^2q(r_i+a_ir,\mu)-(r_i+a_ir)\frac{\partial}{\partial\mu}\left[(1-\mu^2)\phi(r_i+a_ir,\mu)\right], & \text{for spherical geometry}\\ (r_i+a_ir)q(r_i+a_ir,\xi,\chi)+\frac{\partial}{\partial\chi}\left[\eta\phi(r_i+a_ir,\xi,\chi)\right], & \text{for cylindrical geometry,}\end{cases} \tag{2.83}$$

After obtaining the nodal equation, the equation is discretized, and then, the Green's function method is used to find the solution. Finally, the neutron flux in a node is calculated by Legendre polynomial expansion. This method can achieve very high accuracy on coarse spatial meshes and can greatly enhance the computational efficiency of neutron transport in curvilinear coordinate systems.

2. Coarse mesh rebalance method

The coarse mesh rebalance method is an effective and widely adopted acceleration method. The basic idea underlying this method is as follows. Based on the original refined finite difference mesh, a set of regular and nonoverlapping coarse meshes containing several fine meshes is built up with a coincident boundary between the coarse mesh and the fine mesh. After iterations of the fine mesh, a "coarse mesh rebalance factor" is calculated according to the neutron balance principle. The final solution is achieved by multiplying the coarse mesh rebalance factor by the neutron flux obtained by iteration. This approach can force the solution to satisfy the neutron balance relationship in the coarse mesh and accelerate the convergence rate.

3. Diffusion synthetic acceleration method

The diffusion synthesis acceleration method aims to accelerate the convergence of the source iteration for the S_N method. The diffusion equation is a good approximation of the transport equation based on the transport theory. In this method, a modified diffusion equation is obtained by replacing the original leakage term with that calculated by the transport equation. Thus, the solution of the modified diffusion equation has the accuracy of the transport equation. Consequently, the neutron flux obtained in each iteration is rectified by the modified diffusion equation and used to obtain the next iteration source, thereby effectively accelerating the source iteration.

4. Krylov subspace method

The Krylov subspace method has been developed in recent years. Its core idea is to find a subspace k from n-dimensional vector space and then find an approximate solution in subspace k. Assume that the subspace dimension dim $k = m$ and that m constraints are required to derive an approximate solution from subspace k. Usually, m orthogonal conditions are adopted to solve the linear equation sets. In the solution to the eigenvalue problem, an m-order small matrix is usually obtained with the subspace, the eigenvalues of the large matrix are approximately obtained by solving the small matrix, and m coefficients are determined from the eigenvectors of the small matrix to linearly combine the subspaces. In addition to its high computational speed, the Krylov subspace method can solve many eigenvalues and eigenvectors simultaneously and play an important role in solving the high-order harmonics of neutron transport equations.

2.4 MC and Deterministic Coupling Calculations

The MC method can accurately simulate particle transport within complex geometries but requires a long computation time to solve deep-penetration problems. The deterministic method is appropriate for solving deep-penetration problems but defective in treating complex geometries and nuclide distributions. Considering the complicated features of advanced nuclear systems, such as complex neutron spectrum structure and angular distribution, complicated material composition, large spatial span and complex geometry, etc., it is difficult to perform an efficient and accurate neutron transport calculation by a single method. The Monte Carlo–deterministic coupling method, combining the advantages of both the Monte Carlo and deterministic methods, is one of the most efficient and accurate methods for solving transport problems in advanced nuclear systems.

There are three types of MC and deterministic coupling methods of transport calculations: region coupling, energy coupling, and time coupling. This section will introduce the region coupling and energy coupling methods and the time coupling method will be introduced in Chap. 3.

2.4.1 Region Coupling Method

Region coupling between MC and deterministic methods is a commonly used method. In this method, the entire region is divided into several subregions according to the complexity of the nuclide distribution. The MC method is utilized for the subregions with highly complex nuclide distribution, and the deterministic method is utilized for the other subregions. This method can efficiently address the deep-penetration problems in advanced nuclear systems but still requires further improvements.

In the traditional region coupling methods, the secondary particles reflected back into the MC calculation region by scattering are ignored, which results in an underestimation of the interface source (S_0) for the deterministic calculation. A new region coupling method with an adaptive transition region model is introduced by the FDS Team. The spatial integrative neutron transport model was constructed, and the nonlinear impacts on the interface source from the secondary particles are considered to achieve accurate solutions to deep-penetration problems. As shown in Fig. 2.17, according to the geometric and physical complexity of the model, the model is adaptively divided into a strong anisotropy region and a weak anisotropy region. Based on the neutron spectrum at the interface and the physical characteristics of the neighbor region, the thickness of the transition region is calculated by the diffusion model to realize the efficient and accurate solution of a strong anisotropy system. The adaptive partition of the transition region and data exchange between different calculations regions will be introduced in the following part.

1. Adaptive partition of the transition region

The main role of the adaptive partition for the transition region is to obtain the transition region thickness according to the mean free path and the characteristics of the reflected particles. The thickness of the transition region can be deduced from the albedo β, which is expressed as

$$\beta = \frac{J^-(0)}{J^+(0)} \tag{2.84}$$

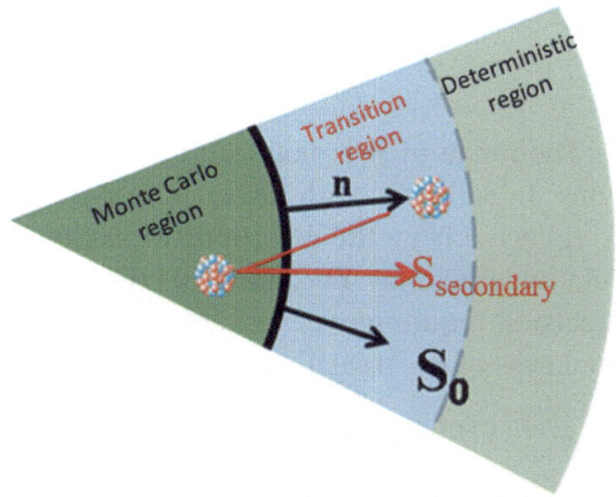

Fig. 2.17 Sketch of the MC and deterministic coupling method based on an adaptive transition region

where $J^-(0)$ is the reflected neutron current density from the deterministic region to the MC region and $J^+(0)$ is the incident neutron current density from the MC region to the deterministic region.

Based on the diffusion approximation and extrapolation boundary conditions, the albedo can be expressed as Eq. (2.85)

$$\beta = \frac{1 - \frac{2D}{L}\coth\left(\frac{a+2.13D}{L}\right)}{1 + \frac{2D}{L}\coth\left(\frac{a+2.13D}{L}\right)} \tag{2.85}$$

where D is the diffusion coefficient in cm, L is the diffusion length in units of cm, and a is the thickness of the transition region in units of cm. The equations for D and L are (2.86) and (2.87), respectively, as follows:

$$D = \frac{1}{\Sigma_s(3 - \frac{2}{A})} \tag{2.86}$$

$$L = \sqrt{\frac{D}{\Sigma_a}} \tag{2.87}$$

where A is the mass number of the medium and Σ_s is the macroscopic scattering cross section in cm^{-1}. In addition, Σ_a is the macroscopic absorption cross section in units of cm^{-1}.

For a medium with infinite thickness, the albedo is expressed as

$$\beta_\infty = \frac{1 - \frac{2D}{L}}{1 + \frac{2D}{L}} \tag{2.88}$$

To take the impact of the thickness of the transition region into account, the reductive albedo $\bar{\beta}$ is introduced as

$$\bar{\beta} = \frac{\beta}{\beta_\infty} = \frac{1 + \frac{2D}{L}}{1 - \frac{2D}{L}} \frac{1 - \frac{2D}{L}\coth\left(\frac{a+2.13D}{L}\right)}{1 + \frac{2D}{L}\coth\left(\frac{a+2.13D}{L}\right)} \tag{2.89}$$

The range of $\bar{\beta}$ is between 0 and 1, and larger $\bar{\beta}$ values indicate thicker transitions. In addition, with larger $\bar{\beta}$ values, fewer numbers of secondary particles will be reflected from the deterministic region to the MC region, resulting in higher calculation efficiency. For practical computations, considering the calculation time and precision requirements, suitable $\bar{\beta}$ values (e.g., 90%) are selected to calculate the thickness of the transition region.

2. Data exchange between different calculations regions

In addition to the albedo of the secondary particles, the data exchange between different calculation regions is also a key problem of MC and deterministic region coupling methods. Take MC-S_N region coupling as an example. It is necessary to convert the MC particle information at the interface into S_N angular fluxes. The S_N mesh cells across the interface are traversed, and the MC particle information is accumulated within each mesh cell. Therefore, the S_N angular flux in the mesh cell (i, j, k) for energy group g and direction m is obtained by using

$$\phi_{i,j,k,m,g} = \sum_n \frac{J_n^{\text{MC}}(\Delta A, \Delta E, \Delta \boldsymbol{\Omega})}{\omega_m N \Delta A |\lambda_n|} \tag{2.90}$$

where $J_n^{\text{MC}}(\Delta A, \Delta E, \Delta \boldsymbol{\Omega})$ is the particle current across area ΔA within the direction range $\Delta \boldsymbol{\Omega}$ and energy range ΔE, which corresponds to the S_N mesh cell (i, j, k) with energy group g and direction m. In addition, ω_m is the weight function of discrete direction $\boldsymbol{\Omega}$, N is the total number of particles in the mesh cell (i, j, k) of the MC simulation, and λ_n is the cosine of the angle between the surface normal and particle trajectory.

After the calculation of the transition region thickness, the data between the different calculations regions are exchanged, and the MC and S_N coupling calculation based on the transition region can be performed. The flowchart shown in Fig. 2.18 includes the following steps:

(1) Perform the transport calculation by using the MC method for the MC region and transition region to obtain the corresponding neutron fluxes and particle information at the interface between the two regions.
(2) Convert the particle information at the interface of the MC and transition regions into S_N angular fluxes. Then, perform transport calculations by using the S_N method in the transition region and S_N region to obtain the corresponding neutron and angular fluxes at the interface between the two regions.
(3) Compare the neutron fluxes obtained from step 1 and step 2 in the transition region. If the deviation is smaller than the given threshold, the final neutron flux can be generated by combining the neutron fluxes from all three regions. Otherwise, set the angular flux at the interface of the transition and S_N regions

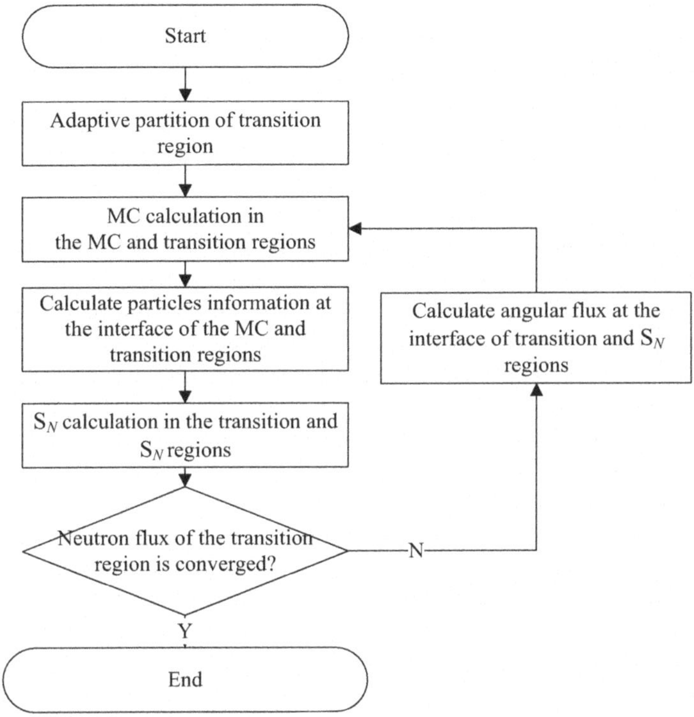

Fig. 2.18 Flowchart of the MC and S$_N$ coupling method based on an adaptive transition region

as the additional source and revise the particle information at the interface of the MC region and transition region. Repeat the above steps until convergence is achieved.

2.4.2 Energy Coupling

The Monte Carlo method uses the continuous-energy cross sections with low convergence rate efficiency, while the deterministic method is faster, but it is difficult to deal with the resonance energy region. According to the characteristics of the Monte Carlo and deterministic methods, the Monte Carlo and the deterministic energy coupling method employs different methods for different energy regions. For example, researchers in Korea proposed a Monte Carlo and MOC energy coupling method [18]. In the method, neutron energy was divided into three regions: fast, resonance, and thermal. Due to the accurate cross-section treatment of Monte Carlo and the fast calculation of MOC, Monte Carlo was used for simulation in the resonance region, and MOC was used for fast and thermal region solution. Scattering sources

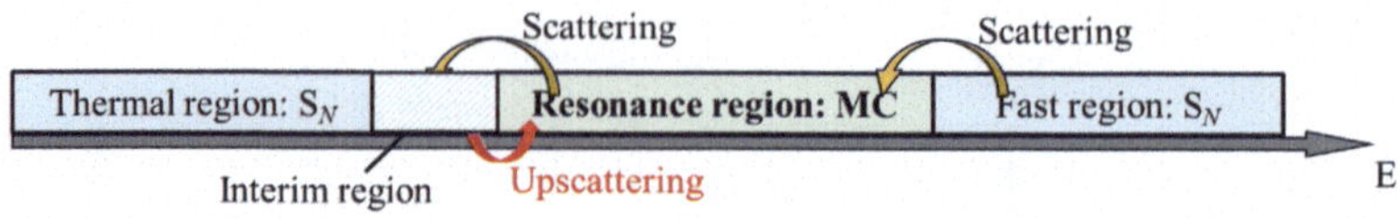

Fig. 2.19 Schematic diagram of the energy coupling method based on the transition region

between energy regions were directly delivered. By this method, neutron transport in the resonance region can be accurately simulated and smaller statistical fluctuation could be obtained, because Monte Carlo was used only for resonance region solution. However, the direction distribution of the scattering source was not considered in the source delivery, and it only applied to the neutron transport calculation of 2-dimensional problems.

This section proposes a Monte Carlo and deterministic energy coupling method based on the transition energy region. The Monte Carlo method is used for the calculation in the resonance region with strong anisotropy (E_{r}, $E_{\mathrm{r+1}}$, …, E_{t}), and the deterministic method is used for the solution in the fast (E_1, E_2, E_3, …, E_{r}) and thermal region (E_{t}, $E_{\mathrm{t+1}}$, …, E_{G}) with weak anisotropy. The neutrons at the lower boundary of the resonance region may have upscattering reactions with a medium and then enter the resonance region. Thus, in addition to the scattering source from the resonance region to thermal region, there may also exist a scattering source from the thermal region to resonance region. This bidirectional scattering source between two regions can be efficiently processed in the hybrid method by the set of transition energy region, as shown in Fig. 2.19. Assuming that the lowest energy that can cause a upscattering reaction into the resonance region is $E_{\mathrm{t+us}}$, the Monte Carlo calculation region is extended from the original (E_{t}, E_{r}) to ($E_{\mathrm{t+us}}$, E_{r}) to fully consider the upscattering effect in the Monte Carlo calculation. In the hybrid method, the mutual source delivery between energy regions can be efficiently processed in the inner iteration to accelerate the convergence of the outer iteration compared with the traditional methods.

Take the criticality calculation as an example. The basic flow of the energy coupling method based on the transition energy region is shown in Fig. 2.20, including the following steps:

(1) According to the input information of geometry, material, and source, the deterministic method is used for the fast region and the distribution of the scattering source into lower energy regions can be obtained;
(2) Based on the scattering source from the fast region, the Monte Carlo method is used for the resonance and transition region and the distribution of the scattering source into the thermal region can be obtained;
(3) With the scattering source from the higher region into the thermal region, the deterministic method is applied in the thermal region. Thus far, one outer iteration calculation has been completed.
(4) According to the results in three regions, the fission source is updated and compared with the previous one to evaluate the convergence. If the fission source

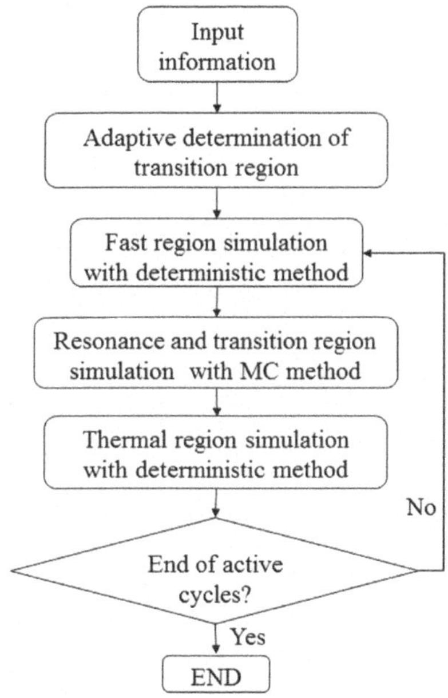

Fig. 2.20 Flowchart of the energy coupling method based on the transition region

is not converged, the updated fission source is delivered to the next iteration; if it is converged, the active cycles begin, and the k_{eff} and flux obtained in each iteration calculation will be counted until the end of active cycles.

2.5 Transport Simulation Codes

2.5.1 Monte Carlo Codes

The Monte Carlo method can handle neutron transport problems of complex geometry, complex neutron spectrum, and anisotropic neutron scattering and has been widely employed in neutron transport calculations of advanced nuclear systems. Table 2.3 shows major Monte Carlo codes, and the typical codes are introduced in the following section, including SuperMC, MCNP, and Serpent.

2.5.1.1 SuperMC

Super Multi-functional Calculation Program for Nuclear Design and Safety Evaluation (SuperMC) [4, 12, 19] is a large-scale integrated software system for neutronics design. Taking the neutron transport calculation as the core, SuperMC supports the whole process of the neutronics calculation containing motion, depletion, radiation

Table 2.3 Major Monte Carlo codes

Code	Developer
FLUKA	European Organization for Nuclear Research (CERN) and Italian Institute for Nuclear Physics (INFN)
Geant4	CERN
JMCT	Institute of Applied Physics and Computational Mathematics, China
KENO	Oak Ridge National Laboratory (ORNL), USA
MC21	United State Naval Research Laboratory (NRL), USA
MCBEND	ANSWERS Software Service, UK
McCARD	Seoul National University, Korea
MCNP	Los Alamos National Laboratory (LANL), USA
MCU	Ulsan National Institute of Science and Technology (UNIST), Korea
Monaco	ORNL, USA
MONK	ANSWERS Software Service, UK
MVP	Japan Atomic Energy Agency (JAEA), Japan
OpenMC	Massachusetts Institute of Technology (MIT), USA
PHITS	JAEA + RIST + KEE, Japan
RMC	Tsinghua University, China
Serpent	VTT Technical Research Centre of Finland Ltd, Finland
SuperMC	FDS Team, China
TRIPOLI	French Alternative Energies and Atomic Energy Commission (CEA), France

source term/dose/biohazard, material activation, and transmutation, and is also in support of the multi-physics coupling calculation of thermo-hydraulics, structural mechanics, chemistry, biology as well. Besides, based on the cloud computing framework, it integrates accurate modeling, visualized analyses, and virtual simulations and comprehensive data libraries as a whole. SuperMC can be used for the design and safety evaluation of nuclear systems, as well as the nuclear technology application field including radiation medicine, nuclear detection, and so on.

SuperMC supports transport simulations of various types of particles, such as neutrons, photons, electrons, and protons. Sources include general sources, critical sources, and user-defined sources. Common physical quantities for nuclear design and analysis, such as the eigenvalue, cell flux, point flux, energy deposition, reaction rate, reactivity coefficient and kinetic parameter, and special quantities, such as k_s for an external source-driven system and the homogenized group parameters for deterministic codes can be calculated in SuperMC. A criticality search aiming for a specified eigenvalue within a prescribed confidence interval is designed exclusively for determining assembly configurations, fuel loadings, and soluble boron concentrations, among others. Considering the multi-physics feedback effect, the functions of structured mesh, unstructured mesh, and continuous tallies, among others, are developed. For shielding calculations, some basic variance reduction techniques, including the weight window, forced collision, and exponential transformation, and adaptive variance reduction techniques, such as GWWG and GWWG coupled UFS, are supported.

For advanced nuclear systems, SuperMC has already been widely applied to nuclear design and analysis, such as ITER, CLEAR, etc.

2.5.1.2 MCNP

MCNP, a General Monte Carlo N-Particle Transport Code, was developed by Los Alamos National Laboratory (LANL) in the USA [20]. The latest version, MCNP6, can be used to perform particles transport computations and has been widely applied in many fields, such as fusion/fission reactor design, radiation protection and dosimetry, medical physics, and nuclear criticality safety. It supports transport simulations of neutrons, photons, electrons, and heavy charged particles and can perform eigenvalue and fixed-source calculations. In the latest version of MCNP6, multi-type descriptions of sources, various tallies and variance reduction techniques are included. MCNP has already been applied in the neutronics analysis of advanced nuclear system, such as ITER, etc.

2.5.1.3 Serpent

Serpent is continuous-energy Monte Carlo reactor physics burnup calculation code developed by the VTT Technical Research Centre of Finland Ltd. [21]. The current version is Serpent 2, which is mainly applied for physics analyses of fission reactors. Serpent 2 supports neutron and photon transport simulations. The flux, reaction rate, energy deposition, and other physical quantities can be calculated. Serpent 2 has been primarily applied to the design and analysis of advanced nuclear systems, such as the lead-based reactor MYRRHA, etc.

2.5.2 Deterministic Codes

The deterministic codes have the advantage of fast calculation speed and have been applied to the neutron transport calculation in advanced nuclear systems. Table 2.4 shows major deterministic codes and the typical codes are introduced in the following section, including WIMS, DOORS, and NECP.

Table 2.4 Major deterministic codes

Code	Developer
ATTILA	Varian, USA
Denovo	ORNL, USA
DOORS	ORNL, USA
ERANOS	European Collaboration on Fast Reactors
NECP	Xi'an Jiaotong University
WIMS	ANSWERS Software Service, UK

2.5.2.1 WIMS

WIMS is a general-purpose reactor physics program developed by ANSWERS software service [22] and the latest version is WIMS11. It can support multiple calculation methods, such as diffusion theory, S_N, collision probability, MOC, etc., and has already been widely applied to the design and operation of nuclear reactors. WIMS can calculate multiple parameters, including poison worth, nuclear heat deposition of neutron and photon, reactivity coefficients, etc., and can be applied to lattice cell calculation and whole core calculation. WIMS has already been applied in the design of advanced nuclear systems, including lead-based reactor, sodium-cooled fast reactor, gas-cooled fast reactor, and high-temperature reactor, and calculation of the eigenvalue, flux, kinetic parameters, reactivity coefficient, etc. For example, WIMS has been used for the reactor physics analysis of advanced nuclear systems, such as the high-temperature reactor HTR-10 and PBMR-400.

2.5.2.2 DOORS

DOORS is a 1D, 2D, and 3D discrete ordinates transport code system for the deep-penetration transport of neutrons and photons developed by ORNL [23], which contains independent codes, such as ANISN, DORT, and TORT. DOORS has been widely applied in nuclear reactor physics and nuclear radiation shielding. In DOORS, the discrete ordinates method, the multi-group approximation, the finite difference method and the nodal method are all used to process the variables of angle, energy, space, respectively. In DOORS, ANISN can calculate the neutron/photon flux and eigenvalue in 1D coordinate system. DORT can calculate the neutron/photon flux and eigenvalue in 1D or 2D systems. TORT can be used to address the deep-penetration problem in radiation shielding and calculate the neutron/photon flux and eigenvalues in 2D and 3D coordinate systems. DOORS has been applied to the design and analysis of advanced nuclear systems.

2.5.2.3 NECP

NCEP is a software package for nuclear engineering developed by Xi'an Jiao Tong University [24], China, and can support multiple calculation methods, such as discrete-ordinates, collision probability, characteristics, etc. NECP can automatically generate multi-group cross sections and can calculate eigenvalue, neutron flux, neutron nuclear heat deposition, and other parameters. It can be applied to the lattice calculation and core calculation.

References

1. Lamarsh JR (1966) Introduction to nuclear reactor theory. Addison-Wesley Publication Company, Massachusetts
2. Wu YC (2017) Fusion neutronics. Springer Nature Singapore Pte, Ltd
3. Lewis EE, Miller WF Jr (1993) Computational methods of neutron transport. American Nuclear Society, La Grange Park, Illinois
4. Wu YC, Song J, Zheng HQ et al (2015) CAD-based Monte Carlo Program for integrated simulation of nuclear system SuperMC. Ann Nucl Energy 82:161–168
5. Song J, Sun GY, Chen ZP et al (2015) Study on Monte Carlo K-effective calculation method. Nucl Sci Eng 35(2):241–245 (in Chinese)
6. Ott K, Neuhold R (1985) Introductory nuclear reactor dynamics. American Nuclear Society, La Grange Park III, USA
7. Haghighat A, Wagner JC (2003) Monte Carlo variance reduction with deterministic importance functions. Prog Nucl Energy 42(1):25–53
8. Zhao JB, Li XM, Wu B et al (2016) An automatic adaptive mesh generation method for weight window in Monte Carlo particle transport. Ann Nucl Energy 91:105–110
9. Wagner JC, Blakeman ED, Peplow DE (2007) Forward-weighted CADIS method for global variance reduction. Trans Am Nucl Soc 97:630–633
10. Davis A, Turner A (2011) Comparison of global variance reduction techniques for Monte Carlo radiation transport simulations of ITER. Fusion Eng Des 86(9–11):2698–2700
11. Zhang S, Yu SP, He P (2016) Verification of SuperMC with ITER C-Lite neutronic model. Fusion Eng Des 113:126–130
12. Wu YC (2018) Multi-functional Neutronics calculation methodology and program for nuclear design and radiation safety evaluation. Fusion Sci Technol 74(4):321–329
13. Chen ZP, Song J, Wu B et al (2015) Optimal spatial subdivision method for improving geometry navigation performance in Monte Carlo particle transport simulation. Ann Nucl Energy 76:479–484
14. Askew JR (1972) A characteristics formulation of the neutron transport equation in complicated geometries. Report AEEW-M 1108. United Kingdom Atomic Energy Establishment, Winfrith, England
15. Bell GI, Glasstone S (1970) Nuclear reactor theory. Van Nostrand Reihold Company, New York
16. Wu YC, Xie ZS, Fischer U (1999) A discrete ordinates nodal method for one-dimensional neutron transport calculation in curvilinear geometries. Nucl Sci Eng 133:350–357
17. Azmy Y, Sartori E (2010) Nuclear computational science: a century in review. Springer, Berlin
18. Lee H, Choi S, Lee D (2015) A hybrid Monte Carlo/method-of-characteristics method for efficient neutron transport analysis. Nucl Sci Eng 180:69–85
19. Wu YC, Team FDS (2009) CAD-based interface programs for fusion neutron transport simulation. Fusion Eng Des 84(7–11):1987–1992

20. X-5 Monte Carlo Team (2003) MCNP-A general Monte Carlo N-particle transport code, Version 5. LA-UR-03-1987, Los Alamos National Library
21. Leppanen J, Pusa M, Vitanen T (2015) The serpent Monte Carlo code: status, development and applications in 2013. Ann Nucl Energy 82:142–150
22. Lindley BA, Hosking JG, Smith PJ et al (2017) Current status of the reactor physics code WIMS and recent developments. Ann Nucl Energy 102:148–157
23. Oak Ridge National Laboratory (1998) DOORS3.2 one, two- and three dimensional discrete ordinates neutron/photon transport code system. CCC-650, Oak Ridge, Tennessee
24. Chen J, Liu ZY, Zhao C et al (2018) A new high-fidelity neutronics code NECP-X. Ann Nucl Energy 116:417–428

Chapter 3
Neutron Kinetics

Neutron kinetics is dedicated to studies of the time-dependent neutron behavior of nuclear systems over a short time range (typically from microseconds to seconds) and the factors that affect this behavior [1]. Neutron kinetics provides neutronics support not only for exploring the conditions of the stable operation of various nuclear systems but also for the prediction and analysis of potential accidents and their consequences. Therefore, the study of neutron kinetics is of great significance for the safe operation of nuclear systems. Compared with traditional nuclear systems, advanced nuclear systems show new characteristics including a complex neutron spectrum, strong coupling effects with an external neutron source, and delayed neutron precursor movement, which will be associated with the new features of neutron kinetics, and pose challenges to traditional neuron kinetics theories and methods [2–6]. This chapter mainly introduces the characteristics of the neutron kinetics, the corresponding theories, and computational methods for advanced nuclear systems.

3.1 Physical Basis of Neutron Kinetics

The transient response is the response of a nuclear system to a rapid change including the loss of flow, the loss of coolants, rod ejection, source trip, and so on. To characterize the transient behavior of the system, safety analysis should be performed, which requires multi-physics and multi-scale simulation that couple neutron kinetics with thermal hydraulics, structural mechanics, and so on. When large positive reactivity is inserted, the system might be out of control. The study of kinetics contributes not only to the development of reliable reactivity monitoring methods but also to the exploration of reactivity change regulation and the corresponding reactivity control approaches. Thus, kinetics theory is fundamental to the safe operation and control of nuclear systems [7]. This section will review the basic concepts of neutron kinetics, the kinetics equations, and the point kinetics model, which is widely used in traditional fission systems.

Y. Wu, *Neutronics of Advanced Nuclear Systems*,
https://doi.org/10.1007/978-981-13-6520-1_3

3.1.1 Basic Concepts

There are several common parameters for describing the kinetics characteristics of a nuclear system. These parameters are delayed neutron fraction β, neutron generation time Λ, reactivity ρ, and reactor period T [8], and they are called as kinetics parameters or integral parameters.

1. Delayed neutron fraction

There are two types of neutrons from fission: prompt neutrons and delayed neutrons. Prompt neutrons are emitted immediately after the fission reaction and account for 99% of the total fission neutrons. Delayed neutrons are produced during the decay process of the fission fragments and account for less than 1% of the total amount of fission neutrons. Delayed neutrons have an average energy of approximately 0.4 MeV, which is lower than the approximate 2 MeV average energy of prompt neutrons. The fission fragments that emit delayed neutrons are called delayed neutron precursors. The delayed release of delayed neutrons slows the transient change of nuclear systems. The delayed neutron fraction β is a ratio of delayed neutron production to total neutron production. The β values of fission nuclides in nuclear fuel vary. For thermal reactors, the β of ^{235}U is high, which makes nuclear systems with this fuel easy to control. The effective delayed neutron fraction β_{eff} is defined as a ratio of the fission times caused by delayed neutrons to the fission times caused by all fission neutrons (see Chap. 2). The β_{eff} value depends on the reactor size and neutron spectrum, which indicates the importance of delayed neutrons. In thermal reactors, delayed neutrons, compared with prompt neutrons, will have higher fission probabilities because they have lower energy. Thus, delayed neutrons have lower leakage risk and higher neutron importance.

2. Neutron generation time

The neutron generation time Λ is the mean time interval between successive generations of neutrons in a chain reaction. The main difference in the dynamic behaviors of the thermal reactors and the fast reactor is their Λ value. If prompt neutrons alone are considered, Λ is rather small (typically 10^{-8} s in a fast reactor). The decay process of delayed neutron precursors enhances Λ greatly, which means there will be less neutron generation and a slower rate of change of neutron flux during the same time interval. Delayed neutrons make nuclear systems easy to control. Another parameter is the neutron lifetime l, which is defined as the mean time interval from the emission of a neutron in fission to the absorption of the neutron somewhere in the reactor. The extension of neutron lifetime induced by the precursors of delayed neutrons will be considered. Dividing l by the multiplication factor k_{eff} equals Λ. The neutron lifetime is smaller than the neutron generation time if the reactor is subcritical, and neutron lifetime is larger than neutron generation time if the reactor is supercritical. For a critical reactor, $k_{\text{eff}} = 1$ (neutron lifetime and neutron generation time are the same).

3. Reactivity

The reactivity ρ can describe the deviation from criticality, and it depends on reactor size, material, and the interactions between the neutrons and nuclei of the material. The relationship between the reactivity and the multiplication factor k_{eff} is shown in Eq. (3.1). In the critical system, the rate of change of the power and neutron flux mainly depends on reactivity. The nuclear systems approach criticality when reactivity is close to zero, which means that the nuclear power will remain steady. Thus, reactivity monitoring and control play a very important role in the safety and operation of nuclear systems.

$$\rho = \frac{k_{\text{eff}} - 1}{k_{\text{eff}}} \tag{3.1}$$

4. Reactor period

The reactor period T is defined as the time required for the neutron flux to change by a factor e (2.718); it reflects the rate of change of the neutron flux in the systems. T is positive when the neutron flux increases, and negative when it decreases. The neutron flux increases faster when T is smaller, and the flux will be stable when T approaches infinity. For nuclear safety, T should not be too small.

3.1.2 *Kinetic Equations*

Starting from the neutron transport equation, which is a function of space and time, and considering the concentration change of delayed neutron precursors, the neutron balance equation at any time t in the phase space ($\boldsymbol{r}$, E, $\boldsymbol{\Omega}$) is

$$\begin{aligned}
&\frac{1}{v(E)} \cdot \frac{\partial \phi(\boldsymbol{r}, E, \boldsymbol{\Omega}, t)}{\partial t} + \boldsymbol{\Omega} \cdot \nabla \phi(\boldsymbol{r}, E, \boldsymbol{\Omega}, t) + \Sigma_{\mathrm{t}}(\boldsymbol{r}, E, \boldsymbol{\Omega}, t)\phi(\boldsymbol{r}, E, \boldsymbol{\Omega}, t) \\
&= \int_0^\infty \mathrm{d}E' \int_{\boldsymbol{\Omega}'} \Sigma_{\mathrm{s}}(\boldsymbol{r}, E') f(\boldsymbol{r}; E' \to E, \boldsymbol{\Omega}' \to \boldsymbol{\Omega}) \phi\left(\boldsymbol{r}, E', \boldsymbol{\Omega}', t\right) \mathrm{d}\Omega' \\
&+ \frac{\chi_{\mathrm{p}}(\boldsymbol{r}, E)}{4\pi}(1-\beta) \iint \upsilon(\boldsymbol{r}, E') \Sigma_{\mathrm{f}}(\boldsymbol{r}, E', t) \phi(\boldsymbol{r}, E', \boldsymbol{\Omega}', t)\, \mathrm{d}E'\, \mathrm{d}\boldsymbol{\Omega}' \\
&+ \frac{1}{4\pi} \sum_i \chi_i(\boldsymbol{r}, E) \lambda_i C_i(\boldsymbol{r}, t) \\
&+ S(\boldsymbol{r}, E, \boldsymbol{\Omega}, t)
\end{aligned} \tag{3.2}$$

$$\frac{\partial C_i(\boldsymbol{r}, t)}{\partial t} = \iint \beta_i \upsilon(\boldsymbol{r}, E') \Sigma_{\mathrm{f}}(\boldsymbol{r}, E', t) \phi(\boldsymbol{r}, E', \boldsymbol{\Omega}', t)\, \mathrm{d}E' \mathrm{d}\boldsymbol{\Omega}' - \lambda_i C_i(\boldsymbol{r}, t) \tag{3.3}$$

where $\chi_i(\boldsymbol{r}, E)$ is the spectrum of the i_{th} group of delayed neutrons in the position $\boldsymbol{r}$; $\chi_{\text{p}}(\boldsymbol{r}, E)$ is the spectrum of prompt neutrons in the position $\boldsymbol{r}$; β is the total fraction of delayed neutrons (β_i is the fraction for the i_{th} group); λ_i is the decay constant of the delayed neutron precursors for the ith group in units of s^{-1}; $\upsilon(\boldsymbol{r}, E)$ is the number of neutrons emitted per fission in the position $\boldsymbol{r}$ in units of $\text{cm} \cdot \text{s}^{-1}$; $C_i(\boldsymbol{r}, t)$ is the concentration of delayed neutron precursors in the position $\boldsymbol{r}$ for the i_{th} group in units of cm^{-3}; and S is the external neutron source in units of $\text{cm}^{-3} \cdot \Omega \cdot \text{s}^{-1}$.

Equations (3.2) and (3.3) represent the complete neutron kinetics equations, which describe how neutron flux changes with space and time in a transient process. In addition to the variables including position, direction, and energy, the complete neutron kinetics equations also depend on time, and the precursors also need to be considered [1, 2], which makes these questions difficult to solve. Thus, it is necessary to adopt proper assumptions that could simplify these equations.

The kinetics analysis of traditional nuclear systems commonly uses point kinetics model with the following assumptions: (1) the one-speed diffusion approximation is applied; (2) spatial effects are not considered in the dynamic process, i.e., the spatial distribution shape of the neutron flux is independent of time; (3) both neutron flux and precursors concentration can be expressed as the product of a space-dependent shape function and a time-dependent amplitude function; and (4) the precursors are immobile because of the solid nuclear fuel.

By using the abovementioned assumptions, and ignoring the external source, the time-dependent one-speed diffusion equation can be rewritten as

$$\frac{1}{v}\frac{\partial \phi(\boldsymbol{r},t)}{\partial t} = D\nabla^2\phi(\boldsymbol{r},t) - \Sigma_a\phi(\boldsymbol{r},t) + (1-\beta)k_\infty\Sigma_a\phi(\boldsymbol{r},t) + \sum_{i=1}^{6}\lambda_i C_i(\boldsymbol{r},t) \tag{3.4}$$

where k_∞ is the multiplication factor for an infinite medium system. The equation for the delayed neutron precursors will be

$$\frac{\partial C_i(\boldsymbol{r},t)}{\partial t} = \beta_i k_\infty \Sigma_a \phi(\boldsymbol{r},t) - \lambda_i C_i(\boldsymbol{r},t) \tag{3.5}$$

Assuming that the neutron flux $\phi(\boldsymbol{r}, t)$ can be expressed as a product of its space-dependent shape function $\varphi(\boldsymbol{r})$ and time-dependent amplitude function $T(t)$, and that the concentration of the delayed neutron precursors $C_i(\boldsymbol{r}, t)$ is the product of its space-dependent shape function $g_i(\boldsymbol{r})$ and time-dependent amplitude function $C_i(t)$, then

$$\phi(r,t) = T(t) \times \varphi(r) \tag{3.6}$$

$$C_i(\boldsymbol{r},t) = C_i(t) \times g_i(\boldsymbol{r}) \tag{3.7}$$

The neutron flux shape function $\varphi(\boldsymbol{r})$ can be expressed as

$$\nabla^2\varphi(\boldsymbol{r}) + B^2\varphi(\boldsymbol{r}) = 0 \tag{3.8}$$

Substituting Eqs. (3.6), (3.7) and (3.8) into (3.4), we have

$$\frac{\mathrm{d}T(t)}{\mathrm{d}t} = \left[(1-\beta)k_\infty - 1 - L^2B^2\right]\Sigma_a vT(t) + \sum_{i=1}^{6}\lambda_i\frac{g_i(\boldsymbol{r})}{\varphi(\boldsymbol{r})}C_i(t) \tag{3.9}$$

where B^2 is the buckling of the system and L is the diffusion length. Assuming the space distribution of the concentration of the delayed neutron precursors and the neutron flux is the same, i.e., $g_i(\boldsymbol{r})/\varphi(\boldsymbol{r}) = 1$, we can obtain

$$\frac{\mathrm{d}T(t)}{\mathrm{d}t} = \frac{(1-\beta)k_{\mathrm{eff}} - 1}{l}T(t) + \sum_{i=1}^{6}\lambda_iC_i(t) \tag{3.10}$$

where l is the neutron lifetime with consideration of leakage. The variables can be denoted as

$$k_{\mathrm{eff}} = \frac{k_\infty}{1+L^2B^2} = \frac{v\Sigma_{\mathrm{f}}/\Sigma_{\mathrm{a}}}{1+L^2B^2} \tag{3.11}$$

$$l = \frac{1/v\Sigma_{\mathrm{a}}}{1+L^2B^2} \tag{3.12}$$

Similarly, the precursor Eq. (3.13) can be derived from Eq. (3.3).

$$\frac{\mathrm{d}C_i}{\mathrm{d}t} = \beta_i\frac{k_{\mathrm{eff}}}{l}T(t) - \lambda_iC_i(t) \tag{3.13}$$

Based on the definition of neutron generation time, Eqs. (3.10) and (3.13) can be expressed as

$$\frac{\mathrm{d}T(t)}{\mathrm{d}t} = \frac{\rho(t)-\beta}{\Lambda}T(t) + \sum_{i=1}^{6}\lambda_iC_i(t) \tag{3.14}$$

$$\frac{\mathrm{d}C_i}{\mathrm{d}t} = \frac{\beta_i}{\Lambda}T(t) - \lambda_iC_i(t) \tag{3.15}$$

Equations (3.14) and (3.15) form the classical point kinetics equations. These equations contain kinetics parameters such as the reactivity ρ, the neutron generation time Λ, and the delayed neutron fraction β, which are able to describe the overall features of the system. If we refer to the point kinetics Eqs. (3.14) and (3.15), it is apparent that for the system to be critical on prompt neutrons alone, we would require $\rho = \beta$. For $\rho < \beta$, the delayed neutron contribution is needed for criticality, and hence, the time response of the system will be determined to a large extent by the time delay characterizing the precursor β-decay. In general, the range $0 < \rho < \beta$ is referred to as delayed critical, while $\rho \geq \beta$ is referred to as prompt critical or prompt supercritical. Obviously, it is very important to avoid this latter situation in reactor operation, since the reactor time response would be very rapid if prompt criticality were exceeded.

The point kinetics model is widely used for the safety analysis of nuclear power plants. The point kinetics model regards the neutron flux distribution with time in the reactor as synchronous, which is suitable for nuclear systems that are not far from the critical state and do not have fast reactivity disturbance. In other words, the space-dependent dynamic effects cannot be described in this model. Thus, when integral parameters are measured in physics experiments, appropriate correction methods based on the point kinetics model are often adopted to solve the problems of spatial effects. Moreover, ignoring the change of the neutron flux shape function leads to inaccurate reactivity estimation, which may cause safety risks. Therefore, the influence of special-spectral effects should be considered in the transient safety analysis. For example, when a fuel assembly accidentally falls into a near-critical nuclear system, the neutron flux in and near the fuel assembly increases, as the fission neutron source will be enhanced locally. If the point kinetics model is used, the change of local neutron flux cannot be simulated, which might underestimate the predicted increment of positive reactivity and the supercritical risk. Thus, the point kinetics model is not applicable in all cases [9, 10].

3.2 Characteristics of Neutron Kinetics in Advanced Nuclear Systems

Compared with traditional nuclear systems, advanced nuclear systems exhibit new characteristics in neutron kinetics, and they pose new problems and challenges to traditional kinetics theories and methods as well as the neutronics design and experiments.

1. The spatial effects owing to the coupling of external source and subcritical reactor

In the critical systems, the system eigenstate corresponding to the effective multiplication factor represents the steady-state condition. In subcritical systems, the neutron flux distribution at steady state strongly depends on the external source. The eigenstate and the spatial distribution of the neutron flux are different from those in the critical system. The influence of this difference on the kinetics characteristics of subcritical systems is not considered in the traditional point kinetics model.

Taking an infinite plate reactor as an example, the neutron flux without external neutron source in this system is

$$\phi(x,t) = \sum_{n=1}^{\infty} A_n \cos\frac{(2n-1)\pi}{a} e^{(k_n-1)t/l_n} \tag{3.16}$$

where A_n is a coefficient to be dsetermined, and $k_{\rm n}$ is the eigenvalue with $k_1 = 1 > k_2 > k_3 \ldots$. The spatial distribution of ϕ is expanded as a linear combination of the wave functions. The wave function corresponding to k_1 is the fundamental mode, and the wave functions corresponding to other $k_{\rm n}$ are the high-order modes. At steady state ($t \to \infty$), except for the maximum eigenvalue k_1, the corresponding exponential terms of k_2, k_3 … are all attenuated, which means the fundamental mode alone can be used to describe the characteristics of the system.

To solve the equation of the infinite plate reactor under subcritical state, the external source term S should be considered. Assuming that the spatial distribution $f(r)$ of fission neutrons is the same as that of independent external neutron sources, the neutron flux, delayed precursor concentration, and external source can be derived by using the separation of time and space.

$$\phi(r,t) = f(r)\phi_0(t), \quad C_i(r,t) = f(r)C_0^i(t), \quad S(r,t) = f(r)S_0(t) \tag{3.17}$$

Substituting Eq. (3.17) into kinetics equations with an external source, we have Eqs. (3.18) and (3.19) based on the point kinetics model.

$$\frac{\mathrm{d}n(t)}{\mathrm{d}t} = \frac{\rho(t)-\beta}{\Lambda}n(t) - \sum_{i=1}^{6}\lambda_i C_i(t) + S(t) \tag{3.18}$$

$$\frac{\mathrm{d}C_i}{\mathrm{d}t} = \frac{\beta_i}{\Lambda}n(t) - \lambda_i C_i(t) \tag{3.19}$$

Considering all the components of harmonics, the neutron flux distribution considering external source for subcritical systems is

$$\phi(x,t) = \sum_{n=odd}^{\infty} \cos\frac{n\pi}{a}x\left[c_n \mathrm{e}^{(k_{\rm s,n}-1)t/l_n} + \frac{S_n}{\Sigma_a(1+L^2B_n^2)(1-k_{\rm s,n})}\right] \tag{3.20}$$

where c_n is an undetermined coefficient, and $1 > k_{\rm s,n}$ (n = 1, 3, 5, …) [11]. The eigenfunction consists of both exponential term and source term. Because all exponential terms correspond to k decay with time, both fundamental and higher order harmonics need to be considered. However, the source term will not decay, and both fundamental and higher order modes are independent of the time variable. The external neutron source enhances the contribution of higher order modes, and the space of neutron flux includes a decaying term and a stable term, which is more complex than that of the critical system.

The above conclusion is based on the simple infinite plate reactor with the point kinetics model. It should be noted that the spatial distribution of external source neutrons is usually quite different from that of fission neutrons in the source-driven subcritical system. The spatial distribution of neutron flux in the system is extremely anisotropic and the neutron energy spectrum is wide and complex. In fact, the spatial distribution cannot be simply characterized by a single unified $f(r)$. The external neutron source term exited the higher order modes. It is hard to obtain the analytical solution without approximation, which makes analyzing the kinetics of complex systems difficult. Typically, for example, reactivity measurements based on the point kinetics model are strongly influenced by spatial effects at deep subcritical levels, and there is a large discrepancy between theory and experiment. To solve these problems, especially the spatial effects with external sources, many methods have been studied. Golovkina improved the point reactor equations from the perspective of power redistribution [12]. Henrice introduced the incomplete gamma function to describe the insertion of reactivity and solve the point kinetic equations [13]. Goncalves defined new weighting functions based on generalized perturbation theory and developed the point kinetics equations [14]. Espinosa proposed a new fractional diffusion model to replace the original model [15]. Nevertheless, there seems to be no mature theory for solving the spatial effects problem caused by external source until now.

2. Kinetics models improvement for fluid fuels

The position of fuel in some special advanced nuclear systems is constantly changing during the operation, resulting in new kinetic characteristics. The molten salt reactor with fluid fuel, for example, uses liquid nuclear fuel. The delayed neutron precursors flow in and out of the core uninterruptedly with the flow of fuel, resulting in the reduction of the effective delayed neutron fraction inside the core. The general point kinetics equations do not consider the characteristics of fluid fuel. Therefore, the traditional theory is not applicable to the mobile fuel system. To characterize the flow mechanism of the delayed neutron precursors, the point kinetics equations are revised as follows:

$$\frac{\mathrm{d}n(t)}{\mathrm{d}t} = \frac{\rho(t) - \beta_{\mathrm{eff}}}{\Lambda} n(t) + \sum_{i=1}^{6} \lambda_i C_{\mathrm{c},i}(t) \tag{3.21}$$

$$\frac{\mathrm{d}C_{\mathrm{c},i}}{\mathrm{d}t} = \frac{\beta_i}{\Lambda} n(t) - \left(\lambda_i + \frac{1}{\tau_{\mathrm{c}}}\right) C_{\mathrm{c},i}(t) + \frac{1}{\tau_{\mathrm{l}}}\left(\frac{V_{\mathrm{l}}}{V_{\mathrm{c}}}\right) C_{\mathrm{l},i}(t - \tau_{\mathrm{l}}) \tag{3.22}$$

$$\frac{\mathrm{d}C_{\mathrm{l},i}}{\mathrm{d}t} = -\left(\lambda_i + \frac{1}{\tau_{\mathrm{l}}}\right) C_{\mathrm{l},i}(t) + \frac{1}{\tau_{\mathrm{c}}}\left(\frac{V_{\mathrm{c}}}{V_{\mathrm{l}}}\right) C_{\mathrm{c},i}(t) \tag{3.23}$$

where $\beta_{\mathrm{eff}} = \beta_{\mathrm{static}} - \beta_{\mathrm{loss}}$; β_{static} is the delayed neutron fraction when the fuel is not flowing; β_{loss} is the delayed neutron fraction that is lost due to fuel flow; τ_{c} equals $M_{\mathrm{c}}/w(t)$ and describes the residence time of the precursors in the core in units of s; τ_{l} equals $M_{\mathrm{l}}/w(t)$ and describes the residence time of the precursor nucleus in the loop in units of s; M_{c} is the mass of the fluid fuel in the core in units of kg; M_{l} is the mass of the fluid fuel in the loop in units of kg; $w(t)$ is the mass flow rate for mobile fuels in units of kg/s; V_{c} is the volume of the fluid fuel in the core in units of cm^3; V_{l} is the volume of the fluid fuel in the loop in units of cm^3; C_{c} is the concentration of the precursors inside the core in units of $1/\mathrm{cm}^3$; and C_{l} is the concentration of precursors in the loop in units of $1/\mathrm{cm}^3$.

Equation (3.22) represents the balance equation of the delayed neutron precursor concentration in the core. The first term on the right represents the term from fission, the second represents the loss term caused by the decay of delayed neutron precursors and drift flow, and the third represents the term where delayed neutron precursor in the loop returns to the core. Equation (3.23) describes the delayed neutrons in the loop. The first term on the right represents the decay and drift loss of the delayed neutron precursors in the reactor core, and the second term on the right represents the contribution of the delayed neutron precursor from the reactor core to the loop. For the characteristics of liquid fuels, it is necessary to study the theory and computational method of the distribution of delayed neutron precursors and carry out experimental validation [5, 6].

3. The variation of kinetic parameters affects transient response behavior

In advanced nuclear systems, especially hybrid nuclear systems (i.e., fusion or accelerator-driven subcritical systems), the average neutron energy is higher and the energy spectrum is more complex than those of traditional nuclear systems. The kinetics parameters are also different, which affects the transient response characteristics of the nuclear system. To understand the physical performance of the systems and establish operation regulation accurately, the related kinetics behavior should be studied. For example, the neutron generation time of fast reactors is usually several orders smaller than that of thermal reactors, which will not result in differences in neutron kinetics when the induced reactivity is small. However, when the induced reactivity exceeds 1β (approximately 600 pcm in a reactor with UO_2 fuel), the rise period of neutron flux or power will be much shorter than that of the thermal reactor, and the nuclear system reaches prompt criticality. Advanced systems with the capability of transmutation are another example. With the amount of MA increasing and the amount of ^{238}U decreasing, the neutron spectrum hardens and, simultaneously, the β_{eff} and fuel Doppler effects decrease. Subsequently, the transient response characteristics of the system will change. It is possible that the system cannot induce sufficient negative reactivity when the temperature increases, and safety risks increase. Therefore, in the design of critical systems with a transmutation function, it is necessary to pay special attention to the change of neutron kinetics response characteristics.

The delayed effect of delayed neutrons on the neutron flux change is not evident in the subcritical systems when compared with critical system after same perturbation. This delayed effect is also not evident in fast system compared with thermal system. With the increase of subcriticality, the importance of the delayed neutrons will be greatly reduced, and the importance of external neutrons will be strengthened. In addition, the response characteristics of the system to reactivity transient or feedback are different. For example, if $+5\beta$ reactivity was inserted instantaneously into a critical system, the nuclear system would reach an instantaneous supercritical state and might be uncontrollable; if $+5\beta$ reactivity is inserted into a core with subcriticality of -10β, the power will only increase and the system will tend to reach a stable state at another power level. In subcritical systems, both source-dominated and feedback-dominated effects should be considered in the transient process, which is different from the considerations of critical systems. The study of the kinetics of these two kinds of effects plays a very important role in system control strategy development [7, 8].

3.3 Computational Methods for Kinetics Problems in Advanced Nuclear Systems

Neutron kinetics in advanced nuclear systems is complex, with new characteristics, such as the movement of the delayed neutron precursors and the spatial effects, among others. It is difficult to apply the traditional point reactor kinetics model for advanced nuclear systems. Space–time kinetics methods that consider all variables including position, direction, energy, and time, should be adopted although new adaptive methods need to be developed. In this section, three kinds of space—time kinetics computational methods, i.e., the direct numerical method, the modal expansion method, and the factorization method, are introduced for solving neutron transport equations.

3.3.1 *Direct Numerical Method*

To solve the neutron transport equation, the basic idea of the direct numerical method is to discretize all variables including time, space, energy, and angle directly. This is one of the most computationally intensive methods in space–time neutron transport calculation. In general, this method is used to provide accurate numerical solutions for other approximation methods and to verify the results, or to make final calculations for the design of nuclear devices of great importance [16].

The direct numerical method mainly involves processing the time variable based on the full implicit method, and its calculation scheme is

$$y^{n+1} = y^n + \Delta t_n f(t_{n+1}, y^{n+1}) \tag{3.24}$$

By substituting the above equation into Eqs. (3.2) and (3.3), the time-dependent neutron transport equation is obtained

$$\begin{aligned}
&\frac{\phi_{n+1} - \phi_n}{v(E)(t_{n+1} - t_n)} + \nabla \cdot \boldsymbol{\Omega}\phi_{n+1} + \Sigma_t(\boldsymbol{r}, E, \boldsymbol{\Omega}, t_{n+1})\phi_n = S(\boldsymbol{r}, E, \boldsymbol{\Omega}, t_{n+1}) \\
&+ \iint \Sigma_s(\boldsymbol{r}, E' \to E, \boldsymbol{\Omega}' \to \boldsymbol{\Omega}, t_{n+1})\phi(\boldsymbol{r}, E', \boldsymbol{\Omega}', t_{n+1})\,\mathrm{d}E'\,\mathrm{d}\boldsymbol{\Omega}' \\
&+ \frac{\chi_p(\boldsymbol{r}, E)}{4\pi}(1 - \beta) \iint \upsilon(\boldsymbol{r}, E')\Sigma_f(\boldsymbol{r}, E', t_{n+1})\phi(\boldsymbol{r}, E', \boldsymbol{\Omega}', t_{n+1})\,\mathrm{d}E'\mathrm{d}\boldsymbol{\Omega}' \\
&+ \frac{1}{4\pi}\sum_i \chi_i(\boldsymbol{r}, E)\lambda_i C_i(\boldsymbol{r}, t_{n+1})
\end{aligned} \tag{3.25}$$

where ϕ_{n+1} means $\phi(\boldsymbol{r}, E, \boldsymbol{\Omega}, t_{n+1})$ and ϕ_n means $\phi(\boldsymbol{r}, E, \boldsymbol{\Omega}, t_n)$. The equation for the delayed neutron precursors is

$$C_i(\boldsymbol{r},t_{n+1}) = \frac{\Delta t_n \beta_i}{1+\lambda_i \Delta t_n}\iint \upsilon(\boldsymbol{r},E')\Sigma_f(\boldsymbol{r},E',t_{n+1})\phi_{g'}(\boldsymbol{r},E',\boldsymbol{\Omega}',t_{n+1})\,\mathrm{d}E'\mathrm{d}\boldsymbol{\Omega}' + \frac{1}{1+\lambda_i \Delta t_n}C_i(\boldsymbol{r},t_n) \tag{3.26}$$

The above time-dependent neutron transport equation can be further expressed as

$$\begin{aligned}\nabla\cdot\boldsymbol{\Omega}\phi(\boldsymbol{r},E,\boldsymbol{\Omega},t_{n+1}) &+ \left[\Sigma_t(\boldsymbol{r},E,\boldsymbol{\Omega},t_{n+1}) + \frac{1}{v(E)\Delta t_n}\right]\phi(\boldsymbol{r},E,\boldsymbol{\Omega},t_{n+1})\\ &= \frac{1}{v(E)}\frac{\phi(\boldsymbol{r},E,\boldsymbol{\Omega},t_n)}{\Delta t_n} + \iint \Sigma_s(\boldsymbol{r},E'\rightarrow E,\boldsymbol{\Omega}'\rightarrow\boldsymbol{\Omega},t_{n+1})\phi(\boldsymbol{r},E',\boldsymbol{\Omega}',t_{n+1})\,\mathrm{d}E'\mathrm{d}\boldsymbol{\Omega}'\\ &\frac{\chi_p(\boldsymbol{r},E)}{4\pi}(1-\beta)\iint \upsilon(\boldsymbol{r},E')\,\Sigma_f(\boldsymbol{r},E',t_{n+1})\phi(\boldsymbol{r},E',\boldsymbol{\Omega}',t_{n+1})\,\mathrm{d}E'\mathrm{d}\boldsymbol{\Omega}'\\ &+\frac{1}{4\pi}\sum_i \chi_i(\boldsymbol{r},E)\lambda_i C_i(\boldsymbol{r},t_{n+1}) + S(\boldsymbol{r},E,\boldsymbol{\Omega},t_{n+1})\end{aligned} \tag{3.27}$$

The equation is similar to the steady-state neutron transport equation in form; however, the total cross section, the external source term, and the fission source term must be updated with time. The flux distribution of the time t_{n+1} could be derived by substituting the flux distribution of the previous time t_n into the equation. By iteration with proper time steps, the flux distribution of each moment can be derived.

3.3.2 Modal Expansion Method

The basic idea of the modal expansion method is that a series of trial functions are selected in advance (or determined by auxiliary calculation), then the neutron flux can be expanded by these trial functions, and then the ordinary differential equations are obtained by the weighted residual method or the variation method to determine the expansion coefficients of these trial functions [17].

The basic formula for expanding the neutron flux by trial functions is as follows:

$$\phi(\boldsymbol{r},\boldsymbol{\Omega},E,t) = \sum_i T_i(t)\phi_i(\boldsymbol{r},\boldsymbol{\Omega},E) \tag{3.28}$$

where $\phi_j(\boldsymbol{r},\boldsymbol{\Omega},E)$ is the i_{th} trial function, and $T_j(t)$ is the undetermined coefficient.

The modal expansion Eq. (3.28) is substituted into Eq. (3.2), which will then be multiplied by weighting function $W_k(\boldsymbol{r},\boldsymbol{\Omega},E)$. The integration is done in the phase space ($\boldsymbol{r}\times\boldsymbol{\Omega}\times E$), and finally Eq. (3.29) is derived. Similarly, Eq. (3.26) is multiplied by $W_k(\boldsymbol{r},\boldsymbol{\Omega},E)\times x_i(\boldsymbol{r},E)$ and then integrated in the same phase space, and Eq. (3.30) is derived

$$\sum_j \Lambda_{kj} \frac{\partial T_j(t)}{\partial t} = \sum_j \left[\rho(t) - \beta'(t)\right]_{kj} T_j(t) + \sum_i \lambda_i C_{ik}(t) + Q_k(t) \tag{3.29}$$

$$\frac{\partial C_{ik}(t)}{\partial t} + \lambda_i C_{ik}(t) = \sum_j \left[\beta_i'(t)\right]_{kj} T_j(t) \tag{3.30}$$

where

$$\Lambda_{kj} \equiv (W_k, \frac{1}{v}\phi_j) \tag{3.31}$$

$$\left[\rho(t) - \beta'(t)\right]_{kj} \equiv (W_k, \left\{-\Omega \cdot \nabla\phi_j - \Sigma\phi_j + S(t)\phi_j + \frac{\chi_p}{4\pi}(1-\beta)F(t)\phi_j\right\}) \tag{3.32}$$

$$\left[\beta_i'(t)\right]_{kj} \equiv (W_k, \frac{\chi_i}{4\pi}\beta_i F\phi_j) \tag{3.33}$$

$$C_{ik}(t) \equiv (W_k, \frac{\chi_i}{4\pi} C_i(r, t)) \tag{3.34}$$

$$Q_k(t) \equiv (W_k, Q) \tag{3.35}$$

If the first-order modal expansion is considered and W_k is ϕ^*/F (ϕ^* is the adjoint flux), then Eqs. (3.29) and (3.30) are the same as the point kinetics equations. If neutron flux is expanded with I order modal, then I order equations can be obtained. These equations can be written in a more simplified way as

$$\Lambda \frac{\partial \boldsymbol{T}(t)}{\partial t} = (\rho - \beta')\boldsymbol{T}(t) + \sum_i \lambda_i \boldsymbol{C}_i(t) + \boldsymbol{Q} \tag{3.36}$$

$$\frac{\partial \boldsymbol{C}_i(t)}{\partial t} + \lambda_i \boldsymbol{C}_i(t) = \beta_i' \boldsymbol{T}(t) \tag{3.37}$$

where Λ, $\rho - \beta'$, and β_i' represent the matrices formed by the elements of Λ_{ki}, $|\rho - \beta'|_{kj}$, and $|\beta_i'|_{kj}$. Obviously, the two equations above are general point kinetics equations. The $\phi(\boldsymbol{r}, \boldsymbol{\Omega}, E)$ can be obtained after the coefficient $T_j(t)$ is determined by using iteration based on Eqs. (3.31) to (3.37).

3.3.3 Factorization Method

The basic idea of the factorization method is to decompose the neutron flux into the product of shape function and amplitude function, and introduce additional conditions to the shape function by using the arbitrariness of the factorization, so that the neutron transport equation can be simplified. Its characteristics are that the time-varying characteristics of neutron flux are mainly reflected in the time-varying characteristics of amplitude function, while the shape function is to reflect the slowly varying space and energy distribution of neutron flux [1, 18].

The basic equation of the factorization method is

$$\phi(\boldsymbol{r}, E, \boldsymbol{\Omega}, t) = T(t) \times \varphi(\boldsymbol{r}, E, \boldsymbol{\Omega}, t) \tag{3.38}$$

Substituting the above equation into the original neutron transport Eq. (3.12), the shape function of neutron flux can be derived as

$$(L_0 + L_\mathrm{f})\varphi + \frac{S_\mathrm{d}\{T\varphi\}}{T} + \frac{S}{T} = \frac{1}{v}\left[\frac{1}{T}\frac{\mathrm{d}T}{\mathrm{d}t}\varphi + \frac{\partial\varphi}{\partial t}\right] \tag{3.39}$$

where

$$L_0\varphi = -\boldsymbol{\Omega}\cdot\nabla\varphi(\boldsymbol{r}, E, \boldsymbol{\Omega}, t) - \Sigma_\mathrm{t}(\boldsymbol{r}, E, \boldsymbol{\Omega}, t)\varphi(\boldsymbol{r}, E, \boldsymbol{\Omega}, t) + \iint \Sigma_\mathrm{s}(\boldsymbol{r}, E' \to E, \boldsymbol{\Omega}' \to \boldsymbol{\Omega}, t)\varphi(\boldsymbol{r}, E', \boldsymbol{\Omega}', t)\,\mathrm{d}E'\mathrm{d}\boldsymbol{\Omega}' \tag{3.40}$$

$$L_\mathrm{f}\varphi = \frac{\chi_\mathrm{p}(\boldsymbol{r}, E)(1-\beta)}{4\pi}\iint \upsilon\Sigma_\mathrm{f}(\boldsymbol{r}, E', \boldsymbol{\Omega}', t)\varphi(\boldsymbol{r}, E', \boldsymbol{\Omega}', t)\,\mathrm{d}E'\mathrm{d}\boldsymbol{\Omega}' \tag{3.41}$$

$$S_\mathrm{d} = \frac{1}{4\pi}\sum_i \lambda_i C_i(\boldsymbol{r}, t)\chi_i(\boldsymbol{r}, E) \tag{3.42}$$

$$S = S(\boldsymbol{r}, E, \boldsymbol{\Omega}, t) \tag{3.43}$$

Multiplying the two sides of the shape function equation by the adjoint neutron flux $\phi^*(\boldsymbol{r}, E, \boldsymbol{\Omega})$, integrating over phase space $(\boldsymbol{r}, E, \boldsymbol{\Omega})$ and using the following condition:

$$\frac{\partial I}{\partial t} = \frac{\partial}{\partial t}\iiint \frac{\phi^*(r, E, \Omega)\varphi(r, E, \Omega)}{v}\,\mathrm{d}E\mathrm{d}\Omega\mathrm{d}V = 0 \tag{3.44}$$

the equation of amplitude function is obtained as follows:

$$\frac{\mathrm{d}T(t)}{\mathrm{d}t} = \frac{\rho(t) - \beta(t)}{\Lambda(t)}T(t) + \sum_i \lambda_i c_i(t) + s(t) \tag{3.45}$$

Similarly, the time-dependent equation for the delayed neutron precursors can be derived as

$$\frac{\mathrm{d}c_i(t)}{\mathrm{d}t} = \frac{\beta_{\mathrm{eff},i}(t)}{\Lambda(t)}c_i(t) - \lambda_i c_i(t) \tag{3.46}$$

where

$$\frac{\rho(t)}{\Lambda(t)} = \frac{k-1}{k}\frac{\iiint \phi^*(\boldsymbol{r}, E, \boldsymbol{\Omega})\left(L_{\mathrm{f}} + L_{\mathrm{d}}^{stable}\right)\varphi(\boldsymbol{r}, E, \boldsymbol{\Omega}, t)\mathrm{d}E\mathrm{d}\boldsymbol{\Omega}\mathrm{d}V}{I} \tag{3.47}$$

$$\frac{\beta_{\mathrm{eff}}(t)}{\Lambda(t)} = \frac{\iiint \phi^*(\boldsymbol{r}, E, \boldsymbol{\Omega})L_{\mathrm{d}}^{stable}\varphi(\boldsymbol{r}, E, \boldsymbol{\Omega}, t)\mathrm{d}E\mathrm{d}\boldsymbol{\Omega}\mathrm{d}V}{I} \tag{3.48}$$

$$L_{\mathrm{d}}^{stable} = \sum_i \lambda_i C_i(\boldsymbol{r}, t)\chi_i(\boldsymbol{r}, E) \tag{3.49}$$

$$\beta_{\mathrm{eff}}(t) = \sum_i \beta_{\mathrm{eff},i}(t) \tag{3.50}$$

$$c_i(t) = \frac{\iiint \phi^*(\boldsymbol{r}, E, \boldsymbol{\Omega})C_i(\boldsymbol{r}, t)\chi_i(\boldsymbol{r}, E)\,\mathrm{d}E\mathrm{d}\boldsymbol{\Omega}\mathrm{d}V}{I} \tag{3.51}$$

$$s(t) = \frac{\iiint \phi^*(\boldsymbol{r}, E, \boldsymbol{\Omega})S(\boldsymbol{r}, E, \boldsymbol{\Omega}, t)\,\mathrm{d}E\mathrm{d}\boldsymbol{\Omega}\mathrm{d}V}{I} \tag{3.52}$$

Equations (3.45) and (3.46) are a set of amplitude function equations in the usual point kinetics form. The difference between these equation and traditional point kinetics equations is that the kinetic parameters $\rho(t)$, $\Lambda(t)$, and $\beta(t)$ remain the time-dependent features. These time-varying characteristics can be obtained after solving the neutron flux shape function Eq. (3.39) by the direct numerical method.

In the factorization method, there are several approximations for solving the neutron flux shape function equation, such as improved quasi-static approximation, quasi-static approximation, and adiabatic approximation. The calculation time step of the neutron flux shape function can be set according to the needs of users. A large time step is used when the shape change is small and no recalculation is needed, while small time steps are adopted when the shape function changes greatly. In this way, the computational time can decrease and the high accuracy of the results can be kept.

The above three types of time-dependent computational methods belong to deterministic methods. The Monte Carlo method is applicable to problems of complex geometry and complex material isotope distribution. However, it is time-consuming to solve time-dependent neutron kinetics problems, especially those with delayed neutrons, by Monte Carlo methods. In recent years, attempts have been made using time-dependent Monte Carlo methods. For example, the Monte Carlo improved quasi-static method embeds the Monte Carlo methods into the improved quasi-static computational framework. According to the improved quasi-static method, the neutron space–time kinetics problem with Monte Carlo methods can be divided into three parts: (1) solving the point reactor kinetics equation; (2) obtaining the adjoint flux and attaining the kinetics parameters by a Monte Carlo adjoint solution method such as the iterative fission probability method; and (3) and calculating the neutron flux shape in the transient process by solving the Monte Carlo steady-state fixed-source problem. Kinetics parameter calculation and steady-state fixed-source calculation are two key technologies of the improved quasi-static Monte Carlo method. Another example is the time-dependent direct Monte Carlo method, which has several principal features: (1) the delayed neutron source is given using the data from the file from the preprocessing step; (2) time step by time step computation is used; and (3) prompt and delayed neutrons are simulated separately.

Among these three methods, the direct numerical method has the fewest assumptions and the highest accuracy but the highest computational cost. A preprocessing step for neutron flux has been developed to obtain reasonable and economic results using the other two methods. Unlike the modal expansion method, the factorization method does not calculate the trial functions (shape function) beforehand. The update of the shape function at the end of the step is affected by the amplitude function $n(t)$ in the same time interval. The expansion method is free to choose trial functions. Therefore, it is difficult to compare the factorization method with the modal method in terms of accuracy, but the advantages and disadvantages of the two methods can be qualitatively compared. The advantages of the factorization method are as follows: (1) it is especially suitable for dealing with nonlinear problems; (2) the results are not affected by the appropriateness of the trial function; and (3) and it is easy to verify the accuracy of the results by specifying the convergence criteria. The advantages of the modal method are: (1) it provides the flexibility of selecting appropriate expansion functions; and (2) the computational cost required is low [18].

Keeping in mind the above discussion, every method has its own advantages. Researchers should select a method according to specific features of the problem. Meanwhile, the space–time kinetics computational methods should be developed for advanced nuclear systems. For instance, there are also high-order harmonic problems and external source effects in hybrid nuclear systems, thus, the corresponding computational methods need to be further studied [19–21].

3.4 Simulation Codes for Neutron Kinetics

Several neutron kinetics calculation programs have been developed for advanced nuclear systems. The representative programs include SuperMC, PARCS (Purdue Advanced Reactor Core Simulator), TDMCC (Time-Dependent Monte Carlo Code), ERANOS (European Reactor Analysis Optimized System), etc. SuperMC has been introduced in the previous section. This section only focuses on the remaining programs.

1. PARCS

PARCS is a three-dimensional deterministic neutronics software developed for the American Nuclear Regulatory Commission (NRC) by Purdue University [22]. PARCS solves the steady-state and time-dependent, multigroup neutron diffusion and SP3 transport equations in orthogonal and non-orthogonal geometries. Time discretization is performed using the direct numerical method with an exponential transformation of the group fluxes (Theta method). Spatial discretization is performed with a nodal method, such as the analytical nodal method or the nodal expansion method. PARCS has been coupled with thermo-hydraulic calculation programs and applied for the study of the kinetics of lead–bismuth eutectic and gas-cooled XADS systems.

2. TDMCC

TDMCC is a three-dimensional time-dependent Monte Carlo simulation program developed by RFNC-VNIIEF (The Russian Federal Nuclear Center—All-Russian Scientific Research Institute of Experimental Physics) [23]. In this program, a direct simulation method other than the improved quasi-static Monte Carlo method is implemented. The prompt and delayed neutrons are simulated separately time step by time step. The program can be executed independently and coupled with thermal-hydraulic programs and is designated not only for calculating time-dependent reactor core neutron kinetics problems but also for transient analysis of various accidents.

3. ERANOS/KIN3D

ERANOS is developed jointly by the European Community and validated to provide a suitable basis for the reliable neutronics calculation of traditional and advanced fast reactors [24]. KIN3D is a kinetics/perturbation calculation module of ERANOS, in which both the point kinetics solver and the space–time kinetics solver (direct and factorization method) are included. KIN3D has been applied in kinetics behavior analysis of both critical (the French Superphoenix Reactor) and subcritical systems (PDS-XADS: Preliminary Design Study-Experimental Accelerator-Driven System).

References

1. Ott K, Neuhold R (1985) Introductory nuclear reactor dynamics. American Nuclear Society, La Grange Park III, USA
2. Wu YC (2017) Fusion neutronics. Springer Nature Singapore Pte. Ltd
3. Wu YC (2018) Multi-functional neutronics calculation methodology and program for nuclear design and radiation safety evaluation. Fusion Sci Technol 74:321–329
4. Rineiski A, Maschek W (2003) On application of quasistatic and point-reactor schemes for sub-critical systems with external neutron source. In: Nuclear mathematical and computational sciences: a century in review, Tennessee, 6–11 April
5. Zhang DL, Qiu SZ, Su GH (2009) Development of a safety analysis code for molten salt reactor. Nucl Eng Des 239(12):2778–2785
6. Wang SS, Rineiski A, Maschek W (2006) Molten salt related extensions of the SIMMER-III code and its application for a burner reactor. Nucl Eng Des 236(14–16):1580–1588
7. Maschek W, Rineiski A, Suzuki T et al (2005) SIMMER-III and SIMMER-IV safety code development for reactor with transmutation capability. In: Mathematics and computation, supercomputing, reactor physics and biological applicants, Palais de Papes, France, 12–15 September 2005
8. Dulla S, Paolo P, Ravetto P et al (2011) Integral parameters in source driven systems. Prog Nucl Energy 53:32–40
9. Bahman Z (2017) Neutronic analysis for nuclear reactor analysis. Springer Nature Singapore Pte. Ltd
10. Alan EW, Donal RT, Pavel VT (2012) Fast spectrum reactors. Springer Science

11. Wei SC (2009) Study on the methodology for sub-criticality monitoring of accelerator driven subcritical system. Master thesis, Shanghai Jiao Tong University
12. Golovkina AG, Kudinovich IV (2015) Two-point kinetic model for coupled subcritical reactor core. St Petersburg State University, pp 53–60
13. Henrice JE, Goncalves AC, Palma DA (2013) Analytical solution of point kinetic equations for sub-critical systems. In: 2013 international nuclear Atlantic conference, Brazil, 24–29 October 2013
14. Goncalves WDC, Martinez AS, Silva FCD (2015) Point kinetics equations for subcritical systems based on the importance function associated to an external neutron source. Annal Nucl Energy 79:1–8
15. Espinosa-Paredes G (2016) Fractional-space neutron point kinetics (F-SNPK) equations for nuclear reactor dynamics. Annal Nucl Energy 107:136–143
16. Azmy Y, Sartori E (2010) Nuclear computational science: a century in review. Springer, Berlin
17. Bell GI, Glasstone S (1970) Nuclear reactor theory. Van Nostrand Reihold Company, New York
18. Hetrick DL, Jarvis RG (1971) Dynamics of nuclear reactors. University of Chicago Press
19. Cacuci DG (2002) On perturbation theory and reactor kinetics: from Wigner's pile period to accelerator driven systems. In: PHYSOR 2002, Seoul, Korea
20. Rineiski A, Maschek W et al (2005) Kinetics models for safety studies of ADS. Annal Nucl Energy 32:1348–1365
21. Hassanzadeh M, Feghhi SAH, Khalafi H (2013) Calculation of the neutron importance and weighted neutron generation time using MCNIC method in accelerator driven subcritical reactor. Nucl Eng Des 262:404–408
22. Joo HG, Barber DA, Jiang G et al (1998) PARCS: a multidimensional two-group reactor kinetics code based on the non-linear analytical nodal method. University of Purdue Report PU/NE-98-26
23. Bakanov VV, Zhitnik AK, Motlokhov VN et al (2004) TDMCC Monte-Carlo package coupled with STAR-CD thermal-hydraulic code. Transac American Nucl Soc 91:250
24. Ruggieri JM, Tommasi J, Lebrat JF et al (2006) ERANOS 2.1: international code system for GEN IV fast reactor analysis. In: 2006 international congress on advances in nuclear power plants, Nevada, 4–8 June 2006

Chapter 4
Neutron-Induced Transmutation

Transmutation is defined as the conversion of one nuclide into another, along with the corresponding change in the number of neutrons or protons in the resulting nucleus. Transmutation can be realized via nuclear reactions or radioactive decays. In nuclear systems, neutron-induced transmutation includes the processes of burnup, nuclear waste transmutation, nuclear fuel breeding, and material activation. Research on burnup mainly focused on studying the consumption of nuclear fuel isotopes and its impact on nuclear system performance. During the process of fuel burnup, a large amount of long-lived nuclear wastes with high level of radiation will be produced, the improper handling of which will cause substantial radiological hazards. In advanced nuclear systems, especially hybrid nuclear systems, long-lived high-level radionuclides in nuclear wastes can be converted into short-lived nuclides or stable nuclides via nuclear waste transmutation. The fissile fuel can only be used by existing nuclear systems for decades as estimated by the International Energy Agency (IEA) [1], while tritium, which is used by fusion systems, is virtually nonexistent in nature. Nuclear fuel breeding refers to the conversion of fissionable nuclides into fissile nuclides or the conversion of other nuclides into tritium via nuclear reactions to satisfy the nuclear fuel demands for the long-term stable application of nuclear energy. During the operation of nuclear systems, neutrons react with nuclei to produce radionuclides, and the materials are activated. The radionuclides that are produced typically decay and emit α, β, and γ rays, which poses potential radioactive hazards to nuclear systems, workers, and the environment.

In this chapter, first, the main principles of transmutation are introduced, and the process of transmutation is described, including burnup, nuclear waste transmutation, nuclear fuel breeding, and material activation. Then the in-core fuel management and fuel cycle of a nuclear system will be discussed. Finally, the calculation methods and common calculation codes for transmutation are presented; their application is introduced in Chaps. 9 and 11.

Y. Wu, *Neutronics of Advanced Nuclear Systems*,
https://doi.org/10.1007/978-981-13-6520-1_4

4.1 Principles of Transmutation

Neutron-induced reactions are an important pathway for including transmutation. However, not all types of neutron nuclear reactions can induce transmutation. For example, the (n, n) and (n, n′) reactions do not result in changes in the numbers of protons or neutrons in the nucleus. Table 4.1 lists the main types of neutron nuclear reactions that can induce transmutation. Radioactive decay is another important pathway for inducing nuclear transmutation. Table 4.2 lists the main types of decay that can induce transmutation. Neutron nuclear reactions and radioactive decays are often combined to form transformation relations, for example, to form a long transmutation chain or multiple transmutation chains. A transmutation chain is used to represent the transformations of nuclides during the transmutation process.

In transmutation, the change rates of the nuclide densities in the medium are mainly affected by the neutron nuclear reactions and decay processes. Under neutron irradiation, the change rate of density for any type of radionuclide is equal to the production rate minus the consumption rate. The production rate and the consumption rate can be calculated by using the reaction rate and the decay constant. According to the conservation relationship, the corresponding differential equation

Table 4.1 Main reaction types of neutron-induced transmutation

Reaction type	Description	Reaction type	Description
(n, γ)	Neutron absorption followed by emission of gamma ray	(n, d)	Neutron absorption followed by emission of a deuteron
(n, 2n)	Neutron absorption followed by emission of two neutrons	(n, h)	Neutron absorption followed by emission of a helium-3 nucleus
(n, α)	Neutron absorption followed by emission of an alpha particle	(n, 2p)	Neutron absorption followed by emission of two protons
(n, p)	Neutron absorption followed by emission of a proton	(n, 4n)	Neutron absorption followed by emission of four neutrons
(n, t)	Neutron absorption followed by emission of a triton	(n, 2nα)	Neutron absorption followed by emission of two neutrons and an alpha particle
(n, nα)	Neutron absorption followed by emission of a neutron and an alpha particle	(n, f)	Neutron absorption followed by emission of a light particle through nuclear fission
(n, 3n)	Neutron absorption followed by emission of three neutrons		

Table 4.2 Main decay types that induce transmutation

Decay type	Description
(β^-)	Spontaneous decay emits β^-
(β^+)	Spontaneous decay emits β^+
(α)	Spontaneous decay emits α
($\beta^-\alpha$)	Spontaneous decay emits β^- and α

can be established as Eq. (4.1), which is known as the Bateman equation. This is the basic equation for describing the evolution of nuclides in materials due to neutron-induced transmutation:

$$\frac{dN_i}{dt} = -N_i(\lambda_i + \sigma_i\phi) + \sum_{j \neq i} N_j(\lambda_{i,j} + \sigma_{i,j}\phi) + S_i \tag{4.1}$$

$$S_i = \sum_k N_k \sigma_k^{\mathrm{f}} \phi Y_{i,k} \tag{4.2}$$

where N_i is the number of nuclide i at time t, in units of cm^{-3}; λ_i is the decay constant of nuclide i, in units of s^{-1}; $\lambda_{i,j}$ is the decay constant of radionuclide i that is generated by the decay of nuclide j, in units of s^{-1}; σ_i is the total cross section for the reaction of nuclide i with a neutron, in units of cm^2; $\sigma_{i,j}$ is the production cross section of nuclide i that is produced by the reaction of nuclide j with a neutron, in units of cm^2; σ_k^{f} is the fission reaction cross section of nuclide k, in units of cm^2; ϕ is the neutron flux, in units of $\mathrm{cm}^2 \cdot \mathrm{s}^{-1}$; S_i is the number of nuclide i that are produced by fission; and $Y_{i,k}$ is the yield of nuclide i from the fission reaction of nuclide k.

In nuclear systems, where many nuclides are being produced and decaying, Eq. (4.1) leads to sets of nonlinear partial differential equations that are often difficult to solve. Thus, the solution space is divided into several independent small spatial regions, and in each region, N_i and ϕ are assumed to change slowly in space (considered as constants). Then, the equations are converted into time-dependent first-order linear ordinary differential equations. The matrix representation of the Bateman equations can be expressed as

$$\frac{d\boldsymbol{N}(t)}{dt} = \boldsymbol{A}\boldsymbol{N}(t) \tag{4.3}$$

where $\boldsymbol{N}(t)$ is the vector of the numbers of nuclide N_i, and $\boldsymbol{A}$ is a coefficient matrix.

The solution of Eq. (4.4) is

$$\boldsymbol{N}(t) = e^{\boldsymbol{A}t}\boldsymbol{N}(0) \tag{4.4}$$

According to the Bateman equations, the element $\boldsymbol{A}_{i,j}$ of the matrix $\boldsymbol{A}$ represents the conversion rate of nuclide j to nuclide i when $i \neq j$ and the reduction rate of the nuclide i when $i = j$:

$$A_{i,j} = -\lambda_i^{\text{eff}}\delta_{ij} + b_{i,j}^{\text{eff}}\lambda_j^{\text{eff}}(1 - \delta_{ij}) \tag{4.5}$$

where λ_i^{eff} is the effective decay constant of nuclide i, and $b_{i,j}^{\text{eff}}$ is the effective branching ratio of nuclide j that has converted to nuclide i, which can be calculated as

$$\lambda_i^{\text{eff}} = \lambda_i + \sigma_i\phi \tag{4.6}$$

$$b_{i,j}^{\text{eff}} = \frac{\lambda_{i,j} + \sigma_{i,j}\phi}{\lambda_j^{\text{eff}}} \tag{4.7}$$

4.2 Nuclear Transmutation Process

This section will highlight the processes, along with the corresponding parameters, which are related to the nuclear transmutation process, namely, burnup, nuclear waste transmutation, nuclear fuel breeding, and material activation for advanced nuclear systems.

4.2.1 Burnup

During the operation of nuclear systems, fission reactions can occur between neutrons and fissile nuclides, in which approximately 200 MeV of energy and two to three neutrons are released on average, and several fission products are produced. This process is called burnup. The depletion of nuclear fuel directly affects the reactivity of a nuclear system. However, some of the nuclides and their decay products in the fission product have large thermal neutron absorption cross sections, which will induce fission product poisoning.

4.2.1.1 Fission Fuel Depletion

In fission systems and hybrid nuclear systems, fission fuel is continuously consumed to produce energy, which directly affects the reactivity and refueling cycle of the systems. In the following section, various typical parameters that characterize fuel depletion, such as the burnup depth, energy multiplication factor, and fission rate, are described.

1. Burnup depth

Burnup is a process in which fission fuels are consumed in fission systems. The burnup depth is represented by the total energy that is released per unit mass of fuel, and it indicates the depletion level of the fuel. The burnup depth is calculated via the following equation:

$$BU = \int_{0}^{T} P(t)dt \frac{1}{W_U} \tag{4.8}$$

where BU is the burnup depth, in units of MWd/t; $P(t)$ is the power of the fuel at time t, in units of MW; and W_U is the mass of the heavy isotopes (such as uranium) in the fuel, in units of t (tons). Taking UO_2 fuel as an example, the fraction of oxygen should be subtracted in the calculation of $W_{U.}$

The conversion relationship between units of burnup depth is

$$1\,\text{MWd/t} = 86.4\,\text{MJ/kg} \tag{4.9}$$

The burnup depth is directly related to the type of nuclear system. Compared with traditional nuclear system, advanced nuclear system can make more effective use of nuclear fuel, and the burnup depth is higher. For example, the designed burnup depth of a lead-based reactor typically exceeds 100 GWd/t, and the designed burnup depth of a gas-cooled fast reactor could exceed 500 GWd/t, while the highest burnup depth of a current PWR is only approximately 60 GWd/t.

2. Energy multiplication factor

In hybrid nuclear systems, an external source neutron in the subcritical reactor/blanket can induce fission reaction which causes large amount of energy to be released. The energy multiplication factor (M) is defined as the ratio of the total energy deposition that is induced by an external source neutron in the subcritical reactor/blanket to the energy of the external source neutron, as expressed in Eq. (4.10). A larger energy multiplication factor indicates that more fission reactions are induced by an external source neutron in the subcritical reactor/ blanket.

$$M = \frac{E_{\text{fission}}}{E_{\text{Source}}} \cdot \frac{\eta k_{\text{eff}}}{\bar{\nu}(1 - k_{\text{eff}})} \tag{4.10}$$

where E_{fission} is the mean energy that is released in each fission reaction, E_{Source} is the mean energy of external source neutrons, η is the effective external source neutron weighting factor, and $\bar{\nu}$ is the average number of neutrons that are emitted per fission in the system.

3. Fission rate

The fission rate corresponds to the number of fission reactions of fissile isotopes in unit volume per unit time. The fission rate varies spatially in the fission system and the hybrid nuclear system and the total fission rate can be obtained by integrating the fission rate to the whole space. The fission rate is an important parameter in the design of fission and hybrid nuclear systems.

4.2.1.2 Fission Product Poisoning

Various fission products have large thermal neutron absorption cross sections, which are called fission poisons. With changes in reactor power and operating time, the concentrations of these fission poisons change continuously, thereby causing significant reactivity change; this process is called fission product poisoning. The toxicity, which is denoted as P, is typically used to measure the effect of fission poisons on the effective multiplication factor, namely, k_{eff}, which is defined as the ratio of the number of thermal neutrons that are absorbed by fission poisons to the number of thermal neutrons that are absorbed by nuclear fuel:

$$P = \frac{\Sigma_a^P}{\Sigma_a^U} \tag{4.11}$$

where Σ_a^P is the macroscopic absorption cross section of the fission poisons, and Σ_a^U is the macroscopic absorption cross section of the nuclear fuel.

Fission product poisoning is especially important in thermal neutron reactors such as VHTRs. In fission products, ^{135}Xe and ^{149}Sm are especially important because not only do they have large thermal neutron absorption cross sections but also their precursor nuclei ^{135}I and ^{149}Nd have large fission yields. Therefore, ^{135}Xe and ^{149}Sm have a substantial impact on the operation of thermal neutron reactors. Since these fission poisons have small absorption cross sections for fast neutrons, fission production poisoning is typically not considered in fast reactors such as lead-based reactors due to the low thermal neutron flux.

1. Xenon poisoning

In a thermal neutron reactor, ^{135}Xe is mainly derived from beta decay of ^{135}I. For the first loading of the reactor core, the concentration of ^{135}Xe and ^{135}I are 0. After several days of steady-state operation, the production rates of ^{135}I and ^{135}Xe become equal to the corresponding rates of disappearance. Both ^{135}I and ^{135}Xe are close to their equilibrium concentrations, which are related to the core power density. When the core power or core power distribution changes substantially, the concentration of ^{135}Xe changes globally or locally, thereby affecting the core reactivity and power distribution. Iodine pit and xenon oscillation are the typical phenomena that occur.

After the reactor has been shut down, ^{135}Xe will no longer absorb neutrons and only decay. However, as the decay rate of ^{135}I is greater than that of ^{135}Xe, the concentration of ^{135}Xe will gradually increase over a period of time. Then as the decay rate of ^{135}I decreases, the concentration of ^{135}Xe will gradually decrease. The change in the residual reactivity over time is opposite the change in the ^{135}Xe concentration. First, it will reach its minimum value and, subsequently, gradually increase. This phenomenon is typically called "iodine pit". In the iodine pit, if the residual reactivity is still greater than zero, the reactor can also be started by moving control rods. If the residual reactivity is less than zero, the reactor cannot be started.

In a large thermal neutron reactor, the change of reactivity in the local area of the core causes changes in the local power and neutron flux, which, in turn, causes a change in the concentration of ^{135}Xe in the local region. The change in the concentration of ^{135}Xe in the local region causes a corresponding change in the local neutron flux. The interaction and feedback may cause spatial oscillations of the concentration of ^{135}Xe and the neutron flux distribution in the core; this is referred to as xenon oscillation. Xenon oscillation can only occur in large thermal neutron reactors that have high neutron flux. When the size of the core exceeds 30 times the migration length and the thermal neutron flux exceeds 10^{14} $cm^{-2} \cdot s^{-1}$, xenon oscillation must be considered. When xenon oscillation occurs, the total amount of xenon in the whole core does not change significantly, and does not constitute a serious supercritical hazard; however, it may cause the local temperature to rise, which may result in the melting of the fuel element, and the alternating temperature will affect the lifetime of the core material.

2. Samarium poisoning

Among the reactor fission products, the influence of ^{149}Sm on the reactor is second to that of ^{135}Xe. ^{149}Sm is produced via twice beta decay of ^{149}Nd. When the reactor is running, the equilibrium concentration of samarium is not related to the thermal neutron flux; however, the time that is required to reach the equilibrium samarium concentration is closely related to the neutron flux. Even for a reactor that is operating at high neutron flux, the time that is required to reach the equilibrium samarium concentration is at least several hundred hours, which is much longer than the time that is required to reach the equilibrium xenon concentration. The main reason is that the absorption cross section of ^{135}Xe is much larger than that of ^{149}Sm, and ^{135}Xe will disappear due to radioactive decay; hence, ^{135}Xe more easily reaches its equilibrium state. If the reactor has been running for a long time before shutdown and the ^{149}Sm in the reactor has reached the equilibrium concentration, then the concentration of ^{149}Sm after shutdown will increase with time. If the neutron flux density reaches 0.9×10^{14} $cm^{-2} \cdot s^{-1}$, the maximum concentration of ^{149}Sm after shutdown can be up to approximately two times the equilibrium concentration. When the reactor starts again, the excess ^{149}Sm will be consumed quickly, and the balance will be restored. If the neutron flux before shutdown is relatively low, the concentration of ^{149}Sm after shutdown will remain unchanged.

4.2.2 Nuclear Waste Transmutation

The nuclear wastes that are produced by fission nuclear systems include long-lived radionuclides such as the long-lived minor actinides (LLMAs) and the long-lived fission products (LLFPs). The main LLMAs in nuclear wastes include ^{237}Np, ^{241}Am, ^{243}Am, and ^{244}Cm; the main LLFPs include ^{99}Tc, ^{129}I, and ^{135}Cs. These long-lived radionuclides pose potential radiation safety problems for nuclear waste transportation and storage. Therefore, it is essential to find a feasible, safe, and economical way to dispose of the nuclear wastes properly. The transmutation of nuclear wastes, in which the LLMAs and LLFPs in nuclear wastes are transformed into short-lived or stable nuclides, is also related to neutron-induced transmutation. Transmutation is an important approach to reducing the nuclear wastes and consequently reducing the potential radiation hazard that is caused by nuclear wastes to the environment.

There are two approaches for transmuting LLMAs: one involves fission reactions with neutrons, whereas in the other approach, the LLMA is transmuted into fissile nuclides via capture reactions with thermal neutrons and subsequent fission reactions of fissile nuclides with thermal neutrons. Figure 4.1 illustrates the transmutation chains of LLMAs. The capture cross sections of ^{237}Np, ^{241}Am, ^{243}Am, and ^{244}Cm are large. The fission of the LLMAs with neutrons is expressed in Eq. (4.12).

$$n + \text{LLMA} \rightarrow {}^{A1}_{Z1}X + {}^{A2}_{Z2}Y + \nu n \qquad (4.12)$$

where ${}^{A1}_{Z1}X$ and ${}^{A2}_{Z2}Y$ are fission products, and ν is the average number of neutrons that are emitted in the fission processes of the LLMAs.

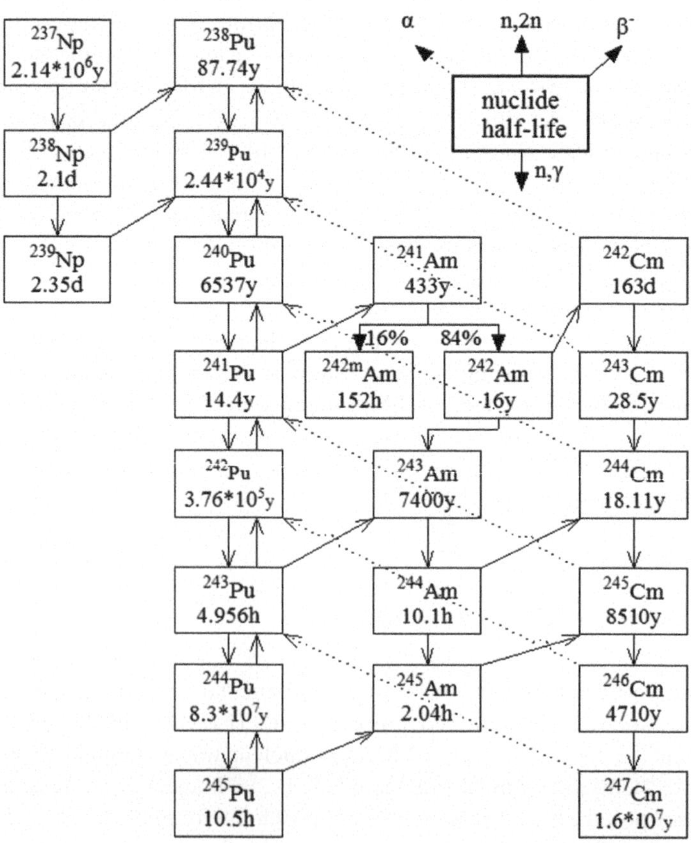

Fig. 4.1 Transmutation of minor actinides via neutron capture and decay

The radioactive fission products in nuclear wastes are mainly medium-mass nuclides, which can be converted into stable or short-lived nuclides via neutron capture. Fission products have a large capture cross section with thermal neutrons; therefore, thermal neutrons are suitable for the transmutation of fission products. The reactions of fission products capturing thermal neutrons are expressed in Eqs. (4.13) and (4.14). Figure 4.2 illustrates the transmutation chains of ^{99}Tc, ^{129}I, and ^{135}Cs.

$$n + {}^{A}_{Z}\text{LLFP} \rightarrow {}^{A+1}_{Z}X + \gamma \tag{4.13}$$

$$n + {}^{A}_{Z}\text{LLFP} \rightarrow {}^{A-1}_{Z}Y + 2n \tag{4.14}$$

where ${}^{A+1}_{Z}X$ and ${}^{A-1}_{Z}Y$ are medium-mass nuclides.

The performance of nuclear waste transmutation can be evaluated from various aspects. The commonly used parameters are as follows.

1. Capture-to-fission ratio

The capture-to-fission ratio is defined as the ratio of the average capture cross section to the average fission cross section and is a quantity of efficiency for transmuting minor actinides. This ratio is expressed as

$$\eta = \frac{\bar{\sigma}_c}{\bar{\sigma}_f} \tag{4.15}$$

where $\bar{\sigma}_c$ is the average capture cross section, in units of barn, and $\bar{\sigma}_f$ is the fission cross section, in units of barn.

A lower capture-to-fission ratio represents a lower probability of minor actinides being converted to transuranic elements via neutron capture, while the probability of fission will be larger; hence, it will lead to a higher transmutation efficiency.

2. Effective half-life

The effective half-life, which is denoted as $T_{1/2}^{\text{eff}}$, is defined as the time interval that is required for the activity of a radioactive nuclide to be reduced to half of its value.

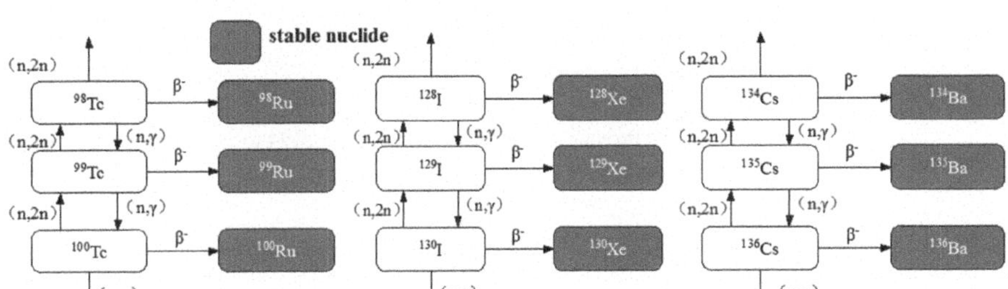

Fig. 4.2 Transmutation chains of ^{99}Tc, ^{129}I, and ^{135}Cs

It is an important parameter for evaluating the transmutation performance of nuclear wastes. The equation for calculating the effective half-life is presented below [2].

$$T_{1/2}^{\text{eff}} = \frac{\ln 2}{3.16 \times 10^7 \times (\lambda_i + \sigma_i \phi)} \tag{4.16}$$

3. Transmutation support ratio

The transmutation support ratio (TSR) is defined as the ratio of the number of minor actinides that are transmuted by hybrid nuclear systems to the number of minor actinides that are produced by a typical PWR with the same power.

4. Transmutation rate

The transmutation rate is defined as the ratio of the number of transmuted long-lived nuclides to the total number of long-lived nuclides.

Transmutation of nuclear wastes is regarded as one of the important objectives for advanced nuclear systems, especially hybrid nuclear systems. The capture-to-fission ratio and effective half-life are used to evaluate the transmutation capabilities of reactors. The effective half-lives of important transuranium nuclides in typical hybrid nuclear systems are listed in Table 4.3.

According to the table above, the effective half-life of ^{237}Np can be reduced from 2.14E+6 years to 41.75 years and 42.16 years in typical fusion-driven subcritical system and typical accelerator-driven subcritical system, respectively, thereby demonstrating the effective transmutation of long-lived transuranium nuclides in hybrid nuclear systems.

Table 4.3 Effective half-lives of important transuranium nuclides in typical hybrid nuclear systems

Nuclides	Natural half-life/a	Effective half-life/a	
		Typical fusion-driven subcritical system	Typical accelerator-driven subcritical system
^{237}Np	2.14E+6	41.75	42.16
^{238}Pu	87.4	12.65	11.69
^{239}Pu	24390	8.19	8.02
^{240}Pu	6580	37.65	36.99
^{241}Pu	13.2	3.94	3.96
^{242}Pu	3.87E+5	52.45	53.12
^{241}Am	433	43.24	45.00
^{243}Am	7370	63.58	66.72
^{243}Cm	32	3.83	4.24
^{244}Cm	17.6	11.50	11.49
^{245}Cm	8300	4.99	5.17

The capture-to-fission ratios of main heavy nuclides in different types of nuclear systems are listed in Table 4.4.

According to the table above, compared to thermal neutron reactors, the capture-to-fission ratios in typical fusion- and accelerator-driven subcritical systems are much lower. For some nuclides, the ratio may be decreased by two orders. Thus, the fusion- and accelerator-driven subcritical systems have stronger transmutation capability and could achieve higher efficiency of nuclear waste transmutation.

4.2.3 Nuclear Fuel Breeding

4.2.3.1 Fission Fuel Breeding

Currently, fissile isotope ^{235}U constitutes approximately 0.72% of natural uranium. There are many fertile nuclei in nature, such as ^{238}U and ^{232}Th. In a hybrid nuclear system and a fast-neutron reactor, the fertile nuclei are converted into fissile nuclei (such as ^{239}Pu and ^{233}U) via neutron capture followed by radioactive decay, and the corresponding system is called a breeding system. The reactions in which ^{238}U and ^{232}Th are converted into fissile isotopes ^{239}Pu and ^{233}U after capturing neutrons are represented by Eqs. (4.17) and (4.18).

Table 4.4 Capture-to-fission ratios of main heavy nuclides in different types of nuclear systems

Nuclides	Typical thermal neutron reactor	Typical fusion-driven subcritical system	Typical accelerator-driven subcritical system
^{237}Np	63	4.90	4.76
^{238}Pu	12	0.69	0.60
^{239}Pu	0.58	0.31	0.29
^{240}Pu	396.6	1.64	1.34
^{241}Pu	0.4	0.19	0.18
^{242}Pu	65.5	2.13	1.68
^{241}Am	100	6.69	6.63
^{243}Am	111	8.26	7.32
^{242}Cm	3.9	2.10	2.00
^{243}Cm	0.16	0.18	0.19
^{244}Cm	16	1.79	1.69
^{245}Cm	0.15	0.18	0.20
^{235}U	0.22	0.31	0.29
^{238}U	8.3	4.26	5.57

$$^{232}\text{Th} + \text{n} \rightarrow {}^{233}\text{Th} \xrightarrow{\beta^-} {}^{233}\text{Pa} \xrightarrow{\beta^-} {}^{233}\text{U} + 5.58\,\text{MeV} \tag{4.17}$$

$$^{238}\text{U} + \text{n} \rightarrow {}^{239}\text{U} \xrightarrow{\beta^-} {}^{239}\text{Np} \xrightarrow{\beta^-} {}^{239}\text{Pu} + 5.77\,\text{MeV} \tag{4.18}$$

The typical parameters for fission fuel breeding, namely, the breeding ratio, fuel breeding ratio, and breeding support ratio, are often used to quantify the fission fuel breeding capability.

1. Breeding ratio

The breeding ratio is defined as the ratio of the number of fissile isotopes that are produced to the number of fissile isotopes consumed. It is typically called the conversion ratio when the breeding ratio is less than 1.

2. Breeding support ratio

The breeding support ratio is the ratio of the number of fissile isotopes that are bred by a hybrid nuclear system to the number of fissile isotopes that are consumed by a typical PWR with the same power, i.e., the number of typical PWRs that can be supplied with the fissile isotopes that are produced by the hybrid nuclear system with the same power.

In advanced nuclear systems, the utilization rate of natural uranium can be increased via fuel breeding. For example, the breeding ratio in a sodium-cooled fast reactor and a thorium-based molten salt reactor are 1.2 and 1.06, respectively, while a higher breeding ratio (of more than 3) can be realized in hybrid nuclear systems [3].

4.2.3.2 Tritium Breeding

A fusion system with a fusion power of 1 GW will consume approximately 55.6 kg of tritium annually to maintain the deuterium–tritium fusion reaction under full power operation [4]. However, the half-life of tritium is 12.31 years, and tritium occurs very rarely in nature. A fusion system [5] must produce the same amount of tritium to maintain tritium self-sufficiency; an effective method for producing tritium is to use neutron nuclear reactions with ^{6}Li and ^{7}Li.

One of the main functions of fusion systems is to breed tritium. Materials that contain lithium are used as tritium breeding materials in fusion systems to produce and recycle tritium fuel for plasma burning. The tritium generation reactions are expressed in Eqs. (4.19) and (4.20). The reaction cross section of neutron-^{6}Li is larger than that of neutron-^{7}Li; thus, the neutron-^{6}Li reaction is the main reaction that is used to produce tritium in a fusion system. The abundance of ^{6}Li is relatively low in nature (approximately 7.6%), and the tritium production cross section of ^{6}Li with thermal neutrons is larger than that with fast neutrons. Therefore, in the design of a tritium breeding blanket, the ^{6}Li abundance must be increased, and the neutron spectrum must be optimized. The threshold energy of the neutron nuclear reaction with ^{7}Li for producing tritium is 3.1454 MeV; if the neutron energy is lower than this threshold energy, the tritium production reaction will not occur.

$$^{6}\mathrm{Li} + \mathrm{n} \rightarrow {}^{3}\mathrm{T} + {}^{4}\mathrm{He} + 4.79\,\mathrm{MeV} \tag{4.19}$$

$$^{7}\mathrm{Li} + \mathrm{n} \rightarrow {}^{3}\mathrm{T} + {}^{4}\mathrm{He} + \mathrm{n} - 2.47\,\mathrm{MeV} \tag{4.20}$$

Due to neutron leakage in fusion systems, neutron multiplication is required when producing tritium via neutron reactions with lithium. A neutron is typically multiplied via (n, 2n) reactions with Be or Pb. At the same time, neutrons in the system are commonly moderated, thereby increasing the probability of neutron–lithium reactions. The neutron nuclear reactions with Be and Pb, in which neutron multiplication occurs are presented below

$$^{9}\mathrm{Be} + \mathrm{n} \rightarrow 2\,{}^{4}\mathrm{He} + 2\mathrm{n} \tag{4.21}$$

$$^{208}\mathrm{Pb} + \mathrm{n} \rightarrow {}^{207}\mathrm{Pb} + 2\mathrm{n} \tag{4.22}$$

The tritium breeding ratio (TBR) is typically used to describe the tritium breeding performance of a fusion system and is defined as the average number of tritium nuclei that are produced by a single fusion neutron in the system. TBR must be greater than 1 to achieve tritium self-sufficiency.

4.2.4 Material Activation

Neutron-induced material activation is the process in which a nucleus absorbs a neutron and changes into a radionuclide. The main activation reaction in the structural material (such as CLAM steel) is the (n, γ) reaction of neutrons with ^{59}Co to produce ^{60}Co, where the reaction processes can be represented as Eqs. (4.23) and (4.24). The activation product ^{60}Co, which has a half-life of 5.27 years, is the main radionuclide in the structural material after the nuclear system has been shut down.

$$^{59}\mathrm{Co} \xrightarrow{(\mathrm{n},\gamma)} {}^{60}\mathrm{Co} \tag{4.23}$$

$$^{59}\mathrm{Co} \xrightarrow{(\mathrm{n},\gamma)} {}^{60m}\mathrm{Co} \xrightarrow{\mathrm{IT}} {}^{60}\mathrm{Co} \tag{4.24}$$

Based on the inventory of radioactive nuclides in materials and nuclear decay data libraries, typical parameters for quantifying the activation of radioactive materials can be calculated. The typical activation parameters that are used in nuclear systems are activity, decay heat, potential biological hazard, clearance index, and dose rate. These parameters can be used for radioactivity level assessments, accident analyses, radiation dose assessments, decommission studies, etc.

1. Activity

Activity is defined as the number of decays of the radioactive material per unit time, which is used to measure the radioactivity of a material. The activity of a nuclide i, which is denoted by A_i, is equal to the product of the number of nuclides, namely, N_i, and its decay constant, namely, λ_i:

$$A_i = \frac{dN_i}{dt} = \lambda_i N_i \tag{4.25}$$

The activity of a material can be calculated by summing the activities of all radioactive nuclides in the material. The variation of a material's activity with time obeys the exponential decay law. The unit of activity is the Becquerel (Bq). One Bq is one nucleus decay per second. Another commonly used unit is Curie (Ci). The conversion between Ci and Bq is

$$1\,\mathrm{Ci} = 3.7 \times 10^{10}\,\mathrm{Bq} \tag{4.26}$$

The specific activity is usually used to characterize the material's radioactivity and is defined as the ratio of the material's activity to the material's mass; it is expressed in units of Bq/kg. The specific activity can be directly measured, and the activity can be calculated as the product of the specific activity and the mass.

2. Decay heat

Once a material is activated, radioactive nuclides in the material will produce heat via decay. Decay heat is the energy that is produced by the decay of radioactive nuclides and is the main heat source for nuclear systems during the shutdown. The decay heat which is denoted as P_i produced by the nuclide i in the material can be expressed as

$$P_i = A_i E_i = \lambda_i N_i E_i \tag{4.27}$$

where E_i is the average amount of energy that is released in each decay of nuclide i.

The material's decay heat can be calculated by summing the decay heats of all radioactive nuclides in the material.

3. Potential biological hazard

The potential biological hazard refers to the impact on the environment of the leakage of radionuclides. The potential biological hazard is described by the inhalation dose and the ingestion dose. International Commission on Radiological Protection (ICRP) typically uses the volumes of water and air that dilute the radionuclide to the maximum allowable concentration to assess the potential biological hazard.

$$BHP_{inh} = \sum_i A_i / DAC_i \tag{4.28}$$

$$BHP_{ing} = \sum_i A_i / MPC_i \tag{4.29}$$

where BHP_{inh} is potential inhalation biological hazard, BHP_{ing} is potential ingestion biological hazard, A_i is the activity of radionuclide i, and DAC_i and MPC_i are the maximum allowable concentration of the activity of radionuclide i in air and water, respectively.

4. Clearance index

The clearance index is a parameter that indicates whether it is possible to dispose of radioactive nuclear wastes [6]. The clearance index is a dimensionless quantity and can be calculated by

$$I_c = \sum_i \frac{A_i}{ML_i} \tag{4.30}$$

where M is the material's mass, in units of kg; and L_i is the activity limitation below which it is possible to directly dispose of the material, in units of Bq/kg. L_i varies with the nuclides that are contained in the material.

As recommended by IAEA, L_i is calculated according to the requirement that the personal annual dose be less than 10 μSv/a. If the clearance index is less than 1, the radioactivity of the nuclear wastes is relatively low and can be disposed of directly.

5. Dose rate

In nuclear systems, the neutrons and α, β, and γ rays from the decay of activation products can cause radiation exposure to the machine and workers. The dose rate is the radiation dose that is received by a machine or a worker per second and corresponds to the radiation level of the radiation field.

To ensure safety, workers cannot enter areas with high radiation dose rates. Therefore, remote handling is required during the maintenance of highly activated components such as a subcritical reactor/blanket, divertor, and vacuum vessel in the ITER machine. For the maintenance of components that have low activation, such as pumps of primary cooling loops in the ITER machine, it is possible to conduct hands-on operation. In addition, the dose rate is one of the bases for classifying the nuclear wastes and determining its disposal strategy. The calculation method of the radiation dose rate and the effect on machines and personnel will be described in detail in Chaps. 5 and 6.

4.3 In-core Fuel Management and Fuel Cycle

In a fission system, fuel management, which is a basic task, refers to the configuration of appropriate fuel assemblies and the corresponding measures of reactivity control for each operating cycle of the core over the life of the nuclear system such that the core can implement the strategy of nuclear power plant operation and meet specified safety requirements. The nuclear fuel cycle (referred to as the fuel cycle) refers to the entire fuel process including the production of nuclear fuel outside the nuclear system, combustion in the nuclear system, and posttreatment. The main objective of fuel management is to operate the nuclear power plant profitably and safety. In fusion systems, the main focus of fuel management and the fuel cycle is the fusion tritium cycle. In hybrid nuclear systems [7, 8], fuel management and fuel cycle of the fission fuel are similar to those in fission systems. For the fusion core of the fusion-driven subcritical system, the fusion tritium cycle is the main focus. The fuel management, the fuel cycle, and the fusion tritium cycle will be briefly discussed in this section.

4.3.1 Fission Fuel Management and Fuel Cycle

4.3.1.1 Fission Fuel Cycle

In this section, the parameters that are typically used in fuel management in nuclear systems will be described: the refueling interval and cycle length, the cycle burnup and discharge burnup, the batch size, and the number of batches. The design and optimization of the fuel management plan will be described in detail in Chap. 9.

1. Refueling interval and cycle length

The refueling interval refers to the time interval between two reactor shutdowns and refueling. To realize the objective of nuclear fuel management, it is necessary to discharge spent fuel within a reasonable interval, to add new fuel and rearrange the fuel position in the reactor, to rationally use burnable poison, to make the fuel assembly reach as close as possible to the maximum fuel burnup depth, and to reduce the neutron irradiation on containers. The reactor completes one cycle of operation for each refueling interval. The time equivalent to the full-load operation of an operation cycle is the cycle length, which is expressed in effective full power days (EFPD).

2. Cycle burnup and discharge burnup

The cycle burnup, which is denoted as BUc refers to the net increase in the average burnup depth of a full reactor of nuclear fuel after the completion of an operating cycle.

$$BUc = (\text{Nominal power} \times \text{Cycle length})/(\text{Load capacity of heavy elements in core}) \quad (4.31)$$

The discharge burnup, which is denoted as BU_{d}, refers to the burnup depth of the fuel from the initial loading into the core to the final discharge after several cycles. BU_{d} is directly related to the economics of a nuclear system and is an important indicator for the design of a nuclear system. In the fuel management of fast reactors, the average discharge burnup, namely, the average burnup depth that is obtained when a batch of fuel is discharged from the core, is often regarded as a target for optimization. To evaluate the performance of a refueling scheme, the average discharge burnup can be used for comparison under the maximum discharge burnup and the maximum line power density of the fuel rod.

3. Batch size and number of batches

A group of assemblies enter and leave the reactor core at the same time. When refueling, only the batch of assemblies with deep burnup is replaced. The number of fuel assemblies that are replaced per refueling is equal to the batch size. Denote the total number of fuel assemblies in the reactor as N_T and the batch size as N, and define $N_T/N = n$ as the number of batches.

4.3.1.2 Fuel Cycle

The nuclear system fuel cycle includes the fission fuel cycle and the fusion tritium cycle. The nuclear fuels that are used in fission systems and hybrid nuclear systems include fissile nuclides such as ^{233}U, ^{235}U, ^{238}U, and ^{239}Pu. Traditional nuclear systems use UO_2 fuel that contains ^{235}U or mixed oxide (MOX) fuel that contains ^{235}U or ^{239}Pu as nuclear fuel. The fuel circulation mode of the traditional nuclear systems can be categorized according to whether it utilizes the once-through or closed thermal reactor cycle.

In the once-through cycle, which is also known as the direct cycle, the spent fuel that is discharged from the thermal neutron reactor is not subject to chemical reprocessing but is disposed of via deep geological disposal after intermediate storage. Considering the economics of power generation, the once-through fuel cycle is cost-effective; however, it will result in the accumulation of large amounts of long-lived radionuclides and the consumption of large amounts of uranium resources. At present, the United States and other countries that have nuclear power plants still utilize the once-through circulation method to directly store spent fuel in deep formation disposal sites.

In the closed thermal reactor cycle, the plutonium and uranium in the spent fuel of the thermal reactor are extracted via reprocessing to produce MOX fuel, which is returned to the thermal neutron reactor for use. The remaining nuclear wastes in the spent fuel is glass-solidified after intermediate storage and eventually stored and disposed of via deep geological disposal. This closed cycle can improve the utilization rate of uranium resources, reduce the consumption of natural uranium, and reduce the amount of nuclear wastes. However, the improvement in the utilization rate of natural uranium resources is limited.

Only ^{235}U, which accounts for less than 1% of the natural uranium, can be utilized by these two types of nuclear fuel cycles, and the natural resources will soon be exhausted. At the same time, a large amount of nuclear wastes will be produced, thereby making it difficult to realize the sustainable development of nuclear energy. Advanced nuclear systems can adopt advanced separation-transmutation closed nuclear fuel cycle models, such as the closed fast reactor fuel cycle and hybrid nuclear system fuel cycles; use fast neutrons that are generated by fast reactor or hybrid nuclear systems to convert ^{238}U to ^{239}Pu, or ^{232}Th to ^{233}U; breed nuclear fuel; and transmute nuclear wastes to realize the sustainable development of nuclear energy.

The closed fast reactor fuel cycle recycles the uranium and plutonium in the spent fuel in thermal neutron reactors via chemical reprocessing and produces mixed oxide fuel elements of uranium and plutonium, which are put into the fast-neutron reactor for recycling. The utilization ratio of natural uranium can be substantially increased up to approximately 60%. After the irradiation of thorium or thorium-based fuel and online or offline reprocessing, the separated and recycled ^{233}U is remade into fuel together with thorium and refueled back to the reactor. A fast-neutron reactor that uses the closed fuel cycle cannot only generate electricity for a long time without adding new fissile materials, but also produce nuclear fuel for new nuclear reactors.

In hybrid nuclear systems (including fusion-driven and accelerator-driven subcritical systems), the external neutron source provides high-current and high-energy neutrons, and the systems have strong nuclear waste transmutation and nuclear fuel breeding capabilities. Spent fuel from the thermal neutron reactor can be transmuted in hybrid nuclear systems. The breeding ratio of a hybrid nuclear system can exceed 3 [3]. The fuel cycle of a hybrid nuclear system can increase the natural uranium utilization rate to more than 95% [9].

4.3.2 Fusion Tritium Cycle

The tritium cycle in fusion systems can be divided into two processes: a combustion cycle and a breeding cycle.

The tritium breeding cycle is the process of tritium breeding in the breeding blanket to supplement the consumption in plasma and the decay and leakage of tritium. In the breeding cycle, at the start-up phase of fusion system, the tritium is initially supplied externally to the fuel storage and supply system and subsequently injected into the Tokamak core by the feeding system. Then, fusion neutrons react with the breeding material in cladding to produce tritium. Next, the tritium extraction system recovers the tritium that was produced in the cladding by blowing helium gas. Then, the helium impurity can be removed by the tritium recovery and purification system to obtain the tritium-containing hydrogen isotope gas. In the isotope separation system, tritium can be accumulated and the isolated products will enter the fuel storage and supply system.

The tritium combustion cycle is the process in which deuterium and tritium are extracted and separated from plasma ash, and transported to the fuel storage and supply system to support the consumption in the core. In the combustion cycle, unburned deuterium and tritium fuel in the core is discharged from the vacuum chamber through the divertor and ash discharge system, enters the tritium recovery and purification system to remove the impurities, and is separated by the isotope separation system to obtain the fuel of D_2, DT, and T_2, which will subsequently enter the fuel storage and supply system. Tritium-containing waste water or wastes gas that is produced in the recycling process is further extracted by a water detritium system and an air detritium system.

4.4 Nuclear Transmutation Calculation

4.4.1 Calculation Methods

The current numerical methods are categorized into two types: analytical and numerical. The numerical methods include the finite difference method of ordinary differential equations and the matrix exponential method.

4.4.1.1 Analytical Method

The analytical method for the Bateman equation is also called the transmutation trajectory analysis (TTA) method, or the linear chains method. In 1910, Harry Bateman proposed the Bateman method for calculating a simple decay chain, based on which the general analytical method was developed by Jerzy Cetnar in 2006 [10]. The TTA method is based on searching and creating transmutation trajectories from the initial nuclide according to its reaction types; the nuclide density is calculated on the transmutation trajectory. Therefore, the process consists of two steps: transmutation trajectory creation and transmutation trajectory solution. The two steps are described below.

1. Transmutation trajectory creation

The main strategy of the TTA method is to decompose the complex transformation web into a set of linear chains that consist of all possible routes or trajectories. Due to the complexity of the nuclide conversion web, it is difficult to consider all possible transmutation chains; therefore, approximations are typically introduced. For a cyclic chain, a linear chain of infinite length is generated, and importance truncation is used to obtain a finite linear chain.

Predefined linear chains are often used in many calculation codes; however, their applicability is not sufficient. The FDS Team has developed a dynamic linear chain creation method that is based on the depth-first search algorithm. With this method, linear chains are created by searching a neutron transport cross section library, decay data library, and fission product yield library. As an example, a tree diagram of nuclide conversions is shown in Fig. 4.3, in which $\boldsymbol{x_1}$ denotes the parent nucleus, and $\boldsymbol{x_2}, \boldsymbol{x_3} \ldots \boldsymbol{x_6}$ denote daughter nuclei. With the depth-first search algorithm, the search is conducted along the reaction chain and decay path of the initial nuclide and the parent nucleus and reaction type are used to determine the type of each daughter nucleus. When the search of the transmutation chain, which started from one nuclide

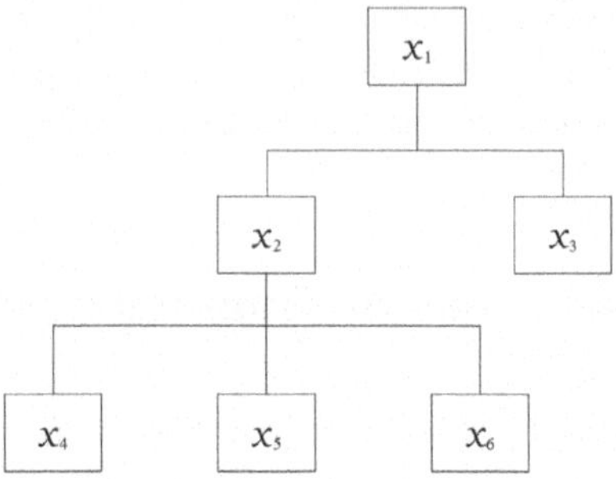

Fig. 4.3 Tree diagram of nuclide conversion

along the reaction chain and decay path, has been completed, the algorithm returns to the upper-level nuclide until the search of all the reaction chains and decay paths has been completed.

2. Transmutation trajectory solution

If only the first nuclide in the transmutation chain has a nonzero initial density, which is denoted as $N_1(0)$, then according to the Bateman solution, the density $N_n(t)$ of the n_{th} nuclide at time t in the linear chain is

$$N_n(t) = N_1(0) B_n \sum_{i=1}^{n} \alpha_i^n e^{-\lambda_i^{\text{eff}} t} \tag{4.32}$$

where

$$B_n = \prod_{j=1}^{n-1} b_{j+1,j}^{\text{eff}} \tag{4.33}$$

$$\alpha_i^n = \frac{\prod_{j=1}^{n-1} \lambda_j^{\text{eff}}}{\prod_{\substack{j=1 \\ j \neq i}}^{n} \left(\lambda_j^{\text{eff}} - \lambda_i^{\text{eff}}\right)} \tag{4.34}$$

When the cyclic chain is linearized, the effective decay constants in the chain become indistinct. This will result in a zero denominator in Eq. (4.34) and cause the failure of the Bateman solution. Thus, the Bateman method is unable to address the linearization of the cyclic chain.

However, the general analytical solution can address the problem of the indistinctiveness of the effective decay constants when the cyclic chain is linearized. For example, for a linear chain of length n, assuming that there are d distinct decay constants, which are denoted as λ_i^{eff}, and each of which is repeated m_i times, according to the general analytical solution, the density of the n_{th} nuclide at time t in the linear chain is

$$N_n(t) = N_1(0)\frac{B_n}{\lambda_n^{\text{eff}}}\sum_{i=1}^{d}\lambda_i^{\text{eff}}\alpha_i e^{-\lambda_i^{\text{eff}}t}\sum_{m=0}^{\mu_i}\frac{\lambda_i^{\text{eff}}t}{m!}\Omega_{i,\mu_i-m} \tag{4.35}$$

where

$$\mu_i = m_i - 1 \tag{4.36}$$

$$B_n = \prod_{j=1}^{n-1} b_{j+1,j}^{\text{eff}} \tag{4.37}$$

$$\alpha_i = \prod_{\substack{j=1 \\ j\neq i}}^{d}\left(\frac{\lambda_j^{\text{eff}}}{\lambda_j^{eff}-\lambda_i^{eff}}\right)^{m_j} \tag{4.38}$$

$$\Omega_{i,j} = \sum_{h_1=0}^{j}\sum_{h_2=0}^{j}\cdots\sum_{h_{i-1}=0}^{j}\sum_{h_{i+1}=0}^{j}\cdots\sum_{h_d}^{j}\prod_{\substack{k=1 \\ k\neq i}}^{n}\binom{h_k+\mu_k}{\mu_k}\left(\frac{\lambda_j^{\text{eff}}}{\lambda_j^{\text{eff}}-\lambda_k^{\text{eff}}}\right)^{h_k}\times\delta\left(j,\prod_{\substack{l=1 \\ l\neq i}}^{d}h_l\right) \tag{4.39}$$

Irrespective of the time step t, the TTA method can quickly address a single transmutation chain with high computational accuracy. However, importance truncation is required when addressing complex transmutation chains such as cyclic chains. In decay problems, there is no cyclic chain; therefore, the TTA method can obtain accurate calculation results. Nuclide transmutation chains under neutron irradiation are usually more complex and the results that are obtained via the TTA method have larger uncertainties while also suffering from lower calculation efficiency.

4.4.1.2 Finite Difference Method of Ordinary Differential Equations

The Bateman equations are time-dependent first-order linear ordinary differential equations, and the nuclide density can be calculated via a time differencing technique, which is called the finite difference method of ordinary differential equations (FDMODE). The time interval of the difference is called the time step. Since the Bateman equations are stiff, it is difficult to achieve convergence using the conventional difference method.

1. Exponential Euler method

The Euler method is a basic numerical method for solving ordinary differential equations. Its main strategy is to replace the change rate of the nuclide density, which is denoted as dN_i/dt, with the forward difference quotient of the nuclide density, namely, $(N_i(t+\Delta t) - N_i(t))/\Delta t$, in a short time interval, namely, $(t, t+\Delta t)$:

$$N_i(t + \Delta t) = N_i(t) + \Delta t \frac{dN_i}{dt} \tag{4.40}$$

Substituting Eq. (4.40) into Eq. (4.1) yields the linear equations of the nuclide density. The nuclide density at time $t + \Delta t$, namely, $N_i(t + \Delta t)$, can be obtained according to the nuclide density at time t, namely, $N_i(t)$. To ensure a highly accurate solution, the Euler method requires step Δt to be minimized. This leads to an increase in the number of steps and makes the calculation time-consuming. In addition, as the number of steps is increased, the calculation error accumulates, thereby resulting in a large error in the final result; therefore, the Euler method is rarely used in practice.

The exponential Euler method is an extension of the Euler method. In the exponential Euler method, the time step is exponentially transformed as a difference interval and associated with the nuclide effective decay constant

$$N_i(t + \Delta t) = N_i(t) + \frac{(e^{\lambda_i^{\text{eff}} \Delta t} - 1)}{\lambda_i^{\text{eff}}} \frac{dN_i}{dt} \tag{4.41}$$

The effective decay constant of a long-lived nuclide is small; therefore, the time step can be relatively large, and the calculation is accurate and fast. For the short-lived radionuclides, since their effective decay constants are large, a small time step is required. In the exponential Euler method, short-lived nuclides are typically ignored hence, only the long-lived nuclides are calculated. The equilibrium assumption is used for the calculation of the nuclide density for short-lived nuclides. Within the time interval Δt, when the product of the effective decay constant of the nuclide and the time step exceeds a specified value, as expressed in Eq. (4.41), the nuclide density will be calculated independently using the equilibrium condition.

$$\lambda_i^{\text{eff}} \Delta t > L \tag{4.42}$$

where L is the equilibrium criterion, which is typically 20 in cases with neutron irradiation and 100 in cases without neutron irradiation.

The change rate of the nuclides in the equilibrium state is zero

$$\frac{dN_i}{dt} = 0 \tag{4.43}$$

Substituting Eq. (4.43) into Eq. (4.1) yields the nuclide density in the equilibrium state in Eq. (4.44)

$$N_i^{\text{equilibrium}} = \frac{\sum_{j \neq i} N_j(\lambda_{i,j} + \sigma_{i,j}\phi) + S_i}{\lambda_i + \sigma_i \phi} \tag{4.44}$$

According to Eq. (4.44), the nuclide density of the equilibrium state is independent of the initial nuclide density.

The exponential Euler method has advantages in terms of accuracy and calculation speed for the nuclide density for long-lived nuclides. However, for the short-lived nuclide density, it is necessary to impose the equilibrium assumption to calculate the nuclide density independently, which may produce errors.

2. Runge–Kutta method

The Runge–Kutta method is an improved Euler method and is widely used in solving ordinary differential equations. In 1989, Yuichi Morimoto applied the Runge–Kutta method to the calculation of transmutation processes [11]. The Runge–Kutta method has many forms, of which the second-order and fourth-order Runge–Kutta methods are the most commonly used. The main strategy of the second-order Runge–Kutta method is to use the gradient of the nuclide density at the midpoint of the burnup steps instead of the change rate. The equations are as follows:

$$\boldsymbol{d}_1 = \frac{d\boldsymbol{N}}{dt}(t^l, \boldsymbol{N}^l)\Delta t = \boldsymbol{A}\boldsymbol{N}^l\Delta t \tag{4.45}$$

$$\boldsymbol{d}_2 = \frac{d\boldsymbol{N}}{dt}\left(t^l + \frac{\Delta t}{2}, \boldsymbol{N}^l + \frac{\boldsymbol{d}_1}{2}\right)\Delta t = \boldsymbol{A}\left(\boldsymbol{N}^l + \frac{\boldsymbol{d}_1}{2}\right)\Delta t \tag{4.46}$$

$$\boldsymbol{N}^{l+1} = \boldsymbol{N}^l + \boldsymbol{d}_2 \tag{4.47}$$

where l is the time interval index. $\boldsymbol{N}^l$ is the nuclide density at time t^l, and $\frac{d\boldsymbol{N}}{dt}(t^l, \boldsymbol{N}^l)$ is the change rate of the nuclide density.

The second-order Runge–Kutta method converts the Bateman equations into linear equations, which are presented as Eq. (4.47), to obtain the nuclide density. The second-order Runge–Kutta method is more accurate than the exponential Euler method because the gradient at the midpoint of the time interval is more accurate than the gradient at the starting point for approximating $\mathrm{d}N_i/\mathrm{d}t$. The fourth-order Runge–Kutta method is similar to the second-order Runge–Kutta method and its equations are

$$\boldsymbol{d}_1 = \frac{\mathrm{d}\boldsymbol{N}}{\mathrm{d}t}(t^l, \boldsymbol{N}^l)\Delta t = \boldsymbol{A}\boldsymbol{N}^l\Delta t \tag{4.48}$$

$$\boldsymbol{d}_2 = \frac{\mathrm{d}\boldsymbol{N}}{\mathrm{d}t}\left(t^l + \frac{\Delta t}{2}, \boldsymbol{N}^l + \frac{\boldsymbol{d}_1}{2}\right)\Delta t = \boldsymbol{A}\left(\boldsymbol{N}^l + \frac{\boldsymbol{d}_1}{2}\right)\Delta t \tag{4.49}$$

$$\boldsymbol{d}_3 = \frac{\mathrm{d}\boldsymbol{N}}{\mathrm{d}t}\left(t^l + \frac{\Delta t}{2}, \boldsymbol{N}^l + \frac{\boldsymbol{d}_2}{2}\right)\Delta t = \boldsymbol{A}\left(\boldsymbol{N}^l + \frac{\boldsymbol{d}_2}{2}\right)\Delta t \tag{4.50}$$

$$\boldsymbol{d}_4 = \frac{\mathrm{d}\boldsymbol{N}}{\mathrm{d}t}(t^l + \Delta t, \boldsymbol{N}^l + \boldsymbol{d}_3)\Delta t = \boldsymbol{A}(\boldsymbol{N}^l + \boldsymbol{d}_3)\Delta t \tag{4.51}$$

$$\boldsymbol{N}^{l+1} = \boldsymbol{N}^l + \frac{\boldsymbol{d}_1 + 2\boldsymbol{d}_2 + 2\boldsymbol{d}_3 + \boldsymbol{d}_4}{6} \tag{4.52}$$

The fourth-order Runge–Kutta method converts the Bateman equations into linear equations, which are presented as Eq. (4.52) to obtain the nuclide density. The gradient that is calculated via the fourth-order Runge–Kutta method is closer to $\mathrm{d}N/\mathrm{d}t$ compared to the second-order Runge–Kutta method. Therefore, the calculation accuracy of the fourth-order Runge–Kutta method is higher; however, it is more time-consuming.

The Runge–Kutta method, with its high accuracy, can address complex transmutation chains. It can quickly solve small-scale transmutation and activation problems. However, the calculation time increases linearly with the number of nuclides. In addition, the Runge–Kutta method requires small time steps, which makes it relatively time-consuming when short-lived nuclides are calculated.

4.4.1.3 Matrix Exponential Method

The matrix exponential method calculates the nuclide density by solving the matrix exponential, namely, e^{At}, in Eq. (4.4). Since many nuclides are produced in transmutation and activation processes, matrix $\boldsymbol{A}$ may be huge. Due to this immense number of matrix elements and the stiffness of the equation, difficulties are encountered regarding the convergence of the direct expansion of the matrix exponential. The following introduces several widely used matrix exponential methods.

1. Taylor series method

The Taylor series method is the most basic matrix exponential method and expands the matrix exponential e^{At} into a polynomial of the matrix $\boldsymbol{A}$ to solve the Bateman equations. The Taylor series expansion of the matrix exponential e^{At} is presented below.

$$e^{At} = \sum_{m=0}^{\infty} \frac{(At)^m}{m!} \approx \sum_{m=0}^{M} \frac{(At)^m}{m!} \tag{4.53}$$

By substituting Eq. (4.53) into Eq. (4.4), a series of linear equations are derived, which can be solved to obtain the nuclide density. Since the decay constant of short-lived nuclides is very large, the norm of the matrix $\boldsymbol{A}$ is also very large. Thus, the convergence of the Taylor series expansion is slow. In addition, the limited numerical precision of computers makes it difficult to ensure improvement in the accuracy of the expansion coefficient, even if the number of expansion terms is increased. This ultimately leads to large numerical errors in the final results.

To improve the accuracy of the Taylor series method, the scale method is often used to reduce the matrix norm. The scale method transforms the matrix exponential as follows:

$$e^{At} = (e^{At/m})^m \tag{4.54}$$

where m is the scaling factor, which is usually chosen as $m = 2^k$.

The Taylor series expansion of $e^{At/m}$ will be more accurate after the scaling transformation; then, the coefficient of the linear equations can be obtained by calculating the m-th power of the expansion. For the Taylor series method to be used, a condition on the time step Δt must be satisfied:

$$\|\boldsymbol{A}\|\Delta t \leq -2\ln(0.001) \tag{4.55}$$

In the Taylor series method, the number of expansion terms is usually determined by the precision requirement, which is typically 0.1% or less. For long-lived nuclides, a precise result can be obtained if a 30th-order Taylor series expansion is used. For short-lived nuclides, even if the number of expansion terms is increased, it is difficult to obtain accurate results. Thus, the Taylor series method is mainly used to calculate the density of the long-lived nuclides, and the density of the short-lived nuclides must be calculated independently. In the Taylor series method, the following condition is typically used to determine whether a specified nuclide must be calculated independently:

$$e^{-\lambda_i^{\text{eff}}\Delta t} < 0.001 \tag{4.56}$$

The treatment approaches for such short-lived-nuclide problems can be categorized into two types: (1) when the short-lived nuclides are initial nuclides, the linear chain method is used, and (2) when the short-lived nuclides are daughter nuclei of long-lived nuclides, the equilibrium assumption is introduced.

2. Krylov subspace method

Akio Yamamoto developed the Krylov subspace method for solving the transmutation problem in 2007 and applied it to solve the transmutation problem of short-lived nuclides [12]. With the Krylov subspace method, the nuclide density at time $t + \Delta t$, which is denoted as $\boldsymbol{N}(t + \Delta t)$, can be expressed as

$$\boldsymbol{N}(t + \Delta t) \cong c_0\boldsymbol{N}(t) + c_1(\boldsymbol{A}\Delta t)^1\boldsymbol{N}(t) + c_2(\boldsymbol{A}\Delta t)^2\boldsymbol{N}(t) + \cdots + c_{m-1}(\boldsymbol{A}\Delta t)^{m-1}\boldsymbol{N}(t) \tag{4.57}$$

In which the expansion coefficient c_k is defined as

$$c_k = \frac{1}{k!} \tag{4.58}$$

The m terms on the right side of Eq. (4.54) form the m-dimensional Krylov subspace:

$$K_m\left(\boldsymbol{A}\Delta t, \vec{N}(t)\right) = \text{Span}\left\{\boldsymbol{N}(t), (\boldsymbol{A}\Delta t)\boldsymbol{N}(t), (\boldsymbol{A}\Delta t)^2\boldsymbol{N}(t), \cdots, (\boldsymbol{A}\Delta t)^{m-1}\boldsymbol{N}(t)\right\} \tag{4.59}$$

In the Krylov subspace method, the expansion coefficients, namely, c_k, are chosen to minimize the square residual of the prediction error for $\boldsymbol{N}(t+\Delta t)$. The choice of the expansion coefficient c_k depends on the matrix $\boldsymbol{A}$. The Krylov subspace method can solve for the eigenvectors that correspond to the maximum eigenvalues of the matrix after an optimization of the expansion coefficient c_k. The maximum eigenvalue of the matrix $\boldsymbol{A}$ corresponds to the short-lived nuclide; therefore, the Krylov subspace method can solve the short-lived-nuclide density properly.

The Krylov subspace method requires relatively small time steps. A longer time step results in a larger calculation error, whereas a shorter time step will result in a longer calculation time. For a stiff matrix, the Krylov subspace method is also time-consuming.

3. Chebyshev rational approximation method

For the complex exponential e^z (the real part of the complex z is negative), the Taylor series method and the Krylov subspace method are also used to expand e^z into polynomial form. To obtain sufficient calculation precision, it is necessary to calculate a high-order expansion if the modulus of z is very large. Another way to expand e^z is to compute its best rational approximation. The Chebyshev rational approximation method (CRAM) converges at a rate of approximately 9.3^{-k} as the order k increases. P. Maria et al. at the VTT applied the CRAM to solve the Bateman equations in 2010 [13]. In their case, the eigenvalues of the coefficient matrix $\boldsymbol{A}$ for long-lived and short-lived nuclides in the Bateman equations were distributed near the negative real axis. Results with relatively high accuracy can be obtained via the CRAM method [14]. The Chebyshev rational function $\hat{r}_{k,k}(z)$ is

$$\hat{r}_{k,k}(z) = \alpha_0 + \sum_{i=1}^{k} \frac{\alpha_i}{z - \theta_i} \tag{4.60}$$

where k is the order of the Chebyshev rational function, α_0 is the limit of the function at infinity; θ_i denotes pole of the function, which is a discrete value; and α_i is the residue at pole θ_i.

Figure 4.4 shows the distribution of the precision when e^z is approximated by $\hat{r}_{k,k}(z)$ with $k = 14$ (the contour distribution of $\left|e^z - \hat{r}_{14,14}(-z)\right|$ in the complex plane). According to the figure, the accuracy can reach 10^{-13} near the negative real axis.

Using the CRAM method, the matrix exponential $e^{\boldsymbol{A}t}$ can be expanded as

$$e^{\boldsymbol{A}t} \approx \hat{r}_{k,k}(\boldsymbol{A}t) = \alpha_0 \boldsymbol{I} + 2Re\left(\sum_{i=1}^{k/2} \alpha_i (\boldsymbol{A}t - \theta_i \boldsymbol{I})^{-1}\right) \tag{4.61}$$

Then, the solution of the nuclide density is obtained as

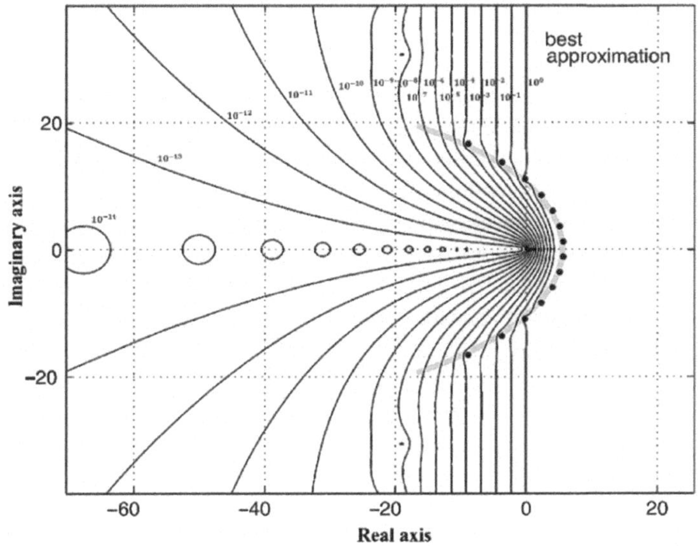

Fig. 4.4 Distribution of the accuracy of the Chebyshev rational function in the complex plane

$$N(t) = e^{At}N(0) = \alpha_0 N(0) + 2Re\left(\sum_{i=1}^{k/2} \alpha_i (At - \theta_i I)^{-1} N(0)\right) \tag{4.62}$$

Equation (4.59) is a linear equation, and matrix inversion of $(At - \theta_i I)$ is required for obtaining the coefficients of the equations. Since the CRAM method can accurately solve the matrix exponential of a stiff matrix, it does not require special treatment for short-lived nuclides. However, matrix inversion is necessary and influences the computational efficiency of the CRAM method.

In an advanced nuclear system, the transmutation and activation processes involve thousands of nuclides and the relevant nuclides vary among problems. Thus, a method with wide applicability is needed. In the analytical method, the importance truncation is necessary for solving the complex transmutation chains, which produces a calculation error. The ordinary differential equations difference method has higher requirements on the time step, and the density of the short-lived nuclides must often be settled independently. The conventional matrix exponential method faces convergence difficulties if the matrix is stiff. The CRAM method can address stiff matrices well and solve problems of transmutation and activation that involve both long-lived and short-lived nuclides with high accuracy.

4.4.2 Simulation Codes

A variety of codes are available that support transmutation simulations such as SuperMC [15], which was developed by the FDS Team in China; FISPACT [16], which was developed by the United Kingdom Atomic Energy Authority (UKAEA); ORIGEN [17], which was developed by Oak Ridge National Laboratory in the USA; CINDER [18], which was developed by Bettis Atomic Power Laboratory in the USA; and Serpent, which was developed by VTT in Finland. In Table 4.5, codes that have transmutation simulation functionalities are listed. SuperMC has been introduced in Chap. 2. In this chapter, the transmutation simulation functionalities of ORIGEN, FISPACT, and CINDER are introduced.

Table 4.5 Codes that have transmutation and activation simulation functionalities

Code	Affiliation	Method
SuperMC	The FDS Team	CRAM
FISPACT	United Kingdom Atomic Energy Authority	FDMODE
ORIGEN	Oak Ridge National Laboratory	Matrix exponential
CINDER	Bettis Atomic Power Laboratory	TTA
Serpent	VTT Technical Research Centre of Finland	CRAM
MC21	Knolls Atomic Power Laboratory	FDMODE
ACT	Japan Atomic Energy Research Institute	Matrix exponential
ANITA	Commission of the European Communities	TTA
MVP-BURN	Japan Atomic Energy Agency	TTA
McCARD	Seoul Nation University	Matrix exponential
FDKR	South Western Institute of Physics	TTA
DEPTH	Tsinghua University	Matrix exponential
JBurn	China Academy of Engineering Physics	CRAM

1. ORIGEN

ORIGEN is a computer system for calculating the burnup, decay, and processing of radioactive materials. It was developed by Oak Ridge National Laboratory and used worldwide. Its decades of development have resulted in several versions, including ORIGEN1.0, ORIGEN2.0, and ORIGEN-S, which is the latest version and is included in the SCALE system. ORIGEN solves burnup equations mainly via the Taylor series method and uses the TTA method and equilibrium approximations to calculate the densities of short-lived nuclides. ORIGEN supports the calculation of activities, biological hazard potentials, and decay gamma spectra and can be applied in nuclear system radioactive source terms analysis and fuel cycle and spent fuel analysis.

2. FISPACT

FISPACT is a transmutation and activation calculation code that was developed by the United Kingdom Atomic Energy Authority. The latest version is FISPACT-II. FISPACT-II can perform transmutation, activation and burnup calculations for neutrons, protons, alpha particles, deuterons, and photons. FISPACT-II solves the Bateman equations with the LSODE ordinary differential equation solver, which was developed by Lawrence Livermore National Laboratory. FISPACT-II supports the calculation of the activity, decay heat, dose rate, biological hazard potential, and decay gamma spectrum, and provides many auxiliary calculation functions, such as pathway analysis, sensitivity analysis, and uncertainty calculation. The nuclear libraries of TENDL-2014, ENDF/BII.1, JENDL-4.0, and JEFF-3.2 [19] are supported in FISPACT-II. FISPACT-II can be applied to advanced fission systems and fusion systems, among others.

3. CINDER

CINDER is a nuclear inventory code that originated at the Bettis Atomic Power Laboratory in 1960s. Its latest version is CINDER'90, which was released by the Radiation Safety Information Computational Center at Oak Ridge National Laboratory. CINDER'90 solves the transmutation and activation problems via the TTA method. The multigroup database that is used in CINDER'90 contains more than 3400 nuclides. CINDER'90 can be widely applied in such areas as reactor burnup, transmutation in accelerator-driven subcritical systems, activation in accelerator systems, and astrophysics.

References

1. NEA (2016) Uranium 2016: resources, production and demand. NEA No. 7301. OECD publishing, Paris. https://doi.org/10.1787/uranium-2016-en
2. Wang MH (2012) Neutronics design and study on subcritical spent fuel burner. University of Science and Technology of China (in Chinese)

3. Zheng SL (2005) Study on neutronics safety characteristic of the fusion-driven subcritical system with the dual-cooled transmutation blanket. Institute of Plasma Physics Chinese Academy of Sciences (in Chinese)
4. Sawan ME, Abdou MA (2006) Physics and technology conditions for attaining tritium self-sufficiency for the DT fuel cycle. Fusion Eng Des 81(8):1131–1144
5. Wu YC, Team FDS (2006) Conceptual design activities of FDS series fusion power plants in China. Fusion Eng Des 81(24–24):2714–2718
6. International Atomic Energy Agency (1996) IAEA TECDOC-855 clearance levels for radionuclides in solid materials: application of the exemption principles. Interim Report for Comment
7. Wu YC, Jiang JQ, Bai YQ (2009) Fusion-fission hybrids driven research in China. Presented at the Fusion-fission research needs workshop, Gaithersburg, Maryland, USA, September 29–October 1
8. Wu YC (2016) Fusion fission hybrid reactor as a bridge between fission and fusion nuclear energy development. In: 2nd international conference on fusion for neutrons and sub-critical nuclear fission waste management and safety, Rome, Italy, 26–28 October
9. Zhan WL, Xu HS (2012) Advanced fission energy program-ads transmutation system. Bull Chin Acad Sci 27(3):375–381
10. Cetnar J (2006) General solution of Bateman equations for nuclear transmutations. Ann Nucl Energy 33(7):640–645
11. Morimoto Y, Maruyama H, Ishii K et al (1989) Neutronic analysis code for fuel assembly using a vectorized Monte Carlo method. Nucl Sci Eng 103(4):351–358
12. Yamamoto A, Tatsumi M, Sugimura N (2007) Numerical solution of stiff burnup equation with short half lived nuclides by the Krylov subspace method. J Nucl Sci Technol 44(2):147–154
13. Pusa M, Leppanen J (2010) Computing the matrix exponential in burnup calculations. Nucl Sci Eng 164(2):140–150
14. Trefethen LN, Weideman JAC, Schmelzer T (2006) Talbot quadratures and rational approximations. BIT Numer Math 46(3):654–670
15. Wu YC (2018) Multi-functional neutronics calculation methodology and program for nuclear design and radiation safety evaluation. Fusion Sci Technol 74:321–329
16. Sublet JC, Eastwood JW, Morgan JG (2012) The FISPACT-II user manual. CCFE-R 11(11)
17. Santamarina A, Bernard D, Blaise P et al (2009) The JEFF-3.1.1 nuclear data library. JEFF report
18. Wilson WB, Cowell ST, England TR et al (2007) A manual for CINDER'90: version 07.4 codes and data. Manual LA-UR 07-8412
19. Hermann OW, Westfall RM (2009) ORIGEN-S: SCALE system module to calculate fuel depletion, actinide transmutation, fission product buildup and decay, and associated radiation source terms. In: SCALE: a modular code system for performing standardized computer analyses for licensing evaluation. Version 6, ORNL/TM-2005/39

Chapter 5
Radiation Dosimetry and Biological Safety

Radiation of nuclear systems mainly comes from neutrons, photons, charged particles, and radionuclides produced by nuclear reactions between neutrons and materials. Radionuclides can migrate from the reactor to the environment, resulting in radiation risks to workers, the public, and the environment. Comparing with the traditional nuclear system, new radionuclides will be produced in advanced nuclear systems because of the differences of coolants between both nuclear systems. This chapter focuses on the radiation effects of these nuclides. Fundamental safety principles need to be followed, and safety requirements need to be met to control radiation risks arising from nuclear systems and to ensure the protection of public health and the natural environment from the harmful effects of ionizing radiation. For workers and the public, it is necessary to estimate the external exposure from neutrons and photons and to evaluate both internal and external exposures to radionuclides. The radioactive source term, nuclide migration, and radiation dosimetry calculations and biological effects of radiation will be described in this chapter.

5.1 Radioactive Source and Nuclide Migration

The radioactive source term refers to the released radioactive materials and their magnitudes, composition, and rates and modes of release.

Determination of the radioactive source terms provides a reference for shielding design, waste evaluation, and environmental impact assessment. After radionuclides are released, they migrate in the nuclear system and the environment (dispersion in the atmosphere, water, and soil, and transfer through the food chain), which may cause radiation hazards to the public and the environment. There are several source terms including fission products and neutron activation products in advanced fission systems, tritium and neutron activation products in fusion systems, and fission prod-

Y. Wu, *Neutronics of Advanced Nuclear Systems*,
https://doi.org/10.1007/978-981-13-6520-1_5

ucts, tritium, neutron activation products, and spallation products in hybrid nuclear systems. This section introduces these radioactive source terms and their migration in the systems and the environment.

5.1.1 Radioactive Source Terms

According to the different generation mechanisms of radionuclides, the radioactive source term mainly includes fission products, neutron activation products, and tritium. New coolants that are applied in advanced nuclear systems will make the neutron activation products of advanced nuclear systems very different from those of traditional nuclear systems.

1. Fission products

Fission products are radionuclides produced by nuclear fission reactions of neutrons bombarding fission materials. According to their volatility and chemical activity, fission products are categorized into fission gases, volatile nuclides, and nonvolatile nuclides [1]. These radionuclides mainly include the following:

(1) Fission gases, such as isotopes of krypton and xenon. When the fuel assembly is damaged or melted, fission gases will neither stick to the surface nor be absorbed by the filter due to their inactive chemical properties, and they will be released through cracks. Fission gases are not easily absorbed by the human body, and they are mainly harmful to people through external exposure.
(2) Volatile nuclides, such as the iodine, caesium, and tellurium group. When the fuel assembly is damaged or melted, volatile nuclides are partially released. Because they are chemically active, these radionuclides combine with others to form aerosols that are released through cracks. Among them, iodine is easily absorbed and accumulated in the thyroid, caesium is deposited in muscle tissues, and strontium is mainly deposited in bones, which may cause radiation hazards to organs or tissues.

Other fission products, such as alkaline earth metals, inert metals, rare earth metals, and actinides, are not volatile and thus are not easily released into the environment in accidents.

2. Neutron activation products

The neutron activation products are mainly derived from the activation products in coolants, structural materials, and cover gases.

(1) Activated products in coolants

The neutron activation products in coolants of different nuclear systems are different; for example, the activation of a supercritical water-cooled reactor produces T, ^{16}N

and ^{17}N; the activation of a lead-cooled reactor produces ^{210}Bi and ^{210}Po. These activation products will generate radiation fields during flow in the coolant loop pipe. When a coolant leak accident occurs, these radioactive products will also leak out and cause radiation exposure to workers.

(2) Activated products in structural materials

The main structural material in nuclear systems is stainless steel, and the radionuclides generated after neutron activation include ^{52}V, ^{51}Cr, ^{56}Mn, ^{59}Fe, ^{57}Co, ^{58}Co, and ^{60}Co. The activated products will corrode the interface between the structural material and the coolant and then dissolve in the form of ions, be suspended as particles in the coolant, or be deposited on the pipe surface or affix to the corrosion surface directly through oxidation, finally forming activated corrosion products. The activated structural materials generate gamma radiation fields around the components and cause radiation hazards to workers.

(3) Activation products in cover gases

In a nuclear system, cover gases can prevent the coolant from contacting the air directly. Generally, argon gas is used as the cover gas. When the system is operating, argon gas will be activated by neutrons leaked from the reactor core, which will generate radioactive ^{41}Ar and ^{37}Zr. When sealing leakage occurs at the top of the reactor vessel, the cover gases will leak out and bring exposure to workers and the public.

3. Tritium

As the main fuel for fusion reactors, tritium is mainly located in the vacuum vessel, hot cell, or tritium plant. In these zones of the ITER, the maximum tritium inventory, which is mainly supplied by external sources, is approximately 4 kg. In addition, breeding blankets are designed to produce tritium in fusion systems. The tritium trapped in the tritium breeding materials of blankets is also a major concern.

Due to its permeation characteristics, tritium can be released into the environment through tritium-related pipe systems under normal conditions, or by rupturing of the containment system during accidents. The human body takes in tritium by breathing, skin contact, drinking water, and ingesting food. Therefore, tritium intake mainly causes internal exposure to the human body.

5.1.2 Radionuclide Migration in Nuclear Systems

The radionuclides that migrate in nuclear systems mainly include fission gas, volatile fission products, tritium, and activated products in the coolant system and cover gas. This section introduces radionuclide migration in fission systems as example. Under normal operating conditions, the radionuclides accumulate in the reactor core. When a leak occurs in the fuel assembly, fission gas and volatile fission products can leak into the coolant and be trapped in the cover gas with the activation products. After

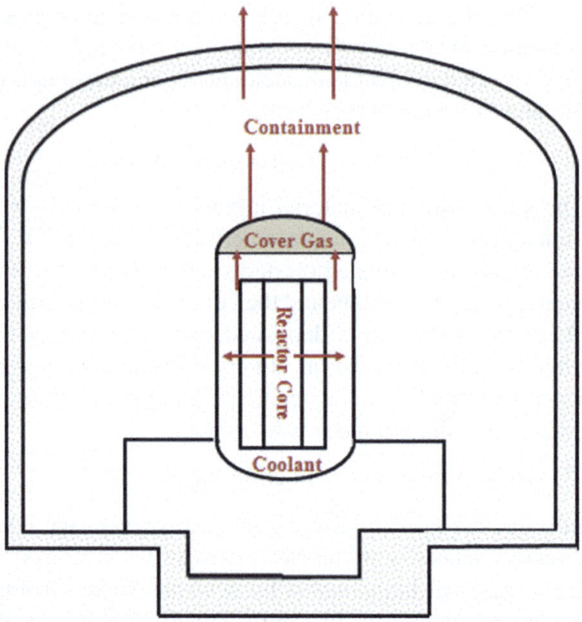

Fig. 5.1 Migration of fission products between systems

fuel assembly damage occurs, the fission products may penetrate through the fuel cladding, the primary system boundary, and the containment system and be released into the environment. The migration of fission products between the systems is shown in Fig. 5.1.

According to the conservation principle of radionuclides, the conversion and migration of fission products between different systems in the nuclear system can be established, as shown in Fig. 5.2. The horizontal arrows represent the conversion of radionuclides within a system, and the vertical arrows represent the migration of radionuclides between different systems. Each circle represents a type of nuclide in a system.

The change rate of any radionuclide in a system is as follows:

$$\frac{dN_{i,j}(t)}{dt} = \delta(i)Y_j + \beta_{i,(j-1)}N_{i,(j-1)}(t) + \sigma_{(i-1),j}N_{(i-1),j}(t) - \gamma_{i,j}N_{i,j}(t) \quad (5.1)$$

where i is the equipment number, which has a the value of 0, 1, 2, …; j is the number of nuclide generations, which has a value of 1, 2, 3, …; $N_{i,j}(t)$ is the number of j generation radionuclides in the ith equipment; $\delta(i)$ is the Dirac delta function, which has a value of 1 when $i = 0$ and a the value of 0 when i $\neq$ 0; Y_j is the generation rate of the jth generation of radionuclides; $\beta_{i,j-1}$ is the generation rate of the jth generation radionuclides in the ith equipment from the $(j-1)$th generation of radionuclides in the ith equipment; $\sigma_{(i-1),j}$ is the generation rate of the jth generation of radionuclides in the ith equipment from the jth generation of radionuclides in

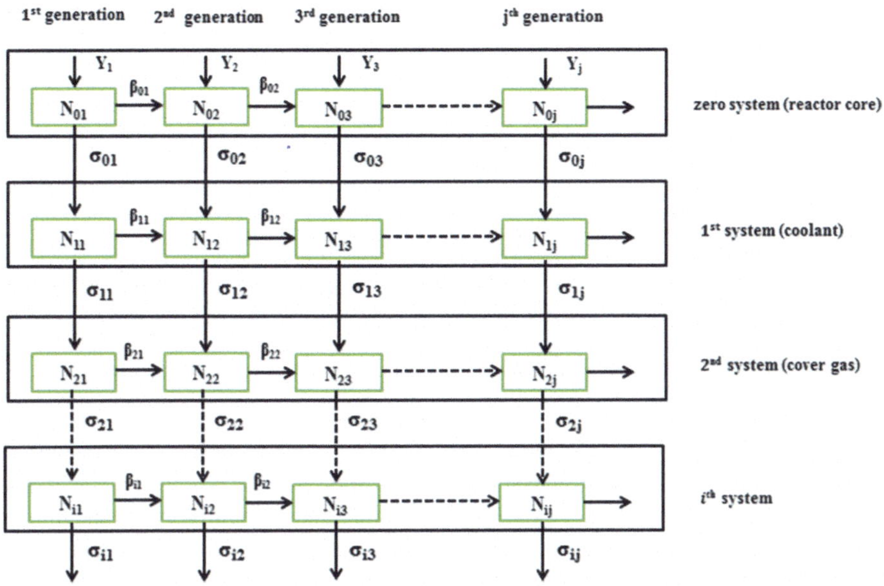

Fig. 5.2 Conversion and migration of fission products between different systems

the $(i-1)$th equipment; and $\gamma_{i,j}$ is the disappearance rate of the jth generation of radionuclides in the ith equipment, which is derived from decay and filtration.

This equation can be applied to calculate the amounts and types of radionuclides in the system due to various factors, such as production, decay, inflows outflows, and removal, and the radioactive source terms in each system can then be obtained.

5.1.3 Radionuclide Migration in the Environment

Radionuclides in the environment are derived from the radioactive materials that are produced during the process of waste disposal, discharged during normal operation, or released during accidents. Airborne radioactive materials include fission gas, volatile fission products, tritium, and gaseous neutron activation products. Radioactive liquid materials include activated corrosion products and some fission products. Solid radioactive materials originate from long-life fission products produced by spent fuel and activation products of structural materials.

The dispersion process of radionuclides in media, including air, water, and soil, and the means of their transfer in food chains are discussed in the following section.

1. Radionuclide dispersion in the atmosphere

The process of radionuclide dispersion in the atmosphere is shown in Fig. 5.3 [2]. A radioactive plume is produced through the discharge of radionuclides from the

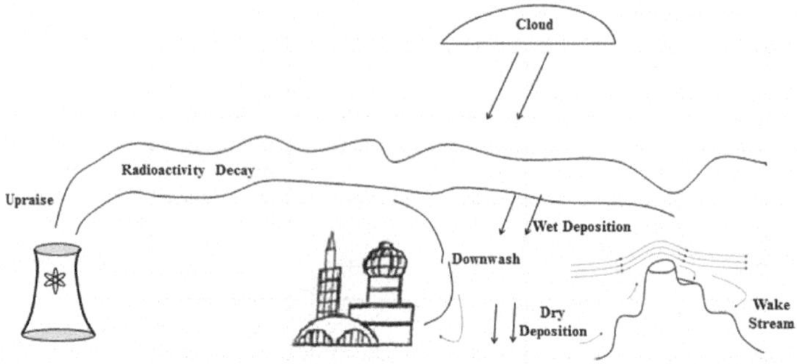

Fig. 5.3 Radionuclide dispersion in the atmosphere

chimney to the atmosphere. The plume rises due to heat conduction and is transported downwind. During its dispersion in atmospheric turbulence, the plume mixes with the surrounding air and continually extends horizontally and vertically. When the plume encounters buildings or hillsides, it sinks in the near downwind area due to airflow perturbations; eventually, it produces a wake stream due to airflow spinning. This process is accompanied by radionuclide decay, dry deposition, and wet deposition. Dry deposition refers to gravity sedimentation of radionuclide aerosols with particle diameters greater than 10 μm. In wet deposition, atmospheric hydrometeors (e.g., raindrops and snow) scavenge the radioactive plume.

Atmospheric dispersion models can be categorized into three types: Gaussian models, Eulerian models, and Lagrangian models [3]. For Gaussian and Eulerian models, a fixed-coordinate system is selected to simulate radionuclide dispersion. The Gaussian models are based on gradient transportation theory and solved the dispersion equation is solved under the assumption of a constant diffusion coefficient. Based on actual meteorological conditions, Gaussian models are modified to account for the mixing layer, radionuclides decay, and dry deposition. According to the wind direction and source release mode, Gaussian models can be categorized into linear Gaussian plume models, segmented Gaussian plume models, and Gaussian puff models. Lagrangian models, which are based on turbulent statistical theory, describe the distribution of radionuclides in a moving coordinate system by taking the center of mass of the puff as the origin. Eulerian models, which are grid models, calculate the radionuclides concentration in each grid based on input data of the wind field or coupling of the flow equation with the equation for the conservation of matter. A comparison of the three kinds of models is shown in Table 5.1.

The linear Gaussian plume model is one of the most widely used atmospheric dispersion models. It is shown below as an example to introduce the radionuclide concentration calculation:

$$C(x, y, z, t) = \frac{Q}{2\pi u \sigma_y \sigma_z} e^{-y^2/2\sigma_y^2} \left(e^{-(z-H)^2/2\sigma_z^2} + e^{-(z+H)^2/2\sigma_z^2} \right) \tag{5.2}$$

Table 5.1 Comparison of atmospheric dispersion models

Dispersion model	Advantages	Disadvantages	Range
Gaussian models	Simple models, fewer input parameters required, simplified meteorological data, and convenient calculation	Difficult to treat complicated terrain and meteorology conditions	<20 km
Lagrangian models	Continual tracing of dispersion trajectory, solving the dispersion problem under conditions of calm wind, light winds, and complicated terrain	Complicated function equation, only applied to radionuclide dispersion simulations under steady and uniform turbulence conditions	Tens to hundreds of kilometers
Eulerian models	Effective treatment of the temporal and spatial variations of the source term, meteorological field, and dynamic pollution dispersion	Difficult to deal with closure problems and numerical diffusion during solving of the transport equation	>200 km

where C is the radionuclide concentration at time t at point (x, y, z) in units of Bq/m^3; Q is the source intensity of the radionuclide discharge rate in units of Bq/s; σ_y and σ_z are the atmospheric dispersion coefficients in the y and z directions, respectively, in units of m; u is the average wind speed in units of m/s; and H is the effective discharge height when a dispersion point is selected as the coordinate point, in units of m.

The typical fission product ^{131}I, easily forms aerosol in the air, and its decay chain is as follows:

$$
\begin{aligned}
{}^{131}_{53}I(8.04d) &\xrightarrow{2.1\%} {}^{131m}_{54}Xe(11.9d) \xrightarrow{IT} {}^{131}_{54}Xe(\text{stable}){}^{131}_{53}I(8.04d) \\
&\xrightarrow{97.3\%} {}^{131}_{54}Xe(\text{stable})
\end{aligned}
$$

Therefore, when calculating the concentration of nuclides, the decay effect of the radionuclides should be considered. In Eq. (5.2), Q_A should be replaced by $Q_A f_F$, in which

$$f_F = \exp\left(-\lambda \frac{x}{u}\right)$$

where λ is the radionuclide decay constant.

Unlike other airborne radioactive materials, tritium gases can be oxidized to tritium water during diffusion. Therefore, the factor of tritium oxidation and reevaporation should be considered in tritium dispersion simulations. To solve this problem, the FDS Team developed a tritium dispersion simulation method, in which tritium oxidation in soil and tritium surface source reevaporation was added [4]. This method can be applied in accurate calculation of tritium concentration.

2. Radionuclides dispersion in water

Models of radionuclide dispersion in water are mainly categorized into Eulerian models and Lagrangian models. One kind of Eulerian model is introduced as an example:

$$\begin{aligned}\frac{\partial C(x,y,z,t)}{\partial t} &= D_x\frac{\partial^2 C(x,y,z,t)}{\partial x^2} + D_y\frac{\partial^2 C(x,y,z,t)}{\partial y^2} + D_z\frac{\partial^2 C(x,y,z,t)}{\partial z^2} \\ &- u\frac{\partial C(x,y,z,t)}{\partial x} - v\frac{\partial C(x,y,z,t)}{\partial y} - w\frac{\partial C(x,y,z,t)}{\partial z} \\ &- \theta(\mathrm{x,y,z,t}) - \frac{0.693}{T}C(x,y,z,t)\end{aligned} \tag{5.3}$$

where $C(x, y, z, t)$ is the radionuclide concentration at time t at point (x, y, z) in units of Bq/m^3; D_x, D_y, and D_z are the water dispersion coefficients in the x, y, and z directions, respectively, in units of m^2/s; u, v, and w are the water flow speeds in the x, y, and z directions, respectively, in units of m/s; $\theta_{x,y,z}$ is the amount of pollution discharged at time t, at point (x, y, z), in units of Bq/(m^3 · s); and T is the half-life of the radionuclides, in units of s.

For an instantaneously released point source, the analytical solution for Eq. (5.3) is

$$\begin{aligned}C(x,y,z,t) &= \frac{Q}{8(\pi t)^{3/2}\sqrt{D_x D_y D_z}} \\ &\exp\left[-\frac{(x-ut)^2}{4D_x t} - \frac{(y-vt)^2}{4D_y t} - \frac{(z-wt)^2}{4D_z t} - \frac{0.693}{T}\right]\end{aligned} \tag{5.4}$$

where Q is the source intensity in units of Bq.

Among the fission products, iodine and caesium are the main nuclides that contribute to the radioactivity in water. Iodine has a large contribution to the radioactivity of water in the short term, whereas caesium can influence the radioactivity in water over the long term.

In the case of a nuclear accident, radioactive material diffuses through the atmosphere and water. For example, some radioactive materials were deposited on the surface of the seawater in the Fukushima nuclear accident, which accelerated the range of influence of the radionuclides. Therefore, the FDS Team developed an air–water coupled radionuclide diffusion simulation method, in which the precipitation correction of radionuclides on the surface of water is added during the atmospheric diffusion simulation of radionuclides [5]. This model can be used to accurately sim-

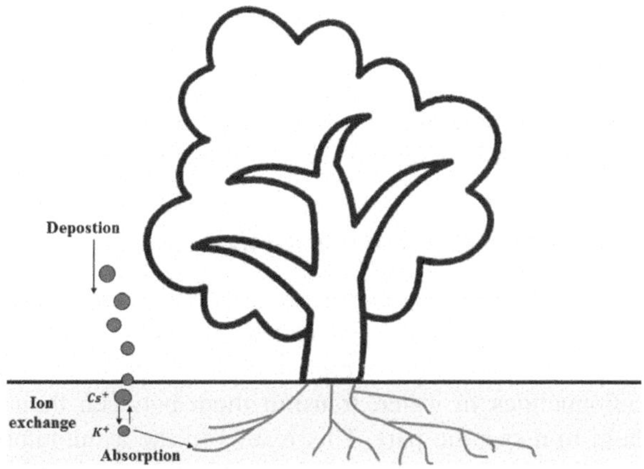

Fig. 5.4 Radionuclide diffusion behaviors in soil

ulate the concentration distribution and diffusion of nuclides at the interface between the atmosphere and water.

3. Radionuclide migration in soil

The radionuclides present in soil are mainly due to the dry and wet deposition of aerosols. The physical and chemical behaviors of these radionuclides in soil include ion exchange and absorption, as shown in Fig. 5.4.

The secondary minerals and humus in the soil mostly exist in the form of colloidal particles and have significant surface energy and adsorptive properties. Therefore, ion exchange is the most important physical and chemical process for the diffusion of radioactive materials in soil. For example, the ion exchange reaction between caesium ions and potassium ions in soil can be expressed as follows:

$$K^+ + Cs^+_{(Liquid)} \leftrightarrow Cs^+ + K^+_{(Liquid)}$$

Radionuclides will enter the soil from surface deposits by cation exchange. Radionuclides precipitate as oxides or hydroxides or combine with soil organic matter. In general, the cation absorption capacity is determined by its valence. The combination of polyvalent particles and clay minerals is relatively firm, and its penetration and cation-exchange capacity are relatively small. This determines the soil-to-plant transfer of radionuclides. According to the ion valence, the selective absorption order of clay minerals is

$$\text{tetravalent ion} > \text{trivalent ion} > \text{divalent ion} > \text{monovalent ion}$$

Among fission products, the size of the hydration particles of alkali and alkaline earth metals decreases as the atomic number increases. Therefore, the selective absorption order is:

$$Cs^+ > Rb^+ > K^+ > Na^+ > Li^+; Ba^{2+} > Sr^{2+} > Ca^{2+} > Mg^{2+}$$

Different plants species have different abilities to accumulate radionuclides. For example, leguminous plants are better at accumulating ^{90}Sr than gramineous plants, but quinoa and calabash plants are much better at accumulating ^{90}Sr than leguminous plants.

4. Radionuclide Transfer in the food chain

The pathways through which radionuclides enter the food chain include absorption of tritium and activated products by plants. Radionuclides are absorbed by terrestrial organisms through their roots and deposited on the surface. Aquatic plants absorb radionuclides in water, transmit them between their tissues, and eventually bind them to a specific part. This results in the accumulation of radionuclides in plant tissues. Radionuclides are then transferred to animals when plants are eaten. Finally, radionuclides are transferred to humans when plants and animals are eaten.

Microelements are essential nutrients for plants, animals, and humans. These elements have similar chemical properties with some fission products. Therefore, fission products released into the environment are easily absorbed by human tissues or organs and cause radiation hazards to human body. For example, the behaviors of nuclides such as ^{45}Ca, ^{89}Sr, ^{90}Sr, ^{240}Ba, and ^{226}Ra are similar to that of ^{40}Ca and these nuclides participate in the entire life process of the human body. Other nuclides, such as ^{22}Na, ^{40}K, ^{86}Rb, and ^{137}Cs, are often involved in food chain transfer along with potassium to maintain the balance of water and pH in the body. The chemical properties of ^{129}I are consistent with those of ^{127}I, and it is easily absorbed by the thyroid gland. The behavior of ^{3}H is the same as that of ^{1}H, and it participates in the circulation of water in the human body.

The transfer processes of radionuclides of the activation products from different types of coolants in the food chain are different. For example, the activation product ^{3}H from fusion systems is released into the environment and then transferred into HTO during the oxidation reaction and isotope exchange reaction. One part of HTO is involved in natural water circulation, and the other part is transferred to the organic tritium after being absorbed by animals and plants. In the environment, inorganic tritium and organic tritium can enter the human body through the respiratory tract and various food chains and are evenly distributed in the human body. The activation product ^{14}C from gas-cooled reactor coolant is oxidized to $^{14}CO_2$ after being released into the atmosphere. $^{14}CO_2$ can be transferred to the human body through the photosynthesis of plants, the food chain, and the respiratory tract and is mainly trapped in the blood, muscle, fat, bone, and liver.

5.2 Radiation Dosimetry Calculations

Radiation can generally be categorized into two types: ionizing and electromagnetic radiation. This section focuses on external and internal exposures to ionizing radiation.

5.2.1 Dosimetric Quantities in Radiological Protection

The interaction between ionizing radiation and matter is a process of energy transfer, in which radiation energy is absorbed by matter, resulting in radiation effects. This section defines the dosimetric quantities used in radiological protection to describe the relationship between the radiation effect and the radiation energy.

1. Absorbed dose

The absorbed dose is a physical quantity representing the ionizing radiation energy absorbed per unit mass of matter. It is the mean energy imparted to matter with mass and is defined as

$$D = \frac{d\bar{\varepsilon}}{dm} \tag{5.5}$$

where $d\bar{\varepsilon}$ is the mean energy imparted to matter with mass *dm* in units of J. The SI unit of the absorbed dose is J/kg, and its special unit is the gray (Gy), where 1 Gy = 1 J/kg. The previous unit was the rad, where 1 rad = 10^{-2} Gy.

The absorbed dose is used to calculate the dose for all types of ionizing radiation for any irradiated material.

2. Dose equivalent

The absorbed dose is insufficient to predict the severity or probability of the deleterious effects on health resulting from irradiation. In radiation protection, it is convenient to introduce a physical quantity for the biological effectiveness of an absorbed dose of ionizing radiation. This quantity, called the dose equivalent, is defined as

$$H = QD \tag{5.6}$$

where H is the dose equivalent in units of Sv; D is the absorbed dose at the point of interest in tissue; and Q is the corresponding quality factor at this point, the value of which is determined by the type and energy of the charged particles passing through a small volume element at that point, which was given in *ICRP Publication 60*.

3. Equivalent dose

The biological consequences of radiation are related not only to the energy deposition per unit mass but also to the type and energy of the radiation. The biological effect

Table 5.2 Radiation weighting factors

Radiation type	Energy range	Radiation weighting factor, ω_R
Photons	All	1
Electrons	All	1
Neutrons	<10 keV	2
Neutrons	10–100 keV	7.4
Neutrons	0.1–2 MeV	20
Neutrons	2–20 MeV	10
Neutrons	>20 MeV	5
Alpha particles	All	20

of the same absorbed dose is different for different types and energy of radiation. Therefore, the equivalent dose is used to assess the damage level of radiation on human organs or tissue. It is defined as

$$H_T = \sum_R \omega_R \cdot D_{T,R} \tag{5.7}$$

where H_T is the equivalent dose of radiation type R on organ or tissue T in units of Sievert (Sv); R is the type of radiation, as shown in Table 5.2; ω_R is the radiation weighting factor for radiation type R, as shown in Table 5.2, given in *ICRP Publication 60*; and $D_{T,R}$ is the average absorbed dose in the specified organ or tissue T due to radiation of type R in units of Gy.

4. Effective dose

The probability of stochastic effects is related not only to the equivalent dose but also to the type of organs or tissues. Therefore, the effective dose is introduced and defined as the sum of the tissue weighting factor multiplied by the equivalent dose of the tissues and organs.

$$E = \sum_T \omega_T H_T \tag{5.8}$$

where ω_T is the tissue weighting factor, which is provided in Table 5.3.

ICRP Publication 103 [6] gives the tissue weighting factors for different tissues and organs, where the sum of all factors is 1. In other words, the value of the effective dose is equivalent to the equivalent dose under uniform gamma ray exposure to all organs and tissues. The unit of the effective dose E is the same as that of the equivalent dose H_T.

5. Committed equivalent dose

The committed equivalent dose $H_T(\tau)$, after intake of radioactive material within a specified time period τ in tissue or organ T, is defined by

Table 5.3 Tissue weighting factor

Tissue or organs	Tissue weighting factor ω_T	Tissue or organ	Tissue weighting factor ω_T
Gonads	0.08	Liver	0.04
Bone marrow (red)	0.12	Oesophagus	0.04
Colon	0.12	Thyroid	0.04
Lung	0.12	Skin	0.01
Stomach	0.12	Bone surface	0.01
Bladder	0.04	Brain	0.01
Breast	0.12	Others	0.12
Salivary glands	0.01	–	–

$$H_T(\tau) = \int_{t_0}^{t_0+\tau} \dot{H}_T(t)dt \tag{5.9}$$

where t_0 is the time of intake of the radioactive material in units of h; $\dot{H}_T(t)$ is the equivalent dose rate in a tissue or organ at time t in units of Sv/h; and τ is the integral time of the intake in units of h.

By default, the integral time is taken to be 50 years for adults, and 70 years is considered for intake by children.

6. Committed effective dose

The committed effective dose $E(\tau)$ is defined by

$$E(\tau) = \sum_T \omega_T \cdot H_T(\tau) \tag{5.10}$$

where $H_T(\tau)$ is the committed equivalent dose in a tissue or organ at integration time τ in units of Sv.

7. Dose equivalent rate

The effective dose cannot be measured directly, so the dose equivalent rate was proposed to provide a reasonable estimate of the effective dose under normal working conditions. The calculations of the contact dose equivalent rate and the dose equivalent rate of the point source are introduced here.

The contact dose equivalent rate, which refers to the dose equivalent rate at the surface for a half infinite flat source, is defined by

$$\dot{H} = C\frac{B}{2}\sum_{i=1}^{N_\gamma} \frac{\mu_a(E_i)}{\mu_m(E_i)} S_\gamma(E_i) \tag{5.11}$$

where C is the conversion coefficient from MeV/(kg · s) to Sv/h, which has a value of 5.76×10^{-10}; B is the buildup factor, which has a default value of 2; N_γ is the number of energy groups in the gamma spectrum history; E_i is the mean photon energy of the ith energy group in units of MeV; $\mu_a(E_i)$ is the mass energy absorption coefficient of air in units of m^2/kg; $\mu_m(E_i)$ is the mass energy attenuation coefficient of the material in units of m^2/kg, $\mu_m = \sum_j f_j \mu_{mj}$, where μ_{mj} is the mass energy attenuation coefficient of radionuclide j, and f_j is the mass fraction of radionuclide j in the material; $S_\gamma(E_i)$ is the rate of gamma emission in units of MeV/(kg · s), $S_\gamma(E_i) = I_i A(t)$, where I_i is the intensity of energy group i of gamma decay in units of MeV; and $A(t)$ is the special activity of the material at time t in units of Bq/kg.

The dose equivalent rate of a point source at a point of interest at distance r in air is defined by

$$\dot{H} = C \sum_{i=1}^{N_\lambda} \frac{\mu_a(E_i)}{4\pi r^2} e^{-\mu(E_i)r} m_s S_\gamma(E_i) \tag{5.12}$$

where $\mu(E_i)$ is the air energy attenuation coefficient in units of m^{-1}, and m_s is the mass of the source in units of kg.

5.2.2 *External Exposure Calculation*

The external exposure dose is an important parameter to assess the radiation hazard for the human body. It is usually described by the dose equivalent of external exposure, which can be calculated as [7]

$$H(\vec{r}) = \int_{E_1}^{E_2} \Phi(\vec{r}, E) F_R(E) \mathrm{d}E \tag{5.13}$$

where $H(\vec{r})$ is the dose equivalent at point $\vec{r}$ induced by radiation R; E is the energy of particle R; E_1 and E_2 are the minimum and maximum energies, respectively; $\Phi(\vec{r}, E)$ is the fluence contributed by particles R with energy E at point $\vec{r}$; and $F_R(E)$ is the fluence-to-dose conversion coefficient for particle R with energy E. The conversion coefficients for neutrons and photons are listed in *ICRP Publication 116* [8].

External exposure in nuclear systems is mainly from fission or fusion neutrons, secondary photons induced by neutron nuclear reaction, and decay photons emitted by radionuclides. This section focuses on calculating the external exposure induced by neutrons and photons.

1. Calculation of the neutron dose

During neutronics design analysis for nuclear systems, considering the complex geometry, the neutron fluence $\Phi(\vec{r}, E)$ is calculated using Monte Carlo methods, and the corresponding neutron dose equivalent is derived using Eq. (5.13).

2. Calculation of the secondary photon dose

In a nuclear system, secondary photons are generated during the interaction of neutrons with materials, and they induce additional external exposures. The fluence of secondary photons can also be calculated during the calculation of the neutron fluence. The dose equivalent of secondary photons is calculated by using Eq. (5.13).

3. Calculation of the decay photon dose

Materials such as structural materials and coolants are activated by fission neutrons, and large amounts of radionuclides are generated. Decay photons emitted by these radionuclides are the primary radiation source after the shutdown of the nuclear system. Two approaches are used to calculate the shutdown dose rate: the rigorous two steps (R2S) method and the direct one-step (D1S) method [9, 10].

In the R2S method, the transports of neutrons and decay photons are simulated separately. First, the model is divided into many small cells or meshes to calculate the spatial distribution of the neutron fluence. Subsequently, the decay photons produced in each cell or mesh at different cooling times are evaluated by activation calculations that are based on the neutron spectrum and fluence, the initial material compositions, and the irradiation scenario. Finally, photon transport is simulated to obtain the decay photon fluence. The corresponding dose equivalent is calculated using Eq. (5.13) with the fluence-to-dose conversion coefficient for photons. The detailed simulation procedure for the R2S method is shown in Fig. 5.5. Using the activation calculations for each cell and each mesh, the R2S method can precisely predict the spatial distributions of the radionuclides and the emitted decay photons. However, the spatial resolution is limited by the size of the cell or mesh in R2S. Therefore, finer cells or meshes are required to improve the accuracy but they may reduce the computational efficiency.

In the D1S calculation, neutron transport is coupled with decay photon transport, as shown in Fig. 5.6. First, a special data library is prepared for the D1S calculation by replacing the nuclear data of the prompt photon with that of the decay photons. The decay photons are assumed to be produced simultaneously when the neutrons interact with materials. Therefore, a time correction factor for each radionuclide is introduced in the D1S method to account for the irradiation scenario and the time effect of radionuclides decay. The time correction factor is calculated for each radionuclide using the ratio of its activity at a certain cooling time to its total number of atoms at a very short irradiation time (usually 1 s). The D1S method has a higher energy and spatial resolutions than the R2S method for the decay photon source. However, the D1S method is applicable to one-step decay, and it is difficult to use to handle multistep decay.

The human body is usually treated as a point for the calculation of external exposure. This treatment makes it difficult to accurately assess the dose of each organ because of the inhomogeneous characteristics of the radiation field and the different radiation sensitivities of organs. Because computational phantoms can describe human tissues or organs, they can be used to assess the external exposure of an organ. The dose assessment method based on computational phantoms provides guidance

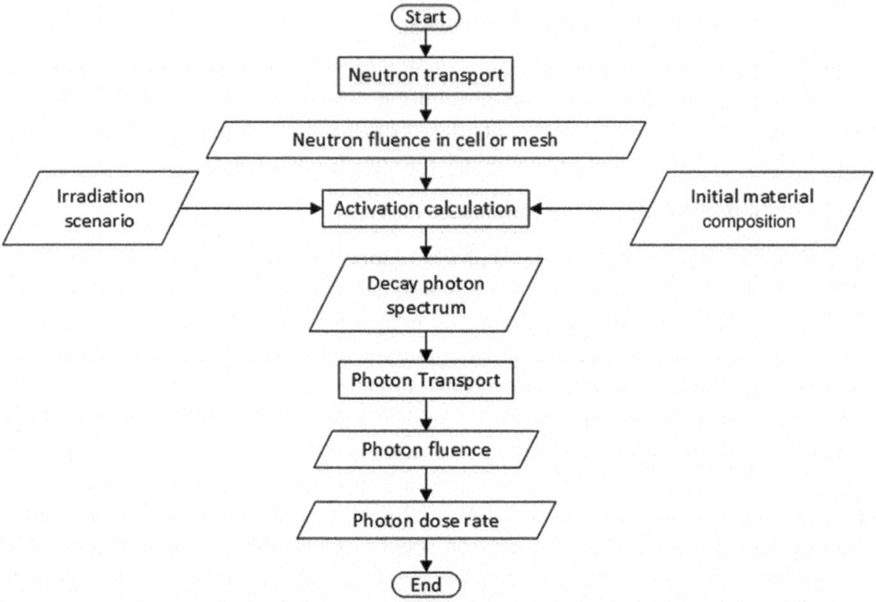

Fig. 5.5 R2S calculation procedure

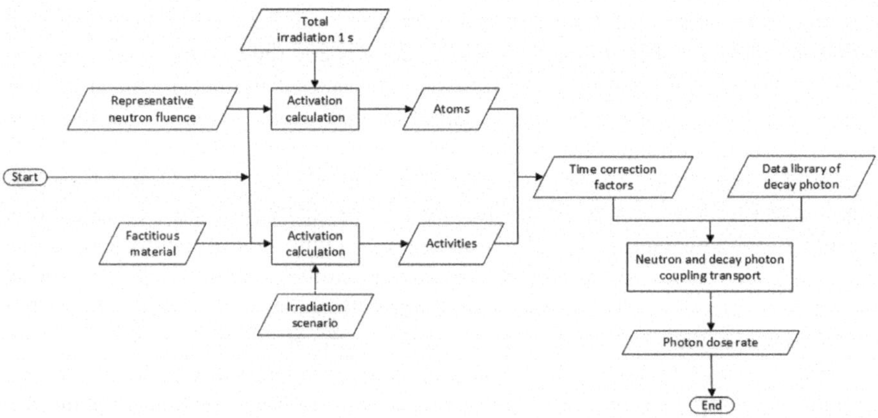

Fig. 5.6 D1S calculation procedure

for the optimal design of nuclear facilities and the optimal scheme for facility maintenance.

Numerous computational phantoms have been developed, ranging from simple mathematical phantoms to complex voxel phantoms. Mathematical phantoms can describe the geometries of human tissues or organs using mathematical formulas and were used to calculate the conversion coefficients in early ICRP publications. Because the number of functions is limited, mathematical phantoms cannot accurately describe complex human anatomy and cannot meet the accuracy requirements for dose calculations. With the development of computer and tomographic imaging technologies, the voxel phantom, which is more similar to the real human body, was developed.

The method to calculate basic data for the radiation field using the voxel phantom was formally proposed in *ICRP Publication 103*, and the reference computational phantom was officially released in *ICRP Publication 110* [11]. Voxel phantoms of different races, genders, ages, weights, and positions (such as sitting, standing, and lying) have been developed and provide important reference data to accurately evaluate human doses.

The ICRP reference computational phantoms were constructed based on images of the Caucasian human body which is different from the Chinese human body. Taking the Chinese body as an example, the computational phantom and its application were introduced. The Rad-Human, which is an accurate whole-body computational phantom of a Chinese adult female, was constructed using high-resolution color digital images from the Chinese Visible Human datasets [12]; a 3D view of the Rad-Human is shown in Fig. 5.7. The Rad-Human construction method is introduced in Chap. 7. The Rad-Human, which includes 46 tissues and organs, has minimum voxel sizes of 0.15 mm $\times$ 0.15 mm $\times$ 0.25 mm for the head and neck regions 0.15 mm $\times$ 0.15 mm $\times$ 0.5 mm for other regions, and it contains more than 28.8 billion voxels. Based on both the Rad-Human and Chinese male computational phantoms, the basal dose data in the radiation field were constructed for the Chinese race, providing reference data for dose assessment in Chinese people.

For voxel phantoms, the calculation process for the organ equivalent dose and the effective dose for external exposure is shown in Fig. 5.8.

The organ equivalent dose induced by gamma rays can be accurately calculated using the following formula:

$$H_T = \frac{1}{n}\sum_{j=1}^{n} H_{T,j} = \frac{1}{n}\sum_{j=1}^{n} \phi_j \times F_R(E) \tag{5.14}$$

where H_T is the equivalent dose of organ T in units of Sv; $H_{T,j}$ is the equivalent dose of the jth voxel in organ T, in units of Sv; ϕ_j is the fluence of the jth voxel, in units of cm^{-2}; $F_R(E)$ is the fluence-to-dose conversion coefficient for particle R with energy E, in units of $\mathrm{Sv} \cdot \mathrm{cm}^2$ which depends on the organ, type of particle, energy group and orientation of the human body in the radiation field (for simplicity,

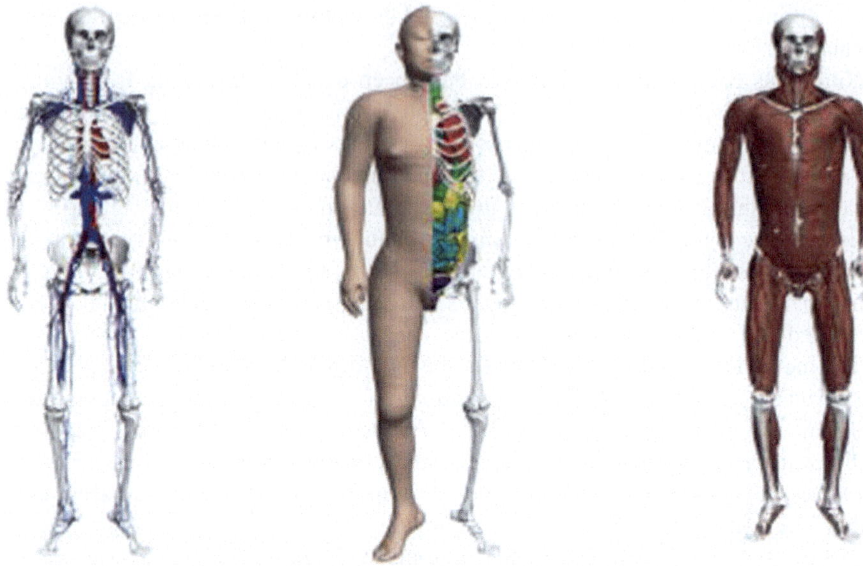

Fig. 5.7 3D view of the Rad-Human

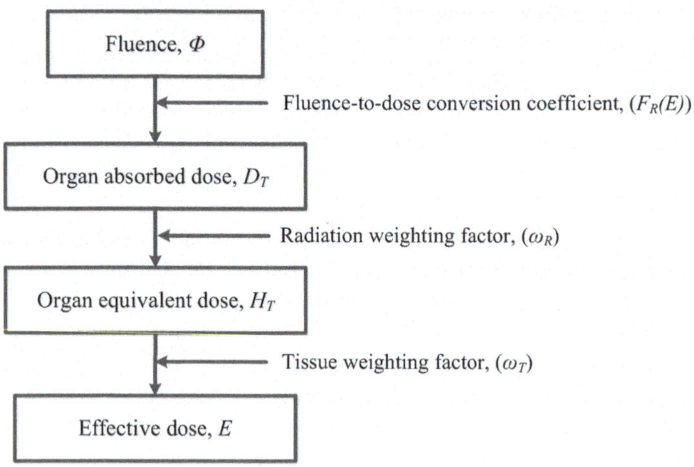

Fig. 5.8 Flowchart for calculating the external exposure dose

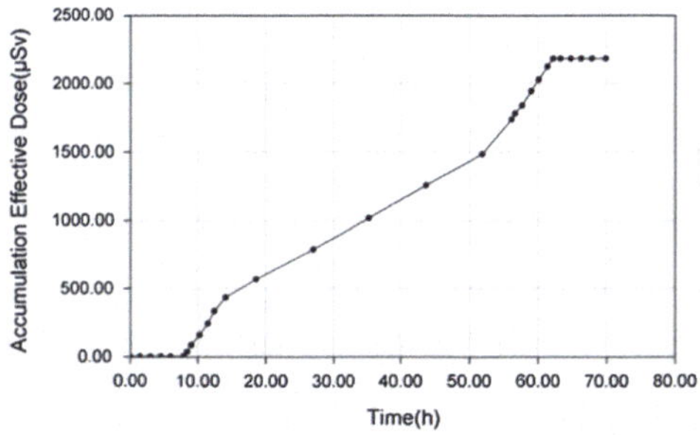

(a) Accumulated effective dose

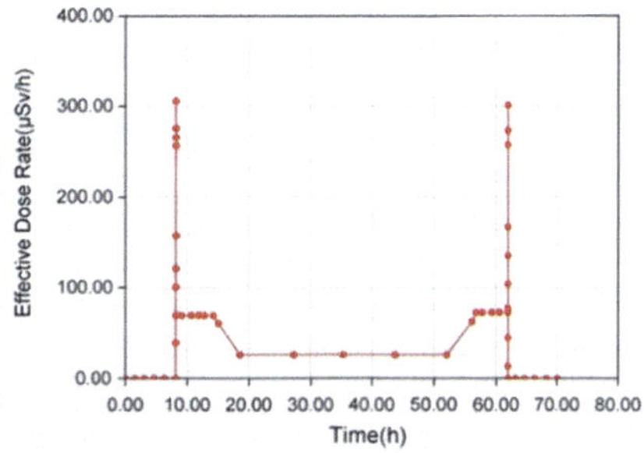

(b) Effective dose rate

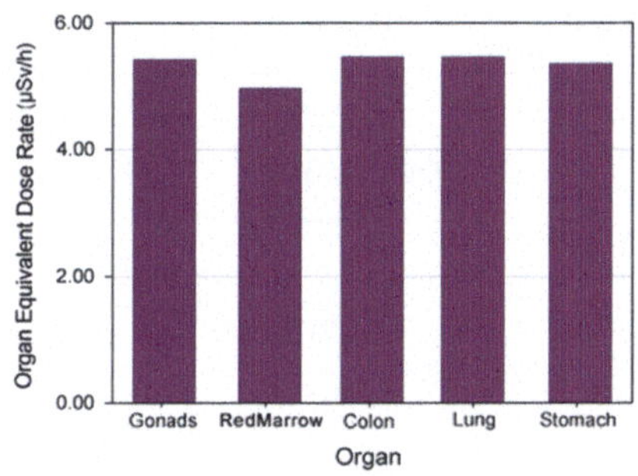

(c) Organ equivalent dose rate

Fig. 5.9 Real-time dose values of human tissues and organs for ITER maintenance plan evaluation

it can be assumed that the human body is in an isotropic radiation field); and n is the number of voxels contained in organ T.

The dose calculation based on voxel phantoms has been applied to the maintenance of fusion systems. Based on the Rad-Human, real-time dose evaluations, including the accumulated effective dose, effective dose rate and organ equivalent dose rate for workers, have been conducted for ITER maintenance plan evaluation. These values can be used to determine whether the accumulated dose in human tissues or organs has exceeded the dose limits of the component maintenance; then, the shielding design can be evaluated and optimized. Figure 5.9 shows the results of the real-time dose to human tissues and organs.

5.2.3 Internal Exposure Calculation

Radionuclides intake can cause internal exposure. Radioactive materials remain in the human body once they enter and continuously induce radiation exposure to organs and tissues. To better simulate the body's metabolic process in mathematical form, the degree of internal exposure is usually evaluated using a biokinetic model.

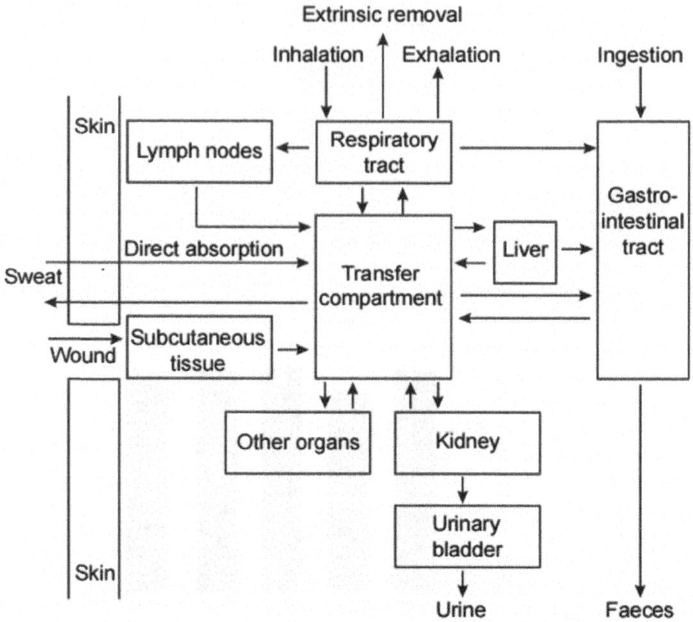

Fig. 5.10 Routes of intake and the metabolism of radionuclides in the human body

A proper model describing the absorption and excretion rates of organs, tissues, and the entire body is applied to calculate the absorbed dose from radionuclides in the human body. In this model, the organs and tissues associated with radionuclide metabolism are assumed to be a series of compartments. The behavior of radionuclides in a compartment is considered to be the same.

Radionuclides may enter the human body through inhalation, ingestion, the skin or wounds. Through retention or transfer via metabolism in the body, radionuclides ultimately cause internal exposure to the human body. The routes of intake and the process of radionuclide metabolism in the human body are shown in Fig. 5.10.

After radionuclides intake, several processes such as deposition, transfer, absorption, and excretion, accompanied by radioactive decay, may occur. These processes are related to the physical and chemical characteristics of the ingested materials, as well as the internal structure of the human body and the functions of the organs and tissues. Radionuclides may be transferred from one organ or tissue to another during the metabolic process, and the metabolic process of radionuclides in each organ or tissue is described by first-order kinetics. Therefore, the variation in the radioactivity of one radionuclide in one compartment with time can be expressed as in Eq. (5.15):

$$\frac{dA(t)}{dt} = \sum_i f_i - \sum_j h_j - \lambda q(t) \tag{5.15}$$

where $A(t)$ is the radioactivity of one radionuclide in one compartment at time t; i and j are the compartment numbers; f_i is the rate of change of radioactivity from either direct intake or transfer from another compartment; h_j is the rate of change of radioactivity transferred to other compartments; and λ is the decay constant of the radionuclide.

The committed effective dose of the intake of one or more radionuclides in the human body can be calculated using the biokinetic model. Due to the complicated calculation process, the dose coefficient is usually defined to simplify the calculation of internal exposure. The radioactivity intake is related to the committed effective dose. The committed effective dose per unit intake of radioactive materials is expressed as $e(g)$, where g refers to a person's age; 50 years is recommended for adults and 70 years is recommended for infants and children. The dose coefficient is divided into the inhalation dose coefficient and the ingestion dose coefficient according to the intake route.

To estimate the internal exposure caused by inhalation of radioactive aerosols for workers and the public, a continuous online monitoring system is usually established in workplaces and near nuclear facilities. Therefore, the concentration of radioactive aerosols can be monitored continuously. For a single radionuclide, the committed equivalent dose of internal exposure can be calculated based on the retention time of workers and the public.

$$E(\tau) = \int_{t_0}^{t_0+\tau} Q(t)Ve(g)\mathrm{d}t \tag{5.16}$$

where $Q(t)$ is the concentration of the radioactive aerosol, which changes over time and is in units of $\mathrm{Bq/m^3}$; V is the exchange rate of air, which has a recommended value of $1.5\ \mathrm{m^3/h}$ according to *ICRP Publication 119*; τ is the retention time in units of h; and $e(g)$ is the dose coefficient for radionuclide intake in units of Sv/Bq.

The dissolution of radionuclides in the blood changes over time during inhalation, in which inhalation can be classified into three categories as shown in Table 5.4. The value of $e(g)$ for all types of radionuclides of different ages and intake types can be obtained from the International Basic Safety Standards. The dose coefficient of internal exposure can be calculated accurately based on the voxel phantom, which provides accurate and reliable data for the human body dose calculation.

5.3 Biological Effects of Radiation

During normal operation or accident conditions, nuclear systems may expose workers and the public to internal or external radiation. The radiation energy deposited in organisms may result in biological changes or radiation damage at either the macro- or micro-level. The in-depth study of the biological effects of radiation is a key part of evaluating the radiation safety of advanced nuclear systems as well as the foundation

Table 5.4 Inhalation form

Inhalation type	Absorbed proportion (%)	Inhalation time
Fast	100	10 min
Medium	10	10 min
	90	140 d
Slow	0.1	10 min
	99.9	1000 d

for radiation protection. This section introduces the biological effects of radiation induced by neutrons, tritium, fission products, and activation products.

5.3.1 Basic Principles

Ionizing radiation can cause various kinds of radiation damage at different levels such as molecules, cells, organs, tissues, and organisms, which are mainly classified into two levels, micro- and macro-level.

1. Micro-level

The micro-level biological effects of radiation refer to radiation damage at the molecular level, such as genes and chromosomes. DNA, a carrier of genetic information, is the main target of radiation damage that can induce a series of physiological and chemical changes in the cell. There are two types of interaction between ionizing radiation and DNA. The first is direct action which refers to the direct interaction between radiation and DNA that may result in rupturing of the chemical bonds of DNA, changes in the molecular structure and loss of biological activity due to ionization and excitation. The other is indirect action, which refers to the interaction between radiation and water molecules in the cell, producing reactive oxygen species (ROS). DNA damage may be caused by ROS. The two types of radiation actions may occur simultaneously on DNA, as shown in Fig. 5.11.

Cells have the ability to self-repair. If the damaged DNA is repaired, the cell will return to normal. However, when DNA repair fails, cells cannot divide normally, leading to distortion of the cells or their daughter cells, and even cell death. Extensive damage to cells can lead to organ or tissue dysfunction, resulting in macro-level biological effects.

2. Macro-level

The biological effects of radiation at the macro-level refer to physiological and functional changes in tissues and organs that ultimately result in radiation damage to living organisms. According to the relationships between the severity or probability and exposure, the biological effects of radiation can be classified into deterministic effects and stochastic effects.

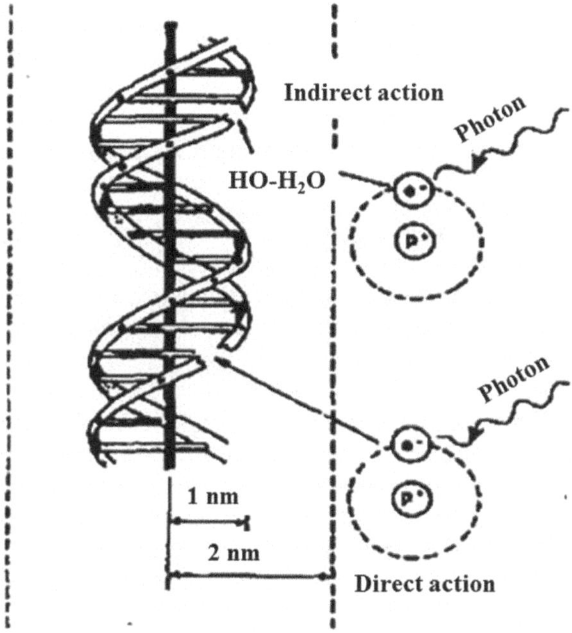

Fig. 5.11 Direct and indirect actions of radiation on DNA

Deterministic effects occur under high radiation doses. The severity of an effect for which a threshold is believed to exist increases with the dose. Cataract formation and permanent hair loss caused by radiation are examples of deterministic effects, and the threshold doses for these two symptoms in a single X-ray or γ-ray exposure are 5 and 7 Gy, respectively. Deterministic effects occur because radiation causes numerous cell deaths and new cells cannot proliferate to compensate, which result in functional obstacles in organs and tissue and ultimately result in visible radiation syndrome of organs and tissues. Intensive studies have been conducted on deterministic effects. The dose threshold values for the deterministic effects for various organs or tissues exposed to different types of radiation with different doses have been reported in ICRP publications.

Rather than the damage severity, the occurrence probability of a biological effect is believed to be a linear function of the dose without a threshold. Genetic effects and cancer are examples of stochastic effects. These effects are caused by radiation energy deposited in key regions of certain cells, which causes gene mutation and leads to the occurrence of cancer or passes on to the offsprings of the exposed individual.

The relationships between the probability of stochastic effects and the dose remain largely unknown. Stochastic effects induced by low-dose radiation are currently a popular research topic, with the main focus on carcinogenesis and hormetic effects. There is no systematic research on the biological effects on human health caused by low-dose radiation. The biological effects of low-dose radiation are mainly derived from extrapolation of the linear no-threshold model. With the rapid development of molecular biology, it is possible to reveal the mechanisms of the biological effects of low-dose radiation at the micro-level.

5.3.2 *Biological Effects of Neutrons*

Neutrons, which are one of the main sources of ionizing radiation in nuclear systems, have the characteristics of electrical neutrality and strong penetrability. They mainly cause external exposure to the human body. This section describes the effects of neutrons, including deterministic and stochastic effects.

1. Deterministic effects of neutrons

The deterministic effect is a kind of "threshold" effect. When the dose that an organ or tissue receives is greater than the threshold, the effect will occur, and the severity is related to the dose, which means that the greater the dose is, the more serious the consequences are. Unlike other types of radiation, the biological effects of neutrons with different energies are different for the same absorbed dose. The threshold values of different tissues or organs to radiation are different. The estimated values of the thresholds for the deterministic effects of several sensitive tissues are shown in Table 5.5.

2. Stochastic effects of neutrons

The carcinogenic effect and hormetic effect induced by low-dose neutron are discussed below.

With in-depth studies of the carcinogenic effects of low-dose radiation at the molecular, cellular, and animal levels, people have become increasingly aware of the important role of DNA damage response and repair, chromosome aberration, and genomic instability in carcinogenic effects. The DNA damage induced by neutrons

Table 5.5 Estimated threshold values for deterministic effects

Organs or tissue	Deterministic effect	Dose threshold/Sv
Testicular	Sperm reduction	0.15
	Permanent infertility	3.5
Ovarian	Permanent sterilization	2.5–6.0
Eye lens	Lens opacity	0.5–2.0
	Visual impairment	5.0
Bone marrow	Blood cells temporarily reduced	0.5
	Fatal reproductive failure	1.5

is mainly DNA double-strand break (DSB) rather than single-strand break (SSB). Neutrons can induce hematopoietic chromosome aberrations in mice and lead to genomic instability. Additionally, they can induce chromosomal aberrations in the peripheral lymphocytes of normal adult males, and the number of chromosomal aberrations increases with decreasing neutron energy. Rossouw studied the relationship between radiation dose and micronucleus formation in human peripheral blood lymphocytes and found that the incidence of micronucleus formation can be regarded as a sensitive indicator of early-stage cancer under neutron irradiation of 0.5 Gy. Other studies found that a neutron dose of 0.08 Gy can significantly induce solid tumors in mice [13, 14]. These results demonstrated that low-dose neutron irradiation can cause DSB and can further lead to chromosomal aberrations, genomic instability, and eventual cancer induction. Figure 5.12 shows solid tumors of female rats induced by low-dose neutron irradiation.

In traditional radiation biology, low-dose radiation can induce DNA damage and gene mutations and increase the risk of cancer. Multiple studies of the biological effects of low-dose radiation have cast light on new low-dose radiation phenomena. For example, low-dose radiation enables cells to fight high-dose radiation. The hormetic effects of low-dose γ-rays and X-rays have been confirmed. Irradiation of animals with low-dose γ-rays or X-rays activates the hormetic effects of the immune system. This indicates that low-dose radiation can activate repair systems, enhance immunity, and alleviate radiation-induced damage to the liver, lungs, and other vital organs. An extensive body of research has shown that the hormetic effects of low-dose radiation exist widely at the molecular, cellular, and individual levels.

The characteristics of the interaction between neutrons and tissues are quite different from those of γ-rays and X-rays. Neutrons mainly interact with nuclei in biological tissues, whereas γ-rays and X-rays mainly interact with extranuclear electrons of atoms. It has been reported that after X-ray or neutron irradiation of hamster

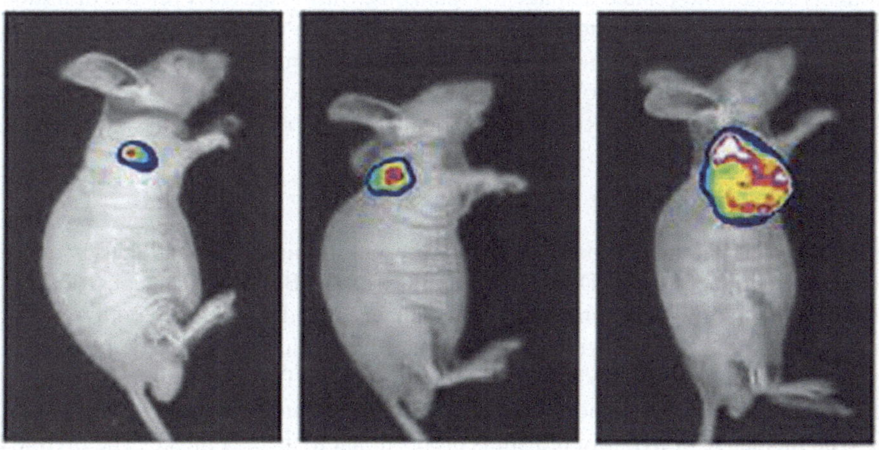

Fig. 5.12 Solid tumors of female rats induced by low-dose neutron irradiation

cells, X-rays below 1 Gy induce hormetic effects in hamster cells, but neutrons do not. However, this does not mean that neutrons cannot induce hormetic effects. More studies are needed to investigate the biological effects of neutrons by applying different end points at various doses. However, cellular experiments carried out with the High Intensity D-T Fusion Neutron Generator (HINEG) have proven that the neutron damage to cells was greater than the gamma damage to cells and that the repairing of DNA double chain fractures induced by neutron damage was significantly slower than that induced by gamma damage. Experiments with nematodes using the HINEG showed low-dose hypersensitivity of neutrons to individual nematodes, demonstrating that low doses of neutrons produce radiation stimulation effects.

The ionizing radiation emitted from nuclear fusion is a potential health risk to humans, including scientists who are currently performing nuclear fusion experiments and the employees at future fusion systems. The DNA damage and its regulation in normal human fibroblasts exposed to fusion radiation were investigated using the High Intensity D-T Fusion Neutron Generator [15]. Heme oxygenase 1 (HO-1), which has been reported to induce anti-inflammatory activity, was upregulated in the irradiated cells. The investigation further clarified that the ataxia telangiectasia-mutated DNA damage response was activated and that it stimulated the phosphorylation of p38 mitogen-activated protein kinase, which was responsible for the upregulation of cOX-2 and HO-1. These results provide novel information about fusion radiation-induced biological effects and potential targets for decreasing the associated health risks.

5.3.3 Biological Effects of Fission Products

Many kinds of fission products are produced in nuclear fission reactions. These products cause time-related damage to organisms. During the early release of fission products, damage is caused to the thyroid by radioactive iodine and to the intestines and stomach by other hard-to-absorb radionuclides. During the later release of fission products, damage is caused to bone tissue by long half-life radionuclides such as ^{90}Sr, ^{137}Cs, ^{147}Pm, ^{144}Ce, and ^{106}Ru. The biological effects of radioactive fission products are described using ^{131}I and ^{90}Sr as examples.

1. Biological effects of ^{131}I

Iodine entering the human body is mainly present in the thyroid and causes radiation damage to the thyroid. Doses of several Gy of 10^5 Bq iodine can cause thyroid damage. The thyroid function was initially enhanced, and sustained hypothyroidism then occurred in some individuals. When the thyroid dose was 30 Gy, permanent hypothyroidism occurred. When the thyroid dose was 100 Gy, the thyroid suffered serious damage, such as glandular follicle atrophy and interstitial and vascular fibrosis. The incidence of hypothyroidism is proportional to the ^{131}I intake. The thyroids of young people are more sensitive to ^{131}I than those of old people.

2. Biological effects of ^{90}Sr

In the human body, most of the absorbed ^{90}Sr is deposited in bones; therefore, it mainly causes radiation damage to bone marrow hematopoietic tissue and bone tissue. ^{90}Sr can be immersed in the formation of salt, which can significantly inhibit the calcification process of bone tissue and induce spontaneous fractures and diffuse osteoporosis. When ^{90}Sr enters the bone tissue, it is mainly deposited in the inorganic substance of the long bones near the bone marrow cavity. Because ^{90}Y (64 h), the daughter of ^{90}Sr, has a short half-life and a large decay energy (2.27 MeV), it will cause strong irradiation to the endothelium, bone marrow, and blood vessels.

The stochastic effect of ^{90}Sr is mainly osteosarcoma, followed by leukemia. Experiments have indicated that the incidence of osteosarcoma induced by ^{90}Sr is related to factors such as the accumulated dose and the animal species bone type. Within a certain dose range, the incidence of osteosarcoma is positively correlated with the cumulative dose. ^{90}Sr-induced osteosarcoma mainly occurs at both ends of long bones.

5.3.4 Biological Effects of Neutron Activation Products

In nuclear systems, the neutron activation products include the activation products of the structural materials, the coolant, and the covering gas. Of these, activated products of structural materials are the most common, including ^{52}V, ^{51}Cr, ^{56}Mn, ^{55}Fe, ^{59}Fe, ^{57}Co, ^{58}Co, and ^{60}Co. The radioisotopes of iron and cobalt are the main contributors to the radioactivity and dose. Radioactive iron causes damage to the liver, spleen, and bone marrow, whereas radioactive cobalt causes damage to the hematopoietic system. The biological effects of the neutron activation products ^{59}Fe and ^{60}Co are described in the following section.

1. Biological effects of ^{59}Fe

Iron is an indispensable element in the human body. It mainly exists in the red blood cells, liver, spleen, and bone marrow. Therefore, ^{59}Fe mainly causes radiation damage to the liver, spleen, and bone marrow. A study was conducted on $^{59}FeCl_3$ with a daily intake dose of 37 kBq/kg over a long period of time. The early results showed that the numbers of red blood cells and hemoglobin increased and that the number of reticular red cells also increased. After 8–9 months, changes in erythrocyte quality and an increase in erythrocyte volume were observed. After 12 months, granulocyte production was inhibited. A study was conducted to increase the daily oral dose of 370 kBq/kg over a long period of time. After 1 month, the number of red blood cells decreased, which increased the hemoglobin content, and there were also qualitative changes. After 2 months, the number of lymphocytes decreased. After 12 months, granulocytes were inhibited.

2. Biological effects of ^{60}Co

When radioactive cobalt enters the body, it is involved in the hematopoietic process. Therefore, serious damage will be caused in the hematopoietic system, which may gradually develop as erythrocyte and lymphocyte dysplasia. A study of rabbits taking 46.25 kBq/kg per day ^{60}Co was conducted. After 2 weeks, there was an increase in the number of red blood cells, which was followed by a significant decrease and a tendency to develop aplastic anemia. After 1 month, the number of lymphocytes began to decrease. Due to the long-term effects of ^{60}Co, the collective resistance is low, and humans often die of pneumonia and other purulent infections. Late in life, the blood sugar levels increase significantly, which may be related to the destruction of the liver's enzyme system.

5.3.5 Biological Effects of Tritium

Tritium is one of the main radiation source terms in a fusion system. It mainly produces internal radiation in the form of tritium water (HTO) and organic tritium (OBT) synthesized in the body. In advanced research of the biological effects of tritium, more attention should be paid to the metabolic characteristics of HTO and OBT in organisms.

OBT is formed by tritium bound to organic compounds, such as proteins, lipids, and carbohydrates, in which hydrogen is replaced by tritium. After entering the human body, HTO rapidly mixes with water. During the metabolic process, some of the HTO is converted to OBT. The amount of OBT incorporated into the tissue is determined by the organism's metabolism and the chemical forms of the OBT. Therefore, the retention time of OBT is longer than that of HTO in vivo, and the biological half-life of OBT is at least several times longer than that of HTO. Thus, continuous irradiation occurs in the body [16].

A report published by the United Nations Scientific Committee on the Effects of Atomic Radiation (UNSCEAR) in 2000 showed that the hazardous impact of OBT is stronger than that of HTO due to the intake and extraction characteristics of OBT [17]. The research was conducted to estimate the proportions of tritium and OBT for workers at a nuclear power plant and residents of a nearby locality. The average OBT was 4.02% of the total tritium in the bodies of the workers, whereas OBT contributed 12.98% of the total dose. A study was performed to determine the biological effects of HTO and OBT in tissues and organs. It indicated that the dose contribution of OBT is nine times higher than that of HTO [18]. OBT and HTO have different accumulation patterns and metabolic routes in the human body; thus, the carcinogenesis of the compounds is different. Therefore, it is essential to perform additional studies to reveal the biological effects of OBT and HTO and their respective mechanisms using advanced biotechnology.

Studies of the biological effects of neutrons, fission products, and neutron activation products are the basis for analyzing the impacts of nuclear systems on workers and the public. They are also the key to biological safety evaluations of nuclear systems. In particular, the damage characteristics of fission products to humans over different periods will be the focus of studies of the biological effects of radiation in future nuclear systems.

References

1. Gianni P (2006). Nuclear safety. Elsevier Butterworth-Heinemann
2. Wu YC (2017). Fusion neutronics. Springer Nature Singapore Pte. Ltd.
3. Bailly BP, Laguionie P, Boust D et al (2012) Estimation of marine source-term following Fukushima Dai-ichi accident. J Environ Radioact 114(12):2–9
4. Nie BJ, Ni MY, Jiang JQ et al (2015) Dynamic evaluation of environmental impact due to tritium accidental release from the fusion reactor. J Environ Radioact 148:137–140
5. Liu P, Dang TQ, He P et al (2016) A multi-medium coupling dispersion model of radionuclide in a nuclear accident. J Radiat Res Radiat Process 34(1):010701 (in Chinese)
6. International Commission on Radiological Protection (2007) ICRP Publication 103: the 2007 recommendation of the international commission on radiological protection
7. Dan GC (2011) Handbook of nuclear engineering. In: Shultis J, Faw R (eds) Radiation shielding and radiological protection, pp 313–1448
8. International Commission on Radiological Protection (2010) ICRP Publication 116: conversion coefficients for radiological protection quantities for external radiation exposures
9. Chen Y, Fischer U (2002) Rigorous MCNP based shutdown dose rate calculations: computational scheme, verification calculations and application to ITER. Fusion Eng Des 63:107–114
10. Villari R, Fischer U, Moro F et al (2014) Shutdown dose rate assessment with the advanced D1S method: development, applications and validation. Fusion Eng Des 89(9):2083–2087
11. International Commission on Radiological Protection (2009) ICRP Publication 110: adult reference computational phantoms
12. Wang W, Cheng MY, Long PC et al (2015) Specific absorbed fractions of electrons and photons for Rad-Human phantom using Monte Carlo method. Chin Phys C 39(07):078203 (in Chinese)
13. Morgan WF (2003) Non-targeted and delayed effects of exposure to ionizing radiation: II. Radiation-induced genomic instability and bystander effects in vivo. Radiat Res 159(5):581–596
14. Rossouw MS, Meehan KA (2005) Micronucleus formation in lymphocytes after exposure to low-dose gamma and neutron radiation. Med Technol SA 19(2):11–15
15. Wang XY, Liu H, Jiang X et al (2018) Cyclooxygenase-2-mediated upregulation of heme oxygenase 1 mitigates the toxicity of deuterium-tritium fusion radiation. Int J Mol Med 42(4):1945–1954
16. UNSCEAR (2016) UNSCEAR 2016 report: sources, effects and risks of ionizing radiation. In: Biological effects of selected internal emitters-tritium
17. UNSCEAR (2000) UNSCEAR 2000 report: sources and effects of ionizing radiation
18. Pietrzak FZ, Radwan J, Judeka L (1978) Tritium in rabbits after ingestion of freeze-dried tritiated food and tritiated water. Radiat Res 76:420–428

Chapter 6
Material Neutron Irradiation Damage

Material neutron irradiation damage refers to the microstructural change and performance degradation of material induced by neutron irradiation [1]. These changes caused by collision displacement, transmutation, and ionization effects can significantly affect the macroscopic physical and mechanical properties of a material. In advanced nuclear systems, the service environment of core components is quite different from that of traditional nuclear systems, and the corresponding materials need to withstand much higher temperatures and higher flux neutron irradiation. Taking the fusion system as an example, 14.06 MeV fusion neutrons will generate more serious irradiation damage than the thermal neutron of the traditional nuclear system. Meanwhile, transmutation reactions, such as (n, p) and (n, α), are more easily generated by high-energy neutrons. In addition, the transmutation gases will interact with the irradiation defects and aggravate the irradiation damage of materials. The degradation of material properties induced by neutron irradiation poses a serious threat to the structural integrity of components, which is a key issue in the development of an advanced nuclear system. Thus, irradiation damage research of materials under harsh environment, such as high operation temperature, high-energy neutron and high-flux neutron irradiation, etc., is an urgent need.

This chapter focuses on the neutron irradiation damage of materials in advanced nuclear systems, including the mechanism of neutron irradiation damage, the influence of neutron irradiation on the microstructure and macroscopic performance of materials, and relevant research methods.

6.1 Mechanisms of Neutron Irradiation Damage

In advanced nuclear systems, the average neutron energy and the total neutron flux are much higher than those of traditional nuclear systems. High-energy neutrons enter the material and interact with the atoms, which will induce changes in the material composition, structure, etc. These changes are the fundamental causes of

Y. Wu, *Neutronics of Advanced Nuclear Systems*,
https://doi.org/10.1007/978-981-13-6520-1_6

irradiation damage in materials. In this section, interactions between neutrons and material atoms and the fundamental principle of irradiation damage will be presented.

6.1.1 Interactions Between Neutron and Materials

Elastic or inelastic collision will occur due to the interactions between neutrons and material atoms. When an inelastic collision occurs, the neutron may be trapped by atomic nuclei. These nuclei may undergo transmutation or electron excitation and ionization. At the same time, the related material atoms will gain some of the kinetic energy of the neutron. In elastic collision, material atoms can obtain part of the kinetic energy of the neutron without changing the atomic type. Generally, the material atoms bombarded by neutrons will leave the lattice position with high kinetic energy, collide with the surrounding atoms, and then trigger a series of collision displacements. The effect induced by the interaction between neutrons and material atoms is the basic reason for the change of the structure and properties of materials.

1. Transmutation reaction

The transmutation reaction is the process that leads to changes in neutron number or proton number in a nucleus and forms another nucleus during the reaction between a neutron and a material nucleus. Meanwhile, hydrogen and helium produced by (n, p) and (n, α) reactions are the important causes of irradiation damage in materials. High-energy neutrons in advanced nuclear systems facilitate the occurrence of transmutation reactions and thus produce much more transmutation gases. The radius of hydrogen atoms is small, enabling them to travel easily to the material surface and escape. Therefore, the irradiation damage caused by hydrogen gases is relatively small. Helium, an inactive element with a larger atomic radius than hydrogen, easily aggregates to form gas bubbles, which can stay in materials for long periods of time and finally aggravate material irradiation damage by promoting material swelling and embrittlement. The production rate of transmutation helium, i.e., He/DPA in appm/dpa, is generally used to evaluate the extent of irradiation damage generated in various materials by different irradiation sources.

In addition, solid transmutation products can also be produced by neutron irradiation. For example, the transmutation of tungsten produces rhenium, tantalum, etc., leading to a change in the material composition that may affect the material properties.

2. Ionization effect

The ionization effect is a phenomenon wherein electrons in atomic orbitals obtain energy transferred from neutrons colliding with material atoms and escape from their orbits. Because the bonding characteristics of metals allow lost electrons in the outer atomic orbitals to be quickly replaced, the ionization effect has little effect on the properties of metallic materials. However, ionization can cleave molecular bonds in polymers and thus degrade their insulating properties. In addition, the ionization effect can also cause the temporary failure of or permanent damage to semiconductors in electronic devices.

3. Collision displacement

Collision displacement is a phenomenon in which energy is transferred from incident neutrons to target atoms of a material, and the target atoms are displaced from their original crystal lattice sites and change the distribution of material atoms. This event consists of the following distinct/sequential processes: (1) the interaction of an incident neutron with a lattice atom; (2) the transfer of kinetic energy from the neutron to the lattice atom, forming a primary knock-on atom (PKA); (3) the displacement of the atom from its lattice site; (4) the passage of the displaced atom through the lattice and the accompanying creation of additional knock-on atoms; (5) the production of a displacement cascade (collection of point defects created by the PKA). During the collision between a neutron and atoms, an atom can be displaced from its lattice site by an incident neutron and leave a vacant site, and if the displaced atom eventually stops in a location between lattice sites, it becomes an interstitial atom. The combination of an interstitial atom and its corresponding vacancy is known as a Frenkel pair. The formation of stable Frenkel pairs indicates the end of a collision displacement event. Notably, this entire process of damage lasts only approximately 10^{-11} s. Collision displacement results in the formation of point defects (vacancies and interstitial atoms) and the aggregation of these defects, which affects the microstructure and properties of a material. The detailed collision displacement process and its calculation method will be addressed in Sect. 6.1.2.

6.1.2 Displacement Damage

Collision displacement under neutron irradiation causes the formation of point defects. Isolated point defects can migrate anywhere in a material, resulting in changes in its microstructure. These changes will further cause degradation in material performance. Meanwhile, the interaction between displacement defects and transmutation gases, also known as the transmutation gas effect, can intensify the irradiation damage in materials. This section concentrates on the collision processes and calculation methods of the displacement damage induced by neutron irradiation. The effect of transmutation gases will be addressed in Sects. 6.2 and 6.3.

1. Primary collision

During the incident process of a neutron, material atoms are impacted and the energy transfers from the neutrons to the atoms. If enough energy is transferred, the atom can escape from its lattice site and form a primary displacement atom in a process known as primary collision. The physical process in a primary collision can be explained from the perspective of classical mechanics by assuming that the crystal effect and interatomic potential can be ignored. When a neutron with a mass of M_{n} and energy of E_{n} interacts with an atom with a mass of M_{T}, the energy transferred from the neutron to the material atom, which is represented by E_T, can be determined according to the equations of energy conservation and momentum conservation in an elastic collision. In this case, although the structure of the nucleus does not change, part of the kinetic energy of the neutron is transferred to the material atom. The expression of energy transferred during the collision between an incident neutron and an atom is

$$E_T = \frac{2M_n M_T}{(M_n + M_T)^2} E_n(1 - \cos\theta) = \frac{1}{2}\mu E_n(1 - \cos\theta) \quad (6.1)$$

where $\mu = 4M_n M_T/(M_n + M_T)^2$. μ is called the neutron energy loss coefficient or energy absorption coefficient of the material atom and is also known as the quality factor, and θ is the scattering angle in the center of the mass coordinate system.

The energy transferred to the atom, E_T, is in the range from 0 to μE_n because of the randomness of the collision between the incident neutron and the target atom (different scattering angle θ). The mean value of the transferred energy for all scattering angles is roughly equal to half of the maximum value. If the transferred energy is too small, then the target atom will not leave its site because of the constraint of the other atoms in the crystal, and the transferred energy will be released in the form of heat. If the transferred energy is large enough, Frenkel pairs will form. The minimum kinetic energy required to produce a stable Frenkel pair is usually defined as the threshold displacement energy, which is denoted as E_d.

The results from calculations and experimental observations revealed that E_d varies with crystallographic orientation. Figure 6.1 shows the variation in the threshold energy of body-centered cubic (BCC) iron with different crystal orientations [2]. The E_d of BCC iron along crystallographic orientation [111] is larger than that of other orientations, and the lowest is along the [100] orientation. E_d can be used to calculate the lowest energy of an incident neutron that leads to lattice atom displacement, i.e., the incident neutron threshold energy. Taking iron as an example, the average E_d is 25 eV, which approximately provides the incident neutron threshold energy of 325 eV according to Eq. (6.1). When iron atoms are impacted by neutrons, primary displacement atoms with high-energy are produced and will continue to collide with other surrounding atoms, causing cascade collisions.

2. Cascade collision

Similar to incident neutrons, the energy of a PKA will transfer to any atom along its path. If the energy of a PKA is larger than the E_d of surrounding atoms, a series of collision will occur in a process called cascade collision. The duration of a cascade collision is dependent on the PKA energy and can last approximately 0.1 ~ 0.3 ps, which is shorter than the period of atomic vibration. The process is repeated unless the transferred energy is lower than E_d.

A cascade collision will lead to the formation of point defects and defect clusters or generate a displacement spike phase transition. Cascade collision induces a severe disturbance of lattice atoms in the crystal defect distribution region and thus forms a low-density and high-temperature melt-droplet-like displacement spike with a shock front. Subsequently, the displacement spike relaxes, and the interstitial atoms escape.

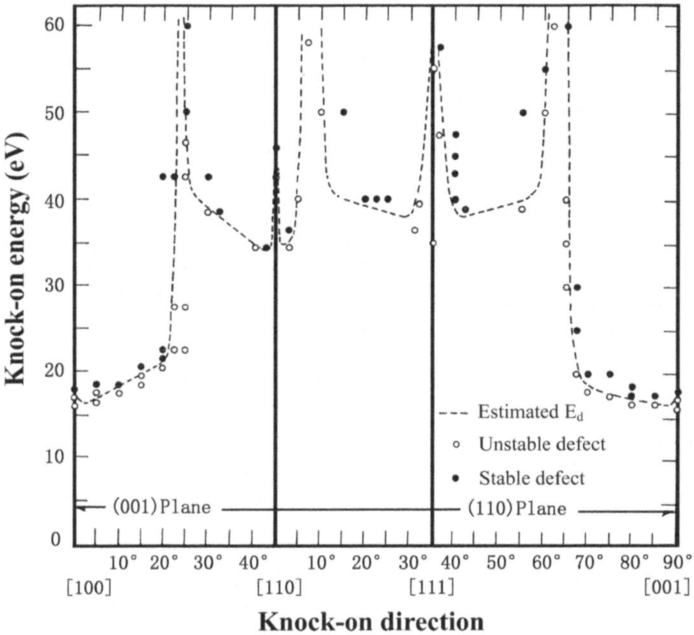

Fig. 6.1 The dependence of the threshold energy of BCC iron on crystallographic orientation [2]

The displacement spike process behaves similarly to a droplet supercooled from a high temperature, leading to the formation of an atom depletion region, point defects or defect clusters. Under high-energy neutron irradiation, a liquid-like displacement spike may form and be cooled rapidly, resulting in the transformation of the ordered crystal structure to a disordered or amorphous phase in a process known as the phase transformation of displacement spikes. The disordered or amorphous phase is preserved by local quenching and expands as the neutron fluence increases until the whole material becomes disordered or amorphous.

3. Displacement damage calculation

The main parameters involved in displacement damage are damage dose, damage rate, energy deposition, and transmutation gas generation rate. The calculation methods of these parameters are presented in this section.

(1) Damage Dose

Damage dose is usually quantified by the average displacement of each atom during the irradiation period, i.e., the displacements per atom (DPA), and the unit is dpa.

The energy distribution function of PKA can be expressed as

$$n\left(E_{\mathrm{p}}\right) = \int_{0}^{T_{\mathrm{irrad}}} \int_{E > E_{\mathrm{d}}/\mu}^{\infty} \phi(E, t)\sigma\left(E, E_{\mathrm{p}}\right) dE\mathrm{d}t \tag{6.2}$$

where T_{irrad} is the irradiation time in units of s, E is the incident energy in units of eV, $\phi(E, t)$ is the neutron flux in units of $cm^{-2} \cdot s^{-1}$), E_p is the energy of PKA in units of eV, E_d is the threshold energy of atom displacements in units of eV, μ is the quality factor, and $\sigma(E, E_p)$ is the energy transfer cross section (i.e., the probability of transferring E_p to the PKA from an incident neutron with an energy of E) in units of barn.

Assuming that the number of displacement atoms generated by each PKA during the cascade collision is $\nu(E_p)$, i.e., the impact damage function for a certain flux of neutrons injected per unit area surface, the number of displacement atoms N_d per unit volume near the surface can be expressed as

$$N_d = \int_{E_d}^{E_{p,max}} n(E_p)\nu(E_p)\,dE_p \tag{6.3}$$

In the study of irradiation damage, N_d/N is generally used to characterize the extent of irradiation damage, i.e., *DPA*

$$DPA = \int_0^T \int_0^\infty \phi(E,t)\sigma_{DPA}(E)\,dE\mathrm{d}t \tag{6.4}$$

where N is the atom density of the irradiated material and $\sigma_{DPA}(E)$ is the displacement damage cross section in units of barn. $\sigma_{DPA}(E)$, the key parameter in *DPA* calculations, can be calculated by the following formula:

$$\sigma_{DPA}(E) = \frac{\kappa}{2E_d}\sigma_{damage}(E) \tag{6.5}$$

where κ is the displacement efficiency (generally 0.8); $\sigma_{damage}(E)$ is the cross section of damage energy in units of eV · barn; and $\sigma_{damage}(E)$ is the probability that an atom is knocked out of its original lattice site due to the impact of a neutron with energy E (described in detail in Chap. 7). The total damage energy E_a can be obtained according to the damage energy cross section

$$E_a = \kappa \int_0^T \int_0^\infty \phi(E,\, t)\sigma_{damage}(E)\,dE\mathrm{d}t \tag{6.6}$$

DPA is not equal to the density of irradiation defects for the following reasons: (1) although an atom can be knocked on many times, only one Frenkel pair form; (2) some Frenkel pairs generated during neutron irradiation will recombine and annihilate.

In addition to the damage dose, the dose rate is another important parameter. The dose rate is the number of displacement atoms generated per unit volume per unit time in material, in unit of dpa/s. The irradiation experiments show that the irradiation damage depends on the damage dose as well as the dose rate. Thus, the dose rate is usually used to compare the irradiation damage generated from different irradiation sources.

(2) Energy Deposition

The energy deposition H in materials can be calculated using the kinetic energy released per unit mass, KERMA factor k

$$H = \sum_i \sum_j \rho_i k_{ij}(E)\phi(E) \tag{6.7}$$

where H is the deposited energy per unit volume of material in units of eV/cm^3; ρ_i is the density of nuclides i in a material in units of n/cm^3; and $k_{ij} \cdot (E)$ is the KERMA factor of nuclide i with respect to nuclear reaction j in units of eV · barn. $k_{ij}(E)$ can be calculated using the following formula:

$$k_{ij}(E) = \sum_i \overline{E}_{ij}(E)\sigma_{ij}(E) \tag{6.8}$$

where $\overline{E}_{ij}$ is the total energy released in nuclear reaction j for nuclide i in units of eV, and $\sigma_{ij}(E)$ is the probability of nuclide i with respect to nuclear reaction j.

6.2 Effects of Neutron Irradiation on Material Microstructures

Under neutron irradiation, there are a lot of point defects, defects clusters formed in material; furthermore, the transmutation reactions can also induce the change in material composition. These radiation defects can diffuse, gather/annihilate, and interact with the transmutation gases, which can cause changes of material microstructure, such as the formation of dislocation loops, stacking fault tetrahedrons, cavities, gas bubbles, and precipitation. Irradiation-induced defects and the transmutation gases are the source of material changes in microstructure and performance. This section will present the effect of neutron irradiation on material microstructure through several typical defects and its evolution mechanisms.

6.2.1 Formation and Evolution of Point Defects

Generation of point defects by neutron irradiation occurs in a very short period that takes approximately 10^{-11} s. Numerous vacancies and interstitial atoms can be generated instantaneously, resulting in a sudden rise in the internal energy of materials and a rapid increase in the lattice configuration entropy, which cause the free energy of the lattice to deviate from the initial low-energy equilibrium state. To return to the equilibrium state, supersaturated point defects may either form stable defect clusters or annihilate at sinks through migration and diffusion. For the remaining isolated

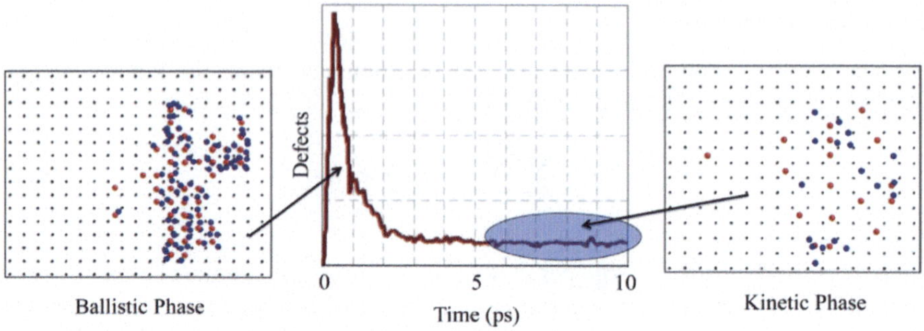

Fig. 6.2 The evolution of irradiation-induced defects in materials over time [3]

point defects, vacancies normally exist as vacancy pairs, whereas interstitials exhibit a dumbbell configuration and are easily aggregated to form interstitial-type stacking faults, which evolve into dislocation loops via dislocation reactions.

Figure 6.2 shows the evolution of point defects in irradiated materials over time [3]. In the initial stage of irradiation, many defects are generated on a timescale of 1 ps in a complex cascade collision process. Subsequently, the energy of the initial energetic particles will disperse into a larger region where no high-energy atoms exist. The local temperature of this region is much higher than that of the surrounding region and ambient temperature. The temperatures in these regions tend to be the same throughout the thermal transmission, and the number of defects in the system also decreases gradually. This phenomenon is considered as a self-healing process. After approximately 1 ns, the system will reach a new thermodynamic equilibrium, after which there is no obvious change in the position of the atoms inside the material. The defects are annealed gradually, and only a small number of defects in the system ultimately remain. Table 6.1 shows the approximate timescale for the production of defects in irradiated materials [4]. The residual defects usually exhibit several specific morphologies, such as dislocation loops, stacking fault tetrahedrons, and cavities, which are the final products of supersaturated irradiation defect clustering. The size, distribution, and morphology of these defects determine the material microstructure, and their formation processes also show the evolution and growth process of the material microstructure.

6.2.2 Dislocation Loops

The clustering of vacancies or interstitial atoms forms circular disks, and then the nearby atoms collapse to form a planar defect called a dislocation loop. Typical dislocation loops are small (3–20 nm) and have a high number density (10^{21}–10^{23} m^{-3}). Figure 6.3 shows the dislocations or dislocation loops in 316LN before and after HFIR neutron irradiation to 3 dpa [5]. At temperature 90 °C, the dislocation loop

Table 6.1 Approximate timescale for the production of defects in irradiated materials

Time (s)	Event	Result
10^{-18}	Energy transfer from the incident neutron	Primary knock-on atoms (PKA)
10^{-13}	Displacements of atoms by the PKA	Displacement cascades
10^{-11}	Energy dissipation, spontaneous recombination, and clustering	Stable Frenkel pairs (self-interstitial atoms (SIA) and vacancies) and defect clusters
$>10^{-8}$	Defect reactions by thermal migration	SIA and vacancy recombination, clustering, trapping, and defect emission

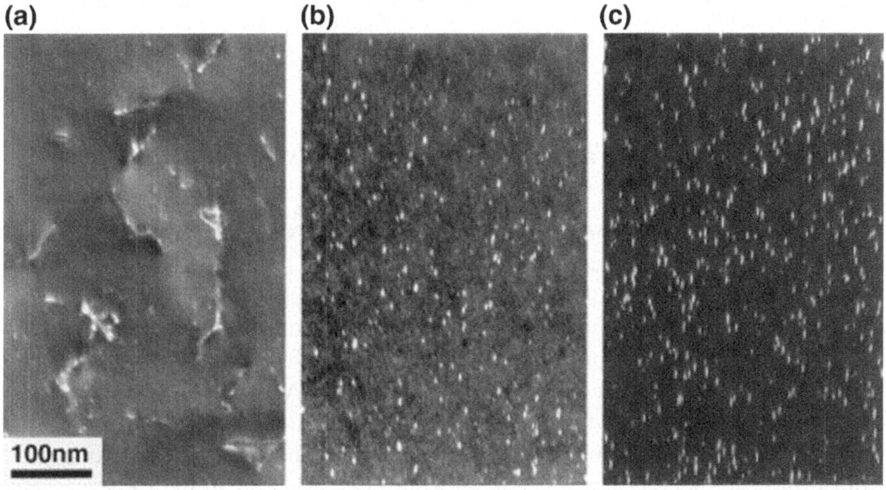

Fig. 6.3 Dislocations or dislocation loops in 316LN before and after HFIR neutron irradiation to 3 dpa [5]. Micrographs were taken with beam direction B close to <1 1 0>. **a** dark field image in unirradiated specimen; **b** and **c** dark field images using streaks arising from faulted loops in the specimens irradiated at 90 and 250 °C

density after irradiation is approximately 3.5×10^{22} m^{-3}, with an average length of 7.6–8.0 nm. During their growth, dislocation loops will interact with other defects, producing new dislocation networks and structures that tend to be stable. These new dislocation structures substantially affect the mechanical properties of a material. In the case of an anisotropic distribution of dislocation loops, permanent deformation such as irradiation creep can be induced.

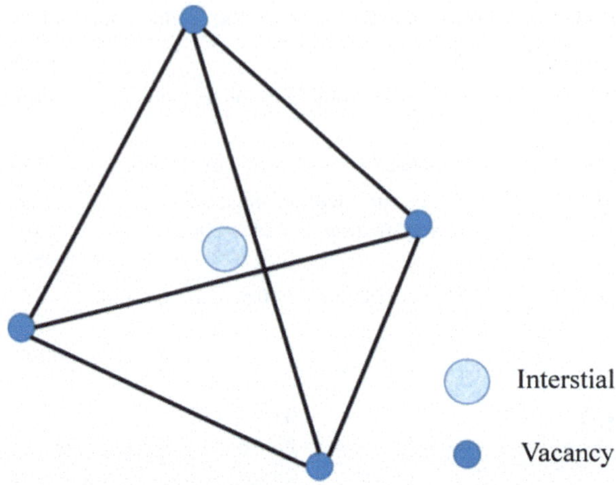

Fig. 6.4 Schematic drawing of a triple-vacancy stacking fault tetrahedron [1]

6.2.3 *Stacking Fault Tetrahedrons and Voids*

The stacking fault tetrahedron is a common crystal defect structure of vacancy clusters in materials with a face-centered cubic (FCC) structure. Figure 6.4 shows a schematic drawing of a triple-vacancy tetrahedron. Three vacancies forming a planar triangle may act as the bottom of a tetrahedron. Atoms in the tetrahedral vertex will descend to a lower position, and a primary tetrahedron lattice defect with four vacancies and an interstitial atom in its center will then be formed. A tetrahedron consisting of four vacancy-type stacking fault rings is called a stacking fault tetrahedron, which has a similar structure to that of a triple-vacancy tetrahedron. In a stacking fault tetrahedron, each face has a piece of stacking fault, whereas in the triple-vacancy tetrahedron, each face is defined by three vacancies. The center of a stacking fault tetrahedron is not a hollow or an atom but a tetrahedral-shaped crystal block. The stacking fault tetrahedron can be generated in many ways, such as quenching, plastic deformation, and irradiation. Typically, the stacking fault tetrahedron is highly stable after being formed. Because of stacking fault energy, the side length of stacking fault tetrahedron is usually shorter than 50 nm. Typically, a stacking fault tetrahedron can be generated in FCC metals with low stacking fault energy because they are generally formed by vacancy-type stacking fault rings.

If the disk of vacancy clusters grows in three dimensions rather than collapsing into dislocation loops, voids will be formed and can be observed in most irradiated materials. The formation of voids can lead to material swelling. Both voids and stacking fault tetrahedrons are secondary defects formed by the clustering of vacancies. Stacking fault tetrahedrons easily form at lower temperatures, whereas voids tend to be formed at higher temperatures.

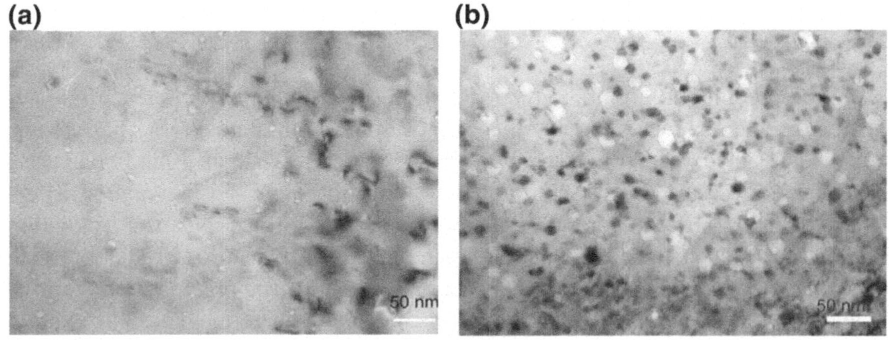

Fig. 6.5 Voids in Fe–18Cr–19Ni after neutron irradiation to 1.7 dpa (**a**) and 20.5 dpa (**b**) [6]

6.2.4 Gas Bubbles and Cavities

High-energy neutron irradiation not only generates defects in materials through displacement damage but also generates transmutation elements, such as hydrogen and helium. These transmutation elements gather to form gas bubbles, which can interact with irradiation defects and affect irradiation damage in materials in a phenomenon called the transmutation gas effect. The production rate of transmutation gases is closely related to the neutron energy. For example, the threshold energies for (n, p) and (n, α) reactions in iron are approximately 1 MeV and 5 MeV, respectively. Therefore, the probability of transmutation reactions in fission systems is quite small because of the lower energy of fission neutrons, whereas these two transmutation reactions are very significant in materials irradiated with neutrons in advanced nuclear systems, especially in fusion systems in which the neutron energy is up to 14.06 MeV. Therefore, the production rates of hydrogen and helium are much higher than those in traditional nuclear systems.

The existence of gas bubbles, which are affected by neutron displacement damage and external stress, enhances the nucleation and growth of voids, leading to the formation of cavities, which contain voids and gas atoms. Figure 6.5 shows the voids features in Fe–18Cr–19Ni after neutron irradiation to 1.7 dpa (a) and 20.5 dpa (b), respectively [6]. The voids in specimens irradiated to 1.7 dpa are small, and not associated with precipitates that exist at rather low density, but they are often associated with dislocations (Fig. 6.5b). As the damage dose increases, the precipitate density also increases. Void–precipitate associations at 20.5 dpa are shown in Fig. 6.5b. Cavity growth causes the irradiation swelling of materials, thus changing the dimensions and density of the structural components and affecting their mechanical properties.

6.2.5 *Irradiation Precipitation*

Irradiation precipitation refers to the irradiation of high-energy particles into materials, causing atoms to deviate significantly from their initial position. These atoms can nucleate and grow at other locations of the material (such as grain boundaries, phase boundaries, surfaces, etc.) through migration, diffusion, and aggregation. Then, a phase of matter different from the material matrix is formed. Neutron irradiation precipitation refers to the precipitation of material caused by neutron irradiation, such as the change in size, density, and distribution of the intrinsic precipitation (carbide in steel, etc.), as well as the new phases precipitated in material. These irradiation precipitates change the microstructure of the material, resulting in the degradation of material properties.

6.3 Effects of Neutron Irradiation on Material Properties

The evolution of neutron irradiation defects may lead to changes in the mechanical and physical properties of materials. This section mainly describes the influence of neutron irradiation on physical properties and mechanical properties.

6.3.1 *Physical Properties*

The physical properties of materials involve many properties, such as density, volume, modulus of elasticity, electrical conductivity, thermal conductivity, magnetic property, and environmental compatibility. For structural materials, the main concern of physical properties is volume change due to neutron irradiation, i.e., irradiation swelling, while for other functional materials, the physical properties concerned differ according to the functional requirements. Here, irradiation swelling of structural materials and the physical properties of several typical fusion functional materials will be introduced.

1. Influence of neutron irradiation on irradiation swelling

As the irradiation dose increases, irradiation swelling increases, which can cause dimensional change in materials and affect the structural stability of components. The nucleation and growth of voids and bubbles occur at an early stage of irradiation. Most vacancies recombine with interstitials or flow into dislocation sinks and have a very low probability to enter into voids and bubbles. In this stage, which is commonly called the incubation period of irradiation swelling, swelling is not prominent, and the growth rate of voids and bubbles is very low.

When the irradiation dose reaches a certain level, the nucleation of voids and bubbles reaches saturation, enabling more vacancies to enter into voids and bubbles and

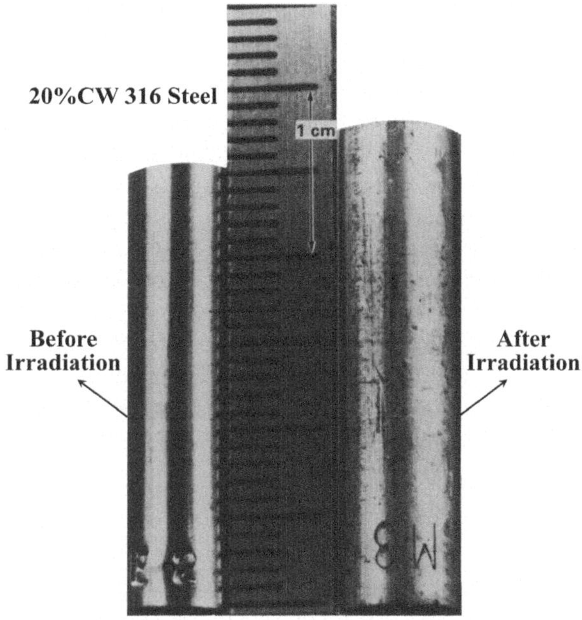

Fig. 6.6 Swelling of austenitic stainless steel following neutron irradiation to 1.5×10^{23} n/cm^2 [7]

rapid growth to occur. Voids and bubbles become large enough to have a remarkable capability to capture vacancies. Meanwhile, the capture of interstitials by dislocations becomes more significant, and irradiation swelling becomes observable. In this stage, i.e., the transition period of irradiation swelling, the swelling rate increases, and then finally stabilized. As the dislocation density increases, the transition period of irradiation swelling can be postponed, which has been confirmed by an irradiation test on stainless steel (as shown in Fig. 6.6).

Irradiation swelling is one of the most significant reasons for material failure induced by high-energy neutron irradiation. The major difference between low-energy and high-energy neutron irradiation is the production rate of transmutation gas (appm/dpa), which indicates the ratio between the concentration of transmutation gas and the irradiation dose. Unlike low-energy neutron irradiation, high-energy neutron irradiation is accompanied by the production of a high concentration of transmutation helium gas, which considerably expands the volume of materials.

In addition, the burnup depth of fuel in advanced nuclear fission systems is high, and irradiation swelling is an important factor hindering its fuel burnup depth. The main reasons are as follows: the total volume of nuclear fuel containing fission products exceeds the volume of pre-fission nuclear fuel (hard swelling) and there are a large number of fissile gas bubbles in fuel. The factors affecting the swelling are the material structure, impurity content, fuel burnup rate and depth, stress state, thermal vibration, and the change of phase during irradiation (phase change, recrystallization, etc.).

2. Influence of neutron irradiation on the electronic properties of electronic devices

Neutron irradiation induces damage in semiconductors, which may cause material performance degradation or even malfunction. Collision displacements break the crystal lattice structure and periodic potential of the material, which produces new electron energy levels in the bandgap and changes the electronic properties of the semiconductor. This property change is mainly illustrated by the decrease in the minority carrier density and in the mobility of the carriers, as well as the carrier removal effect; these effects in turn lead to decrease in the circuit parameters.

For semiconductor materials in electronic devices, transient ionization effects can cause signal and memory errors or single-event latch-ups, which can affect the normal operation of the circuit or even break or burn out the device. In addition, the abundant free electrons and ions generated by ionization accumulate in the oxide layer of the semiconductor surface, resulting in a decrease in the gain of the bipolar transistor, an increase in the reverse leakage current, and an induction of permanent damage in the semiconductor.

3. Influence of neutron irradiation on the thermal properties of tritium breeder materials

The tritium breeders for fusion reactors can be categorized into liquid breeders and solid breeders. While liquid breeders cannot be affected by neutron irradiation, the microstructures and chemical compositions of solid breeders (ceramics) can be changed under neutron irradiation. The most commonly used solid breeder materials include LiO_2, $LiAlO_2$, Li_2TiO_3, Li_2ZrO_3, and Li_4SiO_4. Because an appropriate fusion neutron source is not available currently, the irradiation performance of solid-state breeding materials is evaluated using fission neutron irradiation. Researchers in the EU have conducted irradiation tests on Li-based ceramic breeders in a high-flux reactor (HFR); these tests demonstrated that the content of Li_2SiO_3 in Li_4SiO_4 increases continuously as Li is consumed. Although small fragments were produced from Li_4SiO_4 pebbles after irradiation, the fracture toughness of these pebbles was still high, and the density was almost unchanged, as no new closed pores formed during irradiation. Slight grain growth of Li_4SiO_4 occurred during irradiation, which may affect the mechanical properties of the material. The effects of neutron irradiation on the thermal conductivity of Li_2O are dependent on temperature. The thermal conductivity decreased dramatically below 300 °C, whereas no significant reduction was observed above 300 °C.

4. Influence of irradiation on the superconductivity of superconductors

Magnetic confined fusion systems commonly use superconductors to produce high magnetic fields. Under fusion neutron irradiation, transmutation reactions may change the material composition of these superconductors, and irradiation damage may cause microstructural changes, leading to changes in critical temperature (T_C) and critical current (J_C) and potentially resulting in a loss of superconductivity. Neutron irradiation tests on a YBaCuO superconductor revealed that irradiation slightly increased its T_C but dramatically reduced J_C, causing degradation of its superconductivity.

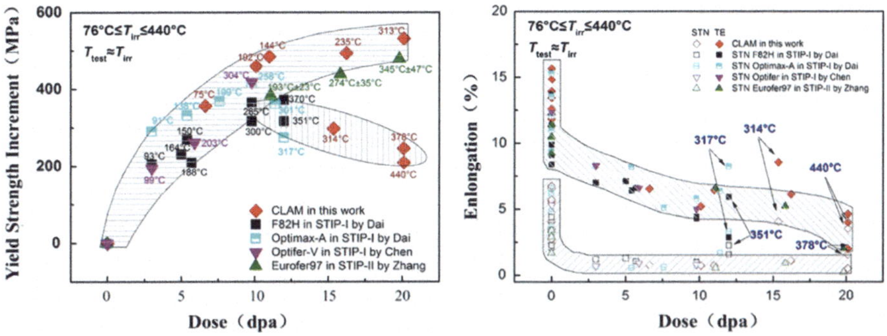

Fig. 6.7 Neutron irradiation hardening and embrittlement of RAFM steel [8]

6.3.2 Mechanical Properties

Microstructure evolutions induced by neutron irradiation may cause an obvious change in material mechanical properties, including strength, plasticity/toughness, creep, and fatigue. The changes in mechanical properties induced by neutron irradiation are of considerable importance for metal structural materials.

1. Irradiation hardening and embrittlement

Generally, the strength of metals increases after neutron irradiation because of irradiation hardening, whereas the elongation decreases and the ductile–brittle transition temperature (DBTT) shifts, leading to the ductility decreasing and causing embrittlement. Because of the formation of irradiation defects, such as dislocation loops, precipitates, and cavities, dislocation gliding is hindered, resulting in an increase in strength and a reduction in elongation. Irradiation hardening and embrittlement are affected by the irradiation dose, temperature, stress, and other environmental factors. Figure 6.7 shows the changes in yield strength and uniform elongation with irradiation dose of RAFMs after neutron irradiation [8]. Irradiation hardening and embrittlement features can be found clearly. In addition, the transmutation helium gas produced by fusion neutron irradiation may accumulate at grain boundaries, leading to the generation of high-temperature helium embrittlement.

2. Irradiation creep

Under a constant load, a material undergoes sustained deformation, and the deformation rate gradually increases over time. This phenomenon, which is called creep, can lead to the creep fracture of a material. The creep process is related to dislocation gliding, atom diffusion, secondary phase precipitation, grain boundary migration, and other factors. Neutron irradiation may dramatically change the microstructure and creep properties of a material.

Irradiation creep includes two types: irradiation-enhanced creep and irradiation-induced creep. For irradiation-enhanced creep, irradiation defects promote disloca-

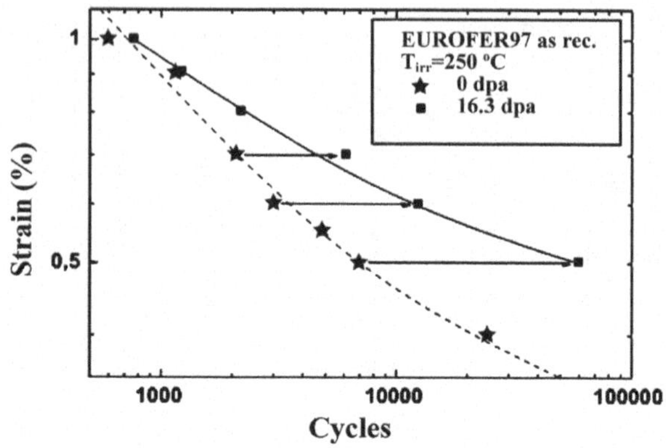

Fig. 6.8 Influence of neutron irradiation on the fatigue life of Eurofer97 [9]

tion gliding, resulting in material deformation. For irradiation-induced creep, dislocations absorb irradiation defects, leading to dislocation gliding and material deformation. Under high-temperature and low-stress conditions, the mobility of defects in materials is large enough that irradiation-enhanced creep plays a dominant role. By contrast, irradiation-induced creep is dominant at low temperature and high stress.

3. Irradiation fatigue

When a periodic load is applied to metal materials under irradiation, cracks initiate and propagate, finally leading to material fracture in a phenomenon known as irradiation fatigue. While the irradiation creep rate is closely related to the production rate of defects during irradiation, irradiation fatigue is dependent on microstructural changes during irradiation. Experiments have shown that fatigue life was reduced under irradiation, which was considered to involve irradiation embrittlement [8]. Figure 6.8 shows the influence of neutron irradiation on the fatigue life of Eurofer97. The fatigue life of material reduced after neutron irradiation, which may be related to neutron irradiation embrittlement. A fusion reactor runs under low-frequency pulsed operation, which establishes thermal loading cycles, and leads to interactions between irradiation creep and fatigue. Irradiation creep influences the stress–strain relation under cyclic loading, which will affect the fatigue life. The effects of irradiation on fatigue, especially when helium is present, require further in-depth investigation.

6.4 Research Methods of Neutron Irradiation and Material Damage

The damage behavior of materials under neutron irradiation in advanced nuclear systems is significantly different from that in traditional nuclear systems. Fission reactors, spallation neutron sources, and particle accelerators are utilized to simulate the neutron irradiation environment of advanced nuclear systems, and multi-scale computer simulations methods are used to study the damage mechanism and establish irradiation damage prediction models. This section describes the experimental investigation and computer simulation methods for the study of neutron irradiation damage.

6.4.1 Experimental Study on Neutron Irradiation Damage

The combination of experimental investigation and computer simulation is often utilized to study changes in the microstructures and properties of materials under neutron irradiation. Neutron irradiation experimental data can be used as the input to a computer simulation or as validation data for established computer simulation models. Neutron irradiation tests and postirradiation examinations are involved in neutron irradiation experiments: the former is carried out by various irradiation facilities, whereas the latter must be performed in a hot cell due to the radioactivity of the samples after neutron irradiation.

The characteristics of material irradiation experiments performed in several different irradiation facilities are shown in Table 6.2. The neutron energy in fission reactors is much lower, thus limiting both the irradiation damage rates and the transmutation gas production rates. One advantage of fission reactors is their large irradiation space. Spallation neutron sources are characterized by a broad neutron spectrum and high average neutron energy, enhancing both the annual irradiation damage dose and the transmutation gas production. However, the available irradiation space in a spallation neutron source is very limited. Generally, fission reactors and spallation neutron sources are used to simulate high-energy neutrons, irradiation damage behavior in materials and investigate the transmutation gas effect of irradiation damage. Meanwhile, during irradiation experiments with spallation neutron sources, small samples are utilized to reduce the radioactive dose and shorten the test cycles of irradiated samples.

Microstructural characterization is essential to the study of neutron irradiation damage. Because the macroscopic property changes of a material after neutron irradiation are determined by the microstructural changes, the mechanism by which the material properties change can be revealed by microstructural characterization. Microstructural characterization techniques such as electron microscopy, small-angle neutron scattering, and positron annihilation spectroscopy are commonly used.

Comparisons between materials before and after neutron irradiation can determine the influence of neutron irradiation on the mechanical properties of those materials and provide an important basis for the evaluation of the service safety of structural materials.

Table 6.2 Characteristics of material irradiation experiments in several irradiation facilities

Irradiation facility	Advantages	Disadvantages
Fusion neutron generator	High neutron energy, high transmutation gas production rate	High activation of irradiated materials, long test cycle, and limited irradiation space
Fission reactor	Large irradiation space, flexible arrangement, and high adaptability	Activation of irradiated samples, long test cycle, and limited transmutation gas production rate
Spallation neutron source	Large damage rate, high transmutation gas production rate, a broader dose, and temperature distribution obtained by one test	Activation of irradiated samples, long test cycle, and limited irradiation space
Ion accelerator	Short test cycle, low cost, and no activation for irradiated materials	Limited incident depth, uneven damage area, and difficulty to obtain the mechanical properties
Electron accelerator	Short test cycle, easy results analysis, and suitable for in situ observation of radiation defects	Small and uneven damage area, and difficulty to obtain the mechanical properties

6.4.2 Numerical Simulation of Neutron Irradiation Damage

In recent years, with the continuous development of computer technology, the calculation of irradiation damage has continuously got breakthrough, and the precision has been continuously improved. Numerical simulation calculations had been playing an increasingly important role in neutron irradiation damage study of materials. In this section, the interaction between neutrons and materials combined with the relevant research method were briefly presented.

The effects of neutron irradiation on the microstructure and properties of a material span a large temporal/spatial scale, increasing the cost and time required for neutron irradiation experiments. As the radiation damage dose of material in some advanced nuclear systems, such as fusion systems, can reach up to hundreds of dpa, it is unrealistic to rely exclusively on neutron irradiation experiments to evaluate the radiation properties of materials. In terms of each process or scale of interaction between neutrons and materials, the simulation methods are adequate, but multi-scale simulation research on neutron irradiation and material damage is rather rare and immature. Currently, models and methods for multi-scale simulation are being developed internationally. Figure 6.9 shows the multi-scale simulation of irradiation damage in materials [10]. Multi-scale simulation methods from small to large in scale include first-principle calculation, molecular dynamics, kinetic Monte Carlo, rate theory, phase field, dislocation dynamics, continuum mechanics, fracture mechanics,

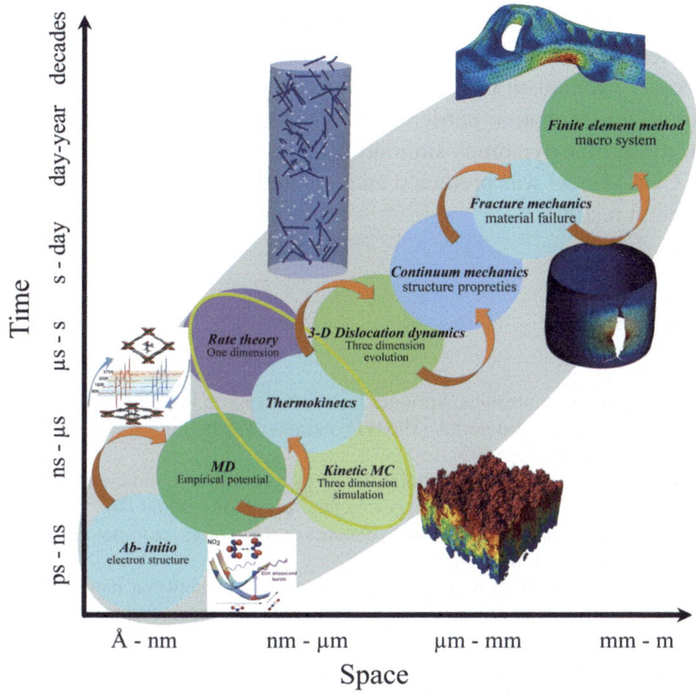

Fig. 6.9 Multi-scale simulation of irradiation damage in materials [10]

and finite element methods, and the reference timescale spans from picoseconds to decades.

In the multi-scale simulation study of material irradiation damage, the parameters of the PKA energy spectrum, transmutation gas generation rate, neutron irradiation damage dose, etc., should first be calculated based on neutron transport theory. These parameters will be used as the input parameters for subsequent multi-scale simulations, and typical computing software includes SuperMC, MCNP, etc. [11–18] On the microscopic scale, the thermodynamic and kinetic characteristic of atomic-scale irradiation defects, including the interaction with the transmutation gas, need to be calculated using the first-principle method, molecular dynamic simulation, kinetic Monte Carlo method, etc., the typical software include VASP, ABINIT, LAMMPS, NAMD, etc. On the mesoscopic scale, the evolution of microstructures such as dislocation loops, voids, bubbles, etc., are studied by the rate theory and phase field method. No general-purpose software has been designed for such calculations. On the macroscopic scale, various mean-field and continuum medium methods (continuum mechanics, fracture mechanics, and the finite element method) based on microstructural changes and empirical relationships between microstructures and properties are used to evaluate the macroscale thermomechanical behavior of materials in the reactor and the safety and reliability of the reactor itself. As for CLAM steel [8,

19], the neutron irradiation experimental data of 20 dpa (annual dose of DEMO) were conducted. Based on these data, the FDS team has preliminarily explained the microscopic mechanism of irradiation hardening and embrittlement. The influence of sputum on irradiation hardening was obtained. Based on the first-principle calculation, molecular dynamics simulation, rate theory, etc., the neutron damage behavior of CLAM steel was predicted and the prediction is in good agreement with the experimental results.

References

1. Wu YC (2016) Fusion neutronics. Springer Nature Singapore Pte. Ltd., Singapore
2. Erginsoy C, Vineyard GH, Englert A (1964) Dynamics of radiation damage in a body-centered cubic lattice. Phys Rev 133:A595–A606
3. Aidhy DS, Lu C, Jin K et al (2015) Point defect evolution in Ni, NiFe and NiCr alloys from atomistic simulations and irradiation experiments. Acta Mater 99:69–76
4. Ullmaier H, Schilling W (1980) Radiation damage in metallic reactor materials. In: Physics of modern materials, vol I. IAEA, Vienna
5. Hashimoto N, Wakai E, Robertson JP (1999) Relationship between hardening and damage structure in austenitic stainless steel 316LN irradiated at low temperature in the HFIR. J Nucl Mater 273:95–101
6. Isobe Y, Sagisaka M, Garner FA et al (2009) Precipitate evolution in low-nickel austenitic stainless steels during neutron irradiation at very low dose rates. J Nucl Mater 386–388:661–665
7. Straalsund JL, Powell RW, Chin BA (1982) An overview of neutron radiation on effects in LMFBR materials. J Nucl Mater 108–109:299–305
8. Ge HE, Peng L, Dai Y et al (2016) Tensile properties of CLAM steel irradiated up to 20.1 dpa in STIP-V. J Nucl Mater 468:240–245
9. Materna-Morris E, Möslang A, Rolli R et al (2011) Effect of 16.3 dpa neutron irradiation on fatigue lifetime of the RAFM steel EUROFER97. Fusion Eng Des 86:2607–2610
10. Wirth BD, Odette GR, Marian J et al (2004) Multiscale modeling of radiation damage in Fe-based alloys in the fusion environment. J Nucl Mater 329–333(Part A):103–111
11. Wu YC, Song J, Zheng HQ et al (2015) CAD-based Monte Carlo program for integrated simulation of nuclear system superMC. Ann Nucl Energy 82:161–168
12. Wu YC (2018) Multi-functional neutronics calculation methodology and program for nuclear design and radiation safety evaluation. Fusion Sci Technol 74(4):321–329
13. Norgett MJ, Robinson MT, Torrens IM (1975) A proposed method of calculating displacement dose rates. Nucl Eng Des 33:50–54
14. Zou J, Zeng Q, Xu DZ (2012) Design and producing of fine-group cross section library HENDL3.0/FG for subcritical system. PHYSOR, 4583–4591. Knoxville, USA
15. Greenwood LR, Smither RK (1985) Specter: neutron damage calculations for materials irradiation. Argonne National Laboratory, ANL-FPP-TM-197: DE85005791
16. Greenwood LR (1994) Neutron interaction and atomic recoil spectra. J Nucl Mater 216:29–44
17. Waters LS (1999) MCPNXTM USER's Manual. Version 2.1.5
18. Shimakawa S, Sekimura N, Nojiri N (2003) Radiation damage calculation by NPRIM computer code with JENDL 3.3. JAERI-Conference, 2003–2006
19. Huang QY (2017) Status and improvement of CLAM for nuclear application. Nucl Fusion 57:086042

Chapter 7
Nuclear Data Libraries

The neutronics simulation of nuclear systems relies on the availability of nuclear data to provide accurate numerical representation of the underlying physical processes. Essential nuclear data include energy-dependent reaction cross sections, the energy, and angular distributions of reaction products for various combinations of incident particles and targets, and the atomic and nuclear properties of excited states as well as their radioactive decay data. Nuclear data are the foundation of both nuclear science and technology research and nuclear engineering design. Advanced nuclear systems have the characteristics of complex neutron spectra and angular distribution, complex material composition, and extreme multi-physics coupling. These issues place higher requirements on the accuracy, the width of the energy region, and the species of nuclides for nuclear data.

A nuclear data library is a dataset of stored nuclear data in a certain format. Neutronics calculations of advanced nuclear systems cover particle transport, nuclide transmutation, neutron radiation damage, radiation dose, biological safety, etc. These calculations require nuclear data that include nuclear reaction cross section data, nuclide transmutation nuclear data, radiation damage nuclear data, and fluence-to-dose conversion factors. Among them, fluence-to-dose conversion factors have relatively few data and do not need to be introduced in this chapter.

7.1 Introduction to Nuclear Data Libraries

According to the data sources and application, nuclear data libraries are categorized into experimental nuclear data libraries, evaluated nuclear data libraries, and application nuclear data libraries. The experimental nuclear data mainly include nuclear reaction cross sections. These data are obtained by direct measurement using experimental methods based on neutron sources and detectors. The evaluated nuclear data are processed from experimental data using theoretical models of nuclear reactions. The application nuclear data are processed by combining the evaluated nuclear data with specific application requirements of nuclear systems. These three categories of

Y. Wu, *Neutronics of Advanced Nuclear Systems*,
https://doi.org/10.1007/978-981-13-6520-1_7

nuclear data libraries are introduced briefly as follows. The application nuclear data library can be applied directly to neutronics calculations and analysis, which will be highlighted in this chapter.

7.1.1 Experimental Nuclear Data Libraries

The experimental nuclear data library EXFOR [1] is the most comprehensive compilation of experimental nuclear reaction data. This library is released by the Nuclear Data Section (NDS) of the IAEA and contains the nuclear data measured in approximately 22,000 experiments. The EXFOR library includes cross sections and other nuclear reaction quantities induced by neutrons, charged particles, and photons. Compilation is mandatory for reaction cross sections of neutrons and light charged particles ($A \leq 12$). The reaction cross sections for heavy ions ($A \geq 13$) and photons are also compiled in the EXFOR library. The types of nuclear data involved in the EXFOR library mainly include total cross section, scattering cross section, absorption cross section, fission cross section, double differential cross section, spectrum, and fission product as well as nuclear structural data.

New requirements for experimental nuclear data are presented by the neutronics calculations for advanced nuclear systems. For example, the cross sections of actinides and minor actinides in the resonance region, which are very important in the performance assessment of transmutation and proliferation for fast reactors and hybrid nuclear systems, remain scarce and have great uncertainties. The fission cross sections with neutron energy up to 200 MeV for key nuclides (Pb, W, etc.) in the ADS spallation target are important for the life assessment of the target, but the experimental data are still deficient. The elastic scattering cross sections of ^{19}F and ^{7}Li (n, γ) cross sections of FLi and BeF_2 coolant in molten salt reactors, the elastic and inelastic scattering cross sections of ^{23}Na in sodium-cooled fast reactors, and the inelastic scattering cross sections of Pb and Bi in lead-based reactors are important to the critical calculation of these reactors. However, these experimental data are still scarce and the uncertainties are high. The (n, p), (n, α), (n, ^{3}He), and (n, t) reaction cross sections of Cu, Fe, Ni, and W are very important in neutron radiation damage estimation in fusion neutronics. These experimental data remain scarce because the reaction cross sections are very small and difficult to measure. Additionally, experimental data for the (n, γ) reaction cross sections of iron and lead, which are important components of shielding materials and can affect the calculation of nuclear heating in superconducting magnets, remain scarce [2]. Currently, the experimental nuclear data cannot meet the requirements of the neutronics calculations for advanced nuclear systems. In the future, it will be necessary to carry out comprehensive measurement on the reaction cross sections with insufficient experimental measurement and high uncertainty.

7.1.2 Evaluated Nuclear Data Libraries

Restricted by the conditions (such as particle sources and detectors) of nuclear experiments, it is difficult to obtain all the nuclear data demanded through only experimental measurements. To fill in the gaps between the measured experimental nuclear data and application requirements, certain physics-based models are used. Nuclear data derived by combing experimental nuclear data with physics-based models that describe the underlying physics of nuclear reactions are known as evaluated nuclear data. Once a full set of isotopes is evaluated with all the incident neutron energies of interest, the evaluated nuclear data library is assembled, representing the backbone of data needed for nuclear design and analysis.

Researchers from the US, China, Europe, Japan, and Russia have been involved in the development and improvement of complete evaluated nuclear data to expand the incident projectile energy range and enrich the isotope types. A complete nuclear data evaluation system has been established. This system includes the theoretical models and calculation programs for nuclear data evaluation, the processing technology of nuclear data, the compilation and benchmark technology of nuclear data, the sensitivity analysis technology, and the adjustment technology of nuclear data.

International nuclear data centers have released several evaluated nuclear libraries, including the ENDF/B [3] library, the CENDL [4] library, the JEFF [5] library, the JENDL [6] library, and the BROND [7] library. In 2018, the ENDF/B-VIII.0 library, which contains evaluated cross sections for 557 nuclides from ^{1}H to ^{257}Fm, was released by the NNDC (National Nuclear Data Center) in the USA. The CENDL-3.1 library was released by the CNDC (China Nuclear Data Center) in 2009. This library contains evaluated cross sections for 240 nuclides from ^{1}H to ^{244}Cm. In 2017, the OECD/NEA released the JEFF-3.3 library, which is a comprehensive, multifunctional data library with evaluated cross sections for 562 nuclides. The JENDL4.0u library was released by the JNDC (JAEA Nuclear Data Center) in 2012 and contains evaluated cross sections for 406 nuclides from ^{1}H to ^{257}Fm. The ROSFOND-2010 library [7] is an evaluated nuclear data library released by the CJD (Russian Nuclear Data Center) in 2010. This library also contains evaluated cross sections for 686 nuclides from ^{1}H to ^{252}Cm. In addition, the TENDL-2015 library was released by the NRG&CEA in 2016 and contains evaluated cross sections for 2800 nuclides from ^{1}H to ^{255}Fm.

The covariance data are released by the nuclear data centers to quantify the reliability and uncertainty of nuclear data. Among them, ENDF/B-VII.1 contains covariance data for 189 nuclides. CENDL-3.1 contains covariance data for six nuclides. JEFF-3.2 contains covariance data for 218 nuclides. JENDL-4.0u contains covariance data for 95 nuclides. ROSFOND-2010 contains covariance data for four nuclides. The development of evaluated nuclear libraries tends to further improve the accuracy and uncertainty quantification for key nuclides, systematization of nuclear data types, diversified application of nuclear data, and specialization of compiling nuclear data.

Advanced nuclear systems have new coolant, fuel, and structural materials compared with traditional reactors. For coolant, the scattering, (n, γ), and (n, 2n) reaction cross sections of lead, bismuth, sodium, lithium, fluorine, and beryllium are very important and could greatly affect the calculation accuracy of neutronics analysis. For fuel, the fission and (n, γ) reaction cross sections of neptunium, americium, curium, and plutonium are very important and could greatly affect the critical and burnup calculation accuracies. The angular distribution, spectra, KERMA factors, and gas production cross sections of iron, lead, tungsten, titanium, tantalum, nickel, and copper are very important and could greatly affect the calculation accuracy in shielding calculations. Therefore, it is necessary to improve the accuracy of the corresponding cross sections and reduce the uncertainty of the data.

7.1.3 Application Nuclear Data Libraries

In neutronics calculations, a nuclear data library that could be directly used by neutronics calculation codes is referred to as an application nuclear data library.

According to the application requirements of neutronics analysis, application nuclear data libraries include nuclear reaction cross section data, nuclide transmutation nuclear data, radiation damage nuclear data, and fluence-to-dose conversion factors. Nuclear reaction cross section data libraries are used for the neutron/photon transport calculations of advanced nuclear systems, which include neutron nuclear reaction cross section data and photon nuclear reaction cross section data.

Nuclide transmutation nuclear data libraries for transmutation calculations of advanced nuclear systems include transmutation reaction cross sections, decay data, and fission yield data.

Radiation damage nuclear data libraries for the radiation damage calculations of advanced nuclear systems include DPA libraries and gas production libraries.

Fluence-to-dose conversion factors libraries for dose equivalent assessment of advanced nuclear systems include fluence-to-dose conversion factors with different incident particles. The ICRP Publication 74 data library [8] has been chosen as the standard fluence-to-dose conversion factor library.

Nuclear reaction cross section data libraries, nuclide transmutation nuclear data libraries, and radiation damage nuclear data libraries are introduced in the following sections.

7.2 Nuclear Reaction Cross Section Data Libraries

According to neutron transport calculation methods, the nuclear reaction cross section data libraries can be classified into two categories. The first category is continuous energy cross section libraries, generally in ACE format [9], which is mainly used for Monte Carlo transport calculations. The other category is multigroup cross section libraries, generally in WIMSD, MATXS, and AMPX format [9], which are mainly used for deterministic transport calculations.

A continuous energy cross section library in ACE format includes almost all the details of the evaluated nuclear data file and is of higher accuracy. However, Monte Carlo transport calculations using continuous energy cross section libraries are extremely time-consuming. A multigroup cross section library describes the evaluated nuclear data based on the group-averaged cross sections. Transport calculations using multigroup cross sections can reduce the calculation time at the expense of calculation precision.

7.2.1 Processing Methods of Nuclear Reaction Cross Section Data Libraries

The cross section processing codes, such as the NJOY [9] code developed by LANL, are used to generate continuous energy cross section or multigroup cross section libraries based on evaluated nuclear data. Continuous energy cross section processing includes resonance cross section reconstruction, Doppler broadening, and thermal up-scatter effect correction. The production of a multigroup cross section library involves resonance cross section reconstruction, Doppler broadening, thermal neutron scattering effect correction, and multigroup processing.

In the processing of a nuclear reaction cross section data library, since the resonance parameter is usually used to describe the resonance phenomenon in the evaluated nuclear data library, it is necessary to generate the resonance cross sections based on the resonance theory for the transport calculation. Among advanced nuclear systems, the temperature distribution gradient of the material in the fusion system is large, and it is necessary to perform accurate Doppler broadening. There are large amounts of thermal neutrons in SWCR, VHTR, and MSR, and therefore thermal neutron scattering effects ought to be accurately considered. Thus, the resonance reconstruction, Doppler broadening, correction for thermal scattering effect, and multigroup processing are introduced in the following sections.

7.2.1.1 Resonance Reconstruction

For heavy nuclei ($A > 100$) and medium-mass nuclei ($40 \leq A \leq 100$), when the energy of an incident neutron is between 1 eV and 1 MeV, the reaction cross sections may oscillate rapidly with the neutron energy; this phenomenon is called resonance [10]. There are many peaks with large cross sections, which are called resonant peaks. Figure 7.1 shows the cross section of ^{238}U in ENDF/B-VII, which includes many resonance peaks ranging from 1 keV to 1 MeV. In addition, especially at the resonance peak, a compound nucleus in a particular quantum state is much more readily formed. Subsequently, the compound nucleus undergoes de-excitation by neutron emission, γ-ray emission, and fission. For light nuclei ($A < 40$), the width

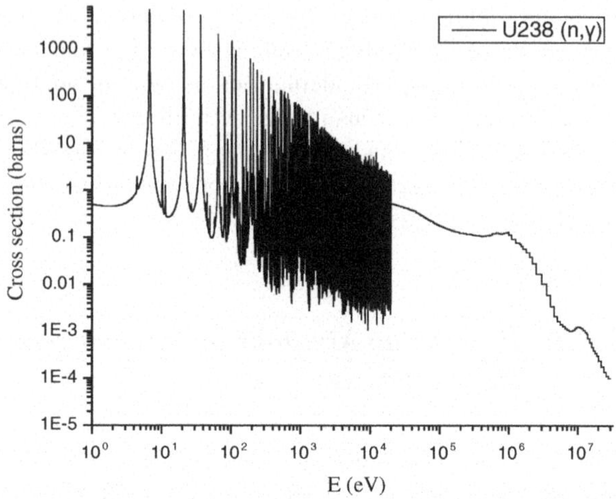

Fig. 7.1 (n, γ) cross section of ^{238}U

and energy intervals of the resonance peaks range from tens to hundreds of keV. Thus, the coherence between the different resonance peaks is weak.

There are distinct resonance peaks that can be clearly distinguished in a certain energy region, and this region is called the resolved resonance energy region [10, 11]. Above the resolved resonance energy region, the resonance peaks overlap and cannot be resolved by detectors, which is called the unresolved resonance energy region [10, 11]. The total capture and fission cross sections of a fertile nucleus show resolved resonances with energies up to several keV.

To reduce the amount of data, resonance parameters are used to describe the resonance cross sections in evaluated nuclear data files. To process nuclear reaction cross section data libraries, the resonance parameters need to be converted into continuous energy cross sections, which is called resonance reconstruction [10].

Several different resonance representations are applied to evaluated nuclear data files, including Single-Level Breit–Wigner, Multi-Level Breit–Wigner, Reich-Moore, and HyBrid R-function. The calculation of elastic scattering cross sections in the resolved resonance energy region represented by the Multi-Level Breit–Wigner formulation is given by Eq. (7.1).

$$\sigma_{\mathrm{n}}(E) = \frac{\pi}{k^2}\sum_l \sum_s \sum_J g_J \left|1 - U_{nn}^{lsJ}(E)\right|^2 \tag{7.1}$$

$$g_J = \frac{2J+1}{2(2I+1)} \tag{7.2}$$

where $\sigma_{\mathrm{n}}(E)$ is the reconstructed capture cross section, in units of barns; other parameters are resonance parameters. l is the orbital angular momentum, S is the orbit spin, J is the total angular momentum, k is the neutron wave number, U is the collision matrix, g_J is the spin statistical factor, and E is the incident energy.

In the unresolved resonance energy region, the resonance parameters are given in the form of statistical parameters. The equation for the elastic scattering cross section is also shown below:

$$\sigma_n(E) = \sigma_p + \frac{2\pi^2}{k^2}\sum_{l,J}\frac{g_J}{\overline{D}}\left[\overline{\Gamma}_n^2 R_n - 2\overline{\Gamma_n}\sin^2\phi_l\right] \tag{7.3}$$

where $\overline{\Gamma_n}$ is the appropriate average width of the resonance capture, R_n is the width fluctuation factor, σ_p is the potential scattering cross section, ϕ_l is the phase shift, and $\overline{D}$ is a statistical parameter.

7.2.1.2 Doppler Broadening

During nuclear reactor operation, cross sections such as fission and resonance capture vary due to the relative changing motions between incident neutrons and target nuclei. At higher temperatures, the relative motion changes intensively, and the width of the resonance peak increases, while the cross sections at the resonance peak energies decrease. This phenomenon is described by the Doppler effect [10]. In processing nuclear data, cross sections with different temperatures are modified according to the Doppler effect, which is called Doppler broadening.

In advanced nuclear systems, the temperatures of different materials are distributed with a large gradient. For example, the temperature of the nuclear fuel and the breeding and energy production blanket could be as high as hundreds of degrees Kelvin (K), whereas a temperature of the superconducting magnet may be as low as 4 K. Therefore, in the processing of nuclear data libraries for advanced nuclear systems, the Doppler effect in the cross section data should be critically considered and corrected accordingly.

According to the conservation of the reaction rate, the broadened cross section $\sigma(v, T)$ at temperature T is described by Eq. (7.4).

$$\sigma(v, T) = \frac{\sqrt{\alpha}}{\sqrt{\pi}}\int_0^\infty \sigma(V)V^2\{\mathrm{e}^{-\alpha(V-v)^2} - \mathrm{e}^{-\alpha(V+v)^2}\}\mathrm{d}V \tag{7.4}$$

where V is the relative speed between the target nuclide and the neutron, v is the speed of the incident neutron, M is the mass of the target nuclide, k is the Boltzmann constant, and $\alpha = M/2kT$.

1. Kernel broadening method

There are several methods to solve Eq. (7.4), and well known among them is the Kernel broadening method, which can be used to accurately calculate broadened cross section at any energy and temperature. With the kernel broadening method, it is convenient to define $x = \sqrt{\alpha}\mathrm{V}$ and $y = \sqrt{\alpha}\mathrm{v}$ in Eq. (7.4). Then, Eq. (7.4) becomes Eq. (7.5).

$$\sigma(y,T) = \frac{1}{y^2\sqrt{\pi}}\int_0^{\infty}\sigma(x)x^2\left\{e^{-(x-y)^2} - e^{-(x+y)^2}\right\}dx \tag{7.5}$$

Equation (7.5) can be simplified as

$$\sigma(y,T) = \sigma^*(y,T) - \sigma^*(-y,T) \tag{7.6}$$

where

$$\sigma^*(y,T) = \frac{1}{y^2\sqrt{\pi}}\int_0^{\infty}\sigma(x)x^2e^{-(x-y)^2}dx \tag{7.7}$$

The integration ranges for $\sigma^*(y)$ and $\sigma^*(-y)$ are $y - 4 \leq x < y + 4$ and $0 < x < 4$, respectively, since the integrands decay with the exponent. Based on the assumption of the linearization of the kernel broadening method, the cross section can be described by Eq. (7.8).

$$\sigma(x) = \sigma(x_i) + s_i\left(x^2 - x_i^2\right), x \in (x_i, x_{i+1}) \tag{7.8}$$

where

$$s_i = \frac{\sigma(x_{i+1},T) - \sigma(x_i,T)}{x_{i+1}^2 - x_i^2} \tag{7.9}$$

Therefore, Eq. (7.7) can be rewritten as

$$\sigma^*(y) = \sum_i\left\{A_i\left[\sigma(x_i) - s_ix_i^2\right] + B_is_i\right\} \tag{7.10}$$

where

$$A_i = \frac{H_2(x_i - y, x_{i+1} - y)}{y^2} + \frac{2H_1(x_i - y, x_{i+1} - y)}{y} + H_0(x_i - y, x_{i+1} - y)$$

$$B_i = \begin{matrix} \frac{H_4(x_i - y, x_{i+1} - y)}{y^2} + \frac{4H_3(x_i - y, x_{i+1} - y)}{y} + 6H_2(x_i - y, x_{i+1} - y) \\ + 4yH_1(x_i - y, x_{i+1} - y) + y^2H_0(x_i - y, x_{i+1} - y) \end{matrix}$$

and $H_n(a,b)$ is defined by Eq. (7.11).

$$H_n(a,b) = \frac{1}{\sqrt{\pi}}\int_a^b z^ne^{-z^2}dz \tag{7.11}$$

This method can provide a solution with high accuracy but low efficiency since the analytical form of Eq. (7.4) introduces the complementary error function (erfc) and a Taylor series expansion.

2. Real-time Doppler broadening method

To improve the efficiency of the kernel Doppler broadening method, many real-time Doppler broadening methods, such as the piecewise Gauss method and the double-exponential method, have been developed.

The piecewise Gauss method uses numerical integration in place of analytical integration. For the integrand $f(z)e^{-z^2}$, the numerical integration form can be described as Gauss–Hermite quadrature.

$$\int_{-\infty}^{\infty} f(z)\mathrm{e}^{-z^2}\mathrm{d}z = \sum_{k=1}^{n} w_k f(z_k) \tag{7.12}$$

where z_k is the abscissa of the Hermite polynomial $H_n(z)$ with n degree and w_k is the corresponding weight.

By defining $x - y = z$, Eq. (7.7) becomes Eq. (7.13).

$$\sigma^*(y, T) = \frac{1}{y^2\sqrt{\pi}} \int_{-y}^{\infty} \sigma(z + y)(z + y)^2 \mathrm{e}^{-z^2}\mathrm{d}z \tag{7.13}$$

The integration interval of Eq. (7.13) is different from that of Eq. (7.12). Since the range of integration is limited to $\mathrm{y} - 4 \leq x < y + 4$, Eq. (7.12) can be directly substituted into Eq. (7.13) when $\mathrm{y} > 4$.

$$\sigma(y, T) = \frac{1}{y^2\sqrt{\pi}} \sum_{k=1}^{n} w_k f(z_k) \tag{7.14}$$

where $f(z) = \sigma(z + y)(z + y)^2$.

When $\mathrm{y} \leq 4$, the numerical integration form of $\int_a^b f(x)\mathrm{d}x$ can be described by a Gauss–Legendre quadrature.

$$\int_a^b f(x)\mathrm{d}x \approx \frac{b - a}{2} \sum_{k=1}^{n} w_k f\left(\frac{b - a}{2}x_k + \frac{b + a}{2}\right) \tag{7.15}$$

Thus, Eq. (7.15) can be described by Eq. (7.16).

$$\sigma(y, T) = \frac{1}{y^2}\sqrt{\frac{1}{\pi}} \sum_i \frac{x_{i+1} - x_i}{2} \sum_{k=1}^{N} w_k f\left(\frac{x_{i+1} - x_i}{2}x_k + \frac{x_{i+1} - x_i}{2}\right) \tag{7.16}$$

where $f(x) = x^2\sigma(x)\left[\exp\left(-(x-y)^2\right) - \exp\left(-(x+y)^2\right)\right]$, x_k are the abscissas of the Legendre polynomial $P_n(x)$ with n degree, and w_k are the corresponding weights.

In the double-exponential method, for integrations of the form $I = \int_a^b f(t)\mathrm{d}t$, the equation can be described by a variable substitution with $t = \psi(x)$.

$$I = \int_a^b f(t)\mathrm{d}t = \int_{-\infty}^{\infty} f(\psi(x))\psi'(x)\mathrm{d}x \tag{7.17}$$

where $\psi(x)$ and $\psi'(x)$ are defined by Eq. (7.18).

$$\begin{aligned} \psi(x) &= \tfrac{b-a}{2}\tanh\left(\tfrac{\pi}{2}\sinh x\right) + \tfrac{b+a}{2} \\ \psi'(x) &= \tfrac{(b-a)\pi}{4}\frac{\cosh x}{\cosh^2\left(\frac{\pi}{2}\sinh x\right)} \end{aligned} \tag{7.18}$$

We define the abscissas x_k and the weights w_k as follows:

$$\begin{aligned} x_k &= \tanh\left(\tfrac{\pi}{2}\sinh kh\right) \\ w_k &= \frac{\frac{\pi}{2}\cosh kh}{\cosh^2\left(\frac{\pi}{2}\sinh kh\right)} \end{aligned} \tag{7.19}$$

The trapezoidal equation with N intervals is applied to solve Eq. (7.17).

$$I_h = \frac{b-a}{2}h\sum_{k=-N/2}^{N/2} f\left(\frac{b-a}{2}x_k + \frac{b+a}{2}\right)w_k \tag{7.20}$$

Accordingly, the Doppler broadening equation can be rewritten as

$$\sigma(y, T) = \frac{y+4}{2y^2\sqrt{\pi}}h\sum_{k=-N/2}^{N/2} f\left(\frac{y+4}{2}x_k + \frac{y+4}{2}\right)w_k \tag{7.21}$$

where $f(x) = x^2\sigma(x)\left[\exp\left(-(x-y)^2\right) - \exp\left(-(x+y)^2\right)\right]$.

The broadening efficiency ratio for generating total cross sections at 600, 900, and 1200 K between the double-exponential and kernel Doppler broadening methods is shown in Fig. 7.2. This figure shows that the average efficiency of the double-exponential method is 10 times higher than that of the kernel Doppler broadening method. Meanwhile, better acceleration is achieved for cross sections at higher temperatures.

In general, cross sections at different temperatures are generated for transport calculations of advanced nuclear systems, which lead to a tremendous amount of computer memory consumption. Therefore, the Doppler broadening method has been developed to calculate cross sections in real time at any required temperature during Monte Carlo transport simulations. The piecewise Gauss method and the

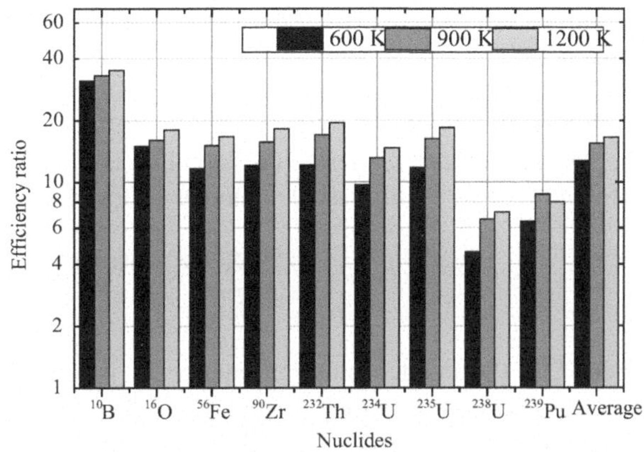

Fig. 7.2 Efficiency ratios between the double-exponential and kernel Doppler broadening methods

double-exponential method can be used in on-the-fly Doppler broadening methods due to their high-efficiency characteristics.

7.2.1.3 Thermal Neutron Scattering

When the energy of incident neutrons is less than 5 eV, the target nucleus cannot be considered stationary because the neutron energy is comparable with the energy of the thermal motion of the atoms in materials.

In addition, there are interactions between the neighboring nuclei of the scattering nucleus in the molecule or in the solid, and it cannot freely recoil when colliding with the neutron, that is, the nucleus is in a bound state. Furthermore, because the De Broglie wavelength of lower energy neutrons is close to the spacing of molecules or crystal cores, interferences may occur. The above phenomena are all thermal neutron scattering effect. When collision occurs between a neutron and a moving target, the neutron can gain energy. Therefore, the energy of the secondary neutron may be greater than the energy of the incident neutron, which is the thermal neutron up-scatter effect [9, 10].

The scattering cross section of a thermal neutron changes with neutron energy and is also related to the temperature, physical and chemical properties of the scattering medium. Thermal neutron scattering cross sections are generally classified into coherent elastic scattering, incoherent elastic scattering, and incoherent inelastic scattering. In practice, the scattering of neutrons can be expressed as the sum of coherent scattering and incoherent scattering. The coherent scattering includes the effects from waves that can interfere with each other, and the incoherent part depends on a simple sum of noninterfering waves from all the N particles. Furthermore, the coherent and incoherent scatterings include both elastic and inelastic parts. In addition, the inelastic part of the scattering contains both interference effects between waves scattered by different particles and direct terms. Coherent and incoherent scatterings are approximately the same in form. The coherent scattering is usually neglected.

1. Coherent elastic scattering

The coherent elastic scattering cross section is mainly for crystalline solids, and the differential cross section is calculated as follows:

$$\sigma(E, E', \mu) = \frac{\sigma_c}{E} \sum_{E_i > E} f_i e^{-2WE_i} \delta(\mu - \mu_0) \delta(E - E') \tag{7.22}$$

where E is the incident neutron energy, E' is the secondary neutron energy, μ is the scattering cosine in the laboratory (LAB) reference system, i is the order of the "Bragg edges", E_i is the so-called "Bragg edges", σ_c is the characteristic coherent cross section for the material, W is the effective Debye–Waller coefficient (which is a function of temperature), f_i is related to the crystallographic structure factors, μ_0 is the cosine of the characteristic scattering angle in the lattice plane, and $\mu_0 = 1 - 2E_i/E$.

2. Incoherent elastic scattering

Incoherent scattering mainly for occurs in hydrogenous solids, and the differential cross section of this type of scattering is as follows:

$$\sigma(E, E', \mu) = \frac{\sigma_b}{2} e^{-2WE(1-\mu)} \delta\left(E - E'\right) \tag{7.23}$$

where σ_b is the characteristic bound cross section.

3. Incoherent inelastic scattering

For incoherent inelastic scattering, the expression of the differential cross section is as follows:

$$\sigma(E, E', \mu) = \frac{\sigma_b}{2kT} \sqrt{\frac{E'}{E}} e^{-\beta/2} S(\alpha, \beta) \tag{7.24}$$

where α is the dimensionless momentum transfer, β is the dimensionless energy transfer, σ_b is the characteristic bound incoherent cross section, and k is the Boltzmann's constant.

Within the incoherent approximation and the Gaussian approximation, the scattering law can be written as

$$S(\alpha, \beta) = \frac{1}{2\pi} \int_{-\infty}^{\infty} e^{i\beta\hat{t}} e^{-\gamma(\hat{t})} d\hat{t} \tag{7.25}$$

The function $\gamma(\hat{t})$ is given by

$$\gamma(\hat{t}) = \alpha \int_{-\infty}^{\infty} P(\beta)\left[1 - e^{-i\beta\hat{t}}\right]e^{-\beta/2}d\beta \tag{7.26}$$

where

$$P(\beta) = \frac{\rho(\beta)}{2\beta\sinh(\beta/2)} \tag{7.27}$$

$\hat{t}$ is the temperature of the scattering nucleus in units of K.

For binding in solids and liquids, the $S(\alpha, \beta)$ for many important moderator materials is available in ENDF/B File 7 format. For the free gas model, it can be expressed as follows:

$$S(\alpha, \beta) = \frac{1}{\sqrt{4\pi\alpha}}\exp\left(-\frac{\alpha^2 + \beta^2}{4\alpha}\right) \tag{7.28}$$

In the ACE format thermal library, thermal scattering cross sections are given at a limited number of temperatures, so calculations can only be accurately performed at these temperatures. For the physical calculation of thermal reactors in advanced nuclear systems, it is necessary to use thermal cross section at different temperatures. The Neville interpolation method is a typical method to generate multi-temperature thermal cross sections online.

The Neville method is an inverse-deducing interpolation algorithm, where each high-order interpolation result can be recursively derived from low-order interpolation. The advantage of this algorithm is that it can directly calculate the approximate value of the function at a certain point without specific expression.

Given a sequence (x_i, y_i) containing $n + 1$ points, the value of each point is not the same. At this time, the polynomial P calculated by the n-time interpolation of the sequence can be expressed as

$$P(x_i) = y_i, i = 0, 1, \ldots, n \tag{7.29}$$

The interpolation polynomial $P_{i,j}(x)$, with the order of $j - i$, is introduced to make it to pass points (x_k, y_k), k = i, i + 1, …, j. Apparently, when j is equal to i, $P_{i,i}(x)$ is equal to $P(x)$.

By defining $P_i(x) = f(x_i) = y_i$, $P_{i,j}(x)$ can be expressed by two points —(x_i, y_i) and (x_{i+1}, y_{i+1}). When $j - i = 1$, $P_{i,j}(x)$ can be described by Eq. (7.30).

$$\begin{aligned} P_{i,j}(x) &= \tfrac{(x_{i+1}-x)}{x_{i+1}-x_i}y_i + \tfrac{(x-x_i)}{x_{i+1}-x_i}y_{i+1} \\ &= \frac{(x_j - x)}{x_j - x_i}P(x_i) + \frac{(x - x_i)}{x_j - x_i}P(x_j) \end{aligned} \tag{7.30}$$

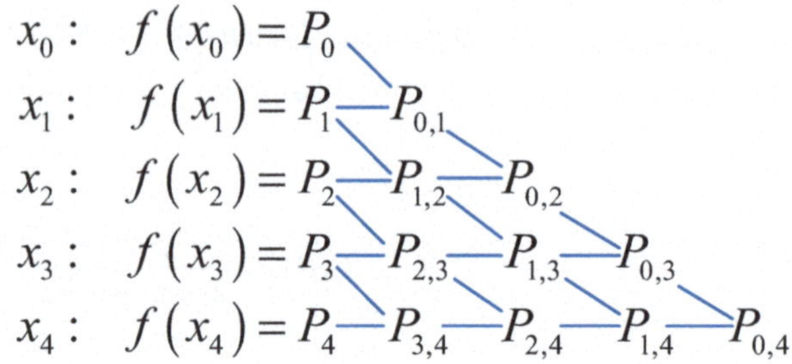

Fig. 7.3 Process of the Neville interpolation method when $j - i = 4$

Equation (7.30) shows that the interpolation polynomial can be represented by the iteration of the previous interpolation polynomial. Therefore, when $j - i > 1$,

$$P_{i,j}(x) = \frac{(x_j - x)}{x_j - x_i} P_{i,j-1}(x) + \frac{(x - x_i)}{x_j - x_i} P_{i+1,j}(x) \tag{7.31}$$

In the equation, $P_{i,j}(x)$ is called a Neville interpolation polynomial. Equation (7.31) shows that each previous interpolation calculation result can be used when adding nodes, so it is not necessary to perform double calculations, that is, the method is adaptive.

To further clarify the adaptability of the method, Fig. 7.3 shows the iterative process of the Neville interpolation method when $j - i = 4$. Figure 7.3 shows that the five reference point data all affect the final interpolation result. The more the original nodes, the higher the accuracy of the Neville interpolation method.

For incoherent elastic scattering and inelastic scattering cross sections, the energy grid does not change with temperature. The inelastic scattering cross section with energy E corresponding to temperature T can be obtained by Neville interpolation of the cross sections with temperatures $T_1, T_2, \ldots, T_n$. For coherent elastic scattering, the number of energy points in the energy grid continuously decreases as the temperature increases. To ensure the accuracy of the cross section, the energy frame with lower temperature and more energy points is preferentially selected. For energy points that are not in the energy grid, the cross sections are obtained using linear interpolation of adjacent energy points.

7.2.1.4 Multigroup Processing

The Boltzmann transport equation in its general form cannot be directly solved, and multigroup methods are the most commonly used technique for numerically solving the equation. The Boltzmann transport equation [10, 11] can be converted into G equations through a multigroup approximation. The multigroup cross sections can be calculated by the following equation:

$$\sigma_g = \frac{\int_{\Delta E} \sigma(E)\phi(E)\mathrm{d}E}{\int_{\Delta E} \phi(E)\mathrm{d}E} \tag{7.32}$$

where $\sigma(E)$ is the microscopic cross section, $\phi(E)$ is the weight function, and σ_g is the cross section of group g.

The energy group structure and weight function are important parameters that influence the accuracy of multigroup cross sections. An appropriate energy group structure and weight function need to be designed according to the characteristics of nuclear systems with large energy spans. The appropriate energy group structure and weight function must be selected for different nuclear reactors in processing multigroup cross sections for advanced nuclear systems. For the thermal reactors including supercritical water reactors, high-temperature gas-cooled reactors, and molten salt reactors, the group in the thermal energy region needs to be subdivided. For LFR, SFR, and GFR, the group in the energy region of the resonance energy and fast neutrons needs to be subdivided. For fusion reactors and fusion–fission hybrid reactors, the group in the fusion energy region (approximately 14.06 MeV) and resonance energy region needs to be subdivided. For the ADS systems, the neutron source coming from the spallation reaction and the neutron energy can reach 1 GeV; therefore, the group in the energy region up to 20 MeV needs to be subdivided.

For hybrid nuclear systems, the fission blanket contains large amounts of actinides and minor actinides. The absorption and fission cross sections in the resonance energy region can cause sharp dips in the spectrum corresponding to each resonance energy region (this is called the resonance self-shielding effect). Therefore, the correction of the resonance self-shielding effect should be considered, based on subdividing the energy group in the resonance energy region. For the thermal reactors, the correction of the thermal scattering effect should also be considered.

The following section introduces multigroup cross section processing methods for the energy group structure, weight function, and correction of the resonance self-shielding effect.

1. Energy group structure

An appropriate energy group structure must be selected for different nuclear reactors in processing multigroup cross sections. The selection of the energy group structure is usually based on the following principles.

(1) The threshold energy of the reactions, including inelastic scattering, fission, (n, 2n) and (n, 3n) reactions, should be considered. The threshold energies of these reactions should be chosen as group boundaries.
(2) For resonance reactions, the energies of the resonance peaks and their overlap should be carefully considered. Different resonance peaks should be located in different energy groups.
(3) The energy group width should be less than the maximum elastic scattering energy loss of most nuclides, since the slowing-down process via elastic scattering is only related to two adjacent energy groups.

(4) The energies, at which the cross section significantly changes, should be group boundaries.
(5) The energy group width is usually designed as a lethargy width.

The above principles are generally used in the design of the energy group structure, while the particularity of advanced nuclear systems should be considered. For the neutronics calculation of advanced nuclear systems, certain systems have plenty of high-energy neutrons (spallation neutrons and fusion neutrons); therefore, the group in the energy region up to 20 MeV needs to be subdivided in the design of multigroup nuclear data libraries. By increasing the number of energy groups, more detailed information about the cross sections could be obtained. The accuracy of multigroup cross sections with a fine energy group structure is higher than that of traditional energy group structures. To satisfy the needs of nuclear analysis for advanced nuclear systems, an ultrafine group cross section library was developed by the FDS Team. In this library, the neutron energy group number is 1200, and the photon energy group number is 42. The neutron energy range is divided into three energy regions: 10^{-5}–4 eV, 4 eV–12.52 MeV, and 12.52–150 MeV. The detailed design of the energy group structure is shown in Table 7.1.

2. Weight function

According to Eq. (7.32) for multigroup cross section calculations, the weight function is important when processing multigroup cross section libraries. The weight function should be designed according to the neutron characteristics of the different energy regions. The Maxwell spectrum is used for the thermal energy region, the fission spectrum is used in the fission neutron energy regions, the D-T fusion spectrum is used for the energy region at approximately 14.06 MeV, and the 1/E spectrum is used for the other energy regions.

Thus, for advanced nuclear systems, the weight function for multigroup cross section libraries is usually based on the VITAMIN-E [9] distribution, as shown in Table 7.2.

The photon energy in fusion systems is generally lower than 50 MeV. The 1/E + roll-offs [9] distribution is used as the weight function.

3. Correction of resonance self-shielding effect

The absorption and fission cross sections in the resonance energy region (heavy nucleus is 4 eV–9.118 keV; medium-mass nucleus is 0.1 keV–1 MeV) for actinide

Table 7.1 Energy group structure of advanced nuclear systems

Energy range	Dividing method	Group number
10^{-5}–4 eV	XMAS 172	80
4 eV–12.52 MeV	Average lethargy decrease (0.016) according to ^{238}U elastic scattering cross section	1055
12.52–55 MeV	FENDL3.0/MG	46
55–150 MeV	HENDL-ADS/MG	19

Table 7.2 Weight function of advanced nuclear systems

Energy range	Weight function
10^{-5}–0.414 eV	Maxwell spectrum
0.414 eV–2.12 MeV	1/E spectrum
2.12–10 MeV	Fission spectrum
10–12.52 MeV	1/E spectrum
12.52–15.68 MeV	D-T fusion spectrum
15.68–20 MeV	1/E spectrum
20–150 MeV	HENDL-ADS/MG spectrum

and minor actinide nuclides become very high, which can cause sharp dips in the spectrum corresponding to each resonance energy region, causing the spectrum to deviate from the standard spectrum. The deviation of the spectrum may result in inaccuracies in the group cross section, which is called the resonance self-shielding effect. In fusion reactors, the resonance self-shielding effect is relatively weak due to the high neutron energy. For fast reactors, ADS systems and fusion–fission hybrid reactors, there are many actinides and minor actinides in the fission blankets that have significant resonance peaks in their cross sections in the resonance regions. Therefore, corrections must be performed for the resonance self-shielding effect. If the resonance self-shielding effect is ignored during neutronics calculations, the k_{eff} of an advanced nuclear system may be underestimated.

The cross sections in the unresolved and resolved resonance energy regions can be treated with the Bondarenko method and the flux calculator method, respectively. In the Bondarenko method, the integral slowing-down equation based on the narrow resonance approximation is solved to obtain the flux. The flux in an infinite homogenous medium is assumed to vary inversely with the total macroscopic cross section [9].

$$\phi_l^i(E) = \frac{C(E)}{\left[\sigma_t^i(E) + \sigma_0^i\right]^{l+1}} \tag{7.23}$$

where $C(E)$ is the standard weight function, l is the Legendre order, and σ_t^i and σ_0^i are the total cross section and background cross section, respectively, of nuclide i.

The background cross section [9] for an isotope i is given by Eq. (7.24).

$$\sigma_0^i = \frac{1}{N_i}\sum_{k \neq i} N_k \sigma_t^k\left(\sigma_0^k\right) + \sigma_e \tag{7.24}$$

where $\sigma_e(E)$ is the escape cross section, used to correct the heterogeneous and Dancoff effects, and N_i and N_j are the atomic densities of nuclides i and j, respectively.

If the background cross section $\left(\sigma_0^i\right)$ is small, then the resonance has a significant impact on the flux, and a large self-shielding effect exists. If σ_0^i is very large, the cross sections of the absorber do not affect the flux, and the flux may be represented as a smooth function of energy.

In the flux calculator method [9], the weight function is computed by solving the slowing-down equation in the resolved resonance energy region.

$$[\sigma_0 + \sigma_t(E)]\phi(E) = C(E)\sigma_0 + \int_{E}^{E/\alpha} \frac{\sigma_s(E')}{(1-\alpha)E'}\phi(E')dE' \quad (7.25)$$

where σ_t is the total cross section, σ_s is the scattering cross section, α is coefficients related to the atomic mass, and ϕ is the flux.

The Bondarenko method, which is based on the narrow resonance approximation, is the most commonly used method in the treatment of the resonance self-shielding effect. However, this method cannot address resonance interference. Resonance interference refers to the way in which cross section resonances from one nuclide affect the resonance absorption or scattering of another nuclide when multiple resonance absorbers exist in the same material. In the creation of a cross section library for fusion–fission hybrid reactors, resonance interference effects cannot be neglected. When the reference background cross section of the nuclide is chosen to be closer to that of the real background, the influence of the resonance interference effects become very weak.

7.2.2 *Typical Nuclear Reaction Cross Section Data Libraries*

Nuclear reaction cross section data libraries can be categorized into general nuclear data libraries and special nuclear data libraries. General nuclear data libraries include the ENDF/B continuous energy cross section library, JEFF continuous energy cross section library, etc. These libraries can be used for different nuclear reactors. Special nuclear data libraries are designed for specific types of application requirements. These libraries include WIMSD [12], FENDL [13], ADS-lib [14], BUGLE-B7, KAFAX-E70, and C4P. The WIMSD nuclear data library was released by IAEA/NDS and is used for fission reactors. The FENDL nuclear data library was released by IAEA/NDS and is used for fusion reactors. The ADS-lib nuclear data library was released by IAEA/NDS and is used for the ADS systems. The BUGLE-B7 nuclear data library was released by ORNL and is used for thermal reactors. The KAFAX-E70 nuclear data library was released by OECD/NEA and is used for fast reactors. The C4P nuclear data library was released by KIT and is used for fission reactors. The SuperMC/HENDL library, developed by the FDS Team based on the need for neutronics calculations for advanced nuclear systems, includes a continuous energy cross section library and a multigroup cross section library.

7.2.2.1 SuperMC/HENDL Nuclear Reaction Cross Section Data Library

The hybrid evaluated nuclear data library SuperMC/HENDL [15–18] was developed by the FDS Team to meet the requirements of nuclear analysis for fusion reactors, fission reactors, fusion–fission hybrid reactors, and ADS systems. The evaluated nuclear data files of SuperMC/HENDL are ENDF/B, JEFF, and JENDL. Based on the simulation results of critical safety experiments and neutron shielding integration experiments, 410 nuclides are selected, therein covering the main nuclides of the reactor transport calculations. SuperMC/HENDL includes the nuclear data for transport calculations, transmutation calculations, and radiation damage calculations. Corrections of resonance self-shielding and Doppler and thermal neutron scattering effects for nuclear systems have been accomplished in the design for SuperMC/HENDL. The library includes different energy structures and weight functions for different reactors. Online group collapse can be realized by using a pre-generated ultrafine group cross section library. Based on the on-the-fly Doppler broadening function of the SuperMC, data at any temperature can be provided.

The SuperMC/HENDL nuclear reaction cross section data library includes the coarse group library HENDL/CG (27 groups of neutrons and 21 groups of photons), the multigroup library HENDL/MG (175 groups of neutrons and 42 groups of photons), and the fine-group library HENDL/FG (315 groups of neutrons and 42 groups of photons), ADS multigroup library HENDL-ADS/MG (366 groups of neutrons and 42 groups of photons), and ultrafine group library HENDL /UG (1200 groups of neutrons and 42 groups of photons). The HENDL/MG and HENDL/FG libraries can be applied in fusion systems and fission systems. The HENDL-ADS/MG can be applied in ADS systems. The HENDL/UG can be applied in fission systems and hybrid nuclear systems. The neutron energy range of the HENDL/MG library is 10^{-11}–20 MeV, while the weight function is the VITAMIN-E distribution, and the photon energy range is 1 keV–50 MeV, while the weight function is given by 1/E + roll-offs. The neutron energy range of the HENDL/FG library is 10^{-11}–20 MeV, while the weight function is the fusion–fission hybrid energy spectrum, and the photon energy range is 1 keV–50 MeV, while the weight function is given by 1/E + roll-offs. The neutron energy range of the HENDL-ADS/FG library is 10^{-11}–150 MeV, while the weight function is the ADS spectrum, and the photon energy range is 1 keV–50 MeV, while the weight function is given by 1/E + roll-offs. The neutron energy range of the HENDL/UG library is 10^{-11}–20 MeV, while the weight function is the fission energy spectrum, and the photon energy range is 1 keV–50 MeV, while the weight function is given by 1/E + roll-offs. HENDL/MC can provide continuous energy cross section data for transport calculation of advanced nuclear systems.

7.2.2.2 ENDF/B Nuclear Reaction Cross Section Data Library

The ENDF/B continuous energy cross section data library was released by the NNDC (National Nuclear Data Center), US and was processed by the NJOY nuclear data processing system. The latest version is ENDF/B-VIII.0, which was released in 2018 and contains continuous energy cross sections with the ACE format for 557 nuclides. This library also contains cross sections at different temperatures including 0.1, 250, 300, 600, 900, 1200, and 2500 K.

The neutron energy range of the cross sections for the actinide/minor actinide nuclides in the ENDF/B-VIII.0 library is 10^{-11}–20 MeV and 10^{-11}–30 MeV. The neutron energy range of the cross sections for the structure material nuclides is 10^{-11}–150 MeV. The neutron energy range of the cross sections for the other nuclides is 10^{-11}–20 MeV. This library is a continuous energy cross section data library. Therefore, this library can be applied to the neutronics calculation and analysis of most reactors.

7.2.2.3 FENDL Nuclear Reaction Cross Section Data Library

The Fusion Evaluated Nuclear Data Library, FENDL, was released by IAEA/NDS to satisfy the neutronics calculation requirements for the ITER, DEMO, and IFMIF systems.

FENDL-3.1, which was released in January 2018, contains cross sections of neutron, proton, photon, and deuteron nuclear reactions with energies up to 200 MeV. FENDL-3.1 contains cross sections for 180 nuclides. The evaluated nuclear data files in FENDL-3.1 are ENDF/B, JEFF, and JENDL/He.

FENDL-3.1 includes the FENDL-3.1/E (an evaluated nuclear data library), FENDL-3.1/MC (a continuous energy cross section library), FENDL-3.1/MG (a multigroup cross section library in MATXS format), and FENDL-3.1/MG (a multigroup cross section library in GENDF format) libraries. The neutron energy range of the FENDL-3.1/MG library is 10^{-11}–55 MeV, and the weight function is the VITAMIN-E distribution. The photon energy range is 1 keV–50 MeV, and the weight function is given by 1/E + roll-offs.

7.2.2.4 WIMSD Nuclear Reaction Cross Section Data Library

The WIMSD (Winfrith Improved Multigroup Scheme-D) library was released by IAEA/NDS to satisfy requirements of neutronics calculations for fission reactors and provide data support to reactor lattice codes. The WIMSD library includes two multigroup cross section libraries, WIMSD-69, XMAS-172, and the evaluated nuclear data files in WIMSD are ENDF/B-VII.1, ENDF/B-VII.0, JEFF-3.1, and JENDL-3.2.

The latest version of WIMSD was released in November 2014, which contains cross sections for 185 nuclides. This library covers the main nuclides in neutronics calculation, including the moderators, actinides/minor actinides, fission products, and structural materials. This library also contains cross sections at different temperatures. For actinides, the temperatures are 293, 600, 900, 1200, 1600, and 2000 K. For minor actinides, the temperatures are 293, 700, 1100, 1500, and 2000 K.

The neutron energy range of the WIMSD-69 library is 10^{-11}–10 MeV, and the weight function is the spectrum of typical fission reactors. The neutron energy range of the WIMSD-172 library is 10^{-11}–20 MeV, and the weight function is the spectrum of typical fission reactors.

7.2.2.5 ADS-Lib Nuclear Reaction Cross Section Data Library

The ADS-lib (ADS Nuclear Data Library) was released by IAEA/NDS to satisfy the neutronics calculation for ADS systems and accelerators. The latest version ADS-2.0, released in November 2008, includes the continuous energy cross section data library ADS-2.0/ACE, the multigroup cross section library ADS-2.0/MATXS and ADS-2.0/GENDF. ADS-2.0 contains neutron cross sections for 156 nuclides, which covers the main nuclides in neutronics calculations, including actinides/minor actinides, fission products, structure materials, and coolant.

The cross section data are available at different temperatures. For actinides, ^{28}Si, ^{29}Si, ^{30}Si, and ^{16}O, the temperatures are 293.6, 600, 900, 1200, 1500, and 1800 K. For the rest of the materials, the temperatures are 293.6, 600, 900, and 1200 K. For C-nat, the temperatures are 293.6, 400, 500, 600, 700, 800, 1000, 1200, 1600, and 2000 K. For ^{1}H (H_2O), ^{2}H (D_2O), the temperatures are 293.6, 350, 400, 450, 500, 550, 600, and 650 K. The correction of resonance self-shielding effects has also been performed in ADS-2.0. For the multigroup cross sections of actinides, 10 background cross sections have been set. For the multigroup cross sections of structure materials, six background cross sections have been set. For other nuclides, the infinite dilution cross section is used as a background cross section. The number of groups for the ADS-2.0 multigroup cross section library is 421, the neutron energy range of the ADS-2.0 library is 10^{-11}–20 MeV, and the weight function is the modified spectrum of a PWR.

To meet the requirements of high-energy neutron cross sections for the neutronics analysis of the ADS system, the ADS-HE library (High-Energy Library for Accelerator-Driven Systems) was released in November 2013. This library contains the cross sections for 10 fertile nuclides. ADS-HE is a continuous energy cross section data library with the ACE format. The highest energy of incident neutrons in the ADS-HE library is 1 GeV.

7.3 Nuclide Transmutation Nuclear Data Libraries

One of the main objectives of advanced nuclear systems is to achieve nuclear waste transmutation and nuclear fuel proliferation. In the design process, accurate calculation, and analysis of burnup, nuclear waste transmutation and nuclear fuel proliferation should be carried out.

To perform these calculations and analyses, nuclear data libraries that include transmutation reaction nuclear data, decay data, and fission yield data are required. These libraries are introduced in this section.

7.3.1 *Transmutation Reaction Nuclear Data Libraries*

Transmutation reaction nuclear data libraries include the energy-dependent transmutation reaction probabilities of particles and materials. The reaction cross sections for (n, γ), (n, xn), (n, p), (n, t), (n, α), (n, d), and (n, f) are particularly important in transmutation calculations for advanced nuclear systems. These libraries are the basis of the study and analysis of the radioactive quantity and rule of change of transmutation products.

The main differences between transmutation reaction nuclear data libraries and nuclear reaction cross section data libraries are the nuclide types and nuclear reaction channels. The transmutation reaction nuclear data library contains the cross section files of thousands of nuclides. Each file has relatively few reaction channels and low accuracy. The nuclear reaction cross section data library contains the cross section files of hundreds of nuclides. Each file has relatively more reaction channels and high accuracy.

The neutron energy in advanced nuclear systems such as fusion systems and ADS systems is as high as tens to hundreds of MeV, which makes it easy to cause certain threshold reactions including (n, 2n), (n, 3n), (n, n + p), etc. Therefore, the accuracy of these threshold reaction cross sections should be improved to satisfy the calculation requirements of high-energy neutron transmutation in advanced nuclear systems.

7.3.1.1 Processing Method for Transmutation Reaction Nuclear Data Libraries

The processing method for transmutation reaction nuclear data libraries is the same as that for nuclear reaction cross section data libraries. To further improve the transmutation calculation speed, the cross sections of each reaction are simply expressed by a single value by taking an average weighted by the flux. This method can be expressed as in Eq. (7.26).

$$\sigma_{\text{eff}} \int\limits_{space} d\vec{r} N_D(r) \int\limits_{G} dE\phi(E, r) = \int\limits_{space} d\vec{r} N(r) \int\limits_{G} dE\sigma(E, r)\phi(E, r) \tag{7.26}$$

where σ_{eff} is the effective one-group cross section, $N_D(r)$ is the average nuclide density, $\phi(E, \vec{r})$ is the flux, $\sigma(E, \vec{r})$ is the cross section, and $N(\vec{r})$ is the nuclide density.

Through discretization of Eq. (7.26), the multigroup cross section can be described as

$$\sigma_{\text{eff}} = \frac{\sum_{j=1}^{cell} N^j \sum_{g \in G} \sigma_g^j W_g^j}{\sum_{j=1}^{cell} N_D^j \sum_{g \in G} W_g^j} \tag{7.27}$$

where

$$N_D^j = \sum_j^{cell} V^j N^j / \sum_j^{cell} V^j$$

$$W_g^j = \int_j d\vec{r} \int_g dE\phi(E, \vec{r})$$

where V^j is the volume of region j.

7.3.1.2 Typical Transmutation Reaction Nuclear Data Libraries

A series of transmutation reaction nuclear data libraries have been released to meet the needs of transmutation calculations. Representative examples include the EAF [19] transmutation reaction nuclear data library provided in the FISPACT [20] code package, the transmutation reaction nuclear data library provided in the ORIGEN [21] code package, the FENDL/A transmutation reaction nuclear data library, and the SuperMC/HENDL transmutation reaction nuclear data library. This section introduces two transmutation reaction nuclear data libraries.

1. EAF transmutation reaction nuclear data library provided in the FISPACT code package

The EAF transmutation reaction nuclear data library was processed by the NJOY code, based on several evaluated nuclear data libraries including EAF, JEFF, ENDF/B, and TENDL. The library contains transmutation reaction information for more than 1900 nuclides. This library contains seven types of transmutation reaction nuclear data sub-libraries, as shown in Table 7.3.

The standard PWR weight function (Maxwell + 1/E + fission spectrum), which is used in the WIMS, XMAS, and TRIPOLI group structures, can be used for fission calculations. The weight functions used to generate the fusion relevant libraries are shown in Table 7.4.

Table 7.3 EAF transmutation reaction nuclear data library

Group structure	Energy range (MeV)	Application
WIMS 69	10^{-11}–20	Fission reactors
GAM-II 100	10^{-11}–20	Fusion systems
XMAS 172	10^{-11}–20	Fission reactors
VITAMIN-J 175	10^{-11}–20	Fusion systems
VITAMIN-J 211	10^{-11}–55	IFMIF, ADS
TRIPOLI 315	10^{-11}–20	Fusion systems
TRIPOLI 351	10^{-11}–55	IFMIF, ADS

Table 7.4 Weight function of the EAF transmutation reaction nuclear data library

Energy range	Weight function
10^{-5}–0.414 eV	Maxwell
0.414 eV–12.52 MeV	1/E
12.52–15.68 MeV	D-T
15.68–19.64 MeV	1/E

Table 7.5 Transmutation reaction nuclear data library provided in ORIGEN-2.0

Group structure	Energy range	Application
218 groups	10^{-11}–20 MeV	Fission/Fusion reactors
27 groups		Fission reactors
238 groups		Fission reactors and fusion systems
44 groups		Fission reactors
HANSEN 16 groups		Fission reactors

2. Transmutation reaction nuclear data library provided in the ORIGEN code package

The evaluated nuclear data files of the transmutation reaction nuclear data library in the ORIGEN 2.0 code package are ENDF/B. The library contains the transmutation reaction nuclear data of 1697 nuclides, including 689 light nuclides, 129 actinide nuclides, and 879 nuclear fission products. The Maxwell spectrum at thermal energies, the fission spectrum at neutron fission energies, and the 1/E spectrum at other energies in the library are combined as the weight function. There are five types of transmutation reaction sub-libraries in the transmutation reaction nuclear data library provided in ORIGEN-2.0, as shown in Table 7.5.

7.3.2 Decay Data Libraries

In transmutation calculation and analysis, decay data, including half-life, decay branch ratio, and decay mode, are required. Short-lived nuclides with half-life less than 1 s have little effect on transmutation calculations, whereas long-lived nuclides have significant effects on the calculated results.

The decay modes include α decay, β^+ decay, β^- decay, isomeric transition, spontaneous fission, neutron release, proton release, and combinations of these decay modes.

The representative decay libraries include EAF-2007/dec, FENDL-2.0/D, TENDL-2010/dec, etc. EAF-2007/dec contains the basic decay information of 2231 nuclides in ENDF-6 format. The data in the library are mainly from the evaluated

Table 7.6 Decay types in the EAF-2007/dec library

Identification	Decay type	Identification	Decay type
(b+)	β^+ decay/e	(b−)	β^- decay
(b − n)	n + β^- decay	(b−b−)	β^- decay + β^- decay
(b + n)	n + β^+ decay	(b −a)	α decay + β^- decay
(b + p)	p + β^+ decay	(b + a)	α decay + β^+ decay
(a)	α decay	(p)	p
(n)	n	(pp)	p + p
(SF)	Spontaneous fission	(IT)	Isomeric transition
(b − SF)	β^- decay followed by sponta-neous fission		

nuclear libraries JEF-3.1 and JEF-2.2. The following decay types in Table 7.6 are mainly considered in the decay data library.

7.3.3 Fission Yields Data Libraries

Fission yields data include the data describing the fission products, fission branching ratios, fission spectra, etc. In fusion–fission hybrid reactors, large amounts of actinides and minor actinides exist in the blanket, so fission yield data are required for the transmutation and burnup calculations.

There are several fission yield libraries, including ENDF/B-nfy, JENDL-nfy, EAF-nfy, etc. The fission yield data in EAF-nfy are usually separated into three energy regions by two energy boundaries (200 keV and 5 MeV). The yields of the thermal energy region (energy under 200 keV), fast energy region (energy between 200 keV and 5 MeV), and high-energy region (energy over 5 MeV) are denoted by $\mathrm{Y_t}$, $\mathrm{Y_f}$, and $\mathrm{Y_h}$, respectively. The EAF fission yield data library does not always give all three energy fission yields, and approximation by extrapolation is generally required. The specific process is [18]

(1) if there is only one fission yield Y, $Y_t = Y_f = Y_h = Y$;
(2) if only Y_t and Y_h, $Y_f = (Y_t + Y_h)/2$;
(3) if not given Y_t or Y_h, $Y_t = Y_f$ and $Y_h = Y_f$.

If the energy spectrums of the three energy regions are denoted by ϕ_t, ϕ_f, and ϕ_h, then the average fission yield Y is

$$Y = (\phi_h Y_h + \phi_f Y_f + \phi_t Y_t)/(\phi_h + \phi_f + \phi_t) \tag{7.28}$$

Thus, the effective cross section of the fission products can be obtained by multiplying the fission cross section by the average fission yield.

7.4 Radiation Damage Nuclear Data Libraries

Radiation damage calculation and simulation is one of the important bases for the material design in advanced nuclear systems, and radiation damage nuclear data libraries are the foundation of radiation damage calculations.

High-energy particles including neutrons, photons and protons in ADS systems, fusion reactors and fusion–fission hybrid reactors, cause material radiation damage.

The radiation damage caused by atomic displacement and transmutation gas in materials induced by the high-energy particles (the energy of neutron source is up to 1 GeV in ADS systems; the energy of neutron source is up to 14.06 MeV in fusion system) in these advanced nuclear systems is much more serious than that of traditional fission reactors. Therefore, the radiation resistance of fusion materials directly determines the safety and lifetime of fusion reactors. The radiation damage calculation for fusion materials is one of the most important bases for the design of structural materials for fusion systems. The radiation damage nuclear data library is the basis of the radiation damage calculation. This section introduces the processing method of the radiation damage nuclear data library and the typical libraries.

7.4.1 Processing Methods of the Radiation Damage Nuclear Data Libraries

The radiation damage nuclear data library contains the gas (helium and hydrogen) production cross sections and displacement damage cross sections (DPA) of materials generated by irradiation.

7.4.1.1 Gas Production Cross Section

Helium and hydrogen produced by neutrons and fusion material result in radiation damage in the material. The gas production cross section of helium and hydrogen can be obtained directly from the helium and hydrogen production cross section.

The helium production cross section is the sum of the ^{3}He and ^{4}He production cross section, given as follows:

$$\sigma_{\mathrm{He}}(E) = \sigma_{\mathrm{n},\alpha}(E) + \sigma_{\mathrm{n,n}\alpha}(E) + 2\sigma_{\mathrm{n},2\alpha}(E) + 3\sigma_{\mathrm{n},3\alpha}(E) + \cdots \tag{7.29}$$

The hydrogen generation cross section is the sum of the (n, p), (n, d), and (n, t) cross sections, given as follows:

$$\sigma_{\mathrm{H}}(E) = \sigma_{\mathrm{n,p}}(E) + \sigma_{\mathrm{n,d}}(E) + \sigma_{\mathrm{n,t}}(E) + \cdots \tag{7.30}$$

7.4.1.2 Displacement Damage Cross Section

In the atomic displacement damage calculations for a material, the entire process of neutron transport in planar targets is tracked until the neutron is absorbed, transmitted, or reflected, and the total number of primary knocked-on atoms (PKA) [22] is recorded. Then, each cascade process of the PKA is simulated, and the total displacements of the material are counted.

The damage energy in the material by PKA with kinetic energy E can be determined by the Lindhard [22] model. The atomic weight and atomic number of the PKA are denoted by A_1 and Z_1, respectively, and the kinetic energy is E. The atomic weight and atomic number of the target nuclide are denoted by A_2 and Z_2, and M_2 is the mass of the target nuclide. According to the Lindhard model, the damage energy T can be calculated.

$$T = \frac{E}{1 + kg(\varepsilon)} \tag{7.31}$$

where

$$g(\varepsilon) = 3.4008\varepsilon^{1/6} + 0.40244\varepsilon^{3/4} + \varepsilon \tag{7.32}$$

$$k = \frac{32}{3\pi}\left(\frac{m_e}{M_2}\right)^{1/2} \frac{(A_1 + A_2)^{3/2} Z_1^{2/3} Z_2^{1/2}}{A_1^{3/2}(Z_1^{2/3} + Z_2^{2/3})^{3/4}} \tag{7.33}$$

$$\varepsilon = [A_2 T/(A_1 + A_2)][a/(Z_1 Z_2 e^2)] \tag{7.34}$$

$$a = a_0(9\pi^2/128)^{1/3}(Z_1^{2/3} + Z_2^{2/3})^{-1/2} \tag{7.35}$$

where a_0 is the Bohr radius, m_e is the mass of an electron, and e is the electron charge.

The cross section of the damage energy can be calculated from the damage energy and energy transfer cross section of the neutron and the PKA [23].

$$\sigma_{damage}(E_i) = \sigma(E_i, E) \cdot T \tag{7.36}$$

where $\sigma(E_i, E)$ is the energy transfer cross section (the probability for transferring energy E from a neutron with E_i to the PKA). T is the normalized damage energy.

According to the NRT (Norgett, Robinson and Torrens) [24] model, the cross section of the damage energy is converted into a displacement damage cross section (DPA cross section).

$$\sigma_{DPA}(E_i) = \frac{0.8}{2E_d}\sigma_{damage}(E_i) \quad (7.37)$$

where E_d is the displacement threshold energy.

The DPA cross section above aims to describe a single-element material. For compound materials, the DPA cross section calculation has the following two treatment methods.

1. Weighted average method

For compound materials, the simple method of taking the weighted average of the isotope-wise DPA cross section can be applied. This approach is accurate when the atomic masses of the components are close, such as 316SS and 304SS stainless steel.

The DPA cross section of a compound material can be obtained from the DPA cross section of a single-element material, according to the component percentage in the compound material.

$$\sigma_{DPA}(total) = \sum_{i=1}^{N} w_i \sigma_{DPA}(i) \quad (7.38)$$

where $\sigma_{DPA}(total)$ is the DPA cross section of the compound material, w_i is the percentage of atom i in the compound material, and $\sigma_{DPA}(i)$ is the DPA cross section of atom i.

2. Collision combination energy transfer method

When the atomic masses of the constituents are not close in compound materials, the DPA cross section cannot be obtained by the weighted average method since the primary recoil atom can collide with other types of atoms with different masses.

In compound materials, the displacement damage associated with each atom depends on the type of the projectile and the target atoms. A method may be devised by considering the different cases of the collision cascade. Taking the collision cascade in SiC as an example, there are four cases, namely, C–C (PKA C collides with C), C–Si (PKA C collides with Si), Si–Si (PKA Si collides with Si), and Si–C (PKA Si collides with C). The damage energy T of the collision combinations is calculated according to the Lindhard model.

$$T_{\text{C-Si}} = \frac{Er(C)}{1 + kg(\varepsilon)}, T_{\text{Si-C}} = \frac{Er(Si)}{1 + kg(\varepsilon)} \quad (7.39)$$

$$g(\varepsilon) = 3.4008\varepsilon^{1/6} + 0.40244\varepsilon^{3/4} + \varepsilon \quad (7.40)$$

$$k = \frac{32}{3\pi}\left(\frac{m_e}{M_2}\right)^{1/2} \frac{(A_1 + A_2)^{3/2} Z_1^{2/3} Z_2^{1/2}}{A_1^{3/2}(Z_1^{2/3} + Z_2^{2/3})^{3/4}} \quad (7.41)$$

$$\varepsilon = [A_2 T/(A_1 + A_2)][a/(Z_1 Z_2 e^2)] \quad (7.42)$$

$$a = a_0(9\pi^2/128)^{1/3}(Z_1^{2/3} + Z_2^{2/3})^{-1/2} \tag{7.43}$$

where $E_r(C)$ and $E_r(Si)$ are the PKA energies of C and Si, respectively, Z and A are the atomic number and weight, respectively, and a_0 is the Bohr radius.

The cross section of the damage energy can be calculated based on the NRT model using the damage energy and energy transfer cross section of the collision combinations in SiC.

$$\begin{aligned} \sigma_{damage}(C/Si) &= \sigma(E_c) \cdot T_{\text{C-Si}} \\ \sigma_{damage}(Si/C) &= \sigma(E_{si}) \cdot T_{\text{Si-C}} \end{aligned} \tag{7.44}$$

The cross section of the damage energy for Si and C in the C–C, C–Si, Si–Si, and Si–C collision combinations can be calculated, where $\sigma(E_c)$ is the energy transfer cross section between a neutron and a C atom, and $\sigma(E_{si})$ is the energy transfer cross section between a neutron and a Si atom. The energy transfer cross section is then converted to the DPA cross section.

$$\sigma_{\text{DPA}}\left(\frac{C}{Si}\right), \sigma_{\text{DPA}}\left(\frac{Si}{C}\right), \sigma_{\text{DPA}}\left(\frac{C}{C}\right), \sigma_{\text{DPA}}\left(\frac{Si}{Si}\right) \tag{7.45}$$

The DPA cross sections of SiC can be obtained from the four combinations.

$$\begin{aligned} \sigma_{\text{DPA}}(\text{SiC}) = {} & 25\% \times \sigma_{\text{DPA}}\left(\frac{C}{Si}\right) + 25\% \times \sigma_{\text{DPA}}\left(\frac{Si}{C}\right) + 25\% \times \sigma_{\text{DPA}}\left(\frac{C}{C}\right) \\ & + 25\% \times \sigma_{\text{DPA}}\left(\frac{Si}{Si}\right) \end{aligned} \tag{7.46}$$

7.4.2 Typical Radiation Damage Nuclear Data Libraries

Most radiation damage nuclear data libraries are released together with radiation damage calculation code, including the radiation damage nuclear data library in SuperMC/HENDL developed by the FDS Team, the radiation damage nuclear data library provided in the NPRIM [25] code package developed by JAEA and the radiation damage nuclear data library provided in the SPECTER [26] code package developed by ANL.

7.4.2.1 Radiation Damage Nuclear Data Library in SuperMC /HENDL

The radiation damage nuclear data in SuperMC/HENDL are from the evaluated nuclear data libraries ENDF/B-VII, JEFF-3.1, and JENDL-4.0. The library includes the DPA cross sections and gas production cross sections of 52 nuclides.

The number of energy groups in the library is 1200, and the energy range is 10^{-11}–200 MeV. The thermal neutron energy region adopts a Maxwell distribution, the fission neutron energy region adopts a fission spectrum distribution, and the others adopt 1/E distributions.

The Lindhard and NRT models are used to calculate and process the DPA cross sections in this library. The gas production cross section is obtained by summing the cross sections of (n, p), (n, d), (n, t), (n, ^{3}He), and (n, ^{4}He).

7.4.2.2 Radiation Damage Nuclear Data Library Provided in the NPRIM Code Package

The NPRIM radiation damage calculation code is Windows and Macintosh-based graphical interface code developed by JAEA. The evaluated nuclear data of the radiation damage library provided in NPRIM are from JENDL-3.3 and JENDL-3.2. In addition, the procedure code is NJOY99.32.

The neutron energy range is 10^{-10}–20 MeV. This weight function combines a Maxwell distribution in the low-energy region, a fission spectrum in the fission energy region, and an l/E function in other energy regions. The library contains the DPA cross sections and gas production cross sections for 47 single-element materials, but does not provide radiation damage nuclear data for compound materials. In this library, the radiation damage nuclear data are calculated mainly by the HEATR module of NJOY, which uses the Lindhard model based on Robinson numerical approximation.

7.4.2.3 Radiation Damage Nuclear Data Library Provided in the SPECTER Code Package

SPECTER is a radiation damage calculation code released by ANL. The evaluated nuclear data of the radiation damage nuclear data library provided in SPECTER are from ENDF/B-V, and the processing code is DISCS. The neutron energy range is 10^{-10}–20 MeV. This weight function combines a thermal Maxwell distribution in the low-energy region, a fission spectrum in the fission energy region, and an l/E function in other energy regions.

The library contains the DPA cross sections and gas production cross sections for 41 single-element materials. This library also includes radiation damage nuclear data for compound materials, such as alloys, insulators, and propagating materials. The NRT model is used to calculate and process the DPA cross section in this library.

References

1. Otuka N, Dupont E, Semkova V et al (2014) Towards a more complete and accurate experimental nuclear reaction data library (EXFOR): international collaboration between nuclear reaction data centers (NRDC). Nucl Data Sheets 120:272–276
2. Ge ZG, Chen YJ (2015) Status and prospects of nuclear data development in China. China Sci Bull 60(3087):3098 (in Chinese)
3. Chadwick MB, Obložinský P, Herman M et al (2006) ENDF/B-VII. 0: next generation evaluated nuclear data library for nuclear science and technology. Nucl Data Sheets 107(12):2931–3060
4. Zhigang G, Hongwei Y, Youxiang Z et al (2007) The updated version of the Chinese evaluated nuclear data library (CENDL-3.1) and China nuclear data evaluation activities. In: International conference on nuclear data for science and technology. EDP Sciences
5. Santamarina A, Bernard D, Blaise P et al (2009) The JEFF-3.1. 1 nuclear data library. JEFF report 22(10.2):2
6. Shibata K, Kawano T, Nakagawa T et al (2002) Japanese evaluated nuclear data library version 3 revision-3: JENDL-3.3. J Nucl Sci Technol 39(11):1125–1136
7. Lemmel HD, McLaughlin PK (1944) BROND-2.2 Russian evaluated neutron reaction data library. International Atomic Energy Agency
8. Lopez Aldama D, Trkov A (2004) FENDL-2.1: update of an evaluated nuclear data library for fusion applications. International Atomic Energy Agency
9. Aldama DL, Leszczynski F, Trkov A (2003) WIMS-D library update. Final report of a co-ordinate research project
10. Forrest RA, Kopecky J, Sublet JC (2007) The European transmutation file: EAF-2007 neutron-induced cross section libraries. EURATOM/UKAEA Fusion Association
11. Jordan WC, Bowman SM, Hollenbach DF (1997) Scale cross-section libraries. Vol. III, Sect. M4 of SCALE: a modular code system for performing standardized computer analysis for licensing evaluation. NUREG/CR-0200 Rev 7(3)
12. Shimakawa S, Sekimura N, Nojiri N (2003) Radiation damage calculation by NPRIM computer code with JENDL3.3. Department of HTTR Project, Japan Atomic Energy research Institute
13. Greenwood LR, Smither RK (2002) SPECTER: neutron damage calculations for materials irradiations. Los Alamos National Laboratory
14. Zou J, Zeng Q, Chen C (2013) Development and application of HENDL-ADS/MG cross section library for ADS system. Am Nucl Soc 108:883–886
15. Zou J, Zeng Q, Xu D et al (2012) Design and producing of fine-group cross section library HENDL3.0/FG for subcritical system
16. Dezheng X, Jieqiong J, Jun Z et al (2009) Integral numerical test of the hybrid evaluated nuclear data library HENDL2.0 for heavy nuclides. Chin J Nucl Sci Eng 29(1):71–75
17. Xu D, He Z, Zou J et al (2010) Production and testing of HENDL-2.1/CG coarse-group cross-section library based on ENDF/B-VII. 0. Fusion Eng Des 85(10):2105–2110
18. MacFarlane RE, Muir DW (1994) The NJOY nuclear data processing system, version 91 (Vol. 7). Los Alamos, NM. LA-12740-M
19. Xie ZS, Deng L (2005) Numerical calculation method of neutron transport theory. Northwestern Polytechnical University Press (in Chinese)
20. Ding DZ, Ye CHT, Zhao ZH X et al (2003) Neutron physics-theory, methods, application. Atomic Energy Press (in Chinese)
21. Sun MP (2015) Development and test of fine group nuclear database for advanced nuclear systems. University of Science and Technology of China (in Chinese)
22. Forres RA, Gilbert MR (2007) FISPACT-2007: user manual. EURATOM/UKAEA fusion association
23. Wang RF (1993) Radiation damage of metallic material. Science Press (in Chinese)

24. Odette GR, Doiron DR (1976) Neutron-energy-dependent defect production cross sections for fission and fusion applications. Nucl Technol 29(3):346–368
25. Zou J, He Z, Zeng Q et al (2010) Development and testing of multigroup library with correction of self-shielding effects in fusion–fission hybrid reactor. Fusion Eng Des 85(7):1587–1590
26. Forrest RA (2007) The European transmutation file: EAF-2007 decay data library. EURATOM/UKAEA Fusion Association

Chapter 8
Comprehensive Neutronics Simulations

Comprehensive neutronics simulations are coupled simulations of neutron and multiple neutron-related physical effects for the entire space and life cycle of nuclear systems. The key goal of comprehensive neutronics simulations is to solve neutron transport problems characterized by strong anisotropy, which exist in the whole-life design and safety operations of advanced nuclear systems. In this chapter, the framework of comprehensive simulation systems is presented. Then, the modeling, calculation, and visual analysis are discussed. Finally, several typical simulation systems are briefly introduced.

8.1 Framework of Comprehensive Simulation Systems

The framework is the basis for comprehensive neutronics simulations of nuclear systems. The main simulation objects, related key problems, and typical system architectures are introduced in this section.

8.1.1 Main Simulation Objects

The main objects of comprehensive neutronics simulations involve neutron motion, nuclide transmutation, energy deposition, and their interactions, which together compose a "whole-process neutron transport simulation." In nuclear systems, neutrons are transported throughout the entire space, inducing transmutation (e.g., nuclear fuel depletion, nuclear waste transmutation, nuclear fuel breeding, and activation of the structural material and coolant). Neutron motion is also affected by the changes in nuclide density due to transmutation. These physical processes are tightly coupled. If the activated radionuclides leak into the environment, they may cause potential biological hazards. In addition, comprehensive neutronics simulations involve the

Y. Wu, *Neutronics of Advanced Nuclear Systems*,
https://doi.org/10.1007/978-981-13-6520-1_8

interactions between neutron transport and other physical processes, such as thermal hydraulics, structural mechanics, chemistry, electromagnetics, and biology, which are known as "coupled multi-physics neutronics simulations" [1].

During the operation of advanced nuclear systems, multiple physical processes interact, which span multiple temporal and spatial scales and include physical synergy effects, making the system extremely complicated. A lead-based reactor is used as an example. Lead-based materials have low moderation and absorption effects; thus, neutrons in the reactor core have higher energies than those in traditional fission reactors. These neutrons can be used to achieve the goals of nuclear fuel breeding and nuclear waste transmutation in addition to releasing energy through nuclear fission reactions. Therefore, it is necessary to consider a whole-process neutron transport simulation, including neutron motion, nuclide transmutation, and so on. Simultaneously, the structural materials are eroded and corroded by the high-temperature and high-density liquid lead-based material, and some of the corrosion products that flow with the coolant are activated under neutron irradiation. Hence, it is necessary to consider the interactions between multiple physical processes, including neutron transport, thermal hydraulics, structural mechanics, etc. in coupled multi-physics neutronics simulations. Comprehensive neutronics simulation theories and methods should be developed to demonstrate the integrated physical and engineering features of advanced nuclear systems throughout their entire space and life cycle and to satisfy the requirements of safety, reliability, economy, and sustainability.

To understand the physical characteristics and evolution behaviors of neutrons in advanced nuclear systems, theoretical and experimental studies and numerical simulations are commonly conducted. For example, the FDS Team developed CAD-based whole-process neutronics modeling and simulation methods. Based on these theories and methods, the FDS Team developed a comprehensive neutronics simulation system for advanced nuclear systems that performs high-resolution whole-process neutron transport and coupled multi-physics neutronics simulations to realize accurate and high-performance three-dimensional nuclear design and safety evaluations of complex nuclear systems. The data flow for a typical comprehensive neutronics simulation is shown in Fig. 8.1. The neutron motion calculation outputs the reaction rate for the nuclide transmutation calculation, which in turn provides feedback about the changes in nuclide density to the neutron motion calculation. In the coupled multi-physics neutronics simulation, the neutron transport calculation provides the nuclear heat for the thermal-hydraulic process, and the thermal-hydraulic calculation provides the temperature and the liquid density as feedback for the neutron transport process.

Based on comprehensive neutronics simulation systems, digital reactors and virtual nuclear power plants are developed to support the comprehensive simulations and behavior predictions for an entire plant and its environment throughout its entire life cycle.

Using information technologies (e.g., high-performance computing, cloud computing, visualization, and middleware) and fundamental physical models, digital reactors can perform comprehensive neutronics simulations and related multiphysics coupling simulations. These simulations can be applied to studies of funda-

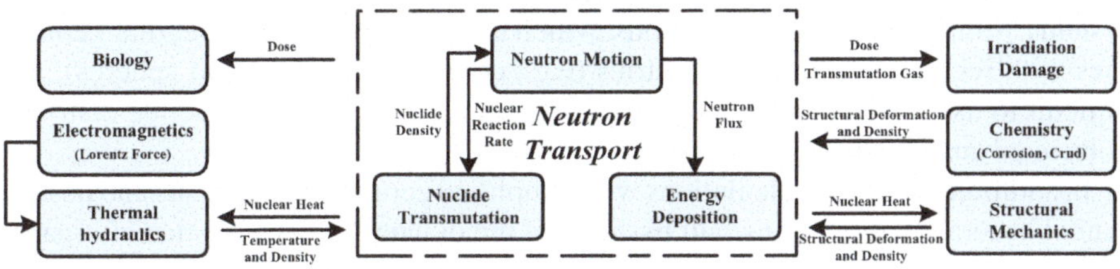

Fig. 8.1 Data flow of the comprehensive neutronics simulation

mental physical problems, the designs, and safety analyses of nuclear reactors and simulations of nuclear reactor operations.

Virtual nuclear power plants provide a comprehensive simulation platform that contains the entire environment within and around the nuclear power plant. Taking a digital reactor as their core, virtual nuclear power plants are deeply integrated with digital technology (including digital earth, digital meteorology, and digital traffic) and adopt advanced information technologies, such as big data, artificial intelligence, mobile internet, cloud computing, and the Internet of Things. Virtual nuclear power plants can be used to simulate multi-operational cases, monitor the evolution of nuclear accident processes, support nuclear emergency decisions, and conduct nuclear emergency exercises.

8.1.2 Key Problems

The complex physical characteristics of advanced nuclear systems cause great challenges for comprehensive neutronics simulations, as described in Chap. 1. For example, in neutron transport simulations of fusion systems, there are two major challenges [2]. First, it is difficult to create complex heterogeneous models, and second, it is nearly impossible to obtain accurate results with good precision in a reasonable time. Furthermore, comprehensive neutronics simulations, which are based on neutron transport and are tightly coupled with other physical processes, face additional challenges.

Generally, the process of comprehensive neutronics simulation includes three steps: modeling, calculation, and visual analysis. Each step has its own difficulties, and the main problems are listed as follows.

1. Large errors caused by simplified modeling methods

An important foundation for accurate nuclear designs and radiation safety assessments of advanced nuclear systems is to build models of the entire space, including all facilities, buildings, and workers.

Advanced nuclear systems contain large numbers of components with complex structures that are tightly arranged with numerous varieties of nuclides with irregular distributions, which causes difficulties in performing accurate modeling of nuclear systems. Simplified modeling methods, which usually approximate irregular geometries with regular and basic geometries (e.g., cuboids, cylinders, and spheres), are difficult to use to accurately describe the complex geometries and nuclide distributions in advanced nuclear systems.

In addition, radiation calculations with simplified geometries of organs and tissues cannot accurately represent small tissues and the dynamic characteristics of organs.

2. Low precision caused by isolated solutions for multiple processes and multiple physical phenomena

Isolated solutions are simply obtained by decoupling strongly related multiple processes and multiple physical phenomena and solving each independently. After one process has been calculated, the results over space, energy, and time are passed on as input parameters for the calculation of the next process. This approach cannot accurately describe the comprehensive features of advanced nuclear systems and results in low precision.

For example, in the coupled neutron motion and nuclide transmutation calculations of fission nuclear systems and hybrid nuclear systems, isolated approaches decouple the neutron motion process and the nuclide transmutation process. The steady-state reaction rate is calculated by solving the Boltzmann equation, and the dependence of the nuclide density with time is calculated by solving the Bateman equations based on the reaction rate obtained in the previous transport calculation. These calculation methods assume that the change in the reaction rate is negligible within a certain time step. Because the reaction rate is closely related to the material nuclide density, errors occur when dividing the time into several steps. There are many similar approaches in traditional simulations, which simply decouple the comprehensive neutronics simulation.

3. Difficulty in effective analysis and utilization of massive multidimensional data

In the analysis and safety evaluation of advanced nuclear systems, massive multidimensional data that represent the features of the advanced nuclear systems under review are generated by simulating various spatial distributions at different times with high-performance computing. Effective analysis and interpretation of the data to optimize the design and safe operation of the system has become the bottleneck of the data analysis.

Traditional visual analysis methods adopt the dimensionality reduction approach, which utilizes a single aspect of the data to assess the global characteristics of the system. For example, three-dimensional data are reduced to one or two dimensions for analysis. This approach has several disadvantages during visual analysis, such as complex processing, intensive labor, and low efficiency. In addition, this approach is poor for visualization, and important details may be lost, resulting in difficulties when determining the global characteristics and variations of the system over space, energy, and time.

8.1.3 Typical System Architectures

Comprehensive neutronics simulations provide an effective approach to perform accurate nuclear designs and safety assessments of advanced nuclear systems. The system architecture of such simulations is important for integrating the modeling, calculation, and visual analysis steps. The typical system architecture of a comprehensive neutronics simulation is shown in Fig. 8.2. The network and modular architecture support flexible flow control and data transfer of different physical processes in the whole-process neutron transport and coupled multi-physics neutronics simulation. Furthermore, a visualized comprehensive simulation platform can be established to enhance the quality and efficiency of the analysis and safety evaluation of advanced nuclear systems. Nonetheless, the difficulties of the modeling, calculation, and visual analysis of comprehensive neutronics simulations require additional attention.

(1) In the modeling step, whole-space and integrated models for whole-process neutron transport and coupled multi-physics neutronics simulations are needed.
(2) In the calculation step, based on the accurate whole-space modeling, the low accuracy and efficiency of isolated solutions must be improved to perform the coupled multi-physics, multi-scale, and multi-process calculations to realize the high-confidence whole-process neutron transport and coupled multi-physics neutronics simulations.
(3) In the visual analysis step, multidimensional and multi-style visual analysis is required to more efficiently explore the connections between different physical parameters (e.g., neutron flux, temperature, and pressure) and reveal the

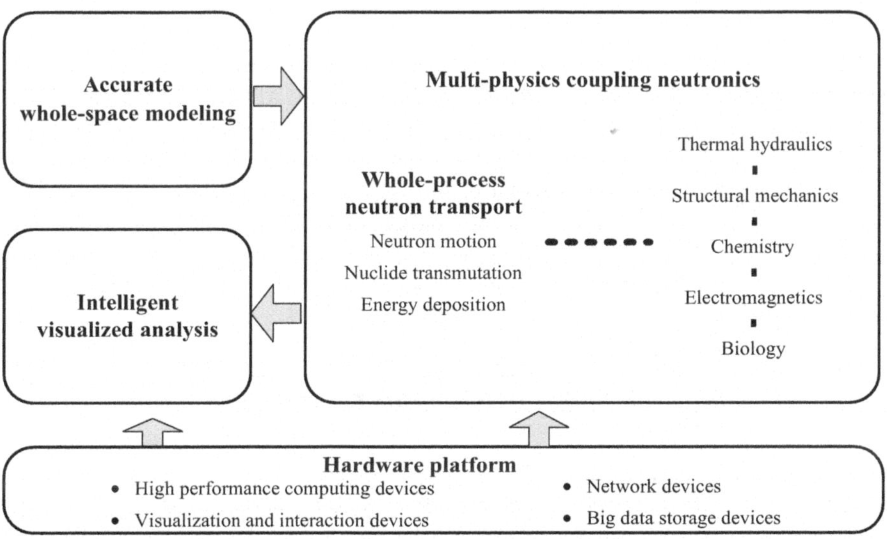

Fig. 8.2 Typical system architecture of a comprehensive neutronics simulation

comprehensive physical characteristics. In addition, a virtual advanced nuclear environment must be constructed with virtual reality technology to simulate the evolution, evaluation, and optimization processes.

8.2 Accurate Whole-Space Modeling

A comprehensive neutronics simulation requires whole-space modeling, including the facilities, buildings, and all of the gaps between them (voids). An accurate and integrated calculation model for the entire space of advanced nuclear systems provides an important basis for comprehensive neutronics simulations.

The complexity of the model in advanced nuclear systems is mainly reflected in two aspects.

(1) The model contains numerous irregular geometric structures with numerous boundary surfaces, which consist of high-order complex surfaces such as B-spline surfaces and swept surfaces. Simplified modeling methods use elements with regular geometries to describe complex irregular boundaries, which are imprecise, time-consuming, unintuitive, and error-prone, resulting in distorted models of complex systems.
(2) Some neutronics models of advanced nuclear systems are characterized by numerous geometric solids, complex topological relations, large spatial spans, and frequent parameter modifications. In these cases, traditional modeling methods have problems of high hardware resource consumption, long model generation time, and redundant information in the generated geometry that lead to low computational efficiency.

Several computer-aided design (CAD)-based neutronics modeling methods have been developed to solve these problems. For example, a CAD-based accurate whole-process modeling method for neutron transport has been proposed by the FDS Team, which can describe irregular geometries using hierarchical structures and hybrid elements (various facets and irregular primitives) and establish a unified model for the whole process of neutron transport, which includes neutron motion, nuclide transmutation, and energy deposition. This section will introduce the CAD-based accurate whole-process modeling methods in detail.

8.2.1 Modeling Methods for Irregular Geometries

The CAD-based accurate irregular modeling methods convert the CAD model described by Boundary REPresentation (BREP) into a neutronics calculation model described by primitives, meshes, high-order surfaces, etc., which can realize the accurate irregular description of a complex geometry by analyzing the topological relations of complex geometries in CAD models. The methods mainly include CSG (constructive solid geometry) modeling, facet modeling, mesh modeling, and voxel-based modeling. Among them, CSG, facet, and mesh modeling are mainly used for facilities and buildings in nuclear systems, and voxel-based modeling is mainly used for human in nuclear radiation environment.

1. CSG Modeling

CSG modeling uses primitives to describe the calculation model. CSG uses Boolean operations to combine basic primitives to compose complex solids. An example of a CSG representation is shown in Fig. 8.3. The representation has a binary tree structure with leaf nodes representing the basic elements, non-leaf nodes representing the Boolean operation results of the elements in the lower level, and a root node representing the generated solid.

A complex solid always consists of several convex solids. A convex solid always falls on one side of each of its surfaces and cannot be divided by its own surfaces. This section introduces accurate whole-space modeling methods based on convex solids [3]. These methods can convert CAD models into CSG models to accurately describe complex structures and nuclide distributions in advanced nuclear systems. The basic flowchart of this method is shown in Fig. 8.4. According to the flowchart, during the traversal of the solids in the CAD model, similarity estimation is performed to check if a solid with the same geometry has been decomposed. If not, face shell shrinking-based decomposition algorithm is used to decompose the complex solid into several convex solids and describe them with intersections of basic elements. Otherwise, the similarity estimation-based decomposition reuse method is utilized

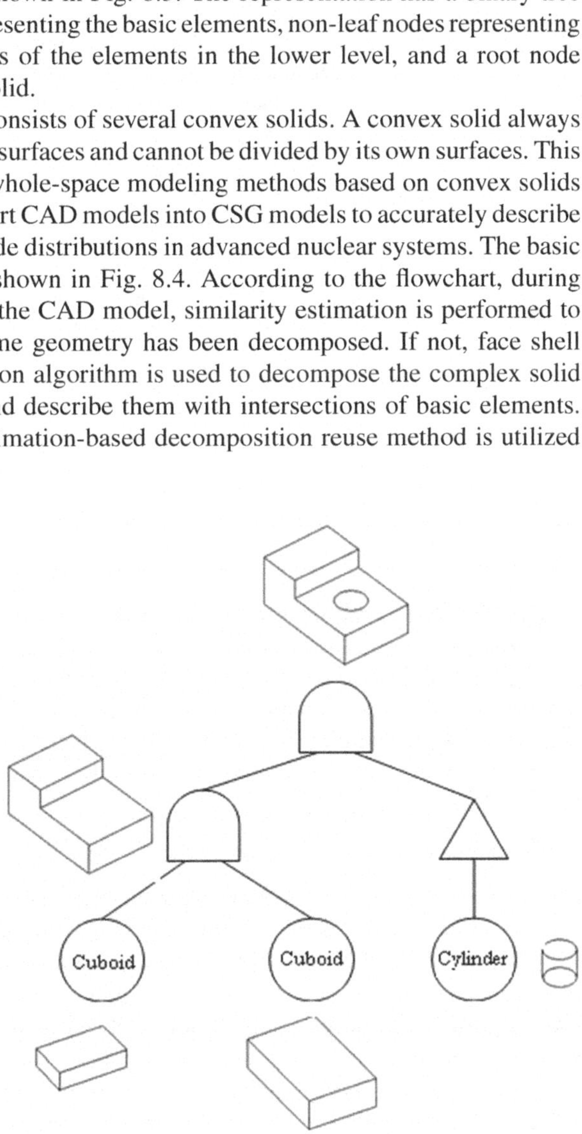

Fig. 8.3 Sketch map of the CSG representation

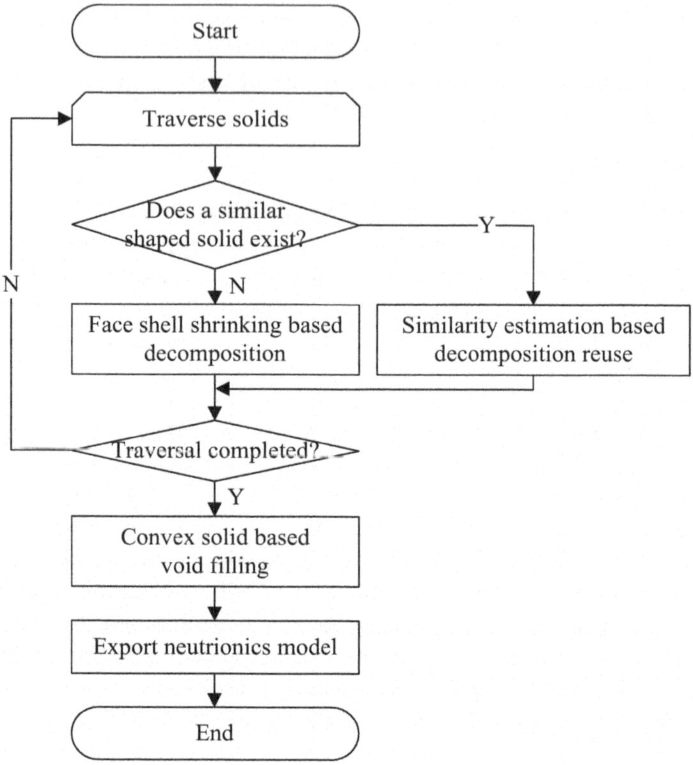

Fig. 8.4 Flowchart of the accurate whole-space modeling method based on convex solids

to reuse the decomposition results of the similar solid. Finally, a convex solid-based void-filling algorithm [4] is used to generate an accurate model for the entire space.

Figure 8.5 shows a cross-sectional view of the accurate whole-space ITER model built based on the modeling methods described above. The voids are hidden for easier observation. This complex and detailed model includes the ITER tokamak machine, tokamak building, tritium factory, diagnostic hall, etc., and it consists of more than 70,000 solids.

2. Facet Modeling

This method uses a set of basic planes or curved surface elements with limited areas (referred to as "facets") to describe the solid boundary and further describe the whole calculation domain. Instead of the entire surface used in the CSG modeling method, abundant small facets are used to approximate the actual boundary of the solid. Theoretically, an accurate calculation model can be generated by using a sufficient number of facets.

In a typical facet model, every solid is surrounded by triangular facets, and each triangular facet is connected to an adjacent facet with shared edges to ensure that the solid surfaces are tightly packed. A facet model can be generated automatically as

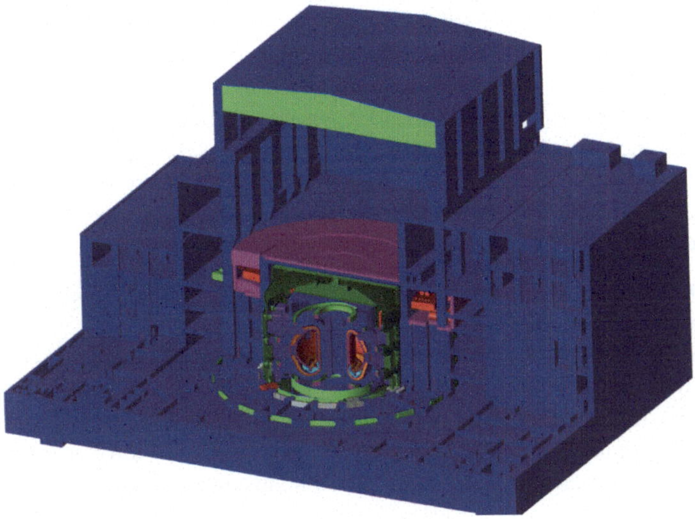

Fig. 8.5 Cross-sectional view of the accurate ITER whole-space model

(a) Solid model (b) Facet model (d=0.1 cm) (c) Facet model (d=0.01 cm)

Fig. 8.6 Facet models with different levels of accuracy

follows. Traverse the surfaces of the solid, and approximate each surface by using triangular facets according to the desired accuracy. Then, record the information of all triangular facets, including the index number, vertex coordinates, topological structure, etc., and finally, generate the facet model by using all the facet solids.

In the facet model, the accuracy is indicated by the maximum distance (d) from the original surface to the facet. Facet models with different levels of accuracy are shown in Fig. 8.6. As shown in the pictures, smaller d values lead to more facets. To accurately describe high-order surfaces, the maximum distance must be sufficiently small, but this leads to a considerable increase in the number of facets, resulting in low calculation efficiency. Therefore, it is necessary to develop a new method to balance the model accuracy and efficiency to optimize the facet modeling method.

3. Mesh Modeling

Mesh modeling uses cuboids, tetrahedrons, or hexahedrons (referred to as "meshes") to describe the whole problem domain. Meshes can be divided into two types: structured meshes and unstructured meshes. Compared to structured meshes, unstructured meshes have flexible node distributions and are easier to fit arbitrary geometries. Therefore, unstructured meshes are more appropriate for advanced nuclear systems with irregular surfaces. The commonly used unstructured meshes are tetrahedrons and hexahedrons. The algorithm used to generate tetrahedrons is more mature than that used to generate hexahedrons and has been widely used in thermal hydraulics, structural mechanics, etc. In addition, it has been applied in neutronics calculations.

This section introduces an automatic modeling method for converting CAD models into mesh models [5]. The basic process is as follows.

(1) Assign the necessary attributes (the material, its importance, etc.) used in the neutronics calculation for the solids.
(2) Analyze the geometry information, physical properties of the solids, and topological relations between the solids to generate the voids between the solids.
(3) Traverse the geometry and material information of all the solids, obtain the geometry boundary, and generate the meshes automatically according to the partitioning scheme to achieve a meshed neutronics model.

Figure 8.7 shows the mesh model of the ITER tokamak. The partition quality of the meshes directly determines the accuracy and efficiency of the solution. In addition, the memory footprint of the unstructured mesh model in the transport calculations is proportional to the number of meshes. When the number of meshes increases, the calculation efficiency decreases. It is therefore necessary to optimize the memory footprint of mesh models to improve their calculation efficiency for high-fidelity comprehensive simulation applications.

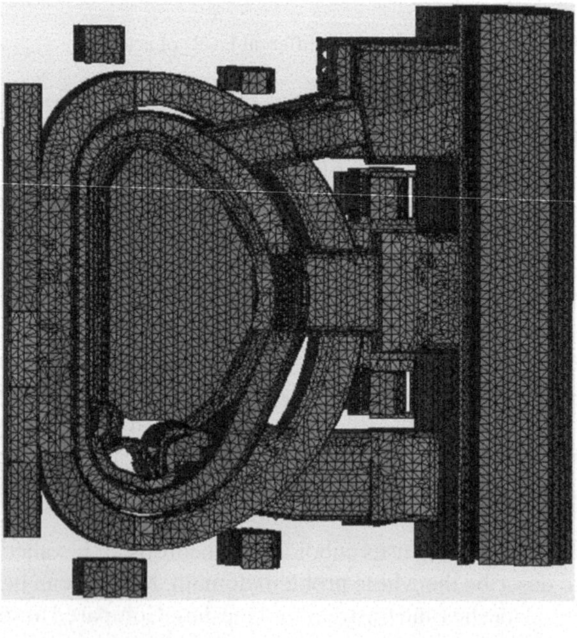

Fig. 8.7 Tetrahedron model of the ITER tokamak

4. Voxel-based Modeling

A voxel contains a value on a regular grid in three-dimensional space. Voxels have been applied to volumetric imaging in medical, scientific data visualization, etc. Based on the anatomical structure of humans, a model of the human body can be constructed using voxels as the basic units. Each voxel represents a certain material with a specific density and nuclide composition, and voxels with the same material represent specific organs or tissues. Figure 8.8 shows a voxel model of a human chest.

With the development of tomographic medical imaging, CT (computed tomography), MRI (magnetic resonance imaging), and section images are now able to accurately describe human organs and tissues. From tomographic images, such as CT and MRI images of live subjects and cross-sectional photographs of cadavers, voxel phantoms can be established to reconstruct the complex anatomical structures of the human body with tissue compositions and dynamic features for accurate dose evaluation. Tomographic medical imaging has become a popular field of research to develop more accurate computational models of the real human body based on such high-precision imaging data.

The main steps in generating a voxel model from medical imaging data are as follows:

(1) Register the images to make the coordinate system of the different images consistent with the space transformation.
(2) Divide the images according to the anatomical characteristics of the human body and draw the 2D contours of the organs and tissues.
(3) Reconstruct the 3D model of the organs and tissues by using 2D contours on each image.
(4) According to the respiratory motion patterns, construct models described by the 3D surfaces for different moments in the entire respiratory cycle.
(5) Generate 3D voxel models from the series of surface-based 3D models.
(6) Assign the material density and nuclide composition for each voxel according to the International Commission on Radiological Protection (ICRP) and Inter-

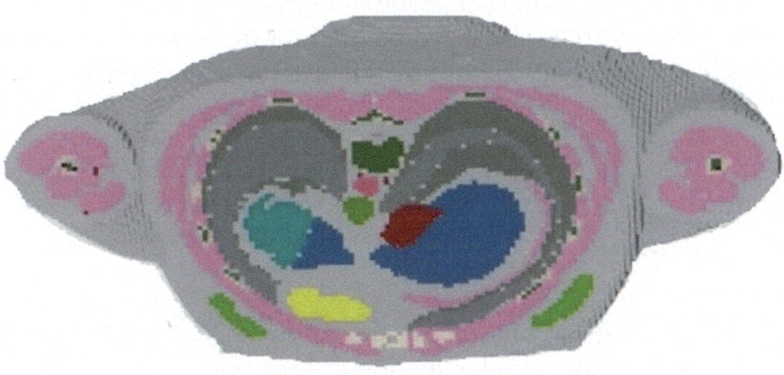

Fig. 8.8 Voxel model of a human chest

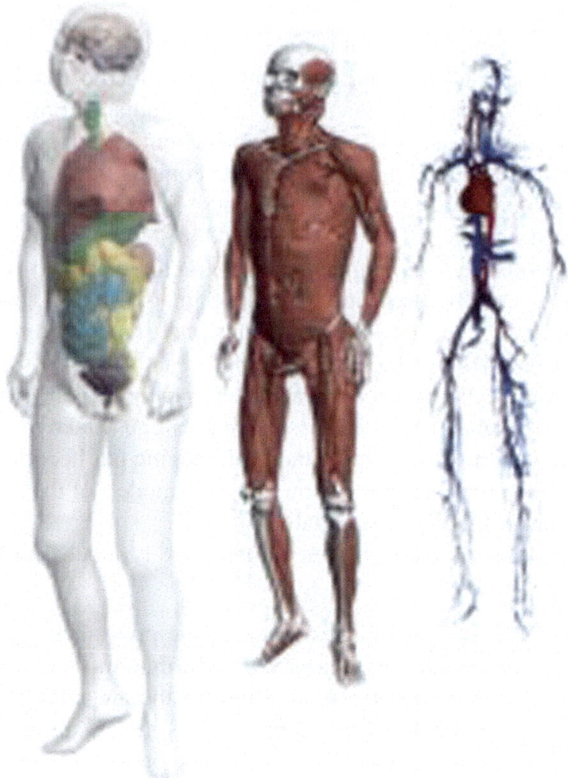

Fig. 8.9 View of the Rad-Human model (Chinese adult radiation virtual human model)

national Commission on Radiation Units and Measurements (ICRU) reports, and generate a four-dimensional voxel model for the final calculations.

Based on high-resolution color digital images of a Chinese adult female from the Chinese Visible Human datasets, the FDS Team constructed the Rad-Human phantom, as shown in Fig. 8.9. This phantom contains 46 organs and tissues with more than 28.8 billion voxels, and it can be used for the evaluation of human body radiation doses in nuclear plants, aerospace, nuclear medicine, radiation therapy, etc.

8.2.2 Modeling Methods for Complex Structure Configuration

When building a whole-space neutronics calculation model characterized by a large number of geometric solids, complex topological relations, large spatial spans, and frequent parameter modifications, traditional modeling methods have several problems, such as large hardware resource consumption and long model generation times. During the process of feature recognition-based whole-space modeling, the definition features are recognized through intelligent analysis of a CAD model, and the whole-space neutronics calculation model is constructed rapidly by utilizing these features to intelligently decompose the complex systems.

In advanced nuclear systems, there are many complex topological relations, such as inclusions, fills, and arrays. For example, fission systems contain an obvious hierarchical feature that the model can be divided into different levels (core → assembly → pin). Each level consists of a large number of similar geometries that are arranged with a certain lattice. In addition, the surface types of the model are quite different in different regions of complex nuclear systems. For example, in fusion systems, the first wall is mainly composed of high-order curved surfaces, and the biological shielding layer is mainly composed of simple planes. Hence, it is important to use these features, such as the hierarchical structure, geometric shape similarity, and surface types, to develop efficient whole-space neutronics modeling methods.

1. Hierarchical modeling method

The parameter visualization-based hierarchical modeling method [6] constructs the model at different independent levels by analyzing the filling relations. The flowchart of this method is shown in Fig. 8.10. All of the geometric parameters are combined into a hierarchical tree through hierarchical relations. The hierarchical tree controls the entire model flexibly, which is convenient for building geometric models with different levels of fineness to satisfy different neutronics calculation requirements. In addition, based on the similarity evaluation and analysis of the topological relations, this method can rapidly create a refined model by defining the geometry with a unique shape and the array rules between the geometric solids.

The neutronics calculation model based on the hierarchical tree maintains the topological relations between the geometries. It can perform iterative modifications of the model parameters based on the constraint relationship between the geometries, and higher computational efficiencies can be achieved by tracking particles in the generated neutronics calculation models with a hierarchical structure. In addition, based on the characteristics of the hierarchy and material properties of the models, a variety of segmented approaches can be constructed, which can realize more efficient interactions and multi-scale visualization of complex models by flexible segmented rendering.

The hierarchical tree contains five kinds of nodes, namely, the hierarchical node (Level), component node (U), subcomponent node (C), single-component filling node (FILL), and multiple-component array-filling node (LAT). Figure 8.11 shows an example of a model of a typical lead-based fast reactor core that was built using the hierarchical modeling method. Consistent with the definitions in the hierarchical tree, U0 represents the entire core model and contains the core fuel (C1), outer coolant (C2), and outer shield (C3) from inside to outside. In the modeling process, the model can be constructed level-by-level. For example, in the core fuel zone, U1 to U6 (fuel assembly, reflection assembly, shielding assembly, absorption rod,

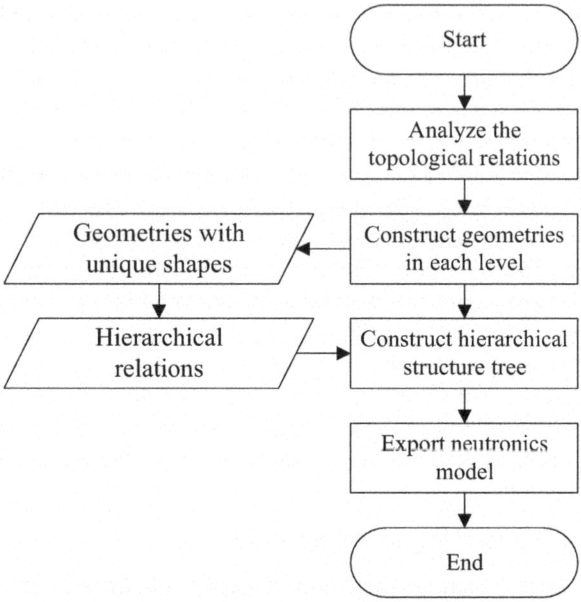

Fig. 8.10 Flowchart of the hierarchical modeling

adjustment rod, and compensation rod, respectively) are arrayed in accordance with the rules of LAT0, and U1 is composed of the fuel (C4), cladding (C5), and coolant (C6); an assembly-level model can then be generated according to the hierarchical relations. If further refined modeling is required, the fuel rod (U7) (composed of the end plug (C7), gas chambers (C8), nuclear fuel (C9 and C10), and cladding (C11)) is arrayed in C4, and a rod-level refined model can be obtained.

2. Adaptive Hybrid Modeling

CSG modeling can address various low-order surfaces, provide a concise representation for the calculation model, and track particle trajectories efficiently. However, it cannot describe high-order complex surfaces and requires time-consuming pre-processing. The facet and mesh modeling can approximate high-order complex surfaces by incorporating numerous facets or meshes without pre-processing, which can save a significant amount of manual time. However, the calculation accuracy and efficiency will decrease when utilizing large amounts of facets or meshes. In addition, mesh modeling has been widely used for thermal hydraulics, structural mechanics, and other physics analyses, whereas voxel-based modeling is always used in the evaluation of irradiation environments on workers.

An adaptive hybrid model established according to the model features can make full use of the advantages of various geometric models, such as CSG primitives, facets, meshes, and voxels, which is important for increasing the accuracy and efficiency of comprehensive neutronics simulations. For example, geometric solids with only analytic surfaces can be described by CSG modeling; geometric solids with high-order complex surfaces can be approximated with a sufficient number of facets; mesh

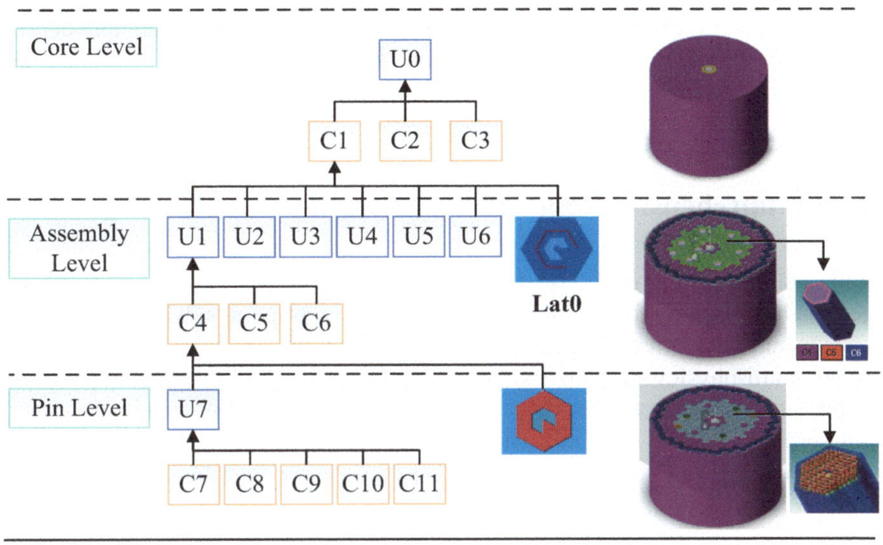

Fig. 8.11 An example of hierarchical modeling based on the core of a typical lead-based reactor

modeling is used in the area of multi-physics coupling calculations, such as thermal hydraulics and structural mechanics; and voxel-based models are used for humans in nuclear radiation environments. In this way, a whole-space calculation model can be established for the whole-process and coupled multi-physics neutronics simulations of advanced nuclear systems.

The adaptive hybrid model of the FFHR (Force-Free Helical Reactor) is shown in Fig. 8.12. Based on the surface types, the outer solids are described with CSG primitives, whereas the inside coils, which consist of high-order surfaces, are approximated

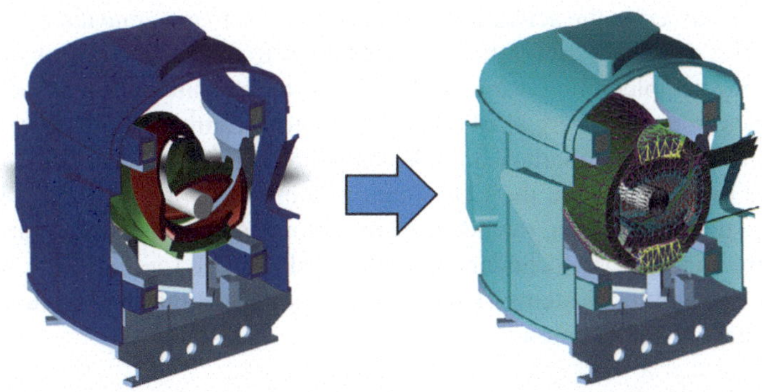

(a) Original CAD model (b) Hybrid model

Fig. 8.12 Construction of an adaptive hybrid model

with facets. Therefore, the adaptive hybrid model can combine the advantages of both CSG primitives and facets to increase the calculation efficiency by decreasing the number of facets.

8.3 Comprehensive Neutronics Calculations

Traditional calculation methods based on isolated processes have many disadvantages, such as oversimplification and poor iterative convergence, which lead to low precision and time-consuming calculations for whole-process neutron transport and coupled multi-physics neutronics simulations. To improve the calculation precision and efficiency, this section is dedicated to introducing the following two topics: whole-process neutron transport calculation and coupled multi-physics neutronics calculation.

8.3.1 Whole-Process Neutron Transport Calculation

Neutron transport describes the motions and interactions of neutrons with materials. Whole-process neutron transport simulation involves neutron motion, nuclide transmutation, and energy deposition. Isolated solutions decouple the multiple closely related processes, solve each process independently, and introduce many assumptions and simplifications, which result in missing several important secondary nuclides and secondary particles.

For example, after the shutdown of nuclear systems, it is necessary to calculate the radiation dose caused by the decay photons released from the activated materials. Isolated approaches decouple the neutron motion process from the activation process. The neutron motion calculation is performed first to calculate the neutron flux, which is passed to the activation calculation codes to generate the decay gamma spectrum. The shutdown dose rate is then obtained from the photon transport calculation with the sampled photon source. Isolated solutions have low accuracy because the actual positions and energies of the decay photons cannot be represented, and the information about the secondary nuclides and secondary particles is lost.

To solve these problems, the FDS Team proposed an integrative physics model that couples neutron motion, nuclide transmutation, and energy deposition based on the response function of secondary nuclides and achieved an accurate neutronics simulation of complex systems. This section first briefly introduces the integrated neutron transport model below:

$$\frac{1}{v}\frac{\partial \phi(\boldsymbol{r}, E, \boldsymbol{\Omega}, t)}{\partial t} + \boldsymbol{\Omega}\nabla\phi(\boldsymbol{r}, E, \boldsymbol{\Omega}, t) + \sum_{i=1}^{N} n_i(\boldsymbol{r}, t)\sigma_{\mathrm{t},i}(\boldsymbol{r}, E)\phi(\boldsymbol{r}, E, \boldsymbol{\Omega}, t)$$

$$= \int_0^{\infty} \mathrm{d}E' \int_{\boldsymbol{\Omega}'} \sum_{i=1}^{N} n_i(\boldsymbol{r}, t)\sigma_{s,i}(\boldsymbol{r}, E')f(\boldsymbol{r}, E' \to E, \boldsymbol{\Omega}' \to \boldsymbol{\Omega})\phi(\boldsymbol{r}, E', \boldsymbol{\Omega}', t)\mathrm{d}\boldsymbol{\Omega}'$$

$$+ \frac{\chi(\boldsymbol{r}, E)}{4\pi} \iint \upsilon(\boldsymbol{r}, E') \sum_{i=1}^{N} n_i(\boldsymbol{r}, t)\sigma_{f,i}(\boldsymbol{r}, E')\phi(\boldsymbol{r}, E', \boldsymbol{\Omega}', t)\mathrm{d}E'\mathrm{d}\boldsymbol{\Omega}'$$

$$+ S(\boldsymbol{r}, E, \boldsymbol{\Omega}, t) \tag{8.1}$$

where $n_i(r, t)$ is the number of nuclide i at position r and time t in units of cm^{-3}. The change in $n_i(r, t)$ with time t can be described by Eq. (8.2)

$$\frac{\mathrm{d}n_i}{\mathrm{d}t} = -n_i(\lambda_i + \sigma_i\phi) + \sum_{j \neq i} n_j(\lambda_{i,j} + \sigma_{i,j}\phi) + \sum_k n_k\sigma_k^{\mathrm{f}}\phi Y_{i,k} \tag{8.2}$$

where λ_i is the decay constant of nuclide i in units of s^{-1}, $\lambda_{i,j}$ is the decay constant of radionuclide i generated by the decay of nuclide j in units of s^{-1}, σ_i is the total cross section for the reaction of nuclide i with a neutron in units of cm^2, $\sigma_{i,j}$ is the production cross section of nuclide i produced by the reaction of nuclide j with a neutron in units of cm^2, σ_k^{f} is the fission reaction cross section of nuclide k in units of cm^2, ϕ is the neutron flux in units of $1/(cm^2 \cdot s)$, and $Y_{i,k}$ is the yield of nuclide i from the fission reaction of nuclide k.

Based on the integrated neutron transport model, an accurate and highly efficient calculation method has been developed for the shutdown dose rate calculation. This method can obtain the distribution of secondary radionuclides under different energies and different neutron irradiation times from activation calculations with full consideration of the interaction between neutrons and the original nuclides/secondary radionuclides and the decay of secondary radionuclides under the real irradiation scheme. The flowchart for this method is shown in Fig. 8.13.

This method obtains the distributions of secondary radionuclides produced by different nuclides under different energies and different neutron irradiation times by activation calculations, fits the relation $f_j(\rho_{i,0}, \phi_{g,0})$ between the amount of radionuclides produced by different nuclides and the initial nuclide density $\rho_{i,0}$ and the neutron flux of energy group g $\phi_{g,0}$, and establishes the corresponding database. For an initial nuclide i with a density of ρ_i, the nuclear density of the produced secondary radionuclide j is $\rho_{i,j,g}$ under irradiation with a given irradiation scheme and a given neutron flux density ϕ_g.

$$\rho_{i,j,g} = \phi_g \times \rho_i \times f_j(\rho_{i,0}, \phi_{g,0}) \tag{8.3}$$

For radionuclide j, decay photons with different energies may be emitted, where the total probability of emitting photons C_j can be obtained by querying the relevant decay database. The decay photon source emitted by radionuclide j with a density of $\rho_{i,j,g}$ is

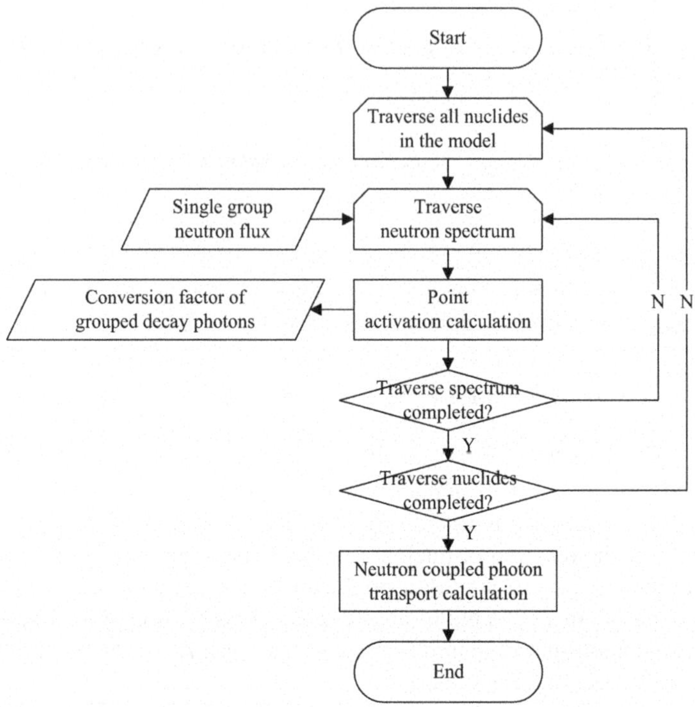

Fig. 8.13 Flowchart of the integrated physics model considering the effect of secondary nuclides

$$I_j = C_j \times \rho_{i,j,g} \tag{8.4}$$

Because the initial nuclides may produce multiple radionuclides under neutron irradiation and each radionuclide may emit photons of multiple energies, the decay photon source generated by nuclide *i* under irradiation with a given irradiation scheme and a given neutron spectrum is

$$I_i = \sum_g \sum_j C_j \times \phi_g \times \rho_i \times f_j(\rho_{i,0}, \phi_{g,0}) \tag{8.5}$$

Based on an accurate distribution of the secondary radionuclide, this method can obtain the decay photon distribution (e.g., energy, intensity) and establish a corresponding database. The decay photon source generated by a material that consists of multiple nuclides with a density of ρ_i under irradiation with a given irradiation scheme and a given neutron spectrum is

$$I = \sum_i \sum_g \sum_j C_j \times \phi_g \times \rho_i \times f_j(\rho_{i,0}, \phi_{g,0}) \tag{8.6}$$

In the calculation of the shutdown dose rate, when a neutron reacts with a nuclide, according to the irradiation plan, the neutron energy, and the nuclide type, the decay photon distribution under the current irradiation conditions can be obtained directly from the database. Finally, the shutdown dose rate is obtained by the photon transport calculation based on the decay photons that are sampled from this distribution.

8.3.2 Coupled Multi-physics Neutronics Calculation

The transient behavior of nuclear systems is highly complex and nonlinear, and it includes several related physical processes, such as neutron transport, thermal hydraulics, and structural mechanics. Understanding the transient behavior is the basis of nuclear design, reactor safety analysis, and operation of nuclear systems and is helpful in improving safety and thus increasing the economic efficiency. In recent years, as the safety and economic requirements for nuclear systems have increased, precise transient simulations of neutronics and non-neutronics feedback effects, namely, coupled multi-physics neutronics simulations, have become necessary.

For example, in the core physics design, an assembly must achieve the design objective of a certain power. To meet this requirement, a certain heat flux must be achieved first, and the highest temperature of the internal part of the assembly and cladding is then calculated. In an isolated thermal-hydraulic design, a conservative estimate of the neutron flux is required to obtain the highest temperature. However, through the coupled simulation, a more reasonable neutron flux can be obtained to accurately calculate the highest temperature, which can prevent excessive design costs due to an overly conservative estimate.

During coupled multi-physics neutronics calculations, the flow control and data transfer of each process should be considered, and the relationships between the processes in terms of space, energy, and time should be accurately represented. This approach can eliminate the inaccuracy and inefficiency caused by isolated solutions and achieves high-performance and high-fidelity simulations. The coupling methods for coupled multi-physics neutronics calculations can be divided into two categories, loose coupling, and tight coupling, based on whether multiple physical processes are calculated simultaneously or not.

1. Loose coupling

In loose coupling calculations, all of the physical processes are calculated separately, and the final results are obtained through iteration. In one iteration step, the output of one process is used as the input to other physical processes. For example, in the loose coupling-based neutron transport and thermal-hydraulic calculations, neutron transport is performed to calculate the power density, which is passed to the thermal hydraulics to generate the temperature distribution. Then, the temperature distribution is used to update the input of the neutron transport to calculate the power density again. These processes are repeated until the power density and temperature distribution converge. This approach is simple and easy to implement using existing codes, and the calculation procedure of one process is a black box to the other processes, so each process can select different codes and methods flexibly. However, errors occur in the decoupling of nonlinear equations. To reduce the calculation errors, the steps are should be very fine, which causes high computational costs.

Common loose coupling methods include the operator splitting (OS) method [7] and the Anderson acceleration method. The OS method only requires the data of the previous iteration for the current step and has a first-order convergence speed. However, it is essentially a fixed-point iteration method, and the calculation results often do not converge and need to be combined with a relaxation iteration. The Anderson acceleration method has better convergence and a faster convergence speed; however, it performs a weighted average of the calculation results of the last several steps for the current step and requires more memory to store the data.

2. Tight coupling

The basic idea of tight coupling calculations is to simultaneously solve the coupled multi-physics equations while considering realistic nonlinear feedback between multiple processes, which can realize an accurate and efficient coupled multi-physics neutronics simulation. For example, in the tight coupling of neutron transport and thermal-hydraulic calculations, the nonlinear equations of neutron transport and thermal hydraulics are simultaneously solved to obtain the power density and temperature distribution in the same time step. The partial differential equations in this approach consider the interactions between multiple physical processes and reflect the physical characteristics of complex systems, resulting in high precision and efficiency with a coarse time step. However, existing codes cannot be used directly with this approach. In addition, it is difficult to provide the equations with the form of residuals, and they are not easy to implement, particularly in cases that the discrete temporal or spatial schemes for multiple physical processes are not consistent.

The Newton family methods are generally used in tight coupling simulations, of which the Jacobian-free Newton–Krylov (JFNK) method is most widely used [8]. In the outer iteration, it uses the Newton iteration method to preserve the good convergence (second order, fast convergence). In the inner iteration, it uses the Krylov subspace method instead of solving the Jacobi matrix required by the Newton iteration method, which can significantly improve the calculation speed and reduce the computational cost. However, the JFNK method needs to simultaneously solve partial differential equations for multi-physics coupling problems; therefore, it is much more complicated when dealing with multi-scale problems.

3. Typical case

This section introduces the coupled method for quasi-static SN neutron transport with thermal hydraulics for the transient safety analysis of nuclear reactors. In this method, the multidimensional discrete-ordinate quasi-static neutron transport kinetics model is adopted for the neutronics calculation, and the Euler fluid dynamic model is adopted for the thermal-hydraulic calculation to achieve multiple velocity fields, multiphase, and multicomponent simulations. During the coupled calculation, physical quantities

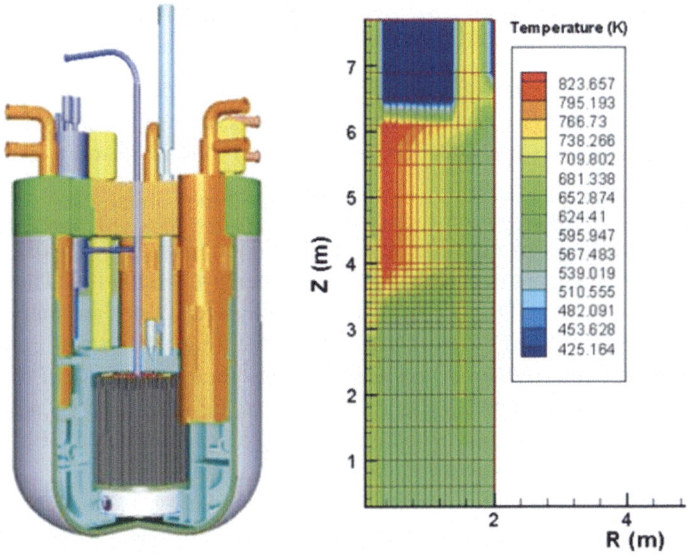

Fig. 8.14 Safety analysis of CLEAR-I based on SuperMC

such as the nuclear heat from the neutronics calculation are utilized as inputs for the thermal-hydraulic calculation, and physical quantities, such as the fuel and fluid temperatures and the fluid density from the thermal-hydraulic calculation, are fed back into the neutronics calculation. The reactor behavior at different time steps can be simulated using this iterative procedure. Based on this method, a coupled neutronics and thermal-hydraulic simulation module has been developed in SuperMC for transient safety analyses of nuclear reactors. Figure 8.14 shows the temperature distribution of CLEAR-I (Lead-based Research Reactor) based on the safety analysis module of SuperMC.

8.4 Intelligent Visual Analysis

With scientific visualization technology, massive amounts of multidimensional data generated by comprehensive simulations can be intuitively analyzed. Based on virtual simulations, work schemes can be evaluated and optimized in a safe and vivid virtual environment. This section introduces advanced methods and technologies for visual analysis and virtual simulation as well as several typical cases.

8.4.1 Multidimensional and Multi-style Visual Analysis

The analysis of calculation results at different times and spatial positions is helpful for guiding system design and optimization (such as identifying the safety risks in component design). Two intelligent visualization methods are introduced as follows.

1. Intelligent visualization of multi-style and multidimensional data

Multidimensional (e.g., space, time, and energy) data, such as flux, dose, temperature, and pressure data, are generated in a comprehensive simulation. These data are usually projected onto a 3D space, in which the physical values are mapped to different colors and visualized by setting the parameters (e.g., space, time, and energy) to enable intuitive and dynamic analysis to reveal their variations in the multidimensional space. Furthermore, a specific region can be visualized by adjusting the range of the data, and the data can be clipped using a programmable GPU (Graphics Processor Unit) to only display the field of interest to the user.

Multi-style data visualization refers to surface rendering, iso-surface rendering, and volume rendering. Surface rendering uses planar elements to approximate the surfaces of a 3D data field. To achieve a high rendering efficiency, only the outer surface of the data field is rendered. Therefore, the distribution of the physical data field over the calculation region is represented quickly. Iso-surface rendering extracts iso-surfaces from the 3D data field and then displays part of the data of interest in 3D. Volume rendering synthesizes colored images from any viewpoint for 3D objects defined in 3D space to visualize the 3D data field over the entire space. Figure 8.15 shows a multi-style visualization of physical data of the CLEAR-I model, in which (a) presents the surface rendering of the neutron flux, (b) presents the iso-surface rendering of the radiation dose, and (c) presents the volume rendering of the nuclear heat.

2. Intelligent-coupled visualization of the data field and geometric model

During neutronics analyses of advanced nuclear systems, due to the complexity of the geometries involved, it is often difficult to efficiently locate results in the data field for an accurate analysis. Intelligent-coupled visualization of the data field and geometric model can directly associate the results with geometric models that analysts are familiar with to effectively help the analysts perform accurate and intuitive analyses and improve the analysis efficiency. Two methods for coupled visualization of the data field and geometric model are described below.

(1) Overlapping visualization of data fields and geometric contours [9]. In this method, the data field and geometric boundary are cut by the same planes to generate geometric contour lines to assist the positioning analysis.
(2) Texture mapping onto geometric surfaces for data visualization. First, using a programmable GPU and 3D textures, the 3D physics data are converted to texture data. The complicated and refined physics data can then be mapped to the model surfaces based on the 3D texture coordinates, and different colors of the model surfaces represent different values of the data.

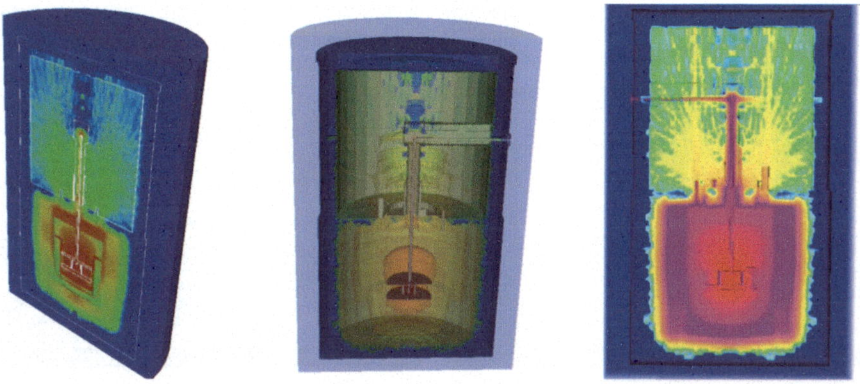

(a) Surface rendering of the neutron flux (b) Iso-surface rendering of the radiation dose (c) Volume rendering of the nuclear heat

Fig. 8.15 Multi-style visualization of physical data

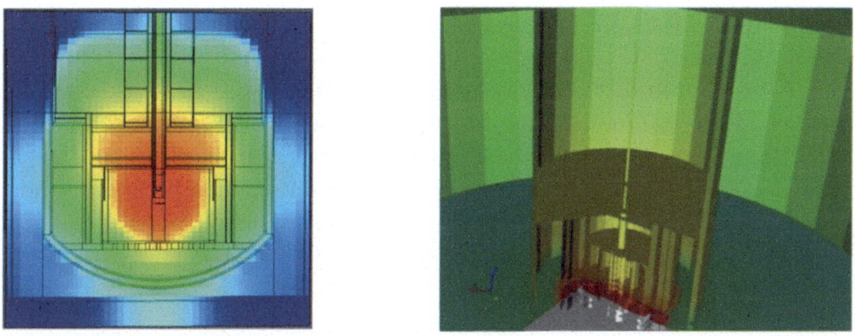

(a) Visualization of the geometric contour lines with the flux on a cut plane (b) Data texture mapped onto the surfaces

Fig. 8.16 Data visualization coupled with the geometric model

Figure 8.16 shows the data visualization coupled with the geometric model, in which (a) presents the flux visualization projected onto a plane and coupled with the contours of the CLEAR-I model, and (b) presents the flux visualization with the flux texture mapped onto the surface of the CLEAR-I model.

3. Typical case

Research on the characteristics of the radiation field in the containment compartment above the reactor cover of CLEAR-I is used as an example of multidimensional visual analysis. In the system, the introduction of an acceleration makes the radiation much more significant, especially above the reactor cover, due to particle leakage through the beam duct and secondary radiation induced by beam losses. Analyzing

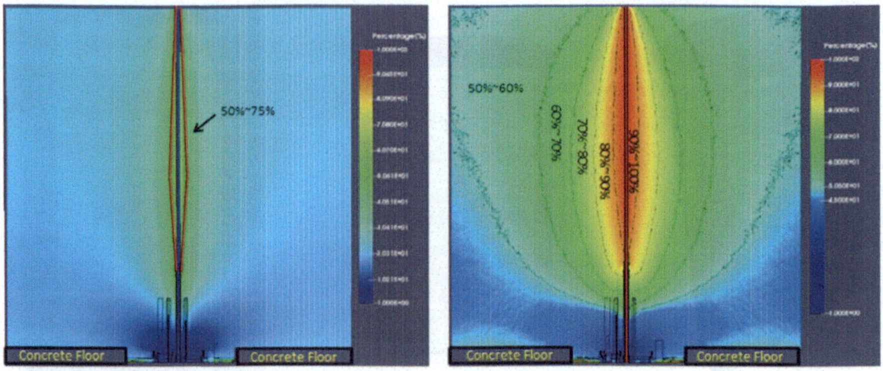

(a) Percentage of neutron flux caused by beam loss (b) Percentage of dose caused by beam loss

Fig. 8.17 Distribution of contributions from beam losses to radiation in the compartment [10]

the radiation characteristics can provide guidance for the shielding design of the upper section of the CLEAR-I system.

Using coupled data field and model visualization, the distribution of the contributions of the neutron flux and dose rate are shown in Fig. 8.17. For the neutron flux, as shown in Fig. 8.17a, the contribution from beam duct leakage dominates except for the area around the upper beam duct. However, for the dose rate, as shown in Fig. 8.17b, the majority of the contribution is from secondary particles induced by beam losses. The visual analysis indicates that the influence of beam losses needs to be carefully considered in the shielding design of CLEAR-I.

8.4.2 *Virtual Simulation*

Based on virtual reality technologies, especially stereoscopic displays and human—computer interactions, virtual simulations enable users to interact with objects naturally within a virtual environment to gain experience with the real system. Nuclear design and radiation safety simulation platforms based on virtual simulations enable design and optimization in safe virtual radiation environments.

Because most advanced nuclear systems are still under development, virtual simulation is one of the most effective approaches to demonstrate their complicated structures and physical processes and determine their design and optimization schemes. Virtual simulations of radiation safety evaluation and their applications in neutronics analysis of advanced nuclear systems will be introduced in this section.

1. Virtual simulation of radiation safety evaluation

Using virtual simulations, which connect theoretical studies with experimental research, work schemes can be simulated and assessed before implementation to improve the efficiency of the implementation and reduce the radiation dose for the

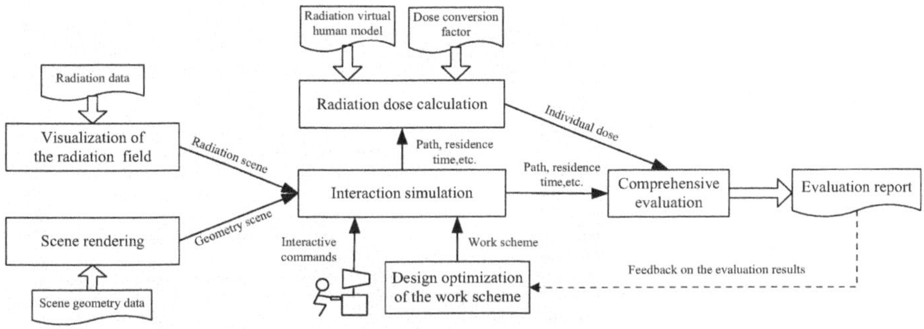

Fig. 8.18 Typical virtual simulation process of a radiation safety evaluation

staff. Virtual radiation safety simulations are mainly used to simulate the working conditions in a nuclear radiation environment, reduce the individual radiation dose, and develop a reasonable work scheme.

A typical procedure of a virtual radiation safety simulation is shown in Fig. 8.18. The operational processes are simulated for a nuclear radiation environment to assess and optimize the work scheme. The radiation data and structure data are used as inputs. Technologies such as radiation field visualization and geometric scene rendering are adopted to optimize the work scheme, interactive simulation, radiation dose calculation, and comprehensive evaluation.

The radiation data described above can be obtained by forward calculations and measurements. In the forward calculation, the radiation source terms (e.g., fission neutrons, photons, and radionuclides) should be determined first; then, the radiation data can be provided by the dose calculation. This approach is mainly used to develop the design scheme and decommissioning plan of nuclear systems. With the measurement results, radiation data can also be obtained by inversion or interpolation, which is usually used in the operations, accident response, and maintenance of nuclear systems. In the virtual simulation of radiation safety evaluations, accurate radiation data are the basis for improving the efficiency of work schemes and decreasing the dose rate of workers.

In addition, due to the increasing demands for improved realism, speed, precision, accuracy, and timeliness in 3D interactive operations, large-scale simulation scenes have become a popular area of research within the virtual simulation technology. To address these challenges, two advanced methods are proposed.

(1) Accelerated rendering technology with view cone clipping based on an adaptive octree. The entire visual scene is divided by an octree to make the number of facets of the leaf nodes meet the requirements of the scene clipping algorithms. In addition, the intersection of the bounding box of the octree nodes and cones is determined in the scene-rendering process to avoid the low efficiency of directly identifying the intersection of huge facets and view cones and to achieve the accelerated rendering of complex systems in the virtual environment.

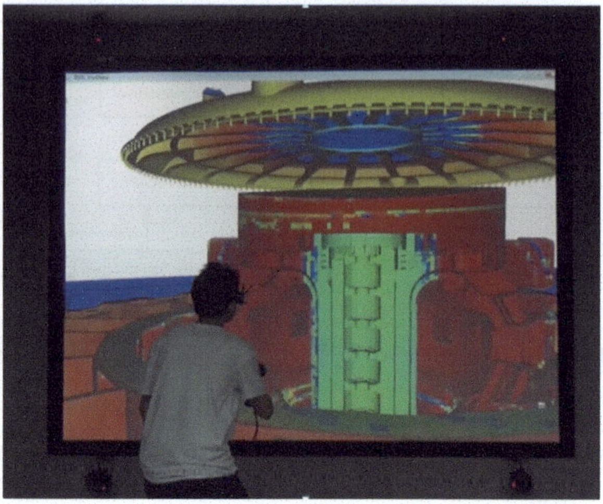

Fig. 8.19 Interactive simulation between a human and a virtual scene

(2) 3D human–computer interaction method based on a multi-degree-of-freedom human motion tracking system [11]. By tracking human motion in real time and superimposing it onto a 3D virtual scene, a person from the real world can interact with virtual objects in a virtual scene in real time. Figure 8.19 shows a 3D interactive simulation between a person and a virtual scene based on a multi-degree-of-freedom human motion tracking system. The realistic experience obtained from the radiation environment and virtual operations can effectively help maintenance workers become familiar with operating procedures and develop basic operating skills.

2. Typical case

As an example case, a virtual simulation of the replacement of the CLEAR-I's spallation target is presented. The spallation target is a key component that connects an accelerator and a subcritical reactor. Spallation reactions occur under the bombardment of high-energy protons, and the generated neutrons are provided for the subcritical reactor. Due to the bombardment of high-energy and high-intensity protons and the irradiation of high-energy neutrons, the target window must be replaced periodically because of its short life. The replacement operation of the spallation target is mainly carried out above the reactor cover. Based on the structural design, the replacement of the spallation target requires extracting the proton beam duct from the target loop and manual interventions, such as the installation of the track and temporary shielding before the replacement operation. Because the activated proton beam duct will produce a large amount of radiation, it is necessary to perform radiation safety simulations of the maintenance process to ensure the smooth implementation of the maintenance and the safety of the workers.

In this case, a 3D interactive simulation of the CLEAR-I's spallation target replacement was performed. Based on the Rad-Human model, the accumulated effective dose, effective dose rate, and equivalent dose for the radiosensitive organs of the

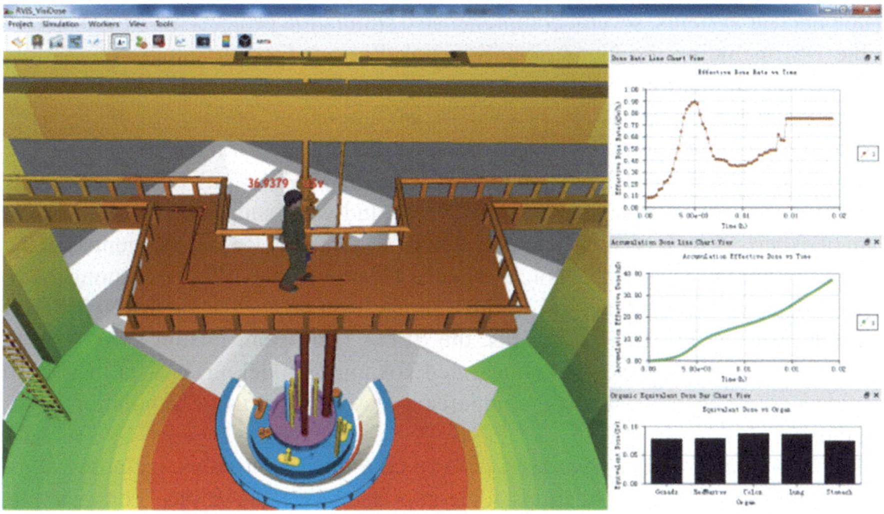

Fig. 8.20 Virtual roaming simulation and real-time organ dose evaluation

maintenance workers were assessed in real time, as shown in Fig. 8.20. Based on the design limits, the accumulated effective dose was 0.25 mSv, and the maximum effective dose rate was 0.69 μSv/h, which is under the design limit of 100 μSv/h. This simulation indicated that for 10 years of operation and a 30-day shutdown, the irradiation environment in the top containment chamber meets the dose limitation requirement for workers to enter before extracting the proton beam duct from the target loop.

8.5 Typical Comprehensive Simulation Systems

Many studies have been carried out to demonstrate the integrated physical behavior of advanced nuclear systems throughout their entire space and life cycle and realize comprehensive neutronics simulations.

The United States, which has the most nuclear power units in the world, has always paid much attention to the development of comprehensive simulation platforms for nuclear systems. For fission systems, the DOE Energy Innovation Hub for Modeling and Simulation of Nuclear Reactors has funded the CASL (Consortium for Advanced Simulation of Light Water Reactors) and NEAMS (Nuclear Energy Advanced Modeling and Simulation) projects to extend the lives of existing reactors and develop new reactors. Among them, VERA (Virtual Environment for Reactor Applications) is a system of physics capabilities being developed and deployed by the CASL. The Idaho National Laboratory and Argonne National Laboratory developed the multi-physics coupling frameworks MOOSE and SHARP, respectively. For fusion nuclear systems, several comprehensive simulation projects have been started in the US, including the Numerical Tokamak Project, First Tokamak Simulation, FSP (Fusion Simulation Project), etc.

The European Union also pays a significant amount of attention to the research and development of comprehensive simulation technology for nuclear systems. For fission systems, the NURESIM (NUclear REactor SIMulation) research projects were proposed for nuclear reactor applications and were supported by a European team of experts, which is consistent with the SRA (Strategic Research Agenda) of the European SNETP (Sustainable Nuclear Energy Technology Platform). Among them, SALOME is a general numerical simulation platform supported by the NURESIM research projects. For fusion systems, the EFDA (European Fusion Development Agreement) established the ITM (Integrated Tokamak Modeling) project in 2003 to provide a comprehensive test and comparison platform for fusion simulation codes and data.

In China, many studies have been performed on these activities, including algorithms, platforms, and applications for comprehensive simulations. For example, focusing on the characteristics of advanced nuclear systems, the FDS Team has been developing SuperMC (Super Multi-functional Calculation program for nuclear design and safety evaluation) and Virtual4DS since 1998. In addition, several nuclear power companies in China have conducted research on the construction of nuclear design software packages for light water reactors, such as the NESTOR software package, which was developed by the Nuclear Power Institute of China National Nuclear Corporation, the COSINE software package, which was developed by State Power Investment Corporation Limited, and the PCM software package "Intelligent Power Station", which was developed by China General Nuclear Power Corporation.

This section provides detailed introductions to typical comprehensive simulation systems, including SuperMC, VERA, and SALOME.

8.5.1 SuperMC

SuperMC [12, 13] is a general, intelligent, accurate, and precise simulation software system that was developed by the FDS Team over more than 20 years of continuous research. Taking whole-process neutron transport as the core, SuperMC also supports multi-physics coupling calculations of thermal hydraulics, structural mechanics, chemistry, and biology. In addition, based on a cloud computing framework [14], it integrates CAD-based accurate modeling of irregular geometries and intelligent visual analysis. SuperMC can be used for the design and safety evaluation of nuclear systems as well as for nuclear technology application fields, including radiation medicine, nuclear detection, and so on. The system architecture of SuperMC is shown in Fig. 8.21.

For the modeling, a CAD-based accurate irregular modeling method that can convert models between CAD systems and multiple calculation codes [15–17], such as MCNP, TRIPOLI, PHITS, TORT, FLUKA, and Geant4, has been developed to

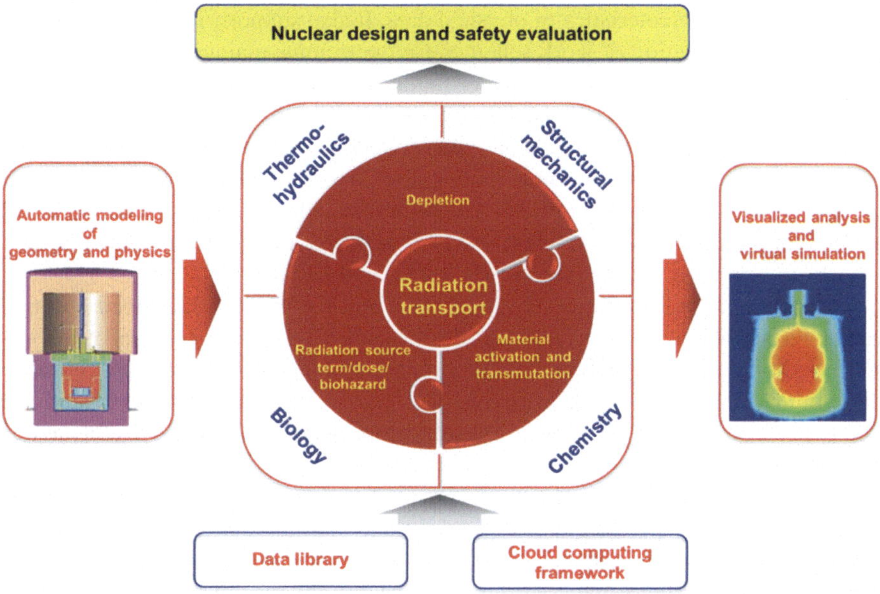

Fig. 8.21 System architecture of SuperMC

provide whole-space and accurate models for comprehensive neutronics simulations. This method also provides parameter-based high-fidelity (pin-by-pin level) complete whole-reactor modeling and interactive modeling with geometric and physical information. In addition, it can convert CT, MRI, and cross-sectional photographs into accurate human computational phantoms, which can be applied to medical physics and radiation protection [18].

For the transport calculations, SuperMC supports MC and deterministic coupling calculations in the spatial, energy, and time dimensions. SuperMC also supports whole-process neutronics calculations, including internally coupled neutron motion, nuclide transmutation, and energy deposition calculations. In addition, SuperMC provides coupled calculations of neutron transport and thermal hydraulics for accurate and efficient 3D nuclear design and safety evaluation for complex nuclear systems, where the MC method or the multidimensional discrete-ordinate quasi-static method can be used for the neutron transport, and the Euler fluid dynamic model is adopted for the thermal-hydraulic calculation to achieve multiple velocity field, multiphase, and multicomponent simulations. In addition, SuperMC provides the functions of intelligent nuclear design, which supports the optimization of the core physics design and radiation protection design and automatically and efficiently obtains the optimal schemes for the materials selection and configuration in the core and shield to satisfy the design objectives of safety, economy, and sustainability.

For the visual analysis, SuperMC developed intelligent visual analysis based on scientific visualization and virtual reality technologies [19, 20]. SuperMC supports multi-style visualization of multidimensional physical data fields to intuitively identify the physical characteristics of physical data. It also supports the evolution, evaluation, and optimization of work schemes for operations, maintenance, decommissioning, and accident emergencies under radiation environments.

SuperMC has been widely used in over 60 countries and more than 40 mega nuclear engineering projects, and it has been collected and publicly distributed by the OECD (Organization for Economic Cooperation and Development)/NEA (Nuclear Energy Agency) Data Bank. SuperMC has been selected as a reference code by ITER and has supported the establishment of ITER 3D basic neutronics models [21–26]. It has been also used in the design of HPR1000 and CLEAR.

In addition, based on SuperMC, the FDS Team has further developed a digital reactor "VisualBUS". This systematic design and safety evaluation software was developed to achieve a visual and whole-process design and 3D whole-space operations simulations for the entire life cycle of a reactor. VisualBUS will eventually be used to perform predictions of reactors with comprehensive multi-physics behaviors with high fidelity.

To address the R&D requirements of advanced nuclear systems, the FDS Team has initiated research on a virtual nuclear power plant "Virtual4DS" based on SuperMC and VisualBUS. In this research, intelligent accident simulations and emergency decision-making systems that consider the impact of the reactor on the environment and society are being developed to support design and operations simulations, accident warning and process simulations, large-scale radionuclide diffusion and environmental impact assessments, nuclear emergency exercises, intelligent decision-making, and social risk assessment.

8.5.2 VERA

VERA (Virtual Environment for Reactor Applications) is a system of physics capabilities being developed and deployed by the CASL [27], which was established in 2010. Based at ORNL, CASL is the DOE's first Energy Innovation Hub—an integrated research center that focuses on a single topic with the objective of rapidly bridging basic research, engineering development, and commercialization.

CASL is a collaboration of US leading scientists, institutions, and supercomputers, with an aggressive 10-year mission to confidently predict the performance of existing and next-generation commercial nuclear reactors through comprehensive, science-based modeling, and simulation. Real world impacts include the following:

(1) Improving efficiency in nuclear power production by reducing unanticipated plant outages and enabling future increases in operating power, resulting in additional power generation.
(2) Lowering costs by better understanding how long fuel can reside in a reactor. With improved reactor and fuel burnup estimates, energy providers could save millions of dollars annually.

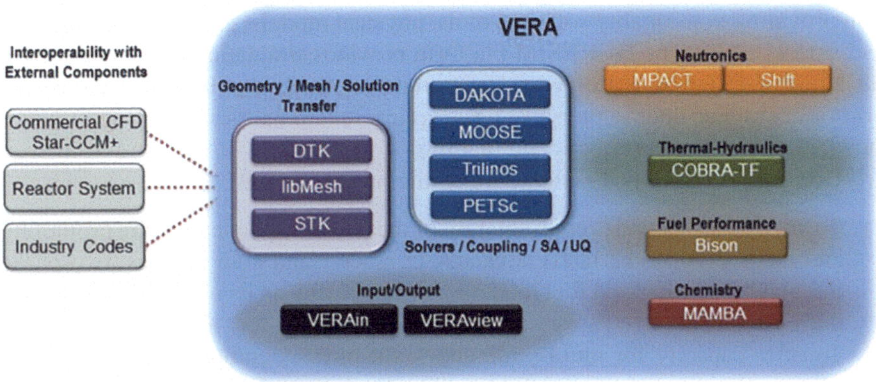

Fig. 8.22 Framework of VERA [28]

(3) Enhancing safety through evaluation of new fuels that can better endure the severe conditions within a reactor.
(4) Extending the life of existing reactors through improved prediction of the lifetimes of key structural components.

VERA includes computational tools and supporting infrastructure that represent the cutting edge in light water reactor modeling and simulation, and it can be used to solve neutronics, thermal hydraulics, fuel performance, and coupled physical problems with advanced uncertainty quantification tools, as shown in Fig. 8.22. VERA integrates physics components based on science-based models, state-of-the-art numerical methods, and modern computational science, and it has been verified and validated using data from operating PWRs, single-effect experiments, and integral tests. VERA is optimized for efficient execution on multiple platforms, including leadership-class computers, advanced architecture platforms now under development, and industrial engineering workstation clusters. In the neutronics calculation, the Monte Carlo code Shift and deterministic code MPACT are used to perform steady-state neutron transport calculations, whereas ORIGEN is used for the nuclide transmutation calculations.

8.5.3 SALOME

SALOME is a general numerical simulation platform supported by NURESIM research projects. NURESIM was developed by 22 organizations in 14 European countries following a roadmap that is consistent with the SRA (Strategic Research Agenda) of the European SNETP (Sustainable Nuclear Energy Technology Platform). The roadmap of NURESIM is generally aimed at improving the safety of light water reactors through deterministic analyses of nuclear power plant events in the scope of the plant design basis. NURESIM intends to be a reference platform providing high-quality software tools, physical models, generic functions, and assessment results. The NURESIM platform provides simulation capacity for light water reactor design basis accidents. Compared to methods used by industry, the platform offers higher fidelity physics and coupling functionalities, resulting in better knowledge about conservatisms and margins.

NURESIM is developed with the following objectives [29]:

(1) To provide an accurate representation of the physical phenomena by promoting and incorporating the latest advances in reactor and core physics, two-phase thermal hydraulics, and fuel modeling (with a focus on its mechanical behavior during accidents).
(2) To offer capacity for multi-scale and multi-physics computations, especially for coupling core physics and thermal hydraulics models for reactor safety; to provide easy coupling of the different codes and solvers through the use of a common data structure and generic functions, for instance, for interpolation between nonconforming meshes.
(3) To provide generic pre-processing and post-processing and supervision functions through the SALOME open source tool.
(4) To validate the individual models, solvers, codes, and the platform through challenging applications corresponding to nuclear reactor situations and including reference calculations, experiments, and plant data; to complement the validation by using quantitative deterministic and statistical sensitivity and uncertainty analyses.

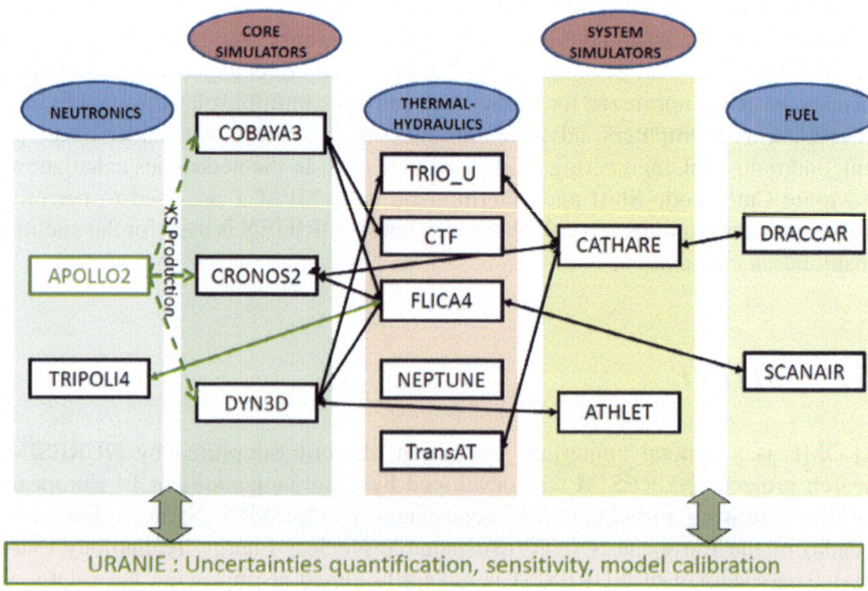

Fig. 8.23 Framework of SALOME [30]

SALOME provides interoperability between CAD, meshing and computing codes, and solvers and shares a pool of generic tools for pre-processing, post-processing, and code coupling, as shown in Fig. 8.23. The core features of SALOME include supporting interoperability between CAD modeling and computational software; facilitating the integration of new components into heterogeneous systems for numerical computation; setting priority to the multi-physics coupling between the computational software; providing a generic, user-friendly, and efficient user interface, which helps to reduce the costs and delays of carrying out studies; reducing the training time to the specific time required to learn the software solution based on this platform; and providing access to all functionalities via the integrated Python console. In terms of the neutronics calculations, that Monte Carlo code TRIPOLI4 and deterministic code APOLLO2 are integrated into SALOME, and several cross section library generation codes are also integrated to realize high-fidelity (pin-by-pin level) simulation.

References

1. Wu YC (2017) Fusion neutronics. Springer Nature Singapore Pte. Ltd
2. Loughlin MJ, Polunovskiy EI, Zheng SL et al (2011) ITER approach to interfacing CAD systems with the nuclear analysis program MCNP. Nucl Technol 175(1):271–275
3. Wu YC (2009) CAD-based interface programs for fusion neutron transport simulation. Fusion Eng Des 84(7–11):1987–1992
4. Yu SP, Cheng MY, Song J et al (2015) Convex-based void filling method for CAD-based Monte Carlo geometry modeling. Annu Nucl Energy 85:380–385
5. Long PC, Zou J, Huang SQ et al (2010) Development and application of SN auto-modeling tool SNAM 2.1. Fusion Eng Des 85(7–9):1113–1116
6. Gan Q, Wu B, Yu SP et al (2016) CAD-based hierarchical geometry conversion method for modeling of fission reactor cores. Ann Nucl Energy 94:369–375
7. Zerkak O, Kozlowski T, Gajev I (2015) Review of multi-physics temporal coupling methods for analysis of nuclear reactors. Ann Nucl Energy 84:225–233
8. Knoll DA, Keyes DE (2004) Jacobian-free Newton-Krylov methods: a survey of approaches and applications. J Comput Phys 193(2):357–397
9. Long PC, Zeng Q, He T et al (2011) Development of a geometry-coupled visual analysis system for MCNP. Prog Nucl Sci Technol 2:280–283
10. Li B, Yang Q, Chang B et al (2016) Preliminary analysis of radiation characteristic at upper section of accelerator driven subcritical system. Ann Nucl Energy 90:410–416
11. Yang ZH, Tao H, Hu LQ et al (2015) Development of high-immersive simulation system for designing maintenance strategy and its application to CLEAR-I. Ann Nucl Energy 83:309–315
12. Wu YC (2018) Multi-functional neutronics calculation methodology and program for nuclear design and radiation safety evaluation. Fusion Sci Technol 74(4):321–329
13. Wu YC, Song J, Zheng HQ et al (2015) CAD-based Monte Carlo program for integrated simulation of nuclear system SuperMC. Ann Nucl Energy 82:161–168
14. http://www.supermc.cn
15. Wang GZ, Xiong J, Long PC et al (2011) Progress and applications of MCAM: Monte Carlo automatic modeling program for particle transport simulation. Prog Nucl Energy 2:821–825
16. Hu HM, Wu YC, Chen ML et al (2007) Benchmarking of SNAM with the ITER 3D model. Fusion Eng Des 82(15–24):2867–2871
17. Zhang JJ, Hu LQ, Zeng Q, et al (2011) Development and application of MC-SN coupled auto-modeling tool RCAM1.0. Fusion Eng Des 86(9–11):2783–2786

18. Cheng MY, Huang SQ, Li J et al (2011) Progress of HUMOP: human automatic modeling program. Trans Am Nucl Soc 104:648–649
19. Luo YT, Long PC, Wu GY, et al (2010) SVIP-N 1.0: an integrated visualization platform for neutronics analysis. Fusion Eng Des 85(7–9):1527–1530
20. Wu YC, He T, Hu LQ et al (2015) Development of virtual reality-based simulation system for nuclear and radiation safety. At Energy Sci Technol 29:77–85
21. Zeng Q, Lu L, Ding AP et al (2006) Update of ITER 3D basic neutronics model with MCAM. Fusion Eng Des 81(23–24):2773–2778
22. Zeng Q, Wang GZ, Dang TQ et al (2012) Use of MCAM in creating 3D neutronics model for ITER building. Fusion Eng Des 87:1273–1276
23. Ying DC, Zeng Q, Qiu YF et al (2011) Assessment of radiation maps during activated divertor moving in the ITER building. Fusion Eng Des 86(9–11):2087–2091
24. Dang TQ, Ying DC, Yang Q et al (2012) First Neutronics analysis for ITER bio-shield equatorial port plug. Fusion Eng Des 87:1447–1452
25. Yang Q, Dang TQ, Ying DC et al (2012) Activation analysis of coolant water in ITER blanket and divertor. Fusion Eng Des 87(7–8):1310–1314
26. Yu SP, Yang Q, Chen C, et al. Shielding design for activated first wall transferring in ITER hot cell building. J Fusion Energy 34(4):887–894
27. http://www.casl.gov
28. Stimpson S, Powers J, Clarno K et al (2018) Pellet-clad mechanical interaction screening using VERA applied to watts bar unit 1, cycles 1–3. Nucl Eng Des 327:172–186
29. (2017) Overview of the NURESAFE European project. Nucl Eng Des 321:1–7
30. Chanaron B, Ahnert C, Crouzet N et al (2015) Advanced multi-physics simulation for reactor safety in the framework of the NURESAFE project. Ann Nucl Energy 84:166–177

Part II
Neutronics Design Principles of Advanced Nuclear Systems

Chapter 9
Neutronics Design of Advanced Fission Systems

Advanced fission systems mainly refer to fourth-generation nuclear systems represented by six typical reactor types: lead-based fast reactors (LFRs), very-high-temperature reactors (VHTRs), sodium-cooled fast reactors (SFRs), supercritical water-cooled reactors (SCWRs), molten salt reactors (MSRs), gas-cooled fast reactors (GFRs), and other unconventional fission systems. These systems are usually designed with characteristics of sustainability, safety, economic efficiency, and effective prevention of nuclear proliferation.

Neutronics design, which is a crucial step in the design of advanced nuclear systems, focuses on the safe and efficient generation, utilization, and protection of neutrons. It is mainly concerned with the safety, economic efficiency, and environmental friendliness of nuclear systems. The neutronics design of advanced fission systems mainly includes the core physics design and the radiation protection design. The main task of the core physics design is to study the neutronics characteristics of the core under different states, to optimize the fuel loading scheme, and to determine methods and strategies for maintaining the critical status and controlling the reactivity while satisfying the main design goals and safety limitations. The radiation protection design includes the radiation source analysis, shielding design, radiation safety analysis, radioactive waste management, etc. It focuses on the study of the characteristics and distribution of neutrons and radioactive rays and radionuclides induced by neutrons. In the radiation protection design, a reasonable arrangement of shielding materials and appropriate measures should be provided to ensure that the equipment, the personnel, the public, and the environment are protected from radioactive hazards.

In this chapter, the neutronics characteristics of advanced fission systems are briefly introduced followed by detailed introductions to their design principles, objectives, and methods. The special design considerations in neutronics for the miniaturization of advanced nuclear systems are introduced in Sect. 9.3, and in the last section, the 10 MW China Lead-based Mini-Reactor CLEAR-M10 is used as an example to illustrate the neutronics design of advanced fission systems.

Y. Wu, *Neutronics of Advanced Nuclear Systems*,
https://doi.org/10.1007/978-981-13-6520-1_9

9.1 Neutronics Characteristics of Advanced Fission Systems

There are several types of advanced fission systems. However, from the view of neutronics, the difference between them is mainly due to the difference in the neutron spectrum in the core. According to the energy of the neutrons inducing the fission reactions of the fuels, the reactors can be classified into thermal neutron spectrum reactors (thermal reactors), fast-neutron spectrum reactors (fast reactors), and intermediate neutron spectrum reactors (epithermal reactors). In thermal reactors, the average neutron energy-inducing fission reactions are less than 1 eV; in fast reactors, the energy is above 0.1 MeV; and in the epithermal reactors, the average energy is between those of thermal reactors and fast reactors. Figure 9.1 shows the spectra of neutron fluxes in these types of reactors.

Of the six typical advanced fission systems, the LFRs, SFRs, and GFRs are usually designed as fast reactors; the MSRs and SCWRs can be designed as thermal reactors or fast reactors; and VHTRs are usually designed as a thermal reactor or epithermal reactor.

The LFRs are liquid-metal-cooled reactors that are usually fueled with metal or ceramic nuclear fuels containing uranium or transuranic elements. Because the neutron slowing-down power and absorption cross section of lead are very small, these reactors are designed as fast reactors to realize energy production, fuel breeding, and transmutation of actinides [1–3]. A large pitch can be adopted to enhance the natural circulation ability of the reactor and to improve the reactor safety. Due to the high density of lead and lead alloys, if the fuel element in the core is damaged, the

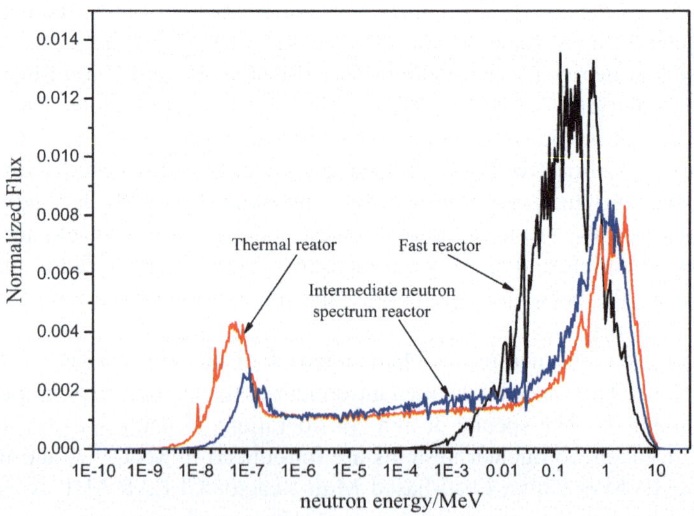

Fig. 9.1 Neutron energy spectra of different types of reactors

leaked fuel will float or pervade, thus avoiding the re-critical safety problem caused by the sinking of leaked fuel in other reactors. The main characteristics of LFRs are the liquid-metal coolant, fast neutron spectrum, and closed fuel cycle.

Similar to the LFRs, SFRs are also a type of fast reactors that are usually fuelled with metal or ceramic nuclear fuels to realize energy production, fuel breeding, and actinide transmutation. However, compared with lead or lead alloys, sodium has a stronger neutron slowing-down power, so the neutron spectrum of SFRs is softer than that of LFRs, and the local void reactivity coefficient may be positive, especially in large SFRs. Sodium is also used as a coolant in the second circuit of SFRs, and the activated products ^{24}Na and ^{22}Na have relatively long half-lives and can be the main radiation source during the maintenance of the secondary sodium system. The main characteristics of SFRs are the liquid sodium coolant, fast neutron spectrum, and closed fuel cycle.

GFRs, which would not have a moderator, usually use helium or CO_2 as coolants. They can operate at high power densities using a direct cycle turbine and achieve highly efficient energy conversion [4]. Metal-clad fuel elements with oxide or carbide fuels are traditionally considered for GFRs. Different core configurations such as hexagonal lattices, plate geometries, prismatic block arrays, and pebble beds, can be adopted. The higher neutron leakage of GFR cores leads to more fuel loadings and smaller fuel Doppler coefficients than those of the LFRs. The main characteristics of GFRs are the gas coolant, fast neutron spectrum, and closed fuel cycle.

MSRs include liquid-fuel MSR and solid-fuel MSRs with LiF, NaF, and other fluoride salts as coolant [4]. In liquid-fuel MSRs, the fluoride salts are also used as the fuel carrier, and uranium, plutonium, and other actinides are melted into the fluoride salts. In MSRs, the fission energy is produced and transferred directly in the fluoride salt without other coolants. Solid-fuel MSRs usually adopt coated fuel particles, and the fission energy is transferred by the circling of fluoride salt. The reactor can be designed as a thermal reactor or a fast reactor based on the usage of a graphite moderator. It can be refueled online, which can prevent a large initial excess reactivity and achieve better resource utilization. The main characteristics of MSRs are the flowing molten salt mixtures (fuel and coolant), thermal or fast neutron spectrum, and closed fuel cycles.

SCWRs are cooled by light water at supercritical pressures (i.e., >22.1 MPa) [4]. Generally, an SCWR can be designed as a fast reactor because of its small coolant flow, low water–uranium ratio, low supercritical water density, and weak moderation ability. The fast core option of SCWRs has advantages of a simplified reactor design and a high power density, but a general safety concern is the positive void reactivity coefficient under accident conditions, which must be avoided by suitable measures. An SCWR can also be designed as a thermal reactor with additional water as a moderator in the core, in which the fuel rods, water rods, and gadolinium rods are usually arranged in tight-pitch lattices. The water rods are used as a moderator, and the gadolinium rods are used to control the reactivity. The main features of SCWRs are the supercritical water coolant, thermal or fast neutron spectrum, and multiple fuel cycle modes.

In VHTRs, graphite is generally used as a moderator and reflector, and hexagonal fuel assemblies or spherical pebble fuels are usually adopted in the core [4]. Helium, which remains in a single phase as well as being neutronically transparent and chemically inert, is the general choice for the reactor coolant and can mitigate the consequences of radioactive coolant release in a Loss Of Coolant Accident (LOCA). Pebble-bed VHTRs can realize online refueling by loading new or spent fuel pellets at the top of the core and removing spent fuel pellets from the bottom. The main characteristics of VHTRs are the helium coolant, thermal neutron spectrum, and once-through fuel cycle.

In general, different types of advanced fission systems have different requirements for material compatibility due to their characteristics and coolant conditions. As a result, there are large differences in the material selections, core structures and configurations, and neutronics design considerations. For example, in MSRs and pebble-bed VHTRs, online refueling is possible, and the initial excess reactivity can be small, making reactivity control easy, so the neutronics design of these systems focuses on the technical feasibility and safety problems caused by the movement and random distribution of fuel elements. However, for advanced fission systems loaded with fixed fuel assemblies, such as LFRs and SFRs, the systems should shut down and refuel regularly, and sufficient initial excess reactivity should be reserved to satisfy the full-power operation during the designed refueling period. Therefore, special attention must be paid to the design and optimization of the reactivity control and fuel management during the neutronics design for these kinds of systems. The neutronics design procedures and considerations for different systems are similar, i.e., to select the appropriate material and optimize the core configuration by coordinating the thermal-hydraulic and mechanical analyses, to achieve the best core design option and realize the goals of energy production, fuel breeding, and waste transmutation.

9.2 Principles and Objectives of Neutronics Design

Compared with traditional fission systems, advanced fission systems have the features of sustainability, safety, economic efficiency, and effective prevention of nuclear proliferation. All of these features are reflected in their design, construction, and operation. Neutronics design is an important part of nuclear system design. This section will introduce the principles and objectives of neutronics design for advanced fission systems.

9.2.1 Core Physics Design

The overall goal of the core physics design is to ensure that the system maintains the fission chain reaction and provides the needed power output in a safe and reliable way. In addition, for the goals of sustainability, economy, and nonproliferation, the following specific principles should be adopted in the design of an advanced fission system.

1. Improvement of the utilization efficiency of nuclear fuels

Improving the utilization efficiency of nuclear fuels means that the advanced fission system should make full use of the limited nuclear fuel resources to generate more energy. In addition, it should reduce the environmental and ecological impacts of the wastes produced during the mining and smelting procedures of uranium and thorium resources to achieve the sustainable development of nuclear energy. In the core physics design, there are many ways to improve the utilization efficiency of nuclear fuels, such as breeding nuclear fuels using the thorium–uranium cycle or uranium–plutonium cycle, reducing neutron leakage to improve neutron economy, improving the fuel burnup and breeding ratio, etc.

2. Reduction of nuclear waste production

The spent fuel discharged from traditional fission systems contains large numbers of long-lived fission products (LLFPs), activation products, and actinides. Most of these nuclides are highly radioactive, have long half-life, and make waste disposal difficult. To reduce the production of nuclear waste, advanced fission systems are usually designed with advanced fuel cycle modes and transmute the nuclear waste through capture and fission reactions between the neutrons and the long-lived fission products or actinides.

3. Reduction of nuclear proliferation risk

Plutonium (Pu) in the spent fuel is an important material for nuclear weapons. In the advanced nuclear system design, the refueling period is extended and the spent fuel of advanced fission systems should have fewer fractions of fissile isotopes and high fractions of spontaneous fission isotopes. This increases the difficulty of extracting fissile Pu from the spent fuel and reduces the risk of nuclear proliferation.

To realize the goals of the advanced fission system, the following principles should be satisfied in the core physics design:

1. Negative reactivity coefficients

The reactivity coefficients are defined as the relative changes in reactivity with respect to parameters such as temperature or power. They are important parameters to evaluate the inherent safety of the nuclear system. The reactivity coefficients include the fuel temperature coefficient, moderator temperature coefficient, coolant temperature coefficient, void coefficient, and so on.

2. Sufficient shutdown margin

Similar to traditional fission systems, reactivity control systems for advanced fission systems should have the ability to compensate for the reactivity changes caused by power variations and to shut down quickly and ensure that the system has a sufficient shutdown margin for safety. This usually requires the reactivity control system to meet the stuck rod criterion; that is, the reactor can be shut down and maintained at a sufficient shutdown margin using any of the control systems when the rod with the largest value is stuck outside the core.

3. Flattening power distribution

In fission systems, the ratio of the local power peak to the average power density is relatively invariant, and the local power peaking factor limits the power generation of the fission system. The power distribution in the fission system should be flattened to be as smooth as possible to improve the economy and power generation to meet the limits of the material and thermal-hydraulic parameters.

4. Sufficient burnup depth

To improve the economy, the fission system is often designed with optimized core management strategies to increase the fuel burnup at the end of life. To ensure the integrity of the fuel elements, the maximum specific burnup of fuel should not exceed the limits of the materials.

In addition to these principles, the fuel breeding ratio and waste transmutation ratio should be as high as possible for advanced fission systems with fuel breeding and waste transmutation capabilities.

9.2.2 Radiation Protection Design

Radiation protection design includes radiation source analysis, radiation shielding design, radiation safety analysis, and radioactive waste management. Its purpose is to prevent harmful deterministic effects, and to reduce the occurrence probability of stochastic effects to be As Low As Reasonably Achievable (ALARA), ensuring the radiation safety of equipment, workers, the public, and the environment. The measures that are usually applied in radiation protection include radiation zoning, radiation shielding design, and radioactive waste management. The detailed objectives and principles for these three main measures are briefly discussed below.

1. Radiation zoning

The purposes of radiation zoning are to control normal exposures and prevent the spread of contamination. The radiation area of the nuclear system should be classified into different zones in accordance with the internal and external exposure risks, and each zone corresponds to a dose rate and has associated time and access conditions [5]. Generally, a zone is defined as a controlled zone when the annual accumulated dose in the area exceeds 30% of the limit and is defined as an unregulated zone when the annual accumulated dose in the area is lower than 10% of the limit. Between 10 and 30%, the zone is defined as a supervised zone.

2. Radiation shielding design

The purpose of radiation shielding design is to shield against neutrons, photons, and other radioactive rays generated in the reactor and ensure that the external exposure to the equipment, workers, public, and environment are lower than the corresponding limits. The principles of radiation shielding design should follow several rules [6]:

(1) As an integral part of the nuclear systems, the shield must be synchronously designed with the system design and cooperate with multiple designs, such as ventilation, drainage, and construction engineering, to fully implement all of the design requirements.
(2) Based on safety and convenience for maintenance and operations, the radiation shielding design should be optimized to minimize the dimensions and reduce the weight and cost.
(3) The shield structure should be as simple as possible and minimize the cavity, pipes, gaps, etc.
(4) Appropriate local shielding with a proper safety margin should be located in or around cavities, pipes, gaps, etc., to compensate for the shielding weakness caused by radiation streaming.
(5) The nuclear shielding materials should be stable, nontoxic, odorless, easy to obtain, easy to transport, and inexpensive. The radiation shielding design should meet the requirements of the reactor structures and their ancillary systems.

3. Radioactive waste management

To reduce the release of radioactive waste into the environment, it is necessary to effectively manage the radioactive waste produced in nuclear systems. In addition, the design procedure should follow the following principles: use low-activation materials to reduce the amount and radioactivity of the waste, ensure the integrity of the containment to decrease the leakage of radionuclides, design waste collection and treatment systems for different waste classifications, recycle the treated wastes and minimize the volume of radioactive waste as much as possible, and ensure that the released wastes meet the national emission standards.

9.3 Methods of Neutronics Design

9.3.1 Core Physics Design

Based on the aforementioned objectives of advanced fission systems, the core physics design mainly focuses on the fuel selection and fuel management design, including selection of core materials, the assembly and core layout design, the reactivity control design, and the fuel management design, etc. As shown in Fig. 9.2, the detailed processes of the core physics design are as follows. First, the fuel cladding and coolant materials are selected based on the objectives of the design, and the assembly and core layout schemes are designed. Subsequently, the detailed core model is developed, and the core physics design is optimized through iterations with the thermal hydraulics, structural mechanics, and neutronics analyses, which

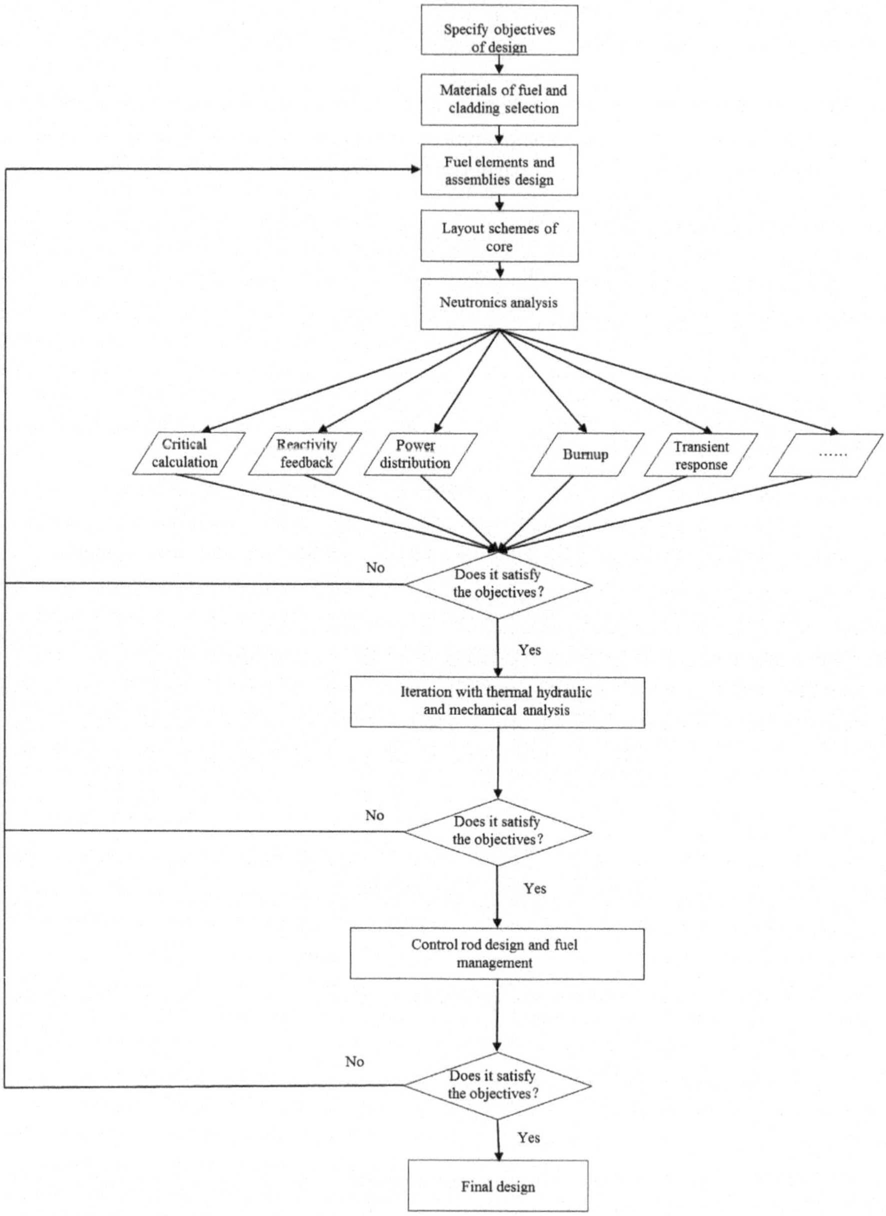

Fig. 9.2 Flowchart of the core physics design

include the reactivity control design, reactivity feedback analysis, power distribution calculation, power flattening optimization, fuel burnup analysis, fuel management, and transient analysis. In addition, reactivity control and fuel management are performed. These activities ensure that the core design meets the various requirements and limitations. Ultimately, the design of the core can be determined.

9.3.1.1 Core Parameters

Table 9.1 presents the main parameters determined in the core physics design. A detailed introduction to the design of these parameters will be presented in the following sections.

1. Materials for the core

The materials in the core mainly include the fuel, cladding, coolant, moderator, reflector, and absorber. The coolant is determined according to the type of reactor.

(1) Fuel types and enrichment

Nuclear fuels can be classified into liquid and solid fuels based on their phase states.

Solid fuels are the most widely used fuel for nuclear systems. They are used due to their good radiation resistance, high melting point, high thermal conductivity, good compatibility with the cladding material and coolant at high temperatures, lack of strong corrosion in contact with the coolant, good processing performance, low manufacturing cost, and so on.

Liquid fuels, including UF_4–LiF–BeF_2–ZrF_4 and ThF_4–UF_4–LiF_4 for MSRs, usually function as a combination of nuclear fuel, coolant, and moderator. Compared with solid-fuel reactors, liquid-fuel reactors have the advantages of a simple refueling system, no manufacturing and handling requirement like solid fuel, and high inherent safety. However, liquid-fuelled reactors also have several problems. For example,

Table 9.1 The main parameters to be determined during the core physics design

Types	Parameters
Materials	Fuel types and enrichment; cladding materials; coolant materials; moderator; reflector; absorber; etc.
Design parameters of the fuel element and assembly	Arrangement of fuel element; dimension of element; thickness of cladding; volume fraction of fuel, moderator and structural material
Design parameters of the core	Arrangement and loading number of fuel assembly; fuel loading; height of core; equivalent diameter of core; structure and positions of reactivity control systems; etc.

molten fluoride salt significantly corrodes structural materials at high temperatures, which directly affects the safety and service life of the reactor.

Based on its chemical composition, solid fuels can be divided into metal fuels, oxide fuels, nitride fuels, and carbide fuels.

Metal fuels have the advantages of high thermal conductivity, high atomic density of the fissionable nuclides, and relatively simple and mature reprocessing technology. Their disadvantages are their low melting point, serious radiation swelling, and poor compatibility with other core materials.

Oxide fuels mainly include uranium oxide fuels and MOX fuels, which have the advantages of high melting points, relatively good radiation resistance, stable chemical properties, and low neutron capture cross sections. However, they also have several problems such as low atomic density of fissionable nuclides and low thermal conductivity. MOX fuels are widely used in closed fuel cycles. They have a larger thermal neutron absorption cross section, fission section, and smaller delayed neutron fraction. Therefore, cores using MOX fuels have higher neutron energy, shorter neutron generation time, and faster rate of power increment at the prompt supercritical state. In addition, MOX fuels are highly toxic and radioactive, and special measures should be taken during manufacturing, transportation, and usage.

Carbide fuels include UC and (U, Pu) C. Because of its high thermal conductivity, the temperature inside the core is flatter, and a higher power density can be obtained. More fissile nuclides can be converted in the core due to higher atomic density of the fissionable nuclides of (U, Pu) C than those of the UO_2. However, these fuels also have disadvantages, such as serious radiation swelling, active chemical properties, easy decomposition at high temperatures, a complex manufacturing process, and immature reprocessing technology.

Similar to the carbide fuels, nitride fuels also have high atomic density of fissionable nuclides, high thermal conductivity, high-temperature and radiation resistance, weak neutron slowing-down power, and good compatibility with liquid metals. Their disadvantage is that ^{14}N should be replaced by ^{15}N to decrease the production of radioactive ^{14}C gas. In addition, complicated dry reprocessing technology is required to manufacture the nitride fuel.

In-core physics design, it is necessary to select the appropriate fuel according to the operating conditions of the nuclear system and compatibility of the core materials as well as technical maturity, economic, and safety objectives. For example, lead-based reactors require that the fuel should have good compatibility with lead-based alloys, high thermal conductivity, high atomic density of the fission nuclides, and high-temperature resistance, and radiation resistance, so nitride solid fuel is the ideal fuel in lead-based reactors. However, considering the technical maturity and reprocessing, most current designs of lead-based reactors mostly use an oxide solid fuel.

After determining the type of fuel, it is necessary to determine the main composition of the fuel, such as thorium, uranium, and plutonium minor actinide (MA), and their proportions. The fuel composition and enrichment will be determined according to such factors as the system function, fuel cycle, fuel cost, core size, fuel burnup, etc. An oxide fuel is often chosen as the fuel for the nuclear system, and the enrichment depends on economic and other considerations. Generally, the fuel enrichment for thermal reactor is 2–4%, whereas that of fast reactors is higher. For example, the enrichment of ^{235}U is 64% for the China Experimental Fast Reactor; and that of the SVBR [7], Russia Modular Lead–Bismuth Reactor, is 16.1% in the first stage. The China Lead-based Mini-Reactor CLEAR-M10 uses UO_2 with an ^{235}U enrichment less than 19.75% as fuel. The Th–U or U–Pu fuel cycle is chosen as the fuel of the nuclear system for fuel breeding, and the proportions of fissile nuclides and fertile nuclides depend on the fuel breeding ratio. In addition, to improve the fuel utilization and reduce the amount of nuclear waste, MA is added to the U–Pu fuel. However, the addition of MA will result in a smaller effective delayed neutron fraction and a smaller reactivity feedback coefficient, which will reduce the inherent safety of the nuclear system. In the design, the proportions of U, Pu, and MA can be adjusted to obtain a core scheme with a high transmutation ratio and high fuel utilization.

(2) Cladding materials

Cladding is the first barrier of fission products. The cladding allows the fuel element to maintain structural integrity, separates the fuel from the coolant, and prevents fission products from leaking into the primary coolant system. The choice of cladding materials must take into account the compatibility between the cladding materials and coolants, the neutron capture capability, and the structural and irradiation properties. The commonly used cladding materials are aluminum alloys, magnesium alloys, zirconium alloys, stainless steel, and nickel alloys. Thermal reactors generally use zirconium alloy as the cladding material because of its small absorption cross section for thermal neutron, good mechanical properties, and corrosion resistance, whereas fast reactors usually use stainless steel as the cladding material because of its high mechanical strength and good radiation resistance.

In addition, there are generally 2–3 coating layers for pellet fuel, which is widely used in pebble-bed VHTRs and solid-state MSRs. The coating layer near the fuel generally uses low-density pyrolytic carbon to hold the gaseous fission products generated by fuel consumption to prevent excessive pressure. The outer coating layer generally uses silicon carbide and dense pyrolytic carbon to bear the pressure and retain the fission products.

(3) Moderator materials

In thermal reactors and epithermal reactors, the average energy of the fission neutrons is approximately 2 MeV, whereas most of the neutrons inducing fission reactions are thermal neutrons with energies far lower than 2 MeV, so moderator materials are needed in these reactors to slow down the fission neutrons. Generally, the moderator should have a high slowing-down power and small absorption cross section for neutrons. Commonly used moderators include graphite, beryllium, light water, heavy water, zirconium hydride, yttrium hydride, etc. Graphite is commonly used as a moderator in pebble-bed MSRs and VHTRs.

(4) Reflector materials

The reflector is generally arranged at the periphery of the fuel zone, and its function is to scatter leaked neutrons back into the core to improve the neutron utilization efficiency. The reflector materials commonly used in fission reactors are stainless steel, graphite, niobium alloys, etc. Stainless steel is commonly used in fast reactors because of its large atomic number, small absorption cross section for fast neutrons, and large scattering angle for neutrons. Lead and lead alloys are also commonly used as coolants and reflector materials in lead-based fast reactors. Graphite, beryllium alloys, and other materials, which have small atomic numbers and strong neutron slowing-down powers, are often used as reflectors in thermal reactors to moderate the leaked neutrons. Some of the leaked neutrons are scattered back to the core by the reflector.

(5) Absorber materials

Absorber materials are used to regulate the reactivity and power distribution. The absorber material is generally required to have a large neutron absorption cross section and good compatibility with the core material. Commonly used absorber materials include boron-containing materials such as boron carbide and boracic acid, as well as Ag–In–Cd, gadolinium, hafnium, europium, tantalum, etc.

2. Fuel element and assembly design

For LFRs, SFRs, GFRs, and SCWRs, it is necessary to determine the dimensions, pitch and layout of the fuel elements, the volume ratio of the core materials, the cladding thickness, and other parameters after determining the core materials.

(1) Determining the fuel element size

The size of the fuel element directly affects the economic efficiency and safety of the nuclear system. If the fuel element is too large, the number of fuel elements could be reduced, which may cause a reduction of the heat exchange surface and an increase in the linear power density. This results in an increase of the fuel center temperature and the cladding temperature, which may exceed the design limits and cause safety problems.

If the fuel element is too small, the resonance absorption effect of ^{238}U increases. Thus, the probability of neutrons escaping from resonance capture is reduced, which results in a decrease of k_{eff}. This is not conducive to improving the neutron economy.

If the power density of fuel elements in the core retains unchanged, a reduction in the fuel element size means that the heat flux on the surface of the fuel elements increase, and the cladding temperature will increase, which increases the risk of fuel cladding damage. If the surface heat flux of the fuel element remains unchanged, a reduction in the fuel element size will reduce the linear power density and the core neutron economy.

In addition, the design of the fuel element should consider the physics and thermal-hydraulic characteristics of the coolant, the fuel breeding capability of core, etc. For example, the diameter of the fuel rod is usually 8–11 mm in thermal reactors, 4–9 mm in sodium-cooled fast reactor, and 8–13 mm in lead-based fast reactors with forced circulation. In lead-based fast reactors with natural circulation, the diameter of the fuel rod is usually greater than 12 mm. For example, the diameter of the fuel rod in SSTAR is 25 mm.

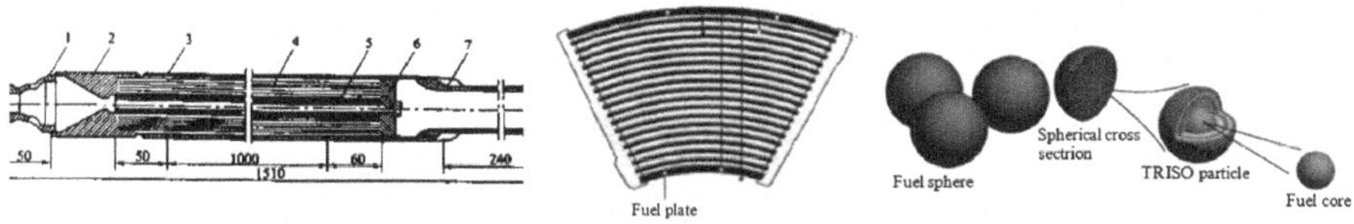

Fig. 9.3 Rod-shaped, plate-shaped, and spherical fuel elements

(2) Determining the fuel element arrangement

The common structures of fuel element include rod-shaped, plate-shaped, and spherical structures, as shown in Fig. 9.3.

For rod-shaped fuels, the fuel rods are usually arranged in square, triangular, or annular arrangement. Square arrangements are commonly used in thermal reactors because they are advantageous for increasing the volume fraction of the coolant or moderator and improving the neutron moderation efficiency. The triangular arrangements are widely applied in fast reactors because they are beneficial for achieving large volume fractions of fuel and improving the power volume ratio. Annular arrangements are often applied in zero-power or low-power experimental reactors, because it facilitates a uniform arrangement of the core and the circular periphery of the active zone.

Plate-shaped fuel elements have many advantages, such as a large heat transmission area per core volume, low central temperature, good shock resistance, and being easy to locate. However, they also have disadvantages, such as difficulty in processing and large volume fraction of structural materials. Currently, the plate-shaped fuel elements are mainly applied in space nuclear reactors and marine reactors.

Spherical fuel elements, such as ceramic-coated fuel pellets, generally have excellent high-temperature performance. In addition, the dispersed distribution of the fuel can reduce the influence of fuel breakage. Currently, spherical fuel elements are mainly used in pebble-bed reactors, such as gas-cooled VHTRs and solid MSRs.

(3) Determining the volume ratio of the core materials

The material volume ratio directly affects the core neutronics, thermal hydraulics, and economic performances. In thermal reactors, such as gas-cooled VHTRs, the volume ratio of the moderator to the fuel is one of the main concerns in the design. Figure 9.4 shows the influence of the moderator–fuel volume ratio on k_{eff} in a typical thermal reactor. For a given fuel enrichment and moderator material, there should be a best moderator–fuel volume ratio at which the core k_{eff} reaches a maximum. In Fig. 9.4, the region in which the moderator–fuel volume ratio is lower than the best value is called under-moderated region, while the right side with volume ratio higher than the best value is called the over-moderated region. In the under-moderated region, as the temperature increases, the slowing-down power of the moderator is weakened due to the changes in atomic density and scattering reaction cross section, and the moderator temperature reactivity of the core is negative. For safety considerations, the thermal reactor is usually designed with under-moderated characteristics. In fast

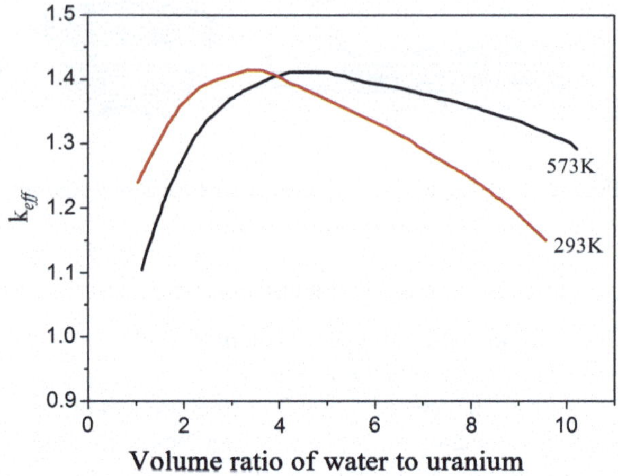

Fig. 9.4 Influence of the water–uranium volume ratio on k_{eff} in a water-cooled reactor

reactors, no moderator is expected in the core, and the coolant–fuel volume ratio is the main concern in the design. From the perspective of reducing fuel loading and improving the economics of the nuclear system, the coolant–fuel volume ratio should be designed to be as small as possible based on the thermal-hydraulic requirements.

(4) Determining the cladding thickness

Generally, the cladding material is selected to have a small neutron absorption cross section and a small influence on the core neutronics performance. In this case, the thickness of the fuel cladding is determined by several factors, including the fission gas pressure in the fuel element, the corrosion of the coolant, the manufacturing tolerances, and the structural strength reduction due to surface defects. The fuel cladding is generally approximately 0.6 mm thick in the PWRs, and between 0.3–0.6 mm thick in fast reactors.

3. Core physics design

Based on the design of the fuel assemblies, the core physics design is used to determine the arrangement of the fuel assemblies, fuel batches and loading, core size, control rod structure and positions, reflector, and shielding assembly.

The arrangement of the fuel assembly mainly depends on the arrangement of the fuel rods. For example, for triangularly arranged fuel rod bundles, the assembly is usually hexagonal with hexagonal arrangement. For square bundles, the assembly is usually rectangular with a square arrangement.

In addition, the core arrangement typically depends on the objectives of the nuclear system. For example, in a fast breeder reactor, to achieve fuel breeding, there are two core arrangements. The outer arrangement is used to arrange the fertile nuclides on the periphery of the core, and the inner arrangement is used to arrange the fertile nuclide and fissile nuclides inside the core. After the arrangement of the fuel assemblies is determined, the core geometry parameters are determined based on the power demands, fuel type, and required fuel cycle period. The specific procedures are as follows: first, the maximum linear power density of the fuel is determined based on the selected fuel type; then, the minimum number of fuel elements is determined based on the required power output, and the approximate equivalent diameter and height of the core are obtained.

Taking a nuclear system with rod-shaped fuel as an example, assuming that the total power is P_T, the total thermal power deposited in the fuel is $P_F = F_U \cdot P_T$, where F_U represents the fraction of power deposited in the fuel, which is generally 0.95–0.98. The total number of fuel elements N in the core can be described as

$$N = \frac{P_F}{\bar{q}H} = \frac{F_U P_T F_Q}{q_{\max} H} \tag{9.1}$$

and the equivalent core diameter D_{eq} is determined by

$$\frac{\pi D_{eq}^2}{4} = \frac{F_U P_T F_Q}{q_{\max} H} A \tag{9.2}$$

where A is the cross-sectional area of a single fuel element, H denotes the length of the core's active zone, $\bar{q}$ is the average linear power density of the fuel element, $q_{\max}$ is the maximum linear power density of the fuel element, and F_Q is the nuclear hot channel factor. Considering the space occupied by other structures, the actual core diameter is slightly larger than the calculated result from Eq. 9.2. In the core physics design, the ratio of the equivalent core diameter to the height should be determined before determining A. For a cylindrical bare reactor, the optimum ratio is 1.08.

Table 9.2 presents the core parameters of several typical advanced fission systems.

9.3.1.2 Power Distribution

The power distribution is an important component of core physics design. An optimum power distribution can reduce the peak power factor and increase the local power density. The power peaking factor F_Q is often used to characterize the power distribution and it can be expressed as

$$F_Q = F_Q^N F_Q^E \tag{9.3}$$

where F_Q^N represents the nuclear power peaking factor, which is the ratio of the local maximum power density to the average power density (for a homogeneous cylindrical bare reactor, $F_Q^N = 2.32 \times 1.57 = 3.638$, where 2.32 is the radial power peaking factor and 1.57 is the axial peaking factor), and F_Q^E is an asymmetrical coefficient caused by several factors, including fuel manufacturing deviation, etc.

The core power distribution is closely related to the arrangement of the assemblies (fuel assemblies, control assemblies, burnable poison assemblies), the arrangement of the reflector, the core operation conditions (e.g., cold zero power, hot zero power,

Table 9.2 Basic parameters of typical advanced fission systems [3, 8]

Typical nuclear system	MSR	VHTR	SCWR	SFR	LFR
System name	MSFR	HTTR	HPLWR	BN-600	CLEAR-M10
Country	France	Japan	Germany	Russia	China
Design organization	CNRS	JAEA	KIT	OKBM	FDS team
Thermal power (MW)	3000	30	2300	1470	35
Fuel type	LiF–(U, Pu)F_3–ThF_4	UO_2	UO_2	UO_2/UO_2–PuO_2	UO_2
Enrichment	–	6%	5, 4%	17, 21, 26%	18, 18.5, 19.75, 17%
Fuel arrangement	Liquid	TRISO-coated fuel particles	Square arrangement	Triangular arrangement	Triangular arrangement
Number of fuel assemblies	–	–	121	369	–
Diameter of the active zone (m)	2.25	2.3	3.5	2.05	1.2
Height of the active zone (m)	2.25	2.9	4.2	1.03	0.79
Diameter of the fuel element (mm)	–	34	8	6.9	16
Coolant	LiF–(U, Pu)F_3–ThF_4	He	Supercritical water	Liquid sodium	Liquid lead
Moderator	–	Graphite	Supercritical water	–	–
Cladding	–	SiC	Stainless steel	Cr16 Ni15 Mo2 + MnTiSi (SW)	15-15Ti (T91 coated by Fe, Cr, Al, Y, etc.)
Thickness of the cladding (mm)	–	0.1	0.5	0.4	0.7
Reflector	–	Graphite	–	Stainless steel	BeO, stainless steel

hot full power, xenon balance status), and fuel burnup. For example, if the core is homogeneously loaded with same enrichment fuel, the power distribution will be flat at the end of fuel cycle because the fuel burnup will be deeper in places with a higher neutron flux (such as the core center), which will result in decrease in the local reactivity and the maximum local power density.

The factors affecting the core power distribution should be analyzed during the design to ensure that the radial and axial power distributions are flattened and that the peaking factors do not exceed the design limits. Generally, the radial power distribution can be flattened by various means, including adjusting the radial distribution of the reactivity control devices, the radial variation of the fuel enrichment, the radial distribution of burnable poison assemblies, etc. Loading fuels with various enrichments is also widely applied for power flattening. Generally, the highly enriched fuel assemblies are usually loaded in the outer zone of the core (i.e., the region with a lower neutron flux density), and the low-enrichment fuels are loaded in the central zone (i.e., the region with a higher neutron flux density). The axial power distribution can be flattened through an axial variation in the fuel enrichment or burnable poison.

The movements of the reactivity control devices are commonly used means for power distribution flattening. For example, at the beginning of the fuel cycle, several control rods in the central region will be inserted to decrease the central neutron flux density. However, special attention should be paid because the insertion of control devices may affect the axial power distribution.

9.3.1.3 Reactivity Control

To meet the requirements of start-up, power regulation, and reactor shutdown, the initial load of fuel should be larger than the critical load with appropriate excess reactivity. Thus, reactivity control devices should be designed to maintain the reactivity at the required level and to prevent reactivity-initiated accidents. Specifically, the basic tasks of reactivity control are emergent shutdown, power regulation, and reactivity compensation.

Reactivity control is usually achieved by changing the fuel loading, neutron moderation, neutron leakage, and neutron absorption. The most commonly used method is to change the absorption of neutrons through neutron absorbers. The absorbers are commonly made of materials with a large absorption cross section (e.g., boron, cadmium, hafnium, Ag–In–Gd alloy). In reactor design, several methods of reactivity control are usually combined.

In fast reactors, the consumption of fissile nuclides can be partially compensated by the breeding of fuel and the absorption cross section of fission products is small, so the initial excess reactivity of fast reactors is usually much smaller than that of thermal reactors with the same power output. It is only possible to control the reactivity with control rods.

In large thermal reactors, the initial excess reactivity may be as high as tens of thousands of pcm (1 pcm = 10^{-5}). If the reactivity is only controlled by control rods, the driving mechanism would be difficult to arrange and many openings in the top cover of the vessel would be needed (which will affect the structural strength of the vessel cover). When numerous control rods are inserted into the core, the neutron flux distribution will become heterogeneous. Therefore, the reactivity of large thermal reactors is often controlled by a combination of control rods, chemical shim control, and burnable poison.

In gas-cooled VHTRs, control rods and absorber balls are both used for reactivity control. The control rod system is used for power regulation, hot shutdown, and long-term cold shutdown of the reactor. The absorption ball device works as the second shutdown system to independently shut down the reactor from hot full power to cold zero power state. In SWCRs, boracic acid cannot be added to the coolant, so the reactivity is controlled by the combination of a large amount of burnable poisons and control rods.

In the miniaturization and modularization design, the reactivity control can be achieved by means of a control drum, sliding reflector, and moving fuel in the advanced fission system.

Moreover, in the design of the reactivity control system, the limit of the control rod withdrawal speed is determined by a safety analysis based on the control rod worth, core dynamics parameters, reactivity coefficients, etc. The neutron generation time of fast reactors is several orders of magnitude smaller than that of traditional fission systems, and the effective delayed neutron fraction is also smaller; thus, the requirements for the reactivity insertion rates in fast reactors are much stricter than those in thermal reactors. For example, the Qinshan Nuclear Power Plant has a maximum reactivity insertion rate of 80 pcm/s, whereas CEFR specifies a maximum reactivity insertion rate of 50 pcm/s.

9.3.1.4 Reactivity Feedback

Reactivity coefficients are important indicators of the inherent feedback behavior of the reactor core. In the core physics design, the reactivity coefficients should be evaluated for each core state and the corresponding fuel management strategy to ensure that sufficient negative reactivity coefficients are provided.

The reactivity coefficients are mainly related to the core operating state, neutron spectrum, materials (i.e., fuel, moderator, and coolant), fuel loading and burnup, core geometric structure, etc. Important reactivity coefficients include the fuel temperature coefficient, coolant temperature coefficient, moderator temperature coefficient, coolant density coefficient, etc. [9].

The fuel temperature coefficient is mainly affected by the variation in the resonance absorption cross section of fuel nuclides with temperature (i.e., the Doppler effect). The fuel temperature feedback is transient because it responds quickly to changes of core power and temperature. The value of the fuel temperature coefficient is mainly related to the fuel temperature, neutron spectrum, fuel nuclide composition, and other factors. Nuclides such as ^{238}U have strong resonance absorption for neutrons with energies between 10 eV and 10 keV, so the absolute value of the negative temperature coefficient of the fuel is larger in thermal reactors than in the fast

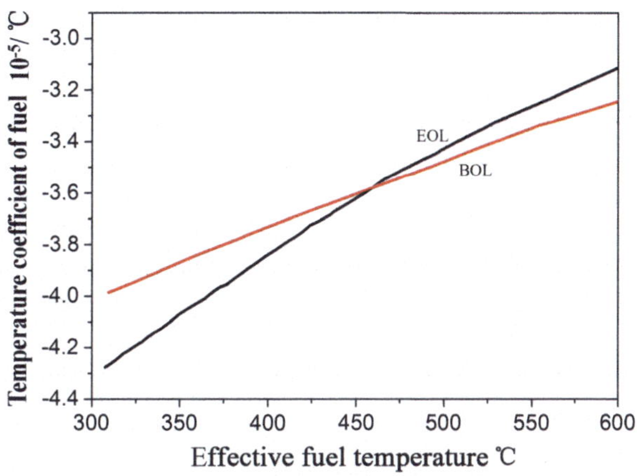

Fig. 9.5 Fuel temperature coefficient versus fuel temperature

reactors. Similarly, the temperature coefficient of ceramic fuel containing oxygen or carbon is greater than that of the metal fuel [8]. Figure 9.5 shows the fuel temperature coefficient of a typical thermal reactor as a function of the fuel temperature. The absolute value of the fuel temperature coefficient decreases as the temperature increases; thus, the negative reactivity feedback due to the fuel temperature effects is more evident at lower operating temperatures.

The moderator temperature effect refers to the change in reactivity induced by the variation of the neutron slowing-down power, which is caused by changes of moderator temperature. For liquid moderators, variation in temperature causes changes in both the neutron reaction cross sections and the density of the moderator nuclides. For solid moderators, when the moderator temperature changes, the change in moderator density is negligible, so the temperature effect of a solid moderator is mainly caused by the variation of the cross section. The moderator temperature effect is a delayed effect, and the coefficient mainly depends on the moderator material and core physics design. In the design, it is generally necessary to ensure that the moderator temperature coefficient is negative to facilitate automatic regulation of the reactor power.

The coolant void coefficient is a relatively important parameter in the design of reactors with liquid coolant. In reactors with liquid coolant, an increase in temperature could generate bubbles, which may cause three effects: (1) a decrease in neutron absorption, which induces an increase in reactivity; (2) an increase of neutron leakage, which decreases the reactivity; (3) and existing bubbles decrease the neutron slowing-down power of the coolant, which causes an increase or decrease in the reactivity depending on the energy spectrum (related to the reactor type). The total coolant void effect is the combination of these effects and the results are related to the locations of the bubbles. Generally, the coolant void coefficient is negative in thermal reactor, whereas positive reactivity feedback may occur in large fast reactors, especially in the central region of the core.

The material density coefficient and the expansion coefficient are also important reactivity coefficients, and their principles are similar to those of moderator temperature coefficient and the coolant void coefficient.

In practice, the temperatures of the materials are difficult to measure in the reactor. Therefore, instead of temperature reactivity coefficients, power reactivity coefficients are often used to characterize the change of reactivity caused by a unit change in power. The power reactivity coefficient is not only related to the nuclear characteristics of the reactor but is also specifically related to thermal hydraulics, structural mechanics, etc., and it is a combination of all of the reactivity effects. To ensure the safety of the reactor, it is necessary to ensure that the power reactivity coefficient remains negative throughout the lifetime of the reactor.

9.3.1.5 Burnup and Poisoning

The burnup analysis is one of the main steps in reactor design. It is the basis for the determination of the fuel cycle and has important impacts on reactor safety and economy. By analyzing the burnup of advanced fission systems, the ability of fuel breeding and transmutation is determined and the refueling cycle can be prolonged by optimizing fuel composition and core layout.

The main content of the burnup analysis includes (1) assessment of the spatial variations of the fuel compositions during operation by tracking various nuclear reaction processes, such as the consumption and breeding of fissile isotopes, and the production and accumulation of fission products (FPs); (2) and determination of the spatial distribution of neutron absorbers for different stages of the fuel cycle, and the refueling scheme at the end of the fuel cycle. The two main burnup processes, the consumption and breeding of fissile isotopes and the production and accumulation of FPs, are introduced below.

1. Consumption and breeding of fissile isotopes

Figure 9.6 shows the variations in the fractions of the uranium and plutonium isotopes except ^{238}U with different burnup depths in a reactor fuelled with low-enrichment uranium. The content of ^{235}U decreases gradually with increasing burnup depth, and the content of ^{239}Pu rapidly accumulates in the initial stage and then becomes constant because of its larger fission cross section. The accumulation of other isotopes of Pu is slow because they are produced by the neutron capture reactions of ^{239}Pu.

In general, the consumption of fissile isotopes is greater than the breeding in thermal reactors. For example, the conversion ratio of fissile isotopes is only ~0.6 in thermal reactors, whereas breeding ratios of ~1.1 can be achieved in fast reactors (e.g., ~1.23 in U.S. Advanced Liquid-Metal Reactor). In addition, the breeding ratio in fast reactors can be improved further by optimizing the fuel loading arrangement and the ratio of fissile nuclides to fertile nuclides in the fuels at the beginning of the fuel cycle. This can reduce the burnup reactivity loss, extend the refueling cycle, increase the fuel burnup level, improve economic efficiency, and effectively prevent nuclear proliferation.

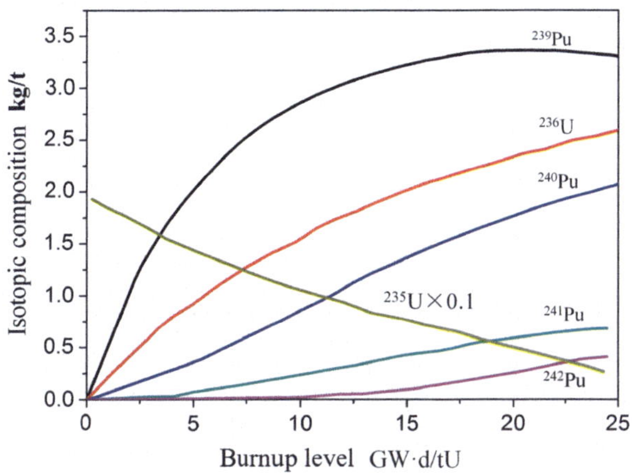

Fig. 9.6 Variations of the uranium and plutonium isotopes fraction with burnup level

2. Production and accumulation of FPs

Some fission products, such as ^{135}Xe and ^{149}Sm, have large reaction cross sections for thermal neutrons, and the production and accumulation of these FPs will cause a decrease of core reactivity (i.e., the FP poisoning). FP poisoning is an important process to be considered during the reactivity control design of thermal reactors, such as VHTRs, MSRs, and SWCRs.

Taking ^{135}Xe poisoning as an example, the equilibrium concentration of ^{135}Xe can be achieved after 30–40 h of continuous operation as shown in Fig. 9.7. When the reactor is shut down, the concentration of ^{135}Xe increases gradually to the maximum during the first 10 h, decreases gradually due to the short decay half-life and is completely decayed after 50–70 h. Due to the concentration of ^{135}Xe, the reactivity decreases gradually after shutdown, reaches the minimum (e.g., iodine well) 10 h after shutdown, and then gradually increases until the poisoning induced by ^{135}Xe is fully released, which may cause the reactor to restart.

The equilibrium concentration of ^{135}Xe is related to the thermal neutron flux in the core. In thermal reactors with neutron fluxes less than 10^{11} n/(cm^2 · s), the effect of ^{135}Xe poisoning can be neglected. The reactivity loss induced by ^{135}Xe poisoning can be up to ~3000 pcm if the neutron flux is approximately 10^{13} n/(cm^2 · s), and up to 5000 pcm when the neutron flux is greater than 10^{15} n/(cm^2 · s).

^{149}Sm is another important fission product for reactivity loss. Figure 9.8 shows the concentration of ^{149}Sm as a function of time. ^{149}Sm requires more time than ^{135}Xe (usually several days) to reach equilibrium. The reactivity loss caused by ^{149}Sm poisoning depends on the ratio of the total fission cross section to absorption cross section, and the equilibrium reactivity loss of ^{149}Sm is usually less than 1000 pcm.

In addition to ^{135}Xe and ^{149}Sm, several other products such as ^{113}Cd, ^{151}Sm, ^{155}Eu, and ^{157}Gd must be considered separately.

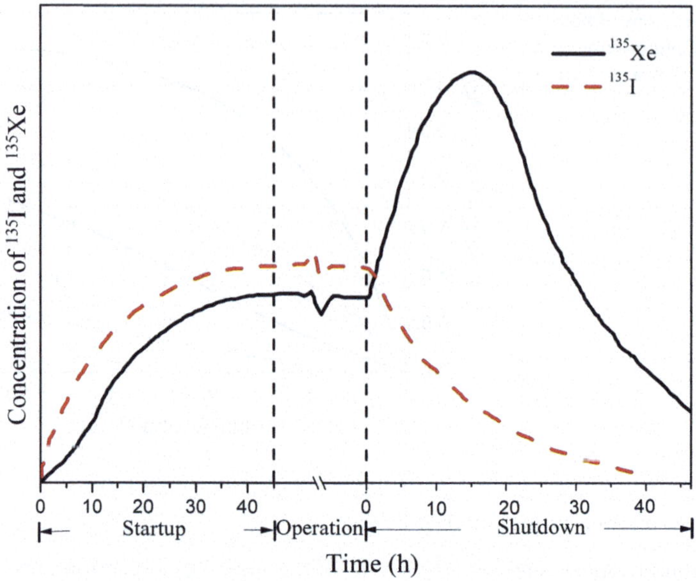

Fig. 9.7 Concentration of ^{135}I and ^{135}Xe as function of time

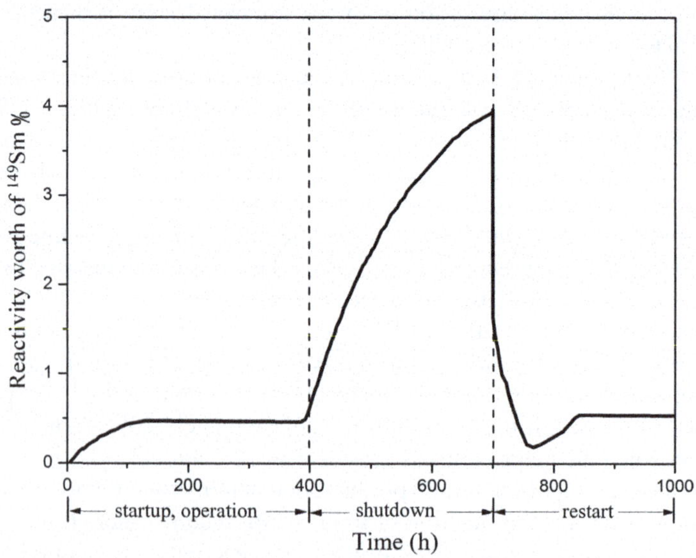

Fig. 9.8 concentration of ^{149}Sm as function of time

9.3.1.6 In-Core Fuel Management

The basic purpose of in-core fuel management is to improve the average discharge burnup level of the fuel to the allowable limits, to increase the utilization efficiency of the fuel, and to decrease the reactor cost per power output by optimizing the loading and refueling scheme. In general, there are usually two alternative refueling schemes, namely, continuous refueling without reactor shutdown and regular refueling with shutdown, depending on the type of reactor and fuel. For MSRs and pebble-bed VHTRs, continuous refueling is used to reduce the excessive reactivity and improve the flexibility of fuel management, regular refueling with shutdown in SFRs, LFRs, SWCRs, etc.

For nuclear systems that use regular refueling, the arrangements of fuel assemblies include uniform arrangements, zoned arrangements, and scattered arrangements based on the fuel enrichments. The loading and unloading patterns include whole batch refueling, batch refueling by moving fuel from center to the edge or from the edge to center for uniform arrangement; the batch refueling is often adopted for zoning arrangement and scattered arrangement. To flatten the power distribution, deepen the burnup level of the fuel, improve the reactor performance and reduce costs, fuel management of advanced fission system usually uses zoned or scattered arrangement and batch refueling.

Figure 9.9 shows the decision process for in-core fuel management. The variables include enrichment of the new fuel ε, cycle time T, batch number n or loading per patch N, power P, loading scheme of the fuel assembly in the core

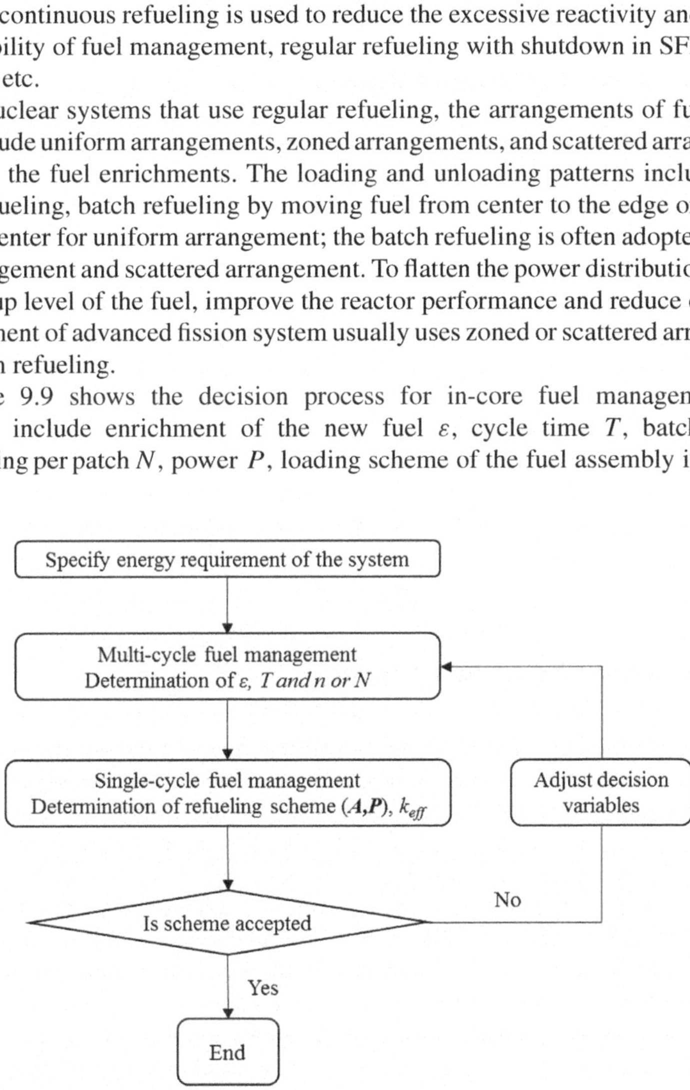

Fig. 9.9 Flowchart for in-core fuel management

$X(i, j)$(i, j is the position of fuel assembly), layout of the burnable poison $BP(i, j)$, and the control scheme. There are close interactions and couplings between these variables. In batch refueling schemes, a fuel assembly usually stays in more than three cycles before it is discharged from the core. In practical engineering calculations, to reduce the number of decision variables, the decoupling method is adopted in the following steps. First, the parameters ε, T, and n/N are determined by multi-cycle fuel management; then, the single-cycle refueling scheme is determined and the optimal fuel management scheme is obtained by iteration.

9.3.1.7 Analysis of the Transient Response

The transient response is the variations of the core parameters due to a sudden disturbance of a stably operating nuclear system. It is important to analyze numerous planned operational transients and potential transients that may cause abnormal or accidental conditions. In particular, power transients are important factors that affect the system safety in accidents.

There are many components of transient response analysis. From the view of core neutronics design, the greatest concern is unexpected changes of reactivity, including abnormal displacements of the control rods, breaks in the main steam pipeline, hydrogen material falling into core, bubble entry, and penetration of fuel assemblies. If the transient response analysis demonstrates that the core parameters (such as the maximum temperature of the coolant, cladding, and fuel) are within the allowable ranges or design limits the transient is determined to be nonmalignant; otherwise, the core needs to be further optimized to assure the safety of the nuclear system.

9.3.2 Radiation Protection Design

The goal of radiation protection design is to ensure the radiation safety of the equipment, workers, the public, and the environment. It includes radiation source analysis, radiation shielding design, radiation safety analysis, and radioactive waste assessment and management. In the design of radiation protection, several characteristics of the radioactive sources, such as the state of matter, type, isotope content, concentration, activity, and spatial distribution, should be calculated and analyzed. Based on the evaluated characteristics of the radiation source, systems such as the shielding system, ventilation system, and radioactive waste treatment system are designed, and the dose and environmental impacts are calculated and assessed. Because the advanced fission systems have high neutron energies, complex energy spectrum, and radiation shielding designs of the system components should pay special attention to reducing the radiation damage caused by neutrons in the radiation protection.

9.3.2.1 Radiation Source Analysis

The radiation in a nuclear system includes the primary radiation and secondary radiation. The primary radiation is released by fission reactions of the fissile nuclides and the decay of fission products, including fission neutrons, prompt γ-rays, delayed γ-rays, α particles, etc. Secondary radiation, such as γ-rays, is caused by interactions between primary radiation and materials.

The distributions of radiation sources are closely related to the nuclear system design. Taking the pool-type reactor as an example, the radiation sources are mainly distributed in the reactor, coolant, auxiliary loops, and spent fuel storage systems. When the reactor is operating, the radiation sources in the reactor system are mainly the various fission products and activated products of the material, which are mainly fission neutrons, prompt and delayed γ-rays, and secondary γ-rays induced by the fission products, and the interactions between the neutron and materials.

After shutdown, the radiation sources are mainly γ-rays released by activated products and fission products. The radioactive substances in the primary coolant system mainly include the activated products of the coolant, the activated corrosion products on the coolant circuit piping and reactor equipment, and the fission products that diffuse from the fuel element when the cladding is damaged. In the secondary coolant system, the coolant flows through the heat exchanger and can be irradiated by neutrons. Therefore, the radioactive substances in this system are mainly activated products of the coolant and its corrosion activation products. The auxiliary circuit system mainly includes the gas system, coolant charging and discharging systems of the primary coolant system, chemical control system, etc. And the composition and characteristics of the radiation source are related to the system design and operating modes. For example, in the gas system of a liquid-metal-cooled reactor, the radioactive substances mainly include the activated products of the reactor cover gas, the gaseous activated product in the coolant, the fission gas, or the volatile nuclides that leak from damaged fuel elements. The radioactive substances in spent fuel storage systems are mainly from the fission products and activation products in the spent fuel assemblies. The activity of each spent fuel assembly discharged from the core is different and is generally related to factors such as the assembly power, irradiation cycle, discharge time, and cooling time in the reactor core. Based on conservative considerations, the spent fuel activity of the maximum power assembly is selected as the source item input in the shield design.

9.3.2.2 Radiation Shielding Design

Based on the assessment of the radiation source, the requirements for radiation zoning and shielding, the appropriate shielding material, and the appropriate shielding arrangement and structure need to be selected in the shielding design. This is to ensure that the radiation damage to materials, temperature, and thermal stress is lower than the limits during the service life, and that the radiations received by workers within the radiation zones are below the specified limits through relevant management systems and control measures. In addition, the distributions of neutron and photon radiation at typical locations under various operating modes must be analyzed to provide a basis for the reactor safety analysis, radiation monitoring system design, equipment maintenance, etc.

Based on the position of the shield, the shielding may include reactor shielding, processing room shielding, the building shielding, etc.

The reactor shielding consists of internal shielding and biological shielding [3, 8]. Taking a pool-type reactor as an example, the internal shielding refers to the shield set inside the primary vessel, which is on the side, top, and bottom of the core. Its main function is to reduce the neutron energy and neutron flux, the neutron radiation damage to reactor components, and the activation of nuclear coolants. Steel, graphite, boron carbide, water, and other materials are commonly used for the internal shielding. For pool-type reactors, the biological shielding generally includes the pit shielding, vessel top shielding, protection platform, etc. Its main function is to further reduce the radiation from the core, which is weakened by the internal shield, to an allowable level. Concrete is the most commonly used shielding material. However, due to the poor thermal conductivity of concrete, it is sometimes necessary to provide a special cooling system in the concrete shield to prevent it from dehydrating or cracking due to heat deposition.

Many systems are involved in the processing room, including primary/secondary circuit, fuel transportation, ventilation and filtration, radioactive waste collection, etc. The distribution of radiation sources in these systems is relatively complex. Usually, the layout of the processing room is optimized according to the intensity of the radiation source. In addition, shielding walls are set up for containment and partition isolation, and ventilation and decreased irradiation times are also used to ensure the radiation safety of the staff.

The shielding of the buildings mainly includes the shielding of the reactor building, auxiliary building, and fuel building. The radiation shielding design of the building should include a calculation of the radiation distribution under the various operating modes of the reactor based on the distribution and intensity of the radiation sources and the arrangement of the equipment and pipelines, and the shielding wall is designed in combination with the shielding requirements and the characteristics of the radiation field.

The radiation sources and shielding schemes are different at different locations, but their design concepts are nearly the same. First, the radiation shielding design limits are determined based on the shielding target and requirements. Second, shielding materials are selected and arranged in combination with environment and space layout according to the intensity and distribution of the radiation sources. Third, the shielding optimization is performed through iterative neutronics calculations with the structural design. Finally, the radiation shielding design scheme is obtained. The design flow for the shielding is shown in Fig. 9.10.

1. Selection and arrangement of shielding material

The selection of shielding materials should consider factors such as the reactor design objectives, weight, volume, price, working environment, and material shielding prop-

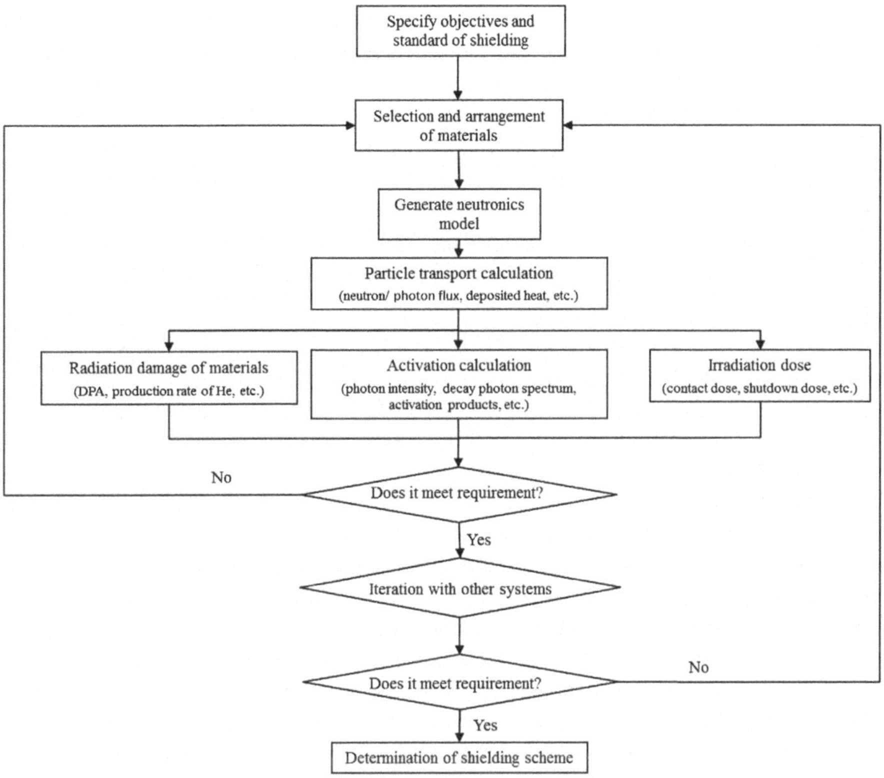

Fig. 9.10 Flowchart for the radiation shielding design

erties. The shielding properties of the materials are generally measured by physical quantities such as the neutron slowing-down power, neutron capture ability, secondary γ-ray released rate, attenuation rate of the flux density, and dose. Different materials have different shielding properties for different types of radiation.

For neutron radiation [6], the commonly used shielding materials in reactors include water, polyethylene, lithium hydride, graphite, boron carbide, concrete, stainless steel, tungsten, copper, etc. Water, polyethylene, lithium hydride, and graphite mainly reduce the energy of neutrons through elastic scattering to achieve the purpose of shielding. Due to its strong slowing-down power, water is often used as a shielding material for the core or spent fuel. Polyethylene has a high hydrogen content and is easy to process and shape, but its melting point is low and it is inflammable. It is usually used in combination with structural materials as an outside shield for the core or a partial shield for the building. Lithium hydride has a high hydrogen content, low density, and a high melting point and boiling point, and it does not release γ-rays after capturing neutrons, but its chemical properties are active. Therefore, it is often used as a shielding material for space reactors. Graphite has excellent high-temperature properties and is not activated by thermal neutrons. It is usually used as a neutron moderator, reflector, and shielding material for the core. Boron carbide and other boron-containing materials mainly utilize the strong absorption of thermal neutrons by ^{10}B for neutron shielding. Boron carbide has a high-density, high-strength, high-temperature stability, and good chemical stability. It is commonly used as a neutron shielding material in reactors. Heavy elements in stainless steels, tungsten, copper, and other materials slow down neutrons through inelastic scattering to achieve shielding purposes. They are often used as shielding materials for fast and high-energy neutrons. Concrete contains a variety of elements and provides good shielding for neutrons and photons. It is commonly used as a shielding material outside the reactor.

Commonly used shielding materials for photon radiation include lead, concrete, water, and so on. Lead has significant shielding ability against low- and high-energy γ-rays, and is an ideal shielding material for γ-rays. Its disadvantages are high cost, low strength, and low melting point; and therefore, it is often used as a γ-ray shield outside the reactor. In lead-based fast reactors, lead alloys are used as coolants, which can effectively shield γ-rays from the core and facilitate the miniaturization of external shields. Iron has weaker shielding performance against γ-rays than lead, but it has the advantages of high mechanical strength, easy processing, and high-temperature resistance. It is often used in the internal shielding of reactors and hybrid radiation fields of neutrons and γ-rays. Concrete has good shielding capability for neutrons and photons, and is commonly used as an external shielding material. Water has poor shielding performance against γ-rays, but it is often used to store solid-state gamma sources in water wells and pools due to its good transparency.

When designing the nuclear system's shielding, based on the shielding characteristics of the nuclear material and the radiation source, a variety of shielding materials are selected and arranged to obtain the optimum results in accordance with the principles of "take moderation into consideration prior to absorption, take the neutron into consideration prior to gamma, and take heavy materials into consideration prior to light materials." For example, CLEAR-M10 adopts a combination of stainless steel and liquid lead as internal shielding, whereas the SVBR uses combination of stainless steel, boron carbide, and lead alloy.

In addition, because of the pipeline arrangement, cable connection, etc., there are numerous penetrations, holes, gaps, etc. in the shield. An improper design may cause radiation leakage, resulting in a higher local radiation field outside the shielding than the limiting level. Therefore, it is necessary to maintain the integrity of the shielding as much as possible, and holes and gaps should be shielded locally throughout the shield design, often using bent pipes, shadow shield, labyrinth design, and so on.

In the case of pipeline shielding, the flux density at any location in the pipeline can be decomposed into three components: direct radiation, escaped radiation that passes through the shield, and reflected radiation from the pipe wall. For a spiral pipe consisting of n pipes with length l, radius a, and rotation angle φ, assuming that the direct radiation and escaped radiation are completely shielded, the flux density at the outlet of the pipe can be approximated as [6]

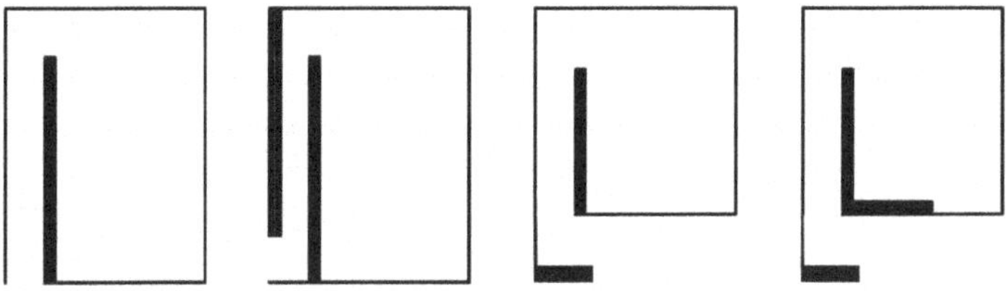

Fig. 9.11 Several typical labyrinth radiation shielding designs

$$\Phi = \frac{N_0}{2^n}\left(\frac{a}{l}\right)^{2n}(A_2\cos\varphi)^{n-1} \tag{9.4}$$

where N_0 is the flux density at the inlet; A_2 is the total albedo. According to Eq. (9.4), the flux density at the outlet decreases rapidly in bent pipe.

To reduce the radiation dose at the entrance and exit of the radiation area, labyrinths are usually designed to shield the direct radiation from the radiation source. The scattered radiation is reduced through multiple scatterings of the radiation by the labyrinth wall. Common labyrinth designs are shown in Fig. 9.11.

2. Shielding calculation optimization

Based on the selection and preliminary arrangement of the shielding materials, a neutron model should be established to perform neutronics calculations, such as neutron/photon transport calculations, irradiation damage calculations, activation calculations, and irradiation dose calculations for evaluating the performance of the shielding scheme. Through iteration with other systems, such as the thermal-hydraulic system, and structural design, a shielding scheme that meets the design requirements can be developed.

Based on the design objectives, shielding calculation is carried out to analyze and optimize the following components.

(1) Flux and nuclear heat deposition

In the shielding calculations, the attenuation of the neutron and photon flux in the shield is usually evaluated to confirm the rationality of the layout. In addition, because various types of radiation are shielded, the radiation energy is deposited in the material in the form of heat, which may cause an increase in the temperature and thermal stress of the material. For example, to prevent the concrete shield from dehydrating or cracking due to heat deposition, the shield design usually requires that the neutron flux density on the inner surface of the concrete should not exceed $5.0 \times 10^9\ \text{cm}^{-2} \cdot \text{s}^{-1}$ and that the photon energy flux should not exceed 4×10^9 $\text{MeV} \cdot \text{cm}^{-2} \cdot \text{s}^{-1}$. Therefore, the nuclear heat deposition in the shield must be evaluated in the transport calculation to ensure that it is below the specified limit.

(2) Irradiation dose

The irradiation dose is an important indicator for determining whether the shielding scheme meets the design requirements. It is necessary to evaluate the neutron, photon, and total dose rate after the shield during operation and shutdown to ensure that the radiation shielding meets the design requirements. The radiation dose should be comprehensively considered to optimize the thickness and arrangement of the neutron/photon shielding materials. To reduce the shutdown dose rate, the radiation shielding design usually arranges neutron absorption materials to reduce the neutron activation (i.e., to reduce the intensity of the gamma-ray source during shutdown), and lead or other materials can be used to shield γ-ray, which reduce the intensity of the γ-rays at the location.

(3) Material radiation damage

Radiation damage to materials is an important indicator for evaluating the radiation safety of reactor equipment and components, and it is also an important parameter for evaluating shielding schemes. Radiation damage to materials is mainly caused by fast neutrons. Several methods can be used to control the fast neutron flux, such as optimizing the arrangement of reflectors to reduce fast neutron leakage of the core or optimizing the selection and arrangement of heavy and light element materials in the shield to slow down fast and high-energy neutrons [3]. In the shielding calculation, it is necessary to evaluate the radiation damage to equipment and components to ensure that it meets the design requirements. Some advanced fission systems have a higher average neutron energy, higher neutron flux, higher neutron leakage rate, and longer refueling period than the thermal reactors. Therefore, the radiation damage to the equipment and materials is more serious than thermal reactors. Special attention should be paid to the radiation shielding design.

(4) Other neutronics parameters

In shielding calculations, neutron parameters such as material activity, residual heat, contact dose rate, etc. should be evaluated to ensure that they meet the reactor safety objectives.

In the design of the radiation shielding, it is necessary to consider the shielding parameters presented above. The optimization algorithm is used to optimize the design parameters of the shielding, including materials, thickness, layout, and position. Finally, the optimal shielding scheme that meets the shielding objectives and the economic and engineering feasibility requirements can be determined.

9.3.2.3 Radiation Safety Analysis

In reactor design, to demonstrate the safety and environmental friendliness of the design, a radiation safety analysis, including an occupational exposure dose assessment, radionuclide diffusion assessment, and environmental impact assessment, must also be performed.

1. Occupational exposure dose assessment for workers [5]

When a worker performs installation and maintenance, he will inevitably be exposed to external or internal radiation from neutrons, photons, and various radioactive rays generated by the nuclear system. In the design and safety evaluation of advanced nuclear system, it is necessary to evaluate the irradiation dose of the worker as he performs various types of operations according to the operating rules. In addition, several precautions must be taken to ensure that the irradiation doses of the worker are as low as possible. For the external irradiation of neutrons and photons, it is necessary to adopt radiation zoning, set up mobile shielding or fixed shielding, and adopt remote operations. For internal irradiation caused by various radioactive volatile nuclides and activated gases, the worker is required to wear protective clothing, and the radioactive area where the worker enters and exits should be fully ventilated.

2. Radionuclide migration and environmental impact assessment

The various radionuclides produced in the reactor will migrate as the coolant and gas flow. When a leakage accident or LOCA occurs, these radionuclides are released into the environment, which affects the environment and the public. In the design and safety evaluation of advanced nuclear systems, calculations and analysis of the migration of radionuclides generated in the reactor are required, and the impacts of radionuclides leakages on the environment are evaluated under normal and accidental conditions to guide the design optimization of the radioactive containment and ventilation systems.

9.3.2.4 Radioactive Waste Assessment and Management

Radioactive waste refers to substances that contains radionuclides or is contaminated with radionuclides but whose radionuclide concentration or activity is greater than the control levels prescribed by the national regulatory authorities. Based on the state of the waste, radioactive waste can be classified into liquid waste, gaseous waste, and solid waste. The liquid waste in the fission system mainly comes from the cleaning water of the primary and secondary coolant equipment, waste resin, the cooling circuit drainage, the radioactive wastewater, etc.; The gaseous waste mainly comes from the normal leakage or discharge of cover gas, the control rod drive, ventilation of radioactive areas and so on. Solid waste mainly comes from various ion exchange waste resins, the failed filter core, contaminated tools, waste activation parts, etc. In the reactor design, the amounts of these radioactive wastes must be calculated based on the neutron activation theory and the nuclide diffusion and migration theory. The use of low-activation materials, radioactive containment, and other means can reduce the total amount of radioactive waste. Waste disposal facilities are designed to collect, process, and store the various wastes generated by the reactor to ensure that the activity of discharged waste meets the related limits.

9.4 Design of Small Modular Reactor

Small modular reactor is an important trend in the development of advanced fission systems because of the huge future market demand. These reactors have the following advantages over large nuclear systems [8, 10].

(1) The single core has a low thermal power and contains a small total amount of radioactive source materials, which can reduce the radiological consequences of accidents conditions.
(2) Small modular systems adopt the concept of an integrated design which simplifies the structure of the reactor system, improves the reliability of the system, and facilitates assembly.
(3) Small modular nuclear systems are designed with passive safety systems. Some of them use multilayered containment or underground layout to enhance radioactive containment, prevent external events, and enhance safety, which reduce the impact on the environment and the public and provides more flexibility in site selection.
(4) A modular design is adopted to reduce construction time, costs, and financing difficulties. They can be built as multistage single or multistage dual units and can be built on demand to achieve construction flexibility.

Many small modular nuclear system concepts have been developed based on various types of reactors, such as light water reactors, high-temperature gas-cooled reactors, liquid-metal-cooled reactors, and MSRs. These systems have higher in safety and economic efficiency, prevent nuclear proliferation, and have long refueling periods through different design concepts and techniques.

Lead-based reactors have outstanding advantages in miniaturization and modularization [4, 11, 12]. For example, the liquid lead-based alloy used as the coolant in lead-based fast reactors has a high boiling point and excellent flow behavior and heat transfer characteristics, such as high thermal conductivity, low viscosity, etc., which increase the core power density and reduce the volume. Moreover, lead-based reactors have a hard neutron flux spectrum and higher neutron yield due to the small neutron absorption and slowing-down power of the liquid lead-based alloy. It is easy to achieve nuclear fuel breeding and a small loss of fuel reactivity, which are beneficial to prolonging the reactor refueling period. In addition, the lead-based metal has a good neutron reflection effect and strong γ-ray shielding ability, so the coolant can serve as both a reflector and a shield, reducing the volume and weight of the shielding system. The lead-based alloy can operate at low pressures and has good compatibility with water, air, structural materials, etc. Therefore, the lead-based fast reactor system has a simple design, and the intermediate circuit and the pressurizer can be omitted. The FDS team designed a small transportable advanced fission system with a liquid lead-based alloy as a coolant. It has technical characteristics of inherent safety, small size, lightweight, long lifetime, and high reliability. It may be a revolution in small high-power energy and can be used in a wide variety of environments, such as islands and oceans, as a power supply for offshore platform, desalinations, power supply in remote areas, and so on.

To meet the diverse demands of the applications, the neutronics design objectives for small modular nuclear systems have higher requirements than those of other nuclear systems. For the core physics, the design of small modular nuclear systems pursues a longer refueling period and an optimum core size after satisfying the thermal-hydraulic, safety, and other requirements. The following aspects should be considered in the design optimization.

(1) Core materials and layout optimization

The fuel type is important for small modular systems. For example, highly enriched uranium nitride is chosen as fuel for the space reactors to minimize the core size and fuel loading. Correspondingly, the weight and volume of the shadow shield are also reduced to realize the light weight of the entire nuclear system. Because small modular nuclear systems usually have small fuel loads, their refueling schemes are relatively simple. Many small modular nuclear systems adopt the no refueling concept in their lifetime. When selecting the core fuel and designing the initial charge, it is necessary to ensure full-power operation over the entire design cycle until refueling.

The neutron spectrum is an important parameter of small modular nuclear systems with long refueling period. Fast reactors have advantages of a small reactivity loss induced by burnup, small space occupied by the reactivity control system, and long-term operation, especially due to the breeding. However, they have a large critical volume. The thermal reactors are the opposite. Therefore, in the core physics design, the optimum neutron flux spectrum can be obtained by adjusting the volume ratio of the moderator to the fuel and optimizing the selection and arrangement of the core materials, which optimize the core size on the thermal hydraulics and safety requirements.

In addition, small modular nuclear systems have greater radial and axial neutron leakage and a nonhomogeneous power factor. This will present challenges for the material and thermal-hydraulic design. In the core physics design, many methods can be used to decrease neutron leakage and improve the fuel economy, such as optimization of the core dimensions, fuel zoning for different enrichments, and using a reflector with a strong reflecting power.

(2) Reactivity control method and its layout optimization

The reactivity control methods used in common nuclear systems, such as control rods and burnable poison, can also be used in small modular nuclear systems. However, the reactivity and power regulation requirements for small modular nuclear systems are stricter because of the more nonhomogeneous distributions of temperature and power. Therefore, there are special considerations for the design and arrangement of the reactivity control system. In addition, the control drum and sliding reflector are used to control the reactivity in small modular nuclear systems.

For radiation protection, the following measures are adopted in the small modular nuclear system to achieve the goals of lightweight, modularization, and transportability.

(1) Shielding at the "source"

The size of the shield is related to its distance from the radiation source. The greater the distance is, the larger the required shield is. To reduce the shielding volume, the shield is usually placed as close as possible to the source (i.e., shielding at the "source").

(2) Multifunctional design of components

Stainless steel, concrete, and lead–bismuth alloys are good neutron and photon shielding materials. They also have good structural mechanics and thermodynamic properties, and can be applied to structural or functional components. In the shielding design, these components can be fully utilized to reduce the weight and volume of the shield.

(3) Full use of multiple shielding modes

To reduce the weight and volume of the shield, shadow shielding, labyrinth shielding, transportable shielding, conformal shielding, and other shielding types are used according to the design objectives. For example, shadow shielding and conformal shielding are often adopted in space reactors to ensure that the fluence is less than 10^{12} n/cm^2 for fast neutrons and accumulated dose is less 10^4 Gy for photons at a position tens meters from reactor. Because the dose plane is far from the nuclear system, the half-cone angle of the shield is usually set to 12°–17°. These measures can effectively reduce the weight and volume of the shield.

In addition, using integrated neutron–photon shielding materials can simplify the design and reduce weight.

9.5 Example of Neutronics Design

This section takes the China Lead-based Mini-Reactor CLEAR-M10 as an example to introduce the core physics design and radiation protection design of advanced fission systems.

CLEAR-M10 is a small modular advanced nuclear system with a 10 MW level electric power supply that was developed by the FDS Team. With its small modular design, ultralong fuel cycle, inherent safety, and combined heat and power (CHP), CLEAR-M10 can flexibly meet various electrical needs, such as power for islands, offshore platforms and remote area, desalination of seawater, and CHP for industrial parks. The reactor structure is shown in Fig. 9.12, and the main parameters of CLEAR-M10 are shown in Table 9.3.

9.5.1 Core Physics Design

To make the loading and unloading of nuclear fuel quick and easy, the core of CLEAR-M10 adopts an integrated design with a triangular arrangement of fuel rods.

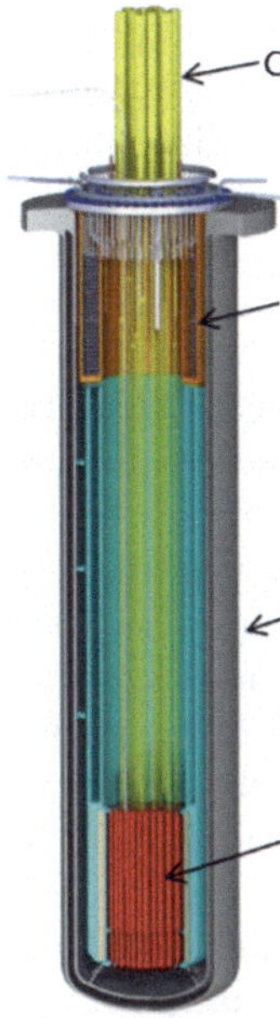

Fig. 9.12 The reactor structure of CLEAR-M10

Table 9.3 Main parameters of CLEAR-M10

Main parameters	Value
Thermal/electric power (MW)	35/14
Refueling cycle (year)	20
Design life (year)	60
Primary coolant	Pb
Fuel	UO_2 with the ^{235}U enrichment $\leq$19.75 wt%
Primary system	Pool type
Primary coolant circulation	Natural
Core inlet/outlet temperature (°C)	375/495
Secondary coolant cycle	Superheated steam
Pressure of secondary circuit (MPa)	13
Heat sink	Air cooler
Height/diameter of vessel (mm)	8500/2200

The lower end of the fuel rod is attached to a grating plate, and the upper end is attached to a lattice frame. The core consists of 31 circles of fuel rods to prevent the local blockage of the coolant channel and to reduce the risk of damage to the cladding. UO_2 is chosen as the fuel, which has enrichments less than 19.75%, and 15-15Ti is chosen as the cladding material. Beryllium oxide (BeO) and stainless steel sandwich panels are the reflectors outside the active zone. To flatten the power and reduce the radial power peaking at the center of the core, the fuel enrichment is reduced in those zones. The fuel enrichments of CLEAR-M10 are 18, 18.5, 19.75, and 17% from the inner core to the outer core. However, because of the moderating and scattering of BeO, the fraction of thermal neutrons in the zone near the reflector is very high, which leads to a high reaction rate near the reflector. To reduce the local power density, the main parameters of the core were determined after iterative optimization with the thermal-hydraulic analysis and structural mechanics analysis. The diameter of the fuel pin is 16 mm, and ratio of the pitch to the diameter ratio is 1.2. The active core diameter is 1200 mm and the active core height is 790 mm, which maintains the power at 35 MW and burns for 20 years. The layout of the core is shown in Fig. 9.13.

According to the preliminary design scheme, a fine model of the reactor was established by SuperMC, and several analyses, such as the power distribution, reactivity control, reactivity feedback, and burnup, were performed.

1. Power distribution

To ensure the integrity of the fuel element, the maximum power density of the fuel element must be below the design limit to ensure that the maximum fuel temperature,

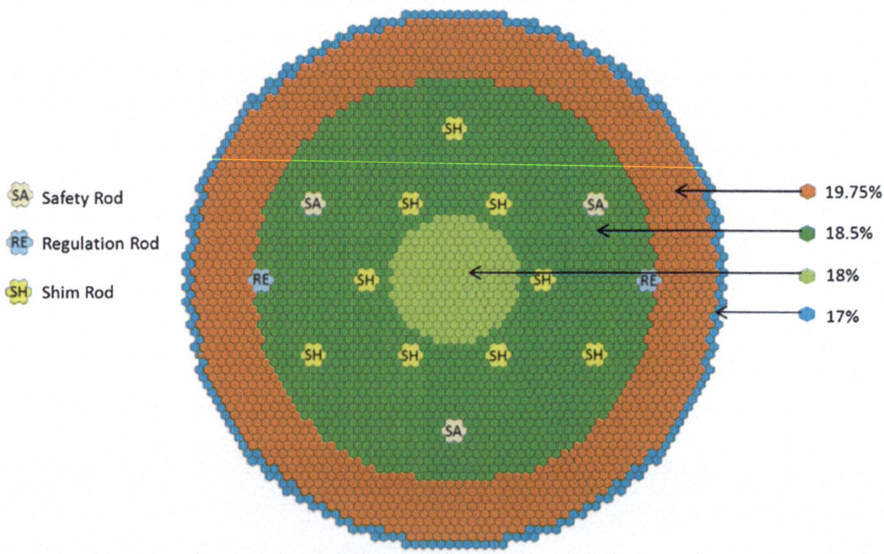

Fig. 9.13 Layout of CLEAR-M10

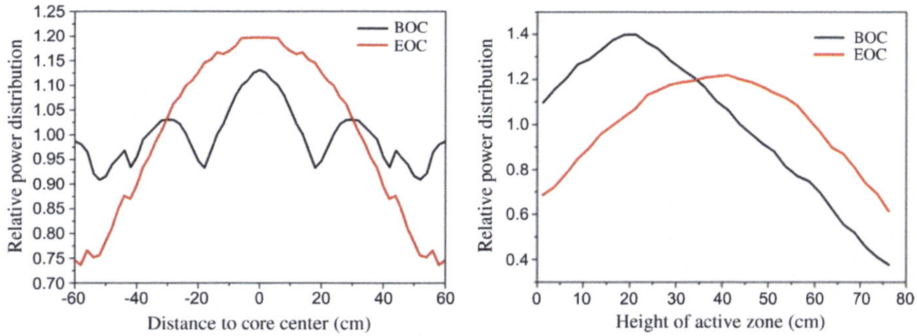

(a) Relative power distribution in the radial direction

(b) Relative power distribution in the axial direction

Fig. 9.14 Power distributions of CLEAR-M10. **a** Relative power distribution in the radial direction. **b** Relative power distribution in the axial direction

maximum cladding temperature, and maximum thermal stress of the cladding meet the material requirements. The rated thermal power of CLEAR-M10 is 35 MW, and its radial and axial power peak factors are 1.13 and 1.40, respectively, at the beginning of cycle (BOC) and 1.20 and 1.22, respectively, at the end of cycle (EOC). The power distributions are shown in Fig. 9.14. The maximum linear power density is 193 W/cm at the BOC and 176 W/cm at the EOC, which are much less than the limit of 400 W/cm and ensures that the maximum cladding temperature is less than 500 °C.

2. Reactivity control

To ensure the critical safety, the reactor must have two independent reactivity control systems, each of which satisfies the stuck rod criterion. The reactivity control of CLEAR-M10 is performed by 14 control rods, which can be divided into two independent control systems. Nine shim rods and two regulation rods comprise the first control system, and three safety rods are in the second control system.

Table 9.4 shows the reactivity balance of the two shutdown systems considering the changing reactivity of the reactor from full power to the cold state. The failure of the control rod with the greatest value is also considered.

Table 9.4 shows that the core could be shut down safely and that each control system satisfies the stuck rod criterion.

3. Reactivity feedback

To ensure the safety of the reactor, the total reactivity feedback must be negative when the reactor is operated at various power levels. The reactivity feedback of CLEAR-M10 includes the Doppler effect, coolant expansion effect, axial expansion effect, and radial expansion effect as shown in Table 9.5. All of the reactivity effects

Table 9.4 Reactivity balance of the control system

Reactivity balance		First control system	Second control system
Reactivity compensation (pcm)		9227	1227
Value of the control system (pcm)	All rods inserted	13,443	3812
	Most valued rod stuck	11,960	2331
Cold shutdown margin (pcm)	All rods inserted	4261	2585
	Most valued rod stuck	2733	1094

Table 9.5 Reactivity feedback

Reactivity effect	Power reactivity (hot zero power–hot full power) (pcm)	
	BOC	EOC
Doppler effect (pcm)	−305	−312
Coolant expansion (pcm)	−92	−95
Axial expansion (pcm)	−384	−363
Radial expansion (pcm)	−446	−454
Total	−1227	−1224

are negative, and the total reactivity effect is negative, which means that the reactor has inherent safety.

4. Burnup

CLEAR-M10 is designed to run at full power for 20 years without refueling, and its reactivity loss is approximately 8000 pcm. At the EOC, the maximum burnup level is approximately 95 MWd/kgU and the average burnup level is approximately 60 MWd/kgU, which are both less than the maximum discharged burnup level of 100 MWd/kgU and more than the average burnup level of current PWRs. The changes in density of several main nuclides with time are shown in Fig. 9.15.

9.5.2 *Radiation Protection Design*

The radiation protection of CLEAR-M10 includes the radiation source analysis, radiation shielding design, radiation safety analysis, and radiative waste assessment and management. This subsection introduces the radiation source analysis and radiation shielding design of CLEAR-M10.

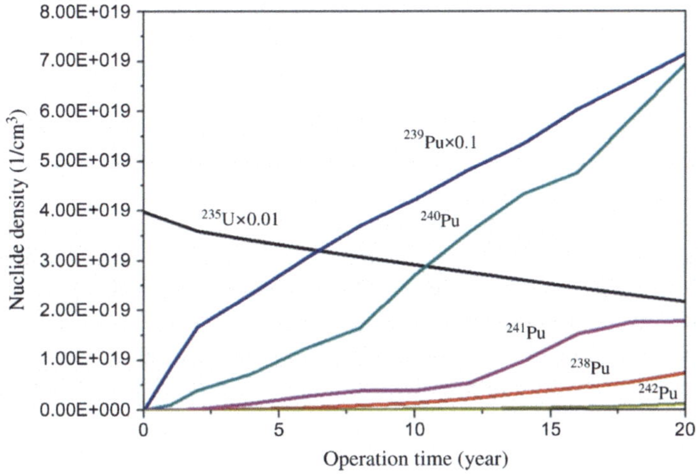

Fig. 9.15 Changes in nuclide density with time

9.5.2.1 Radiation Sources

The radiation sources in CLEAR-M10 mainly come from the core, the primary circuit, structure components, and so on.

1. Core

Neutrons and γ-rays are the main radiation sources during operation of CLEAR-M10. The average neutron energy in the core is approximately 500 keV, and the average neutron flux was approximately 3.0×10^{14} $cm^{-2} \cdot s^{-1}$.

The γ-rays released by fission products (FPs) in the core are one of the main radiation sources after reactor shutdown. When the system runs at full power for a refueling period, the total activity of the fission products is approximately 1.2×10^{19} Bq in the entire core. These FPs exist as inert gases, aerosols, and solids. If fuel claddings fail due to breakage or gas-tightness leakage, some of the FPs will leak into the coolant.

2. Primary system

During operation, the primary coolant is activated by fission neutrons when flowing through the core. The total activity of the lead coolant was approximately 3.0×10^{12} Bq at the EOL of CLEAR-M10. The main contributor to the gamma source was ^{207m}Pb, which has a short half-life of 0.806 s but high-energy secondary photons of 1.06 MeV. Several corrosion-activated products, such as ^{55}Fe, ^{51}Cr, ^{59}Fe, and ^{58}Co, were present in the coolant due to the corrosion of the lead coolant by stainless steel.

Moreover, several gaseous and solid FPs can leak into the coolant and diffuse in the reactor due to breakage or gas-tightness leakage of the fuel assembly .

3. Structural components

In CLEAR-M10, the main structural components, such as the internal equipment, main vessel, and safety vessel, were made of 316L stainless steel. During reactor operation, the structural materials will be activated by neutrons and produce radioactive nuclides. The γ-rays emitted by these nuclides during their decay process are one of the main gamma sources during shutdown.

At the EOL of CLEAR-M10, the total activity of the reactor structural material was approximately 2.5×10^{16} Bq, and the main contributors were ^{56}Mn and ^{51}Cr, whose half-lives are relatively short. One year after reactor shutdown, the main radionuclides were ^{60}Co and ^{55}Fe.

The radiation source of the second circuit is not the main source for the shielding design because the neutron flux of the second circuit is much lower than that of the core and primary circuit.

9.5.2.2 Radiation Shielding Design

In this subsection, the reactor shielding is taken as an example to introduce the radiation shielding design of CLEAR-M10.

The reactor shielding of CLEAR-M10 consists of internal shielding and external shielding as shown in Fig. 9.16. The internal shielding of CLEAR-M10 makes full use of the shielding function of the beryllium oxide–stainless steel reflector and lead coolant above the core. To ensure that the irradiation damage to the vessel does not exceed the limit values during its lifetime, stainless steel is set up around the reflector layer to reduce the irradiation damage. The irradiation damage at the bottom of the vessel is reduced by boron carbide and the lead coolant. The reactor cover is filled with boron carbide to decrease the thermal neutron flux above the cover. On the side and bottom of the reactor pit, borated graphite rods with stainless steel cladding and ordinary concrete are deployed from the inside to the outside as external shielding to ensure that the dose rate outside the external shielding meets the requirements.

1. Distributions of the neutron flux and photon energy flux

Figure 9.17 shows the distributions of the neutron flux and photon energy flux at full-power operation of CLEAR-M10. Because of the natural circulation of CLEAR-M10, the lead coolant, which is 6 m above the core, can effectively shield leaked neutrons and photons from the core. The axial attenuation of the neutron flux is shown in Fig. 9.16a. The neutron flux is less than $3.0 \times 10^4\ \mathrm{cm}^{-2} \cdot \mathrm{s}^{-1}$ through the shielding of the lead coolant and cover.

As shown in Fig. 9.16b, the gamma energy flux attenuates rapidly along the axis and is less than the 20 $\mathrm{MeV} \cdot \mathrm{cm}^{-2} \cdot \mathrm{s}^{-1}$. This indicates that the lead coolant has good shielding ability against gamma energy.

2. Irradiation damage induced by neutrons

The main vessel is one of the key components in the reactor and acts as the primary coolant system's boundary to contain the coolant and radionuclides in the primary coolant system and the cover gas in the reactor. To protect it against radiation damage,

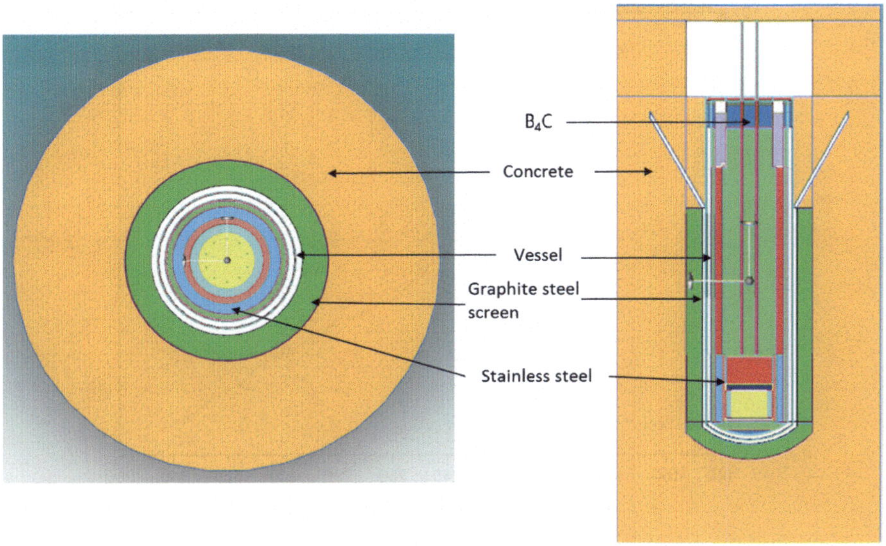

Fig. 9.16 The shielding of CLEAR-M10

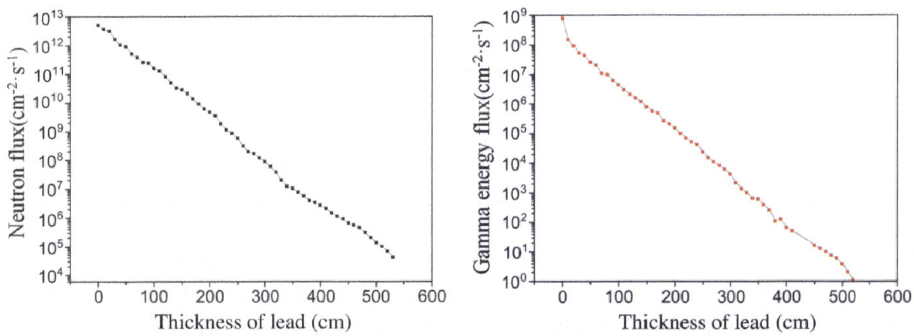

(a) Distribution of the neutron flux (b) Distribution of the gamma energy flux

Fig. 9.17 Distributions of the neutron flux and gamma energy flux of CLEAR-M10 under full-power operation. **a** Distribution of the neutron flux. **b** Distribution of the gamma energy flux

CLEAR-M10 uses stainless steel to protect the container. The maximum neutron flux at the internal surface of main vessel is approximately $4.0 \times 10^{12}\ \mathrm{cm^{-2} \cdot s^{-1}}$, and the total fluence is approximately $8.0 \times 10^{21}\ \mathrm{cm^{-2} \cdot s^{-1}}$, lower than $2.0 \times 10^{22}\ \mathrm{cm^{-2}}$ within the 60 years of full-power operation. The displacement per atom on the main vessel is approximately 0.9 dpa, which is less than the irradiation damage limit of 3 dpa.

3. Nuclear heating and radiation dose

To prevent the production of gaps and the loss of water due to nuclear heating from the interaction between neutrons/photons and the nuclides in the concrete, a 40-cm-thick

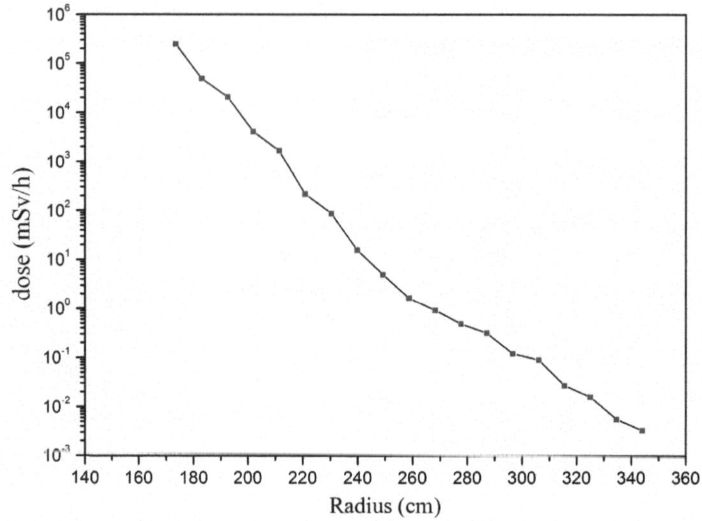

Fig. 9.18 Radial distribution of the dose rate on the side of reactor pit

borated graphite steel screen consisting of graphite rods and steel cladding is located on the inner surface of the CLEAR-M10 concrete pit. Its function is to moderate the fast neutrons and shield the photons leaking from the vessel. The shielding analysis shows that the maximum neutron flux and photon energy flux were approximately $3.0 \times 10^9\ \mathrm{cm}^{-2} \cdot \mathrm{s}^{-1}$ and $4.0 \times 10^9\ \mathrm{MeV} \cdot \mathrm{cm}^{-2} \cdot \mathrm{s}^{-1}$, respectively. These are lower than the limits of $5.0 \times 10^9\ \mathrm{cm}^{-2} \cdot \mathrm{s}$ and $4.0 \times 10^{10}\ \mathrm{MeV} \cdot \mathrm{cm}^{-2} \cdot \mathrm{s}^{-1}$, respectively, indicating that the side radiation shielding design satisfies the requirements of the concrete.

Figure 9.18 shows the attenuation of the dose rate along the radial shielding of the pit during full-power operation. The total dose rate at the side of the pit was decreased from 2.3×10^5 mSv/h to 3.1×10^{-3} mSv/h by the 180 cm radial concrete shielding.

The analysis presented above demonstrates that the CLEAR-M10 radiation shielding design meets the design requirements.

References

1. Wu YC, Bai YQ, Song Y et al (2016) Development strategy and conceptual design of China lead-based research reactor. Ann Nucl Energy 87:511–516
2. Wu YC, Team FDS (2016) Design and R&D progress of China lead-based reactor for ADS research facility. Engineering 2(1):124–131
3. IAEA (2016) Fast Reactor database 2006 update. IAEA-TECDOC-1531
4. Pioro I (2016) Handbook of generation IV nuclear reactors. Woodhead Publishing

5. IAEA (1996) International basic safety standards for protection against ionizing radiation and for the safety of radiation sources. International Atomic Energy Agency
6. Schaeffer NM (1973) Reactor shielding for nuclear engineers. Inc. Fort Worth, Tex (USA)
7. Zrodnikov A, Toshinsky G et al (2011) Fuel cycle for reactor SVBR-100. J Mater Sci Eng B 1:929–937
8. Waltar AE, Todd DR, Tsvetkov PV (2012) Fast spectrum reactors. Springer
9. Doderlein C, Cetnar J et al (2013) Definition of the ELFR core and neutronic characterization. LEADER project deliverable DEL005-2011
10. Carelli MD, Ingersoll DT (2014) Handbook of small modular nuclear reactors. Woodhead Publishing
11. Wu YC (2016) CLEAR-S: an integrated non-nuclear test facility for China lead-based research reactor. Int J Energy Res 40(14):1951–1956
12. Wang M, Lian C et al (2015) Preliminary conceptual design of a lead–bismuth cooled small reactor (CLEAR-SR). Int J Hydrogen Energy 40(14):15132–15136

Chapter 10
Neutronics Design of Fusion Systems

A fusion system is a facility that could utilize the energy released via a fusion reaction in a controllable and peaceful way. There are many facilities based on different methods to achieve the fusion reaction, including magnetic confinement fusion (MCF), inertial confinement fusion (ICF), etc. Among these facilities, the tokamak is the well-developed candidate for a commercial fusion system. In this chapter, the tokamak is used as a representative system to introduce the neutronics design of a fusion system.

In a fusion system, the neutron carries most of the fusion energy (the kinetic energy of the neutron accounts for ~80% of the energy released by the D-T fusion reaction), is the key to attain the tritium self-sufficiency, and is also the source of radioactivity in the system. Thus, the neutronics design is a crucial step in fusion system design and is concerned with the feasibility, safety, economy, and environmental friendliness of the system. The neutronics design for the fusion system is focused on the blanket, tokamak machine, and corresponding buildings, related to the whole lifecycle of systems, including the procedures of design, licensing, operation, and decommission. Compared to fission energy systems, fusion systems have a more complex geometry and a harsher service environment for in-vessel components, which creates great challenges for neutronics design and analysis. In this chapter, our discussion on fusion systems will be as follows: (1) the principles, features, and typical conceptual designs of fusion systems; (2) the neutronics design principles, requirements, and methods; and (3) taking the Dual-cooled Lead Lithium (DLL) blanket adopted in FDS-II and the tokamak machine and buildings in the ITER as examples to illustrate neutronics design for the blanket, the tokamak machine, and the corresponding buildings, respectively.

Y. Wu, *Neutronics of Advanced Nuclear Systems*,
https://doi.org/10.1007/978-981-13-6520-1_10

10.1 Introduction to Fusion Systems

10.1.1 Basic Principles

In tokamak-type fusion systems, the plasma is heated in various ways (such as Ohmic heating, electromagnetic wave heating, neutral beam injection, etc.) and is confined by the specific shape of the magnetic fields produced by coils to ensure the sufficiently high fuel temperature, fuel ion density, and long energy confinement time for plasma. The product of fuel temperature, fuel ion density, and energy confinement time is usually defined as fusion triple product. For example, considering the α-particle heating, the ignition condition for plasma in the D-T fueled fusion system approximately requires the triple product $>3 \times 10^{21}\ \mathrm{m}^{-3} \cdot \mathrm{keV} \cdot \mathrm{s}$ in the temperature of range of 10–20 keV.

The D-T fusion reaction releases one neutron and one ^{4}He nuclide (a.k.a., α particle), together with 17.6 MeV of energy, ~80% and ~20% of which are contributed by the kinetic energy of the neutron and the α particle, respectively. The fusion neutrons penetrate from the core plasma and interact with the materials of the components in the system to achieve energy conversion (neutron kinetic energy $\rightarrow$ thermal energy $\rightarrow$ other forms of energy) and the breeding of nuclear fuel tritium (tritium is a radionuclide with the half-life of 12.3 years. It is naturally very scarce and there is no artificial production available to support fusion systems as external sources. Thus, it is necessary to breed the tritium by relying on the fusion neutrons to react with Li.), thereby realizing the utilization of fusion energy.

The tokamak-type fusion systems are composed of the tokamak machine and external auxiliary systems. The tokamak machine is generally composed of blankets, divertor cassettes, a vacuum vessel (VV), magnets, thermal shields, cryostats, etc. Among these components, the blanket plays a key role in energy conversion and tritium breeding. The external auxiliary systems include plasma heating and current-driven systems, diagnostic systems, control systems, cryogenic systems, power supply systems, heat transfer and transport systems, power conversion systems, etc. [1, 2].

Converting energy and generating the electricity, and attaining tritium self-sufficiency are the two primary functions of fusion systems, the working principles of which are described in this section.

1. Energy conversion and electricity generation

The thermonuclear fusion reactions occurring in the plasma release abundant energy in the forms of kinetic energy of particles (such as neutrons and alpha particles) and thermal radiation. The particles interact with the materials of the components (mainly blankets), converting their kinetic energy into nuclear heat. Additionally, the thermal radiation from the plasma is also an important heat source for the first wall (FW) of the blanket and divertor. The heat in the blanket is taken away by coolant and used for power generation. The schematic diagram of the fusion energy conversion to electricity is shown in Fig. 10.1.

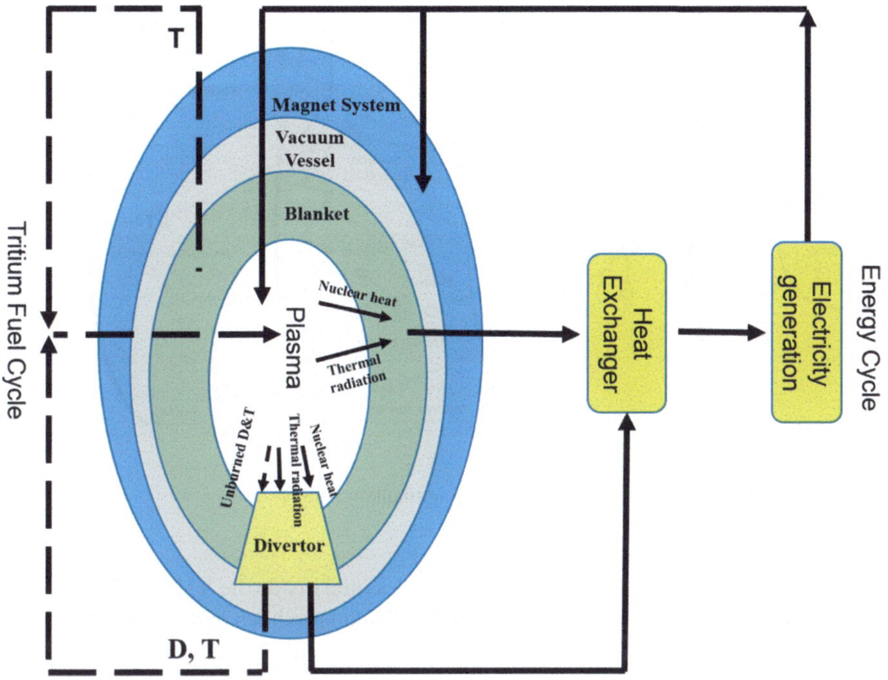

Fig. 10.1 Schematic diagram of a fusion system

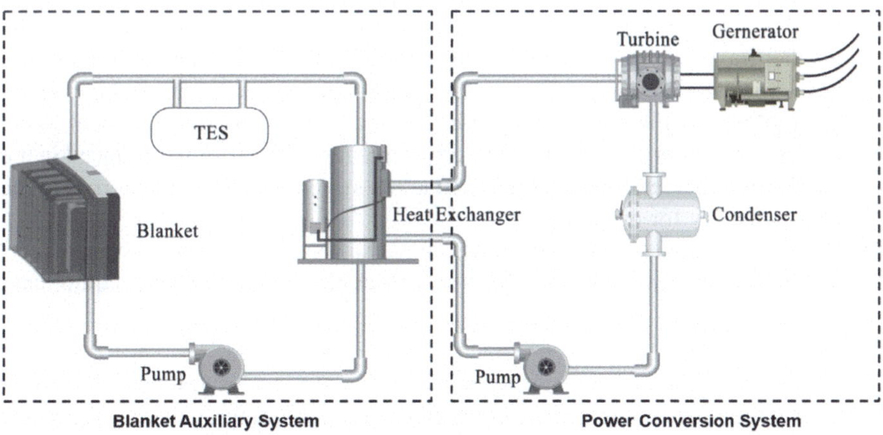

Fig. 10.2 Schematic diagram of a fusion blanket system

The blanket systems, comprising the blanket modules, blanket auxiliary systems, power conversion systems, and corresponding power generation and transmission facilities, are key systems for the extraction and conversion of fusion energy, as shown in Fig. 10.2.

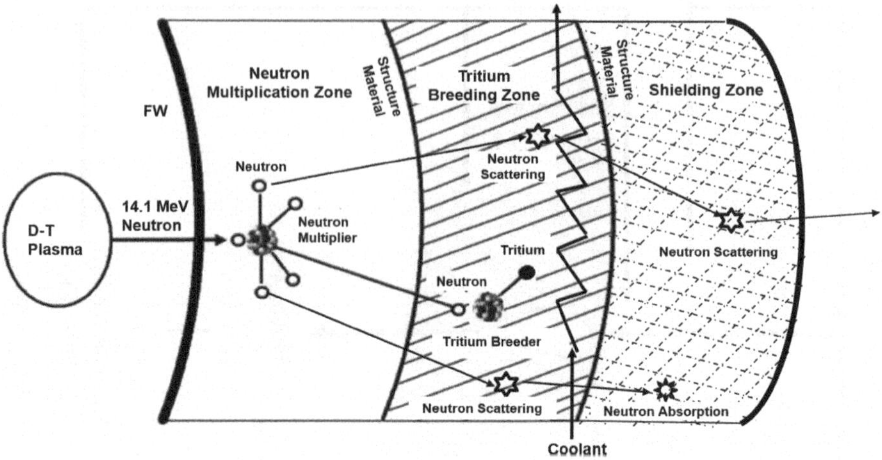

Fig. 10.3 Schematic diagram of a tritium breeding blanket module

(1) Blanket module: The blanket modules are located inside the VV and directly facing the plasma. The main functions of the blanket modules include tritium breeding, energy conversion and transfer, radiation shielding, etc. A blanket module consists mainly of the first wall (FW), tritium breeding zone (TBZ), neutron multiplication zone, cooling channels, and shielding structures. A schematic diagram of a blanket module is shown in Fig. 10.3.
(2) Blanket auxiliary system: The blanket auxiliary system is connected with the blanket modules; this system consists mainly of a tritium extraction system, heat exchangers, pumps, and a coolant purification system. The coolant, which is circulated by the pumps, flows through the blanket and transfers heat to the power conversion system for electricity generation. The tritium produced in the blanket is collected by the tritium extraction system and then transferred to the tritium factory for purification.
(3) Power conversion system: The power conversion system converts the heat that is transferred by the coolant into electricity through turbines and other equipment.

2. Tritium fuel cycle

The tritium fuel cycle of the fusion system consists of two closed loops, a breeding cycle that extracts the tritium bred in the blanket and burning cycle that processes the exhaust gases from the D-T fusion exiting in the tokamak, as shown in Fig. 10.4. The tritium system consists of a tritium extraction system (TES), a fuel purification system (FPS), a tokamak exhaust processing (TEP) system, an isotope separation system (ISS), and a storage and delivery system (SDS) and a fuelling system.

At the beginning of operation, the D and T fuels are supplied from external sources and stored in the SDS for reactor start-up. Because of the low burnup fraction, a considerable amount of D and T fuel is exhausted from the tokamak to the TEP system. The tritium bred in the blanket is extracted by the TES. In both closed

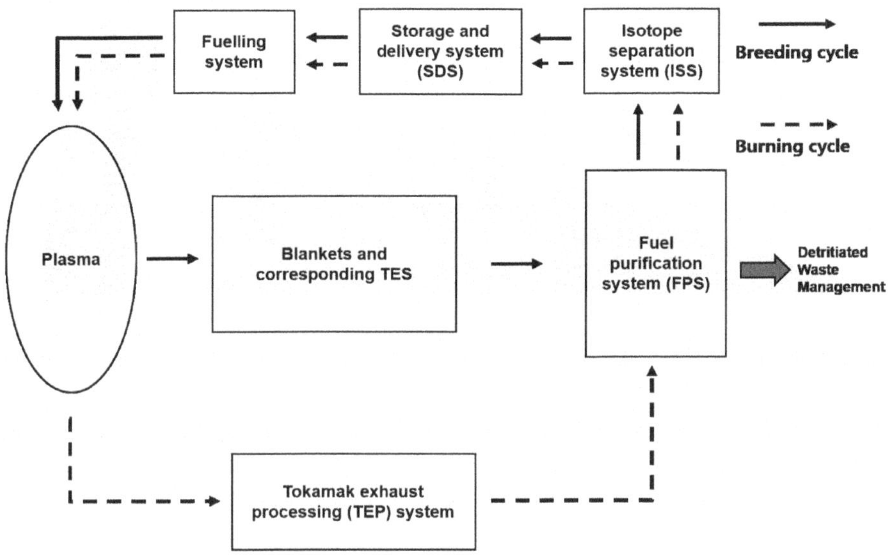

Fig. 10.4 Schematic diagram of the tritium fuel cycle in the fusion system

cycles, the hydrogen isotopes from either the TES or the TEP system will remove impurities through FPS to obtain the purified hydrogen isotopes including tritium and deuterium. Subsequently, these purified hydrogen isotopes will be sent to the ISS to separate tritium and deuterium. Finally, these separated purified fuels will be transformed to the SDS and re-injected into the plasma for burning through fuelling system.

10.1.2 Technical Features

Fusion systems are usually operated under extreme conditions such as strong magnetic fields and extreme temperature variations and pressure gradients among the components, as shown in Fig. 10.5. For example, the magnetic coils used to confine the plasma generate a very strong magnetic field (10–20 T) in the VV. The temperature of the plasma can reach 10^8 °C, and the highest temperature of key components (such as the blanket and divertor) can be as high as 1000 °C, whereas the superconducting magnet needs to be operated at ~ 4 K. Plasma operation requires a vacuum environment with a very low pressure of 10^{-6}–10^{-5} Pa, whereas the pressure in the coolant pipeline can be as high as 10 MPa [3].

To meet the requirements of commercial applications, the fusion system should not only generate stable energy but also ensure high safety, reliability, economical performance, and environmental friendliness. The design of the fusion system should possess the following characteristics:

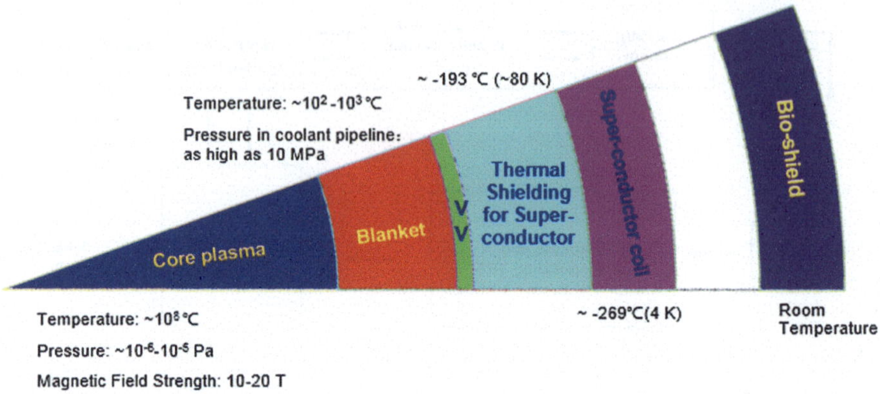

Fig. 10.5 Schematic diagram of physical fields (temperature, pressure, and magnetic field strength) distribution in the fusion system

1. High fusion power and energy gain ratio

The fusion system usually needs 50–100 MW of plasma auxiliary heating or current-driven power. Considering energy conversion efficiency to achieve a substantial electricity output, the fusion energy gain ratio (Q) should generally be greater than 30. Consequently, the NWL on the FW can be as high as 2–5 MW/m^2, and the peak heat load on the divertor target plate can be up to 10–40 MW/m^2. This presents higher requirements for materials in the irradiation resistance and the heat resistance.

2. Steady-state operation and high availability

To stabilize the energy output of fusion system and ensure its controllability, steady-state operation of the plasma is required. Simultaneously, the fusion system needs to achieve high availability (80% or more) to ensure its economic competitiveness. The availability depends on the advancements in fusion plasma physics, industrial manufacturing technology, materials technology, and other factors.

3. High energy conversion efficiency and tritium self-sufficiency

Approximately, 80% of the energy produced in the fusion reactions is carried by neutrons and is converted into heat by the blanket surrounding the plasma. Tritium is continuously consumed because of the D-T reactions in the plasma, the decay of tritium, and its retention in the material of the system. Therefore, neutrons must continuously react with lithium in the blanket to produce more tritium to compensate for these tritium losses in sustaining the tritium fuel supply.

Because there is no practical external source of tritium for the fusion system beyond ITER, the fusion nuclear energy system must rely on itself to achieve self-sufficiency. In addition, the fusion nuclear energy system operates in a variety of extreme physical environments (neutron/photon radiation, temperature, pressure, electromagnetic, etc.), so that the components in the system face a harsh service environment (high irradiation damage, high heat load, etc.). Meanwhile, high-intensity neutron radiation results in a higher level of component activation. Therefore, it becomes challenging in the neutronics design of fusion systems to choose the appropriate functional materials and their layout to achieve tritium production, ensure the safety of materials in harsh service environments, and reduce the activation level of activated components.

10.1.3 Typical Conceptual Designs

To verify and demonstrate the feasibility of fusion energy utilization from the view of physics and technology and to improve safety, economic performance, and reliability of the fusion system, a variety of fusion systems such as International Thermonuclear Experiment Reactor (ITER), Power Plant Conceptual Study (PPCS) and DEMOnstration fusion plant (DEMO) program in European, ARIES (Advanced Reactor Innovation and Evaluation Study) series reactor in the US, SSTR (Steady-State Tokamak Reactor) in Japan and FDS series reactor in China have been proposed and designed in the world. The typical concepts of fusion systems will be introduced in this section. The main plasma parameters and blanket design parameters of these concepts are listed in Table 10.1.

1. ITER

ITER is the largest magnetic confinement fusion experimental reactor under construction, and its operation will be an important milestone for the development of fusion energy. The first plasma discharge of ITER is expected to be achieved in 2025 [4]. The overall programmatic objective of the ITER is to demonstrate the scientific and technological feasibility of fusion energy for peaceful application. Its technical objectives are (1) to achieve a 300–500 s inductively driven plasma with energy gain ratio (Q) higher than 10 for a range of operating scenarios and to demonstrate the steady-state operation using a noninductive current drive with Q-value of at least 5 and controlled ignition; (2) to demonstrate the availability and integration of technologies essential for a fusion reactor (such as superconducting magnets and remote maintenance), to test components for a future reactor (such as systems to exhaust power and particles from the plasma), and to test the tritium breeding module concepts that would lead to a future reactor tritium self-sufficiency and the extraction of high grade heat for electricity production; and (3) to demonstrate the safety characteristics of a fusion system [5].

2. PPCS and DEMO

The PPCS was started in Europe in 2001, with the objective of assessing the feasibility of fusion energy applications in terms of safety, environmental friendliness, and economic competitiveness [6]. There are four blanket design schemes for the PPCS series fusion systems: a Water-Cooled Lithium Lead (WCLL) blanket, a Helium-Cooled Pebble-Bed (HCPB) blanket, a Dual-Coolant Lithium Lead (DCLL) blanket,

Table 10.1 Main design parameters of typical fusion systems

Parameter		ITER	FDS		PPCS				ARIES		A-SSTR
			FDS-II	FDS-III	PPCS-A	PPCS-B	PPCS-C	PPCS-D	ARIES-AT	ARIES-ST	
Major radius (m)		6.2	6.0	5.1	9.55	8.6	7.5	6.1	5.2	3.2	6
Plasma current (MA)		15	15	16	30.5	28	20.1	14.1	13	29	12
Toroidal magnetic field (T)		5.3	5.9	8.0	7.0	6.9	6.0	5.6	5.8	2.2	11
Bootstrap current fraction		~0.5	0.69	0.65	0.45	0.43	0.63	0.76	0.91	0.96	0.8
Ave. NWL (MW/m^2)		≥ 0.5	2.72	4	2.2	2	2.2	2.4	3.2	4.1	6
TBR		–	1.29	1.10	1.06	1.12	1.15	1.12	1.10	1.10	1.20
Auxiliary heating power (MW)		50	80	–	246	270	112	71	35	28	60
Fusion power (GW)		0.5	2.5	2.6	5	3.6	3.41	2.53	1.7	2.86	4.5
Q		≥ 10	31	–	20	13.5	30	35	50	108	75
Blanket	Structural materials	SS316L(N)-IG[a]	CLAM	CLAM	RAFM	RAFM	RAFM	SiC_f/SiC	SiC_f/SiC	SiC_f/SiC	RAFM
	Tritium breeder	–	LiPb	LiPb	LiPb	Li_4SiO_4	LiPb	LiPb	LiPb	LiPb	Li_2O
	Coolant	–	LiPb/He	LiPb/He	H_2O	He	LiPb/He	LiPb	LiPb	LiPb/He	H_2O
	Inlet/Outlet temperature (°C)	–	480/700	400/1000	285/325	300/500	480/700	700/1100	645/1100	350/500	–
Efficiency, η		–	0.47	–	0.31	0.36	0.42	0.6	0.59	0.3	0.35

Note [a]In ITER, the blankets can be classified into two types by their function: shielding blanket designed by ITER IO and test blanket module (TBM) designed by TBM program participants. The structural material is SS316L(N)-IG for shielding blanket and is varied for TBM due to different TBM participant design schemes and a Helium-Cooled Lithium Lead (HCLL) blanket. These blanket designs differ in terms of the device size, fusion power, and material selection as well as economic performance, safety, and environmental impact. The designs cover a variety of concepts ranging from older concepts, which utilize conservative technology and extrapolations of plasma physics, to advanced physical and technical concepts.

Based on the PPCS results and the R&Ds that were followed, Europe initiated a comprehensive design study of a DEMO in 2014 with the aim of generating several hundred MWs of net electricity and operating with a closed tritium fuel cycle around the middle of the century [7]. Like PPCS, there are also four types of blankets concept, including HCPB, HCLL, DCLL, and WCLL. These blanket concepts are the status of design updating at present.

3. ARIES

The ARIES program is a national research project funded by the US Department of Energy in 1990 with the participation of numerous US fusion research institutions. The purpose of the ARIES program is to conduct advanced fusion system research and to explore the potential for fusion development. The ARIES program has produced a series of conceptual designs for fusion systems over the past two decades [8], including ARIES-I, ARIES-II, ARIES-III, ARIES-V, ARIES-RS, ARIES-ST, ARIES-AT, and ARIES-ACT. ARIES-AT is an advanced tokamak fusion system that is characterized by high safety, environmental friendliness, and economical performance, based on the appropriate extrapolation from the existing technology.

4. SSTR

The SSTR is a steady-state tokamak fusion system designed by the Japanese Atomic Energy Research Institute in the 1990s. Its blanket concept uses water as the coolant and is based on the mature technology of light water reactors. The Advanced Steady-State Tokamak Reactor (A-SSTR), the latest conceptual design in the SSTR series, is based on advanced physical and engineering designs. The A-SSTR has a more compact core size and an efficient auxiliary system to improve the economic performance of the plant [9].

5. FDS series reactors

The FDS Team has developed a series of original conceptual designs to address future fusion energy needs [10, 11]. FDS-II is designed to exploit and evaluate the potential of fusion energy for electricity generation, based on advanced plasma parameters [12]. The plasma parameters can be extrapolated from the successful operation of ITER. FDS-II can satisfy various requirements regarding tritium self-sufficiency, safety margin, operational economic performance, environmental friendliness, etc. FDS-III, which can be widely used for hydrogen production, uses a super-high-temperature LiPb blanket with an outlet temperature of approximately 1000 °C. FDS-III has several important features such as an innovative coolant flow channel design and the use of relatively mature steel as a structural material [13]. The Fusion-Driven Subcritical reactor based on a Spherical Tokamak (FDS-ST) was designed with a high energy multiplication factor for economical operation. The center conductor post (CCP) in FDS-ST is a liquid-metal-covered copper conductor that uses liquid metal as both the coolant and the electric-current-carrying medium, which can reduce irradiation damage and improve the TBR [14].

10.2 Principles and Requirements of Neutronics Design

The neutronics design for fusion systems is an essential step in achieving safe and efficient utilization of fusion neutrons, involving the design of blanket, tokamak machine, and corresponding buildings. This section will introduce the general principles and requirements of their neutronics design.

10.2.1 Blanket

To achieve the goal of safe and efficient utilization of fusion neutrons, the design of the blanket should follow such principles as (1) adoption of reasonable overall layout for facilitating installation and maintenance, and simple structures for easy fabrication and high reliability; (2) selection of structural materials with good mechanical performance, low activation, neutron irradiation resistance, corrosion resistance, and high temperature resistance; and (3) achievement of high thermal-power conversion efficiency.

The neutronics design of blankets primarily requires consideration of the following aspects of performance assessment:

1. Tritium self-sufficiency

Tritium self-sufficiency is a prerequisite for maintaining the sustainable operation of a fusion system. The tritium generated in the blanket should compensate for the natural decay of tritium, tritium losses in the fuel system, and tritium consumption via D-T fusion reactions. Therefore, a high tritium breeding ratio (TBR) is usually required in the blanket design.

2. High conversion efficiency of fusion energy

To achieve high-efficiency utilization of fusion energy, the blanket should be designed to make full use of the energy of the fusion neutrons, which means that the energy carried by the fusion neutrons should be deposited in the blanket as much as possible.

3. Sufficient radiation shielding

To ensure the safe operation of fusion system, the blanket should provide sufficient shielding to reduce the radiation loads and material activation of the components outside the blankets (especially, VV and the superconducting magnet systems).

4. Low activation and radiological hazards

To ensure the environmental friendliness of fusion energy from the start, attention should be paid to the following aspects during the blanket design stage. For example, to keep the activation of the blanket as low as possible, appropriate materials (e.g., low-activation materials) should be used, and their arrangement should also be reasonable. In this manner, potential radiological hazards (e.g., tritium and activated products) in the blanket can be minimized.

10.2.2 Tokamak Machine and Buildings

Considering safety and operability during the full lifecycle of the fusion reactor, the basic objectives of neutronics design for the tokamak machine and corresponding buildings are summarized as follows: (1) Ensure the safety of the personnel and the public, and satisfy all of the regulatory requirements; (2) Provide sufficient shielding to ensure that all components function adequately; and (3) Reduce the volume and radiotoxicity of the waste for decommissioning. In this subsection, the objectives and requirements for the neutronics design are discussed by using ITER as an example.

1. Radiation shielding and dose limitations for the personnel and the public

The radiation shielding design follows the "As Low As Reasonably Achievable" (ALARA) principle and should meet the following requirements:

(1) Ensure the dose rate in the operation area to be lower than the design limits.
(2) Facilitate the maintenance tasks, and ensure the occupational dose as low as reasonably achievable.
(3) Add local shielding for the radiation penetrating areas to ensure the radiation levels to be lower than the mandated regulatory limits.
(4) Ensure the occupational and public doses to be lower than the required limits.

According to the requirements acknowledged by the International Commission on Radiological Protection (ICRP), the general safety objectives of ITER are listed in Table 10.2.

2. Radiation protection requirements for components;

(1) Radiation limits for the magnets.

In superconducting magnet systems, there are some radiosensitive components such as the superconductor, stabilizer, insulator, etc. The high-intensity irradiation penetrating the gaps between the blankets or port plugs to the superconductor magnets system may result in the degradations of superconductivity of the superconductor, the electrical resistivity of the stabilizer, and the mechanical strength of the insulator, which may jeopardize the safety operation of fusion systems. Thus, sufficient shielding for these components should be provided, and the radiation loads on them

Table 10.2 General safety objectives for the dose limits in ITER

	For worker	For the public and environment
Situations in design basis		
Normal situations[a]	As low as reasonably achievable, and in any case: maximum individual dose ≤10 mSv/year, average individual dose for workers classified for radiation exposure ≤2.5 mSv/year	Releases less than the limits authorized for the installation, impact as low as reasonably achievable, and in any case: ≤0.1 mSv/year
Incidental situations[b]	As low as reasonably achievable, and in any case ≤10 mSv per incident	Release per incident less than the annual limits authorized for the installation ≤0.1 mSv/year
Accident situations[c]	Consider the constraints related to the management of the accident and post-accident situation	No immediate or deferred countermeasures (confinement, evacuation) <10 mSv No restriction of consumption of animal or vegetable products
Situations beyond design basis		
Hypothetical accidents[d]	No cliff-edge effect, possible countermeasures limited in time and space	

Note [a]Normal situations: events and plant conditions planned and required for ITER normal operation, including some faults and events which can occur as a result of the ITER experimental nature
[b]Incidental situations: event sequences not planned but likely to occur one or more times during the life of the plant but not including normal situations, with frequency greater than 10^{-2}/a
[c]Incidental situations: postulated event sequences not likely to occur during the life of the plant, with frequency ranging from 10^{-4} to 10^{-2}/a
[d]Hypothetical accidents: postulated event sequences not likely to occur during the life of the plant with a very large margin, with frequency ranging from 10^{-6} to 10^{-4}/a

Table 10.3 Limits of radiation loads on superconductor in ITER [17]

Parameters	Limits
Total neutron fluence to epoxy insulator (m^{-2})	1×10^{22} (equiv. to 10^7 Gy)
Peak fast neutron fluence to the Nb_3Sn superconductor (m^{-2})	0.5–1 $\times 10^{22}$
Peak displacement damage to Cu stabilizer between TFC warm-ups (m^{-2})	1–2 $\times 10^{21}$ (equiv. to 0.5–1 $\times 10^{-4}$ dpa)
Peak nuclear heating density in winding pack (W/m^3)	1×10^3
Peak nuclear heating density in cases and magnet structures (W/m^3)	2×10^3

Table 10.4 Thresholds for radiation qualification in ITER [16]

Subsystem	Accumulated dose (Gy)	Neutron fluence ($1/cm^2$)	Neutron flux ($1/(cm^2\ s)$)
Critical system[a] with electronics	1	10^8	10
Noncritical system with electronics	10	10^{10}	10^2
Critical system without electronics	10^3	–	10^4
Noncritical system without electronics	10^4	–	10^5

Note [a]Safety important class (SIC) systems may require a specific radiation qualification

should also be analyzed in detail. Table 10.3 lists the limits of radiation loads on the superconductor in ITER.

In addition, considering the thermal load of cooling system for the superconductor in ITER, the limit of the total nuclear heat of the toroidal field coils (TFCs) is set to 14 kW. To prevent component failure due to local temperature increases, the nuclear heat density in the vacuum vessel should not exceed 0.61 MW/m^3 [15].

(2) Radiation limits for the electronics

Radiation can damage or destroy electronic components and sensors, corrupt signals in analog or digital circuits, corrupt data in memories, etc. In ITER, these effects can appear progressively due to accumulated ionization or accumulated atomic displacements, or instantaneously due to a single highly ionizing particle. The electronic configuration components and systems in ITER are grouped according to their functional requirements and radiation environment. For each group, the radiation thresholds are defined as listed in Table 10.4. If one or more of the three collective unmitigated radiation conditions is above the specified threshold, a radiation mitigation strategy and radiation qualification strategy will be mandatory.

Other than these limits, linear electronic components (such as resistors, capacitors, and inductors) rather than semiconductors are the preferred electronic devices for logic or signal conditioning due to their higher tolerance for radiation.

(3) Radiation limits of re-welding

Components such as blankets, divertors, port plugs, and failed components would have to be replaced during the lifetime of ITER. This replacement would be possible only if the hydraulic weld connections are protected by sufficient shielding to allow for re-welding, also known as weldability.

The weldability of irradiated stainless steel is determined mainly by the production of helium, and the limits for ITER are as follows: <1 appm for thick plate welding and <3 appm for thin plate or tube welding.

The main cause of the helium production in stainless steel is due to the reactions between the neutrons and boron within the steel. The recommended boron contents in steel for different components are as follows: 0.002 wt% in steel for the first wall and 0.001 wt% in steel for the vacuum vessel cooling tubes [15].

(4) Radiation limits for other components

The displacement damage to the components made of steel (such as vacuum vessel) should be limited to prevent the degradation of stainless steel performance. For example, for European DEMO, it is required that the accumulated displacement damage for the first wall and the vacuum vessel, respectively, during their lifetime must not exceed 50 dpa and 2.75 dpa, respectively [17].

3. Decommissioning considerations

To minimize the quantity of waste required for disposal, the neutronics design for decommissioning purposes is related mainly to the impurity specifications of the in-vessel materials, decay heat of the radioactive wastes, and tritium production at the end of the lifecycle of fusion systems. These considerations imply that, except for shielding, the material specifications (e.g., ITER utilizes steels with low cobalt and niobium content) and the use of remote-handling (RH) techniques, which in turn require activation and dose rate estimates, are needed. Taking ITER as an example, the anticipated categories and corresponding limitation amounts are described in detail in Sect. 10.4.2.

10.3 Methods of Neutronics Design

In the design of fusion system, the neutronics design needs to be continually iterated and optimized with the designs of electromagnetism, structural, thermal hydraulics, and other disciplines. These designs jointly determine the system scenario and ensure that the requirements of function and safety are satisfied. In this section, the neutronics design methods for blankets, tokamak machine, and corresponding buildings are discussed.

10.3.1 Design of Blankets

Figure 10.6 summarizes the basic procedures for the neutronics design of the blanket. The designers should first determine the structural layout and pre-select the materials, based on the general design goals and objectives. Then, the design should be optimized, based on the calculation of neutronics parameters, including the TBR, heating deposition, material activity, and irradiation damage. The process should be iterated, until both the design goals and the safety requirements are satisfied.

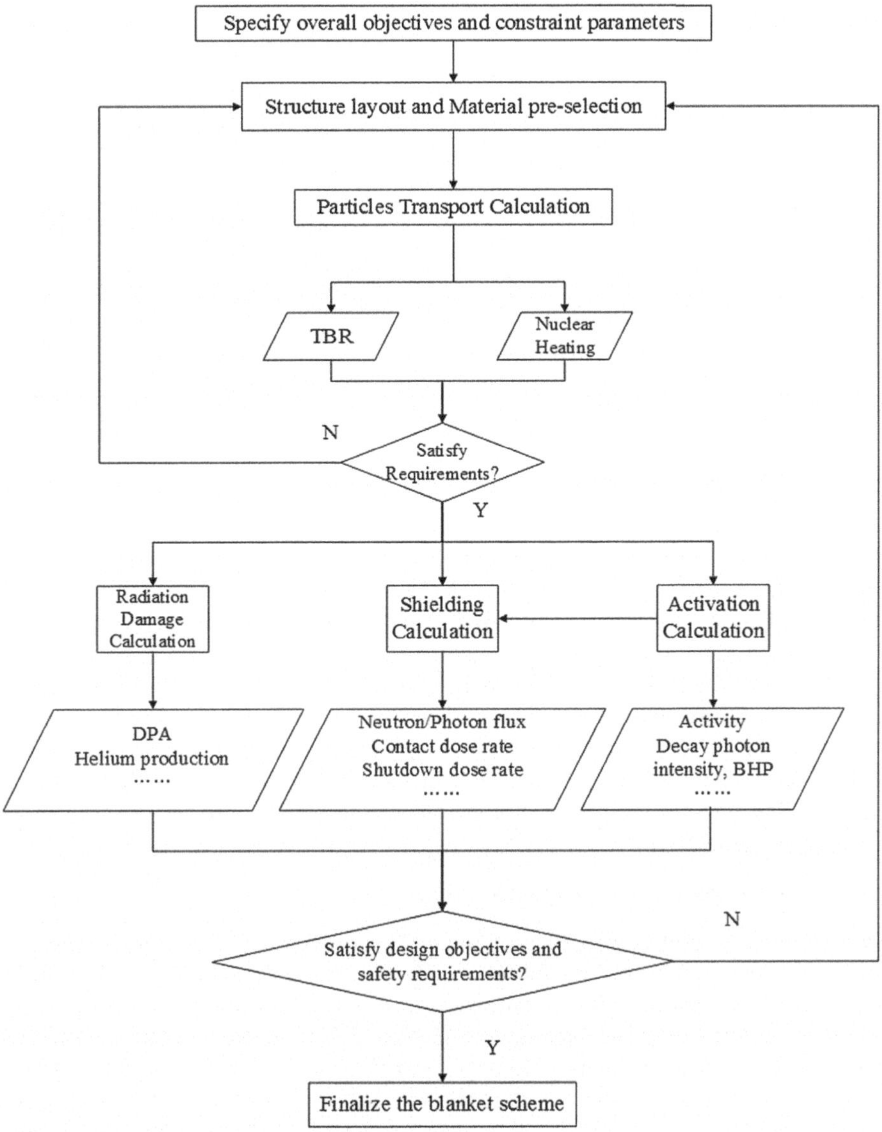

Fig. 10.6 Procedure for the neutronics design of a fusion blanket

Table 10.5 Common materials used in fusion reactors

Type	Candidate
Tritium breeder	Solid: Li_2O, Li_4SiO_4, Li_2Ti_2O3, Li_2ZrO_3, $LiAlO_2$, etc. Liquid: Li, LiPb alloy, FLiBe, etc.
Neutron multiplier	Be, Pb, etc.
Structural material	RAFM steel (such as CLAM steel, EUROFER, F82H, etc.), vanadium alloys, SiCf/SiC, etc.
Coolant	Water, helium gas, liquid LiPb alloy, etc.
Shielding material	Water, steel, graphite, lead, borated materials, etc.

1. Structural layouts and pre-selection of materials

The first step in the neutronics design of a fusion blanket is to determine the layout of the functional zones and the corresponding materials and constrained thicknesses of each zone. Generally, the blanket is divided mainly into an FW, a TBZ, a neutron multiplication zone, and a shielding zone. The materials for each zone should be selected in accordance with the requirements of the service environment, with consideration of the physical performance, compatibility with the coolant, manufacturing feasibility, and other factors. Moreover, the materials used in the blanket should also be safe, cheap, and easy to obtain. During blanket design, the choices made concerning the tritium breeder, neutron multiplier, structural materials, coolant, and shielding materials are vital. Some common materials used in fusion reactors are listed in Table 10.5.

2. Neutronics analysis and optimization

Based on the preliminary structural layout and pre-selection of functional materials, one-dimensional neutronics models (1D on a sphere or a cylinder/slab with infinite length) or 2D models (a cylinder/slab with finite length) are created for performing neutron transport and activation calculations to calculate parameters such as the TBR and the nuclear heat, irradiation damage, and helium gas production rates. These results are fed back into the thermo-hydraulic and structural designs for iterative calculations. The parameters are then compared against the design requirements to optimize the design of the layout, thicknesses, shapes, and materials for each functional zone. Finally, 3D models with detailed structures, holes, penetrations, and slots are created, and detailed neutronics calculations are performed. The details of the neutronics simulation and optimization process are introduced below.

(1) TBR

According to the underlying principle of tritium production, the calculated TBR is determined by the neutron flux, the neutron energy spectrum, and the tritium breeder used in the blanket. Due to a high cross section of tritium production for ^6Li (as high

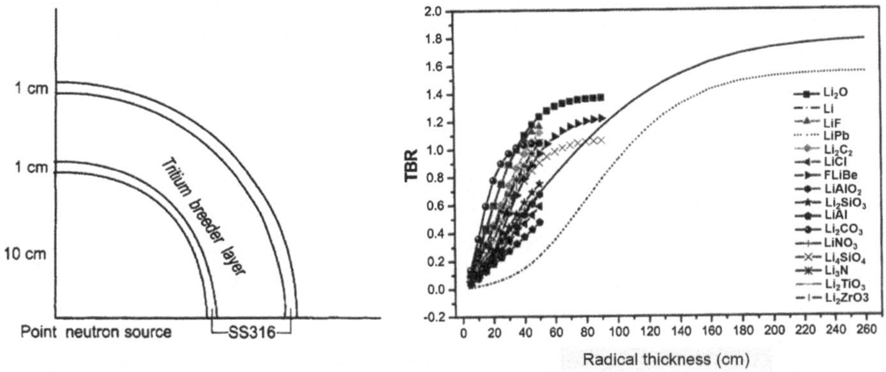

(a) Blanket neutronics model **(b)** TBR versus the radial thickness of the TBZ

Fig. 10.7 Blanket model and TBR versus radial TBZ thickness for various breeders [18]

as 100 b for intermediate- and low-energy neutrons), one can improve the tritium production rate and the utilization of fusion neutrons by increasing the ^{6}Li enrichment and optimizing the materials, layout, and thicknesses of the structural and coolant zones to moderate the neutrons sufficiently before they enter the TBZ. Moreover, one can also improve the TBR by adopting a neutron multiplier such as Be or Pb to multiply the neutrons. Figure 10.7b shows how the TBR depends on the radial thickness of the TBZ for various materials calculated, based on the spherical neutronics model shown in Fig. 10.7a. The results indicate that the TBR increases with increasing TBZ thickness, until it reaches an asymptotically stable value. Different compounds of Li exhibit different saturation thicknesses corresponding to the maximal TBR. Thus, the optimal thickness of the TBZ also differs for different tritium breeders. Suitable tritium breeder materials should be selected in accordance with the specific design requirements.

The uncertainty in the TBR calculations depends mainly on the data libraries used, the simplification of the models, and other assumptions made in the calculations. Considering the losses of tritium in the fuel system and the consumption of the tritium breeders and neutron multipliers, there will be a certain TBR margin in the design of any fusion blanket.

(2) Nuclear heat deposition

Directly affecting the economic performance of a fusion system, the heat deposition in the blanket is one of the most important neutronics parameters. In blanket design, the power density distribution in the blanket is calculated via neutron transport simulations. Based on these calculations, the layout and selection of materials are optimized.

(3) Other neutronics parameters

The shielding performance of the blanket also attracts considerable attention. Thus, neutronics parameters such as the neutron irradiation damage, helium gas production rate, activities of radionuclides, and amount of radioactive waste must be calculated to ensure that the specified safety requirements are satisfied.

10.3.2 Design of Tokamak Machine and Buildings

The neutronics design of the tokamak machine and corresponding buildings in the fusion system is another key task, and the design procedures can be summarized as follows:

(1) to carry out neutronics design for the tokamak machine, based on general design objectives and constraint parameters;
(2) to perform shielding design for radiological buildings and to determine the shielding scenario in accordance with goals of radiation protection;
(3) to conduct safety analysis and radiological environmental impact assessment to evaluate safety and environmental friendliness of neutronics design for the tokamak machine and corresponding buildings;

1. Design of the tokamak machine

The neutronics design of the tokamak machine is the basis of the shielding design of the radiological buildings, safety analysis, and radiological environmental impact assessment. The design can be performed in two stages: (a) to initiate and optimize the design of the tokamak machine based on one-dimensional (1D) calculations; (b) to perform local design and/or optimization for specific areas and/or components based on multidimensional (2D/3D) simulations. The detailed design procedure is summarized as follows:

(1) Preliminary design based on 1D simplified model

Figure 10.8 shows the detailed process of the basic neutronics design based on the neutronics scoping calculations. According to the schematic diagram, the goal of the initial stage is to define the basic layout of the tokamak and select suitable materials based on the objectives and overall parameters specified for the project. Then, the neutronics design and analysis, such as the transport simulation and irradiation damage, activation, and dose rate calculations, are performed to estimate the neutronics parameters during the iterative design process. Once the main requirements for the functionality and safety of each component are satisfied, the preliminary neutronics design of the tokamak is finalized.

(2) Detailed multidimensional (2D/3D) design and optimization

Many components have a complex geometrical structure (such as holes, penetrations, and slots), which is difficult to describe by using a one-dimensional model. With the development of design, it is necessary to perform more detailed neutronics simulation and optimization for some important components based on a 2D or 3D model to

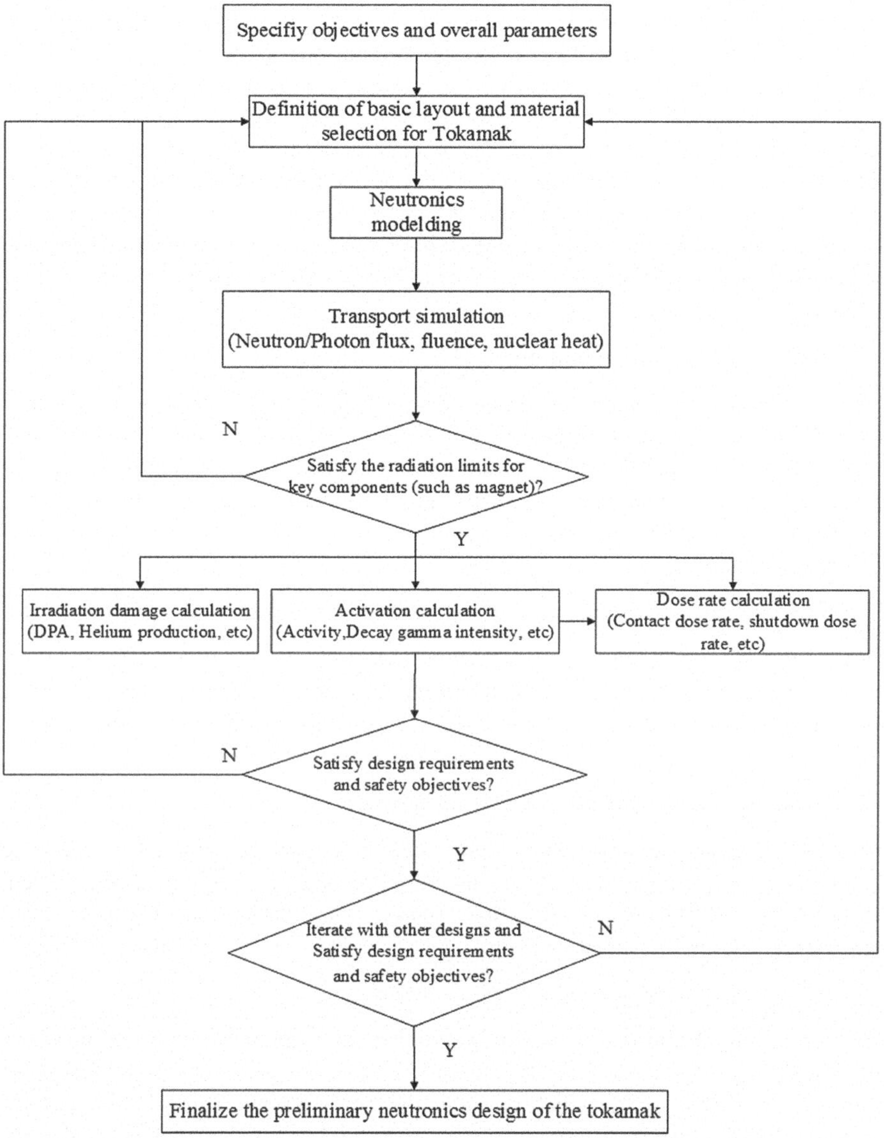

Fig. 10.8 Procedure for the preliminary neutronics design of tokamak machine based on 1D calculation

ensure that the radiation on the components and dose levels in specific regions are within design limits under different operating conditions.

Figure 10.9 shows the detailed procedure of designing and/or optimizing for local components based on 2D/3D calculations. The designer first chooses the candidate materials and puts forward the possible geometrical configurations based on the goals of the design or optimization. Then, several detailed models are built to perform neutronics calculations (such as transport, activation, etc.). Finally, a parametric study is conducted by fitting the calculation results from different cases to propose the potential design and/or optimization scenarios until they satisfy the objectives of design or optimization.

2. Shielding design for the buildings

The shielding design is the main content of neutronics design for radiological buildings. The whole design process contains iterative sequential steps: (1) Determine the shielding requirements and dose rate limits according to the function and layout of the buildings; (2) Calculate the distributions of the neutron and photon fluxes, dose rates, etc.; and (3) Optimize the layouts, materials, and shielding thicknesses of different rooms.

3. Safety analysis and assessment of the radiological impact on the environment

The safety analysis and radiological environmental impact assessment of fusion reactors involve the evaluation of the occupational radiation exposure (ORE) doses, the radioactive waste, the disposal of radioactive wastes, the diffusion of radionuclides, and their radiological impact on the environment, the decay heat, etc.

(1) Evaluation of occupational radiation exposure

For the ORE dose evaluation, the accumulated doses and maximum dose rates are calculated for various operations by occupational workers in radiological areas. Then, the radiation shielding design is optimized or necessary measures are adopted to keep the dose rate below the specified limits.

(2) Evaluation of radionuclide diffusion and environmental impacts

The evaluation of radionuclides diffusion and their environmental impacts are performed primarily based on the methodologies on release and diffusion of radionuclides (in Chap. 5), which provides the diffusion behavior of radionuclides inside/outside the fusion system and environment and sensitivity analysis of the key parameters. With these results, the environmental impacts and the dose to the public induced by the diffusion of radionuclides are evaluated to support the design and safety assessment of the fusion system.

(3) Evaluation of radioactive wastes

The evaluation of the radioactive waste is based mainly on activation calculations, which provides their activity and total inventories according to the specified irradiation scenario. At present, the clearance and recycling are the two most widely used approaches for waste disposal in the world. The clearance method is still subject

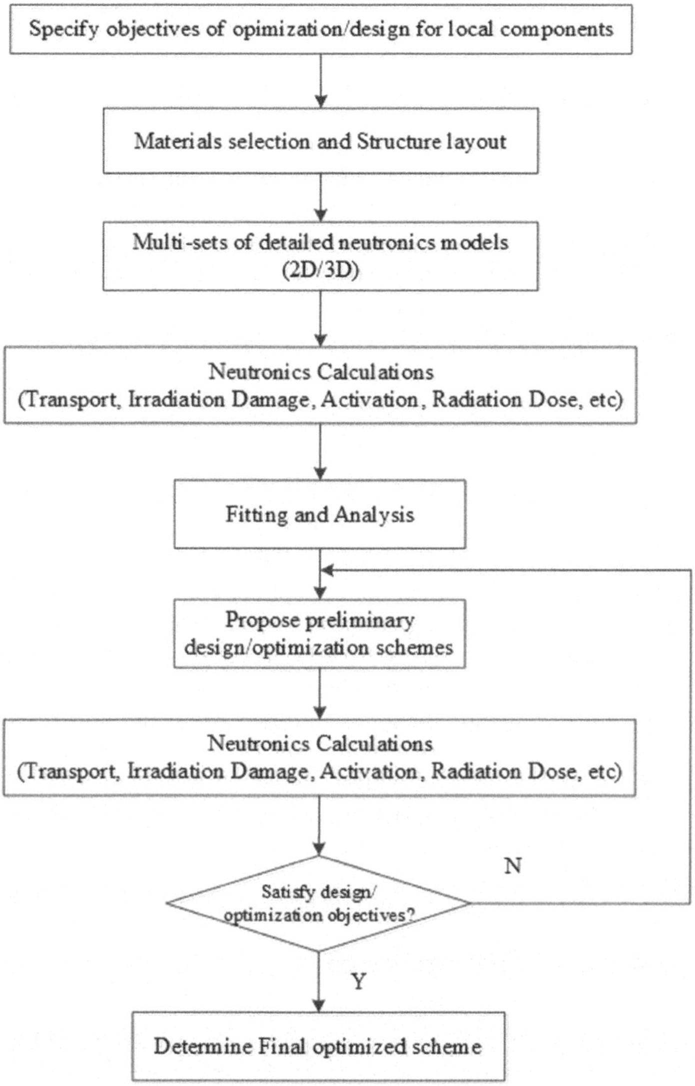

Fig. 10.9 Procedure of multidimensional (2D/3D) design/optimization for tokamak machine

to several problems, such as inconsistent standards among different countries, an absence of standards for certain nuclides, demands for abundant temporary storage, and low public acceptance for clearance materials. For recycling, the problems such as the disposal of large-volume components with high dose rates, the disassembly of complicated components, and the separation and detritiation of materials still need to be solved [19]. Consequently, reduced activation materials should be considered to reduce the activity of the radioactive wastes that must be disposed of during decommissioning. Simultaneously, the structural layout and material selection should be optimized to reduce the total amount of radioactive wastes.

(4) Evaluation of shutdown decay heat

The shutdown decay heat of a fusion system is generated mainly by the decay of the radionuclides that exist in activated materials such as the structural materials and tritium breeders. Compared to ITER, the activation of the materials in the fusion system is much higher because of the higher neutron irradiation, leading to a higher shutdown heat [19]. Therefore, it is necessary to assess the decay heat in the fusion system and to evaluate the temperature increase in its structural materials to ensure that the structural integrity of the reactor can be maintained during shutdown.

10.4 Examples of Neutronics Design

A series of fusion systems such as ITER, PPCS, DEMO, ARIES, SSTR, and FDS series fusion systems have now been proposed and designed worldwide. Among these fusion systems, except ITER which is being constructed as the experimental system, the others remain in the status of conceptual design. Thus, ITER with relatively mature design is taken as an example to illustrate the neutronics design for the tokamak machine and corresponding buildings. As mentioned above, the blanket in fusion systems can be generally classified into two types: the liquid blanket such as Dual-cooled Lead Lithium (DLL), HCLL, WCLL; and the solid blanket such as HCPB. The DLL blanket in FDS-II, which is a typical and attractive blanket option for future fusion power plant, is selected as a typical example to demonstrate the neutronics design for blankets in the fusion systems.

10.4.1 Example of Blanket Design

The designed fusion power of FDS-II is 2.5 GW, with ~1 GW of electrical power, and the power gain Q reaches approximately 31. The NWL and the surface heat flux are designed to be 2–3 MW/m^2 and 0.5–1 MW/m^2, respectively. The operational availability is greater than 80%, and the TBR is ~1.29. Consequently, FDS-II has a high-power density and a high energy conversion.

The liquid LiPb blanket has many advantages, such as an adaptable geometry, manufacturability, high engineering feasibility, high TBR, and online extraction of tritium. Thus, the liquid LiPb blanket concept was adopted in FDS-II. The LiPb alloy has been selected as both the tritium breeder and the neutron multiplier. There are two options for liquid breeding blanket concepts, i.e., the He/LiPb Dual-cooled Lead Lithium (DLL) concept and the He-cooled quasi-Static Lead–Lithium (SLL) concept. The DLL concept has been selected as the main candidate for FDS-II. Figure 10.11 shows the structure of a DLL equatorial outboard blanket module.

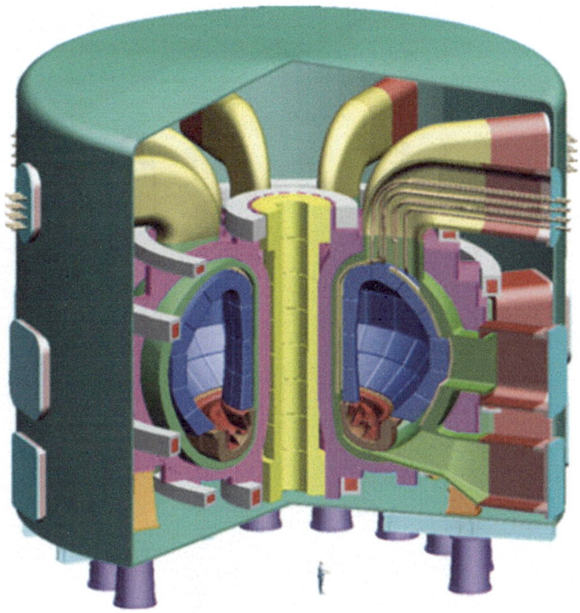

Fig. 10.10 Configuration of FDS-II [12]

The DLL blanket is divided into the FW, a three-segment TBZ, helium manifolds, and a shielding zone in the radial direction, as shown in Fig. 10.10. The flow channel inserts (FCIs) are used to increase the outlet temperature to achieve high thermal efficiency. Moreover, a coating (e.g., Al_2O_3) on the RAFM steel is also under consideration in the design for reducing tritium permeation and preventing corrosion of the RAFM by LiPb. The TBZ is divided into many small segments by the toroidal–poloidal stiffener plates (tpSPs) and radial–poloidal stiffening plates (rpSPs). In the DLL blanket, the helium gas, used as the main coolant for the FW and stiffening plates, enters the U-shaped FW from both sides of the first helium manifold and then flows into the second helium manifold. Then, the helium flows through tpSPs, rpSPs, and the top and bottom covers (Fig. 10.11).

In this section, DLL blanket in the FDS-II is selected as an example to illustrate the blanket neutronics design for a fusion reactor. A simple neutronics model of FDS-II for preliminarily optimizing blanket nuclear performance was created, as shown in Fig. 10.12. The preliminary selections of the materials and thicknesses for the components are listed in Table 10.6.

The DLL blanket consists of two sub-blankets: the working blanket (including the FW, TBZ, and helium manifold) and the shielding blanket. The neutronics parameters of the blanket include the TBR, power density distribution, irradiation damage rate, activation rate, and production rate for radioactive waste. The blanket was designed and optimized , based on these neutronics parameters, as discussed below.

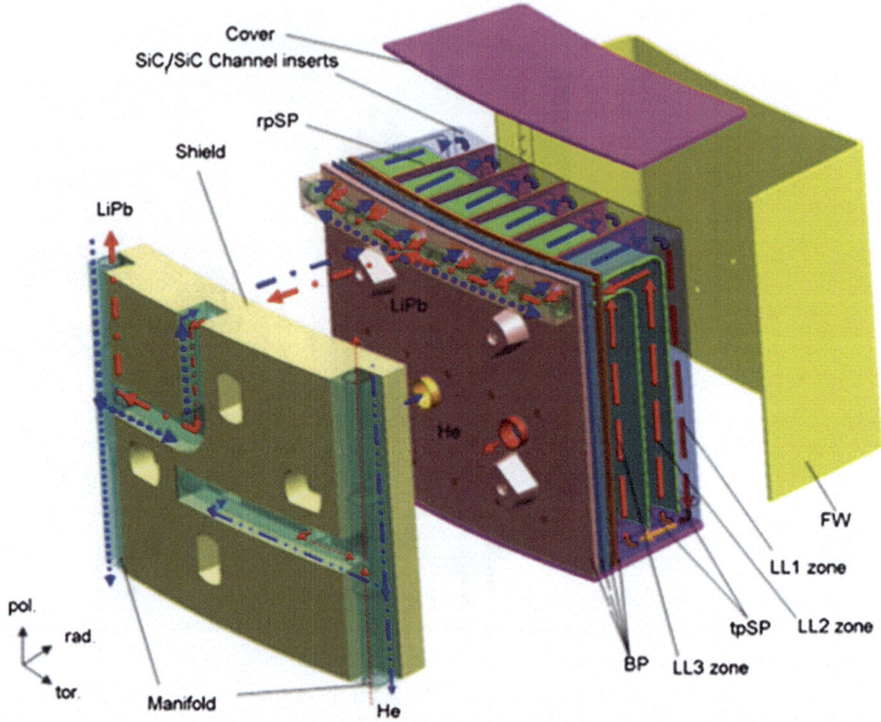

(a) Detail Structure [12]

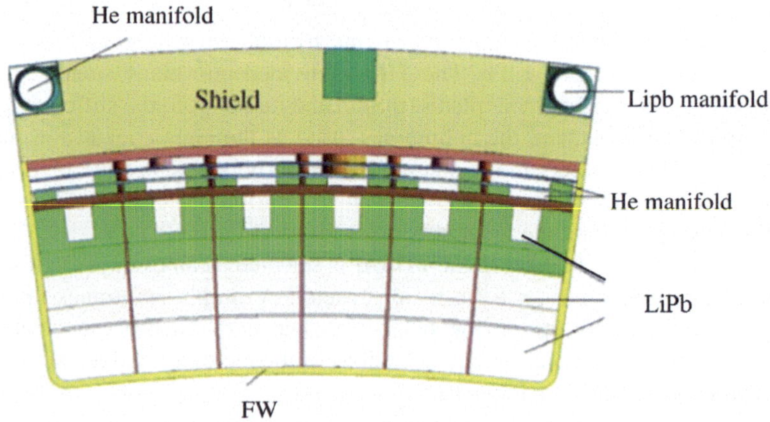

(b) Top View [20]

Fig. 10.11 Schematic structure of the DLL blanket, **a** detail structure [12], **b** top view [20]

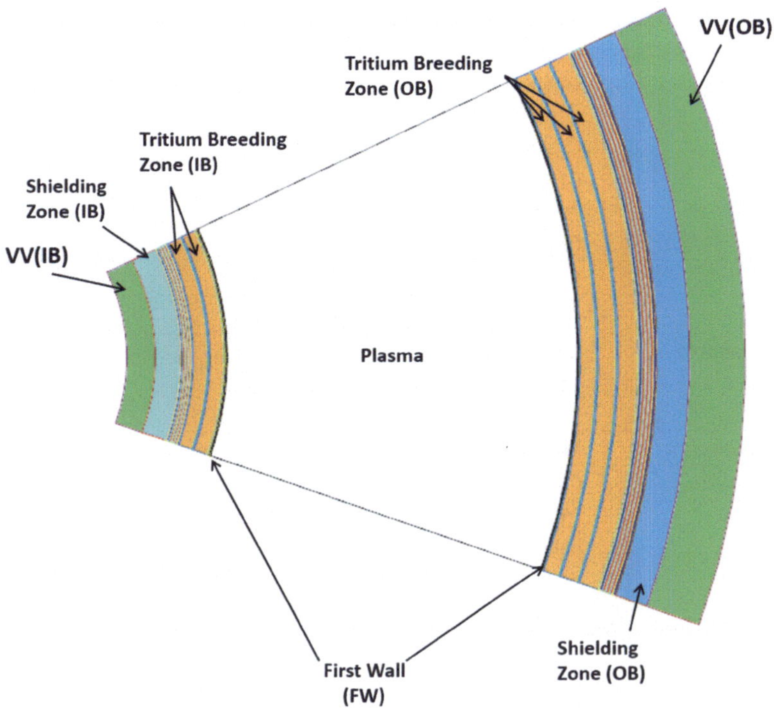

Fig. 10.12 1D neutronics model of FDS-II [21]

1. Design and analysis of the tritium breeding blankets

A commercial fusion reactor must be tritium self-sufficient, which means a TBR of >1.0. In effect, the TBR must be some additional margins in excess of unity to account for some additional impacts (such as tritium losses). Further considering the simplification of the model, it was required that the TBR should be greater than 1.30 in the optimization for FDS-II blankets based on the 1D model. The tritium breeding performance of a blanket depends mainly on the thickness of the TBZ, the fraction of structural materials, and the ^{6}Li enrichment. To improve the tritium breeding performance, the blanket of FDS-II was optimized by considering the following aspects:

(1) TBZ thickness

A comparative 1D analysis was performed to assess the variation of the breeding potential with the TBZ thickness. The impact of the tpSPs on the TBR was considered in the optimization of the TBZ thickness.

The material in the TBZ is the mixture of LiPb and the "grid plate mixture" (a homogenized material containing 50% SiC, 30% RAFM steel, and 20% helium gas) by volumic fraction of ~85 and ~15%. In the optimization, it was assumed that the outboard blanket thickness was fixed at 75 cm, and the inboard blanket thickness

Table 10.6 Material compositions and thicknesses for the main components of FDS-II [22]

Zone	Material composition (volume fraction)	Radial thickness (cm)	Poloidal thickness (cm)
FW	CLAM: 100%	0.4	–
	CLAM: 100%, helium gas: 80%	1.8	–
	CLAM: 100%	0.8	–
TBZ	Variable (LiPb and "grid plate mixture")	IB: 20–60	–
		OB: 20–100	–
Helium manifold	CLAM: 100%	16(IB)/22(OB)	–
VV	CLAM: 80%, water: 20%	35(IB)/70(OB)	
Tungsten (divertor)	W: 100%	–	1
Copper (divertor)	Cu: 100%	–	1
Divertor	CLAM: 25%, helium gas: 75%	–	140

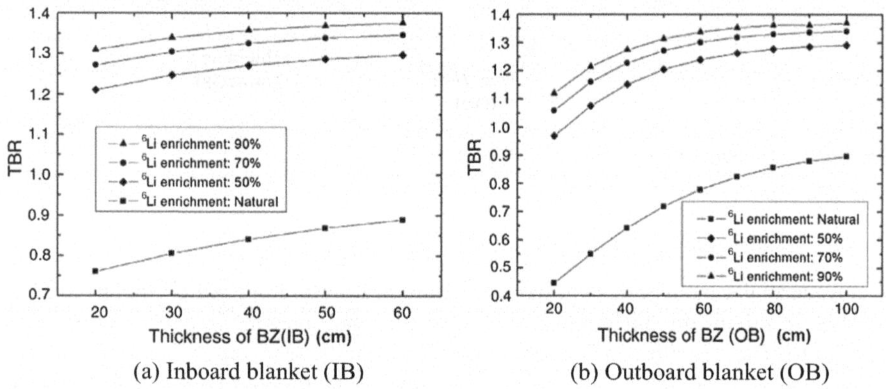

Fig. 10.13 TBR as a function of the TBZ thickness

was gradually increased from 20 to 60 cm. The calculated variation in the TBR with the inboard thickness and the ^{6}Li enrichment level is shown in Fig. 10.13a. Similarly, the results obtained with the inboard blanket thickness fixed at 40 cm and the outboard blanket thickness gradually increasing from 20 cm to 100 cm are shown in Fig. 10.13b. In these two plots, the different curves correspond to 6Li enrichments of 7.4% (natural), 50, 70, and 90% [22].

The TBR increases with the increasing TBZ thickness and tends to saturate (TBR ~ 1.4) when the TBZ thickness reaches 40 cm and 70 cm for the inboard and outboard blankets, respectively. Considering engineering limitations, the TBZ thick-

Table 10.7 TBR values with different "grid plate mixture" fractions and 6Li enrichment levels [22]

"Grid plate mixture" fraction (%)	^{6}Li enrichment		
	50%	70%	90%
5	1.28	1.33	1.36
10	1.25	1.30	1.33
15	1.21	1.27	1.30
20	1.18	1.23	1.27

nesses were chosen to be 41 cm and 75 cm for the inboard and outboard blankets, respectively [22].

(2) Volume fraction of structural materials

In the analysis above, the rpSPs and cover plates were neglected due to the constraints of the 1D geometric model. To investigate the impact of the structural materials on the TBR, a series of calculations with different fractions of "grid plate mixture" in TBZ were performed, based on the optimized TBZ thickness obtained above. Table 10.7 presents the results for the TBR as a function of the "grid plate mixture" fraction and the ^{6}Li enrichment [22].

Considering the decay and leakage of tritium and the calculation uncertainty induced by the model simplification, the TBR calculated using the 1D neutronics model should be greater than 1.30. However, as shown in Table 10.7, the TBR does not reach 1.3 when the "grid plate mixture" fraction is 20% even when the ^{6}Li enrichment is increased to 90%. Therefore, to ensure tritium self-sufficiency, the "grid plate mixture" fraction and the ^{6}Li enrichment were selected to be 10% and 90%, respectively. These selected parameters were used as inputs for subsequent calculations.

(3) TBR uncertainty analysis

The accuracy of the TBR prediction has considerable impact on blanket design. As stated in Sect. 10.3.1, the uncertainties in the TBR are caused mainly by the simplification of the neutronics model, the computational approach used in the calculations, and the tritium losses during the fuel cycle of the designed reactor. The influences of these factors on the TBR are discussed below.

To investigate the impact of the model simplification on the TBR of the DLL blanket, 1D (in Fig. 10.12), 2D (in Fig. 10.14), and 3D (in Fig. 10.15) neutronics models with different refinements were built using SuperMC.

Table 10.8 lists the calculated TBRs based on the different models. The TBR gradually decreases with the stepwise refinement of the models because of the serious neutron leakage that arises in the 3D model with the presence of gaps and openings.

The TBR uncertainties due to the computational approach include uncertainties in the nuclear cross section, calculation errors (e.g., statistical errors in MC simulations), etc. The uncertainties due to the nuclear cross section depend strongly on the blanket

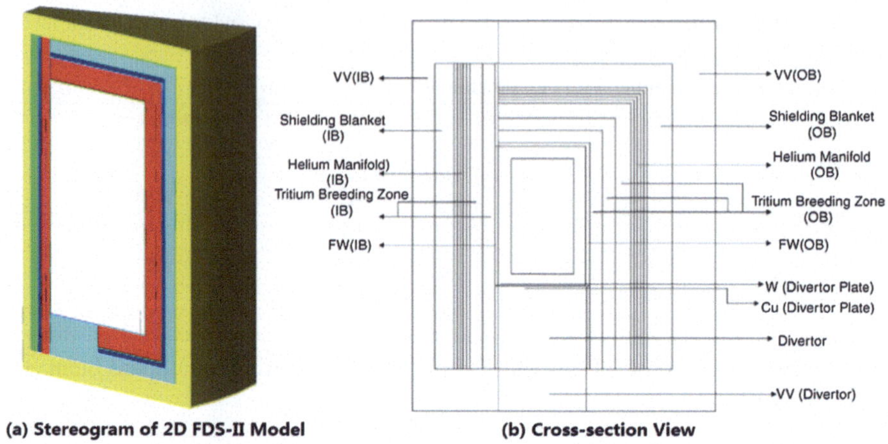

Fig. 10.14 2D neutronics model of FDS-II [21]

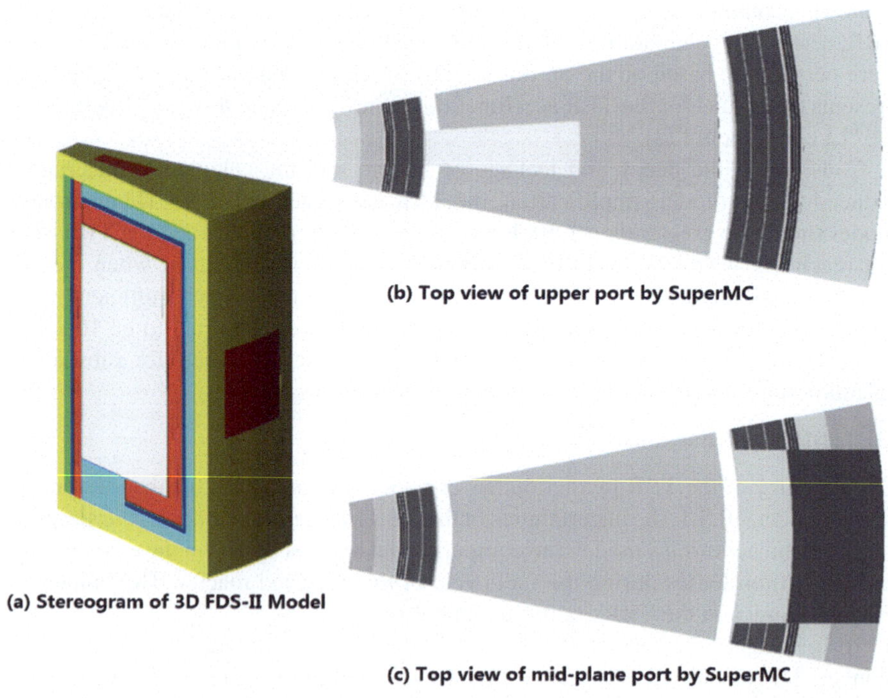

Fig. 10.15 3D neutronics model of FDS-II [21]

Table 10.8 TBRs of the DLL blanket calculated using 1D, 2D, and 3D models [21]

Zone	TBR		
	1D	2D	3D
TBZ (IB/OB)	0.33/1.00	0.23/1.05	0.21/0.75
Shielding zone (IB/OB)	0.04/0.02	0.03/0.02	0.03/0.02
Port	–	–	0.28
Total	1.39	1.33	1.29

concept and materials. For example, an uncertainty of approximately 1–2% should be considered for a liquid metal blanket with LiPb as the breeder; for HCLL and WCLL blankets (solid breeder blankets with beryllium as the neutron multiplier), this uncertainty could be up to 5–10%. In addition, the impact of the burnup of tritium breeders on the TBR throughout the lifetime of the blanket can be neglected for LiPb-based liquid metal blankets, whereas burnup of tritium breeders cannot be ignored for solid breeder blankets. TBR uncertainties arising from statistical errors of less than 0.1% can be neglected.

The TBR uncertainties caused by tritium losses (e.g., the decay, retention, and leakage of tritium) depend strongly on the time window assumed between the generation of the tritium in the blanket and its injection into the plasma. In general, a margin of 5% needs to be added to the TBR when a 1-year delay is assumed regarding the use of the generated tritium for refueling [23].

2. Distribution of nuclear heat

The power density is an important input for the design analysis of the auxiliary systems, such as the cooling system. The deposited power distributions in the blanket based on the DLL concept with average NWL of 2.72 MW/m^2 were calculated using the 2D and 3D models, as displayed in Fig. 10.16. The results show that the nuclear heat power deposited in the DLL blanket is 2330 MW, which accounts for approximately 93% of the total fusion power, and the total nuclear heat deposited in the TBZ of the outboard blanket is 726 MW [21].

According to the objective of the thermo-hydraulic design, the inlet and outlet temperatures of the LiPb coolant in the FDS-II blanket are set to 380 °C and 700 °C, respectively, and 8 MPa helium was chosen as the coolant for the FW and structural material. Considering the temperature limits of the structural materials and the requirements of the power conversion system, the helium inlet and outlet temperatures are set to 300 °C and 450 °C, respectively. The helium flows out of the blankets and exchanges heat with the helium in the secondary loop. In addition, the helium in the secondary loop also exchanges heat with the high-temperature LiPb from the blanket. The helium in the secondary loop can be heated up to 680 °C and then used to produce electricity via the Brayton cycle [12]. Based on the power distribution, the simulation of the thermal hydraulics demonstrates that the design above could transfer the nuclear heat and thermal radiation in the FW away from the blanket and efficiently convert them into electric power.

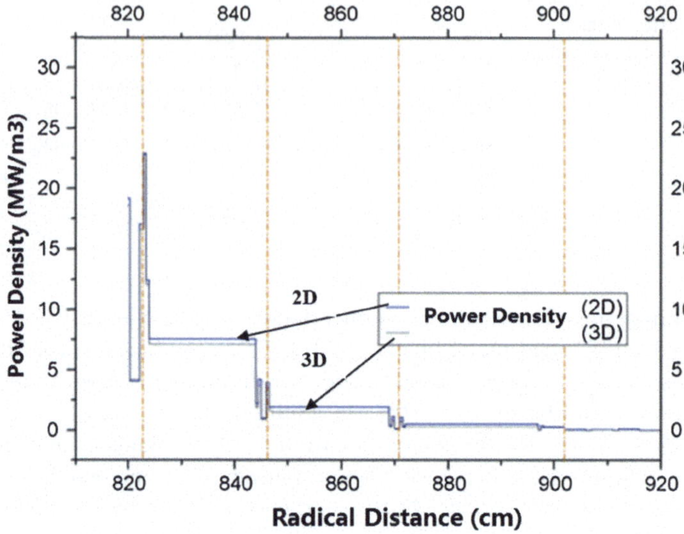

Fig. 10.16 Distribution of power density in DLL blanket [21]

3. Assessment of irradiation damage induced by neutrons

In FDS-II, the FW is subjected to severe neutron irradiation. CLAM steel has been chosen as the structural material for the FW because of its remarkable features of low activation and good irradiation resistance. The neutron spectrum in the FW is shown in Fig. 10.17. The irradiation damage induced by neutrons in the FW is estimated to be ~48.5 dpa after 1 year of full-power operation, thus indicating that the material damage is an important factor affecting the FW lifetime.

4. Assessment of material activation

Considering the service environments of the components in FDS-II, the research on material activation has focused on the activation of the blanket and the corresponding method for the disposal of radioactive waste. Activation calculations were performed to evaluate the activity properties of the materials in different parts of the FDS-II blanket. The irradiation times for the removable and permanent components were assumed to be 5 full-power years (FPYs) and 30 FPYs, respectively. Figure 10.18 shows the decay heat density, specific activity, ingestion dose, and shutdown dose rate as functions of cooling time for different materials.

As shown in Fig. 10.18, compared with other components, the activation properties of the CLAM steel in the FW are always the largest within the first 10 years of cooling. After 30–100 years of cooling, except for the shutdown dose rate, the values of the activation properties of LiPb in the Lead–Lithium-1 (LL1) zone are always larger than the values of the activation properties of the other materials, and the dominant contributor is the residual tritium. The shutdown dose rate of the blanket is less than 2 mSv/h after 100 years of cooling.

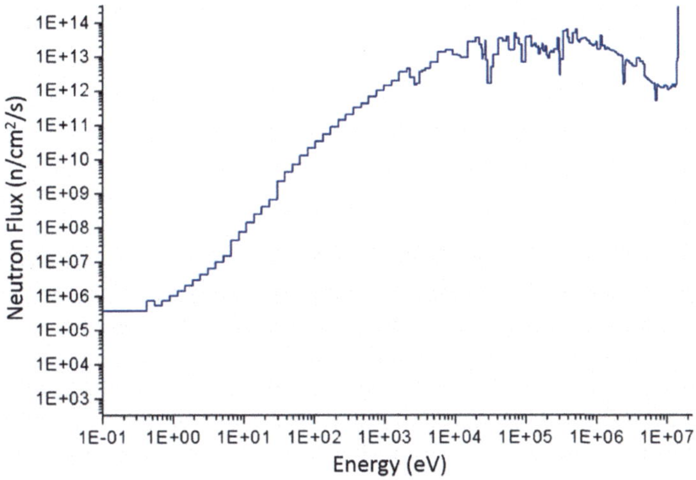

Fig. 10.17 Neutron spectrum in the FW of the DLL blanket

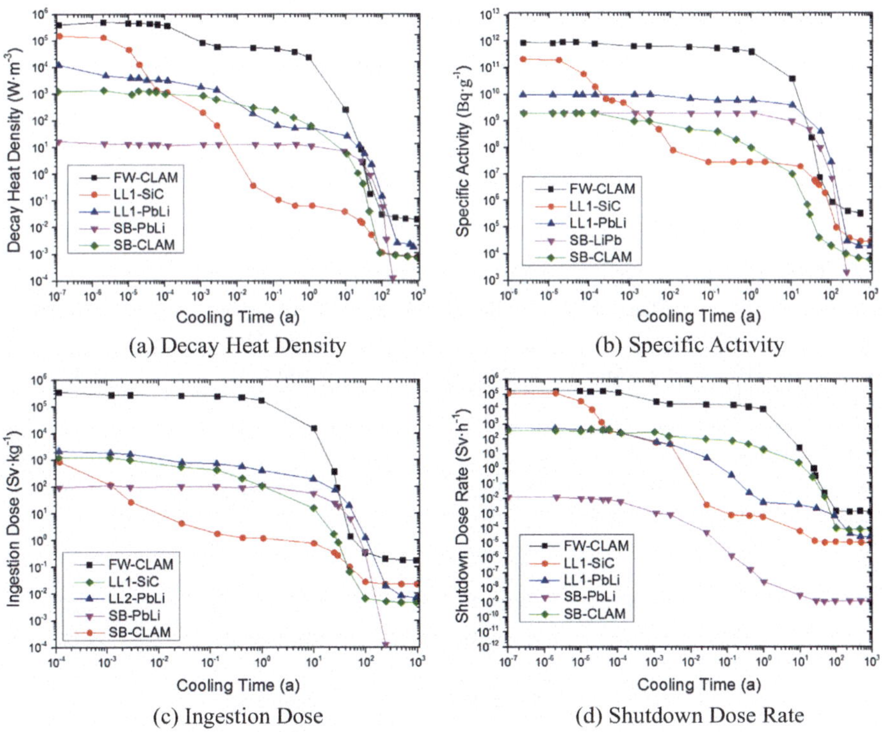

Fig. 10.18 Variation of the activation properties of the DLL blanket with cooling time [24]

5. Management of radioactive wastes

According to the Safety and Environmental Assessment of Fusion Power (SEAFP), the activated materials from fusion systems should be classified for disposal based on the dose rate D, the decay heat density H, and the clearance index I_c, as presented in Table 10.9.

Generally, activated materials should be cooled for 50 years before disposal. However, the cooling time may be extended to 100 years to facilitate the recycling and clearance of radioactive materials to the greatest possible extent. Based on the activation properties of DLL inboard and outboard blanket materials after cooling for 50 and 100 years, the management options for these activated components according to the SEAFP strategies are listed in Table 10.10.

As shown in Table 10.10, among these materials, the shielding and structural materials of the tubes and helium manifolds in the inboard blankets cannot be recycled after only 50 years of cooling. After 100 years of cooling, all materials can be classified as SRM.

Table 10.9 Disposal strategies for activated materials

Category	D (mSv/h)	H (W/m^3)	I_c
PDW (permanent disposal waste)	>20	>10	–
CRM (complex recycle material)	2–20	1–10	–
SRM (simple recycle material)	<2	<1	–
NAW (non-active waste)	–	–	<1

Table 10.10 Management options for activated materials [24]

Components	Material	Cooling for 50 years	Cooling for 100 years	Mass (t)
LL1, wall of TBZ, helium manifold, and shielding blanket of OB	CLAM	CRM (RH)	SRM (RH)	4049
FW, helium manifold, shielding blanket, and tube of IB	CLAM	PDW	SRM (RH)	2595
OB LL3	CLAM	SRM (RH)	SRM (RH)	161
LL2, LL3, shielding blanket	LiPb	SRM (HOR)	SRM (HOR)	2853
LL1	LiPb	SRM (RH)	SRM (RH)	1312
TBZ	SiCf/SiC	SRM (HOR)	SRM (HOR)	812

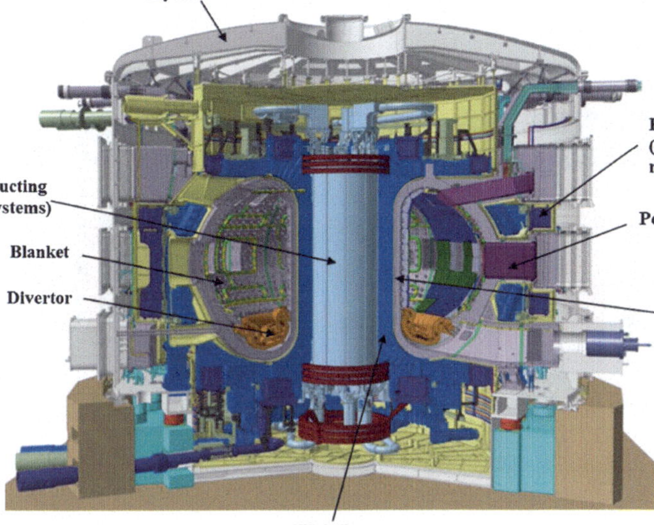

Fig. 10.19 ITER tokamak machine

10.4.2 Example of Tokamak Machine and Buildings Design

The ITER facility consists of the tokamak and some associated buildings. As shown in Fig. 10.19, the ITER tokamak consists of the vacuum vessel (VV), the in-vessel components, the superconducting magnet system, the cryostat, the thermal shield, etc. The in-vessel components include blankets, test blanket modules (TBMs), divertors, etc. The buildings in the ITER site are classified as radiological zones (such as Tokamak Building) and non-radiological zones based on their function and radiation level. In this subsection, ITER is taken as an example to demonstrate the neutronics design for the tokamak machine and buildings.

1. Neutronics design for ITER tokamak machine

(1) Preliminary neutronics design for ITER tokamak machine based on 1D calculation

Based on the basic geometrical configuration and materials in ITER, a 1D neutronics model, as shown in Fig. 10.20, was created to calculate some key neutronics parameters such as NWL, flux, absorbed energy in magnet system, helium production rate, etc.

1) NWL and flux

The average NWLs on the inboard and outboard first wall are 0.43 MW/m^2 and 0.64 MW/m^2, respectively. The fast neutron flux on the first layer of winding pack

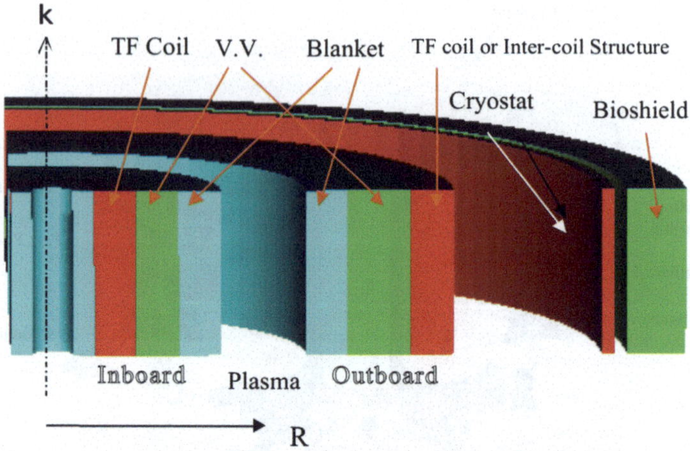

Fig. 10.20 1D calculation model of ITER [15]

and r-Epoxy Insulator in the TF coil inboard leg are 6.91×10^{16} n/cm^2/s and 7.83×10^{16} n/cm^2/s, respectively, which are still lower than design limits even considering the correction factor (>1.0) due to multidimension [15].

2) Absorbed dose in magnet system

The absorbed dose in the inboard TF coil insulator is 0.67 MGy for an average "first wall neutron fluence" (neutron wall loading times irradiation time in MWa/m^2) of 0.5 MWa/m^2. Considering the correction due to the 3D geometry effect (such as gaps), the absorbed dose is found to be 3.2 MGy, which is smaller than the design limit by a factor of ~3.

3) Transmutation helium production

When an average "first wall neutron fluence" of 0.5 MWa/m^2 is assumed, the helium production rates are found to be 0.10 appm and 0.13 appm at the inboard and outboard vacuum vessel surface, respectively [15]. Considering the gap between blankets, the corrected results are 0.6 appm for the inboard side and 0.8 appm for the outboard side, not exceeding the design limits [15].

(2) Detailed multidimensional (2D/3D) design and/or optimization

In this subsection, the Neutral Beam Injectors (NBIs) are taken as an example to illustrate the neutronics design and optimization for local components on the basis of the 2D or 3D model.

Figure 10.21 shows the 2D R-Z model for the neutral beam injector (NBI) port. Based on this model, the nuclear heat changes in the TFCs (P_{TF}) over the port radius (R) with different wall thicknesses (T_w) were evaluated, as shown in Fig. 10.22.

Based on a parameter fitting procedure, the following equation can be derived to describe the relationship between P_{TF}, R, and T within 50 cm < R < 130 cm and 20 cm < Tw < 50 cm.

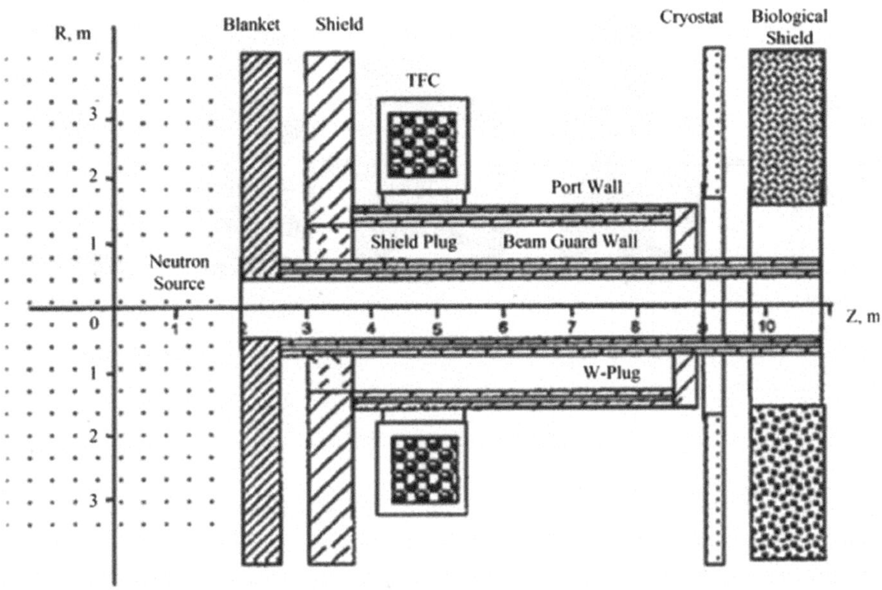

Fig. 10.21 2D RZ model of a neutral beam injector port [25]

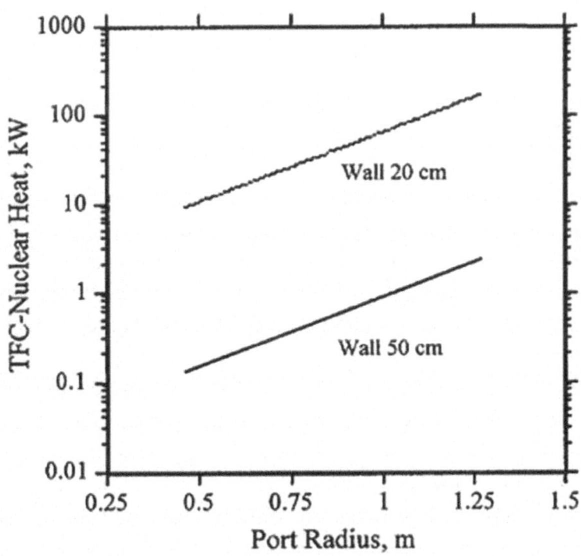

Fig. 10.22 Nuclear heat in the TFCs near an open NBI port versus the port radius [25]

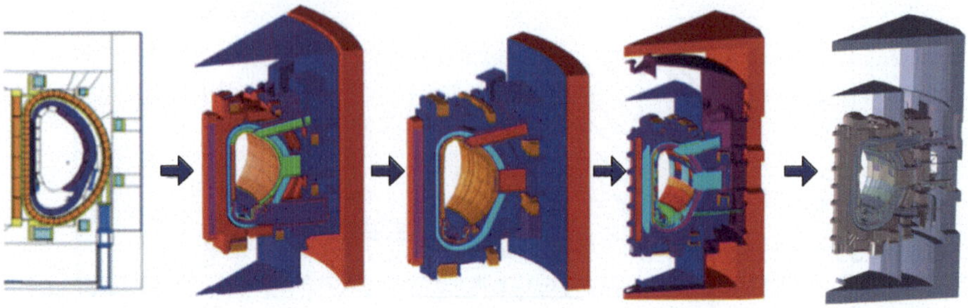

Fig. 10.23 ITER neutronics models

$$P_{TF} = 31e^{(0.036R - 0.14T_w)} \tag{10.1}$$

where P_{TF} is the nuclear heat in the TFC in units of kW, R is the radius of the NBI port in units of cm, and T_W is the wall thickness in units of cm. According to Eq. (10.1) [25], the minimum port wall thickness required to achieve the response design limits for the TFCs adjacent to the NBI is 63 cm, and the required port radius is 117 cm.

An accurate map of the radiation distributions around the tokamak machine relies on the availability of precise and consistent models and design information. For ITER, more detailed 3D models were adopted to perform detailed neutronics calculations for the complex structures and important components of the tokamak.

ITER IO has created and released a series of reference models for its tokamak and containment buildings. As shown in Fig. 10.23, a series of neutronics models for the tokamak, including the so-called Brand, A-lite, B-lite, C-lite, and C-Model models, have been created to enable worldwide consistent analyses and to facilitate to catch up with the frequently updated designs. These neutronics models are distributed in the form of Monte Carlo transport calculation input files, containing information regarding the geometry, materials, neutron source, etc. With the development of CAD-based neutronics modeling technology, CAD models are also distributed along with the calculation input files.

The ITER Brand model was released in 2000 and created manually according to the CATIA CAD model of the ITER FEAT design. The ITER Brand model contains approximately 2900 cells and 2500 surfaces and describes a 20° sector of the tokamak, including the blanket, divertor, vacuum vessel, etc. In the Brand model, all the components are described using extremely simplified geometry. In 2004, the geometrical visualization function of the SuperMC was applied to detect and correct errors in the model, update the blanket description, and expand it from 20° to 40° to contain whole ports.

The ITER A-lite model was released in 2008. It was jointly created using SuperMC by the FDS Team, the Italian National Agency for New Technologies, Energy and Sustainable Economic Development (ENEA), JAEA and ITER IO based on the Brand model. The A-lite model describes all the important components and structures of the tokamak such as the plasma-facing components (PFCs), TFCs, central solenoid (CS), vacuum vessel, upper ports, equatorial ports, and lower ports. Compared with the Brand model, the A-lite model covers a 40° sector and incorporates more details regarding the blanket, divertor, vacuum vessel, etc.

The ITER B-lite model was released in 2011 and jointly created using SuperMC by the FDS Team, the University of Wisconsin, ENEA, CCFE, JAEA, and ITER IO based on the A-lite model and updated ITER engineering design. The B-lite model adopts hierarchical structures, in which simple solids are used to represent boundaries of the subsystems, thus enabling a convenient way to update specific subsystems for specific analyses.

The ITER C-lite model was released in 2013 and was also jointly created by the FDS Team, the University of Wisconsin, ENEA, CCFE, JAEA, and ITER IO, based on the latest engineering design at the time and the B-lite model. In the C-lite model, some components such as the NBI port and cryostat were accurately modeled, and the complexity of the hierarchical structure of the whole model was decreased to two levels to improve the calculation and modeling efficiency. Meanwhile, based on the latest designs at the time, ITER also provided detailed models of the components, which can be used to replace the simplified components in the C-lite model according to the analysis requirements, thereby achieving high accuracy and efficiency.

After the release of the ITER C-lite model in 2013, the ITER engineering design has continuously evolved. To synchronize the neutronics analysis with the latest design, the ITER organization released the latest ITER tokamak 3D neutronics reference model in the 2016—C-Model which was jointly created by the FDS Team, AMECFW (International), UKAEA, ENEA, Fusion for Energy (F4E), IDOM, QST, KIT, UNED, and ITER IO. The C-Model is probably one of the most complex models used in fusion neutronics and has more complexity than ever ITER neutronics reference models. C-Model has reached a high level of configuration and design compliance with a respective increase in model complexity and computational demands. The recently released version features more than 70,000 cells and 100,000 surfaces. Compared with the C-lite model, the C-Model has been improved significantly in terms of specific system model design and level of detail, especially for in-vessel and vessel components [26].

Based on the 3D ITER reference neutronics model (A-lite), a detailed analysis for NBI was performed. In A-lite, the NBI is located in ports number 4 and 5, while port number 4 also contains the diagnostic devices for the NBI. The NBI contains several open channels directly connected to the plasma chamber. The regions closest to these ports are thus subjected to radiation loads caused by the neutrons streaming through these channels. Figure 10.24 shows a detailed neutronics model of the NBI and its integration into an 80° sector model of ITER based on the A-lite model [27].

The related neutronics analysis showed that the maximum radiation dose in the insulator closest to the ITER NBI ports is below 1×10^6 Gy (lower than the design limit of 1×10^7 Gy). The peak nuclear heat in the winding pack and TFC cases is approximately 0.5 kW/m^3, lower than the ITER design limit of 1 kW/m^3. Figure 10.25 shows the dose rate distribution around the NBI ports after a decay period of 10^6 s. The dose rate is approximately 0.5 Sv/h in the space between the vacuum vessel and TF

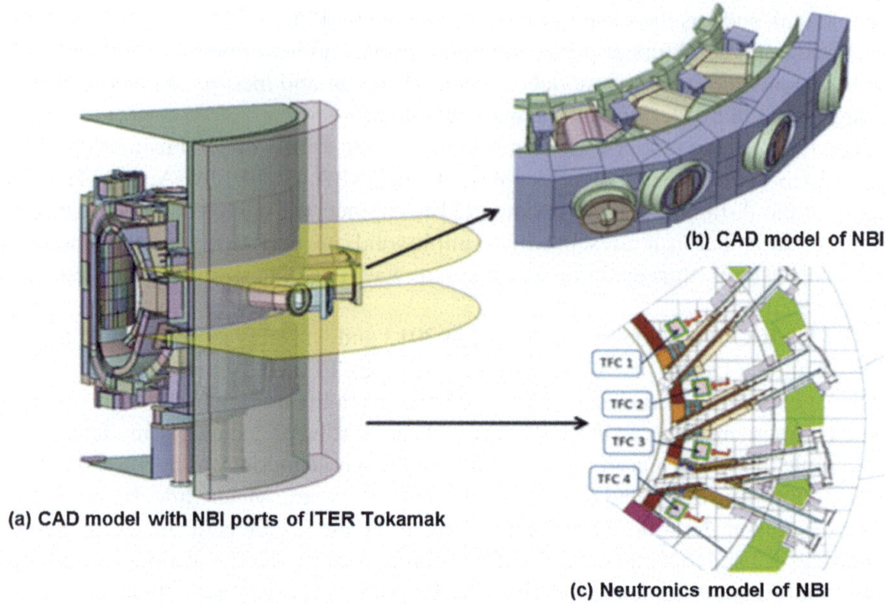

Fig. 10.24 Neutronics model of the NBI [27]

coils. In the vicinity of the duct liner shields, the dose rates also reach approximately 0.5 Sv/h. However, the dose rate outside the vacuum vessel and inside the cryostat exceeds the limit for hands-on maintenance which is 100 μSv/h. Therefore, this region requires special considerations for personnel access during maintenance [27].

2. Shielding design of the buildings

At the ITER site, buildings such as the Tokamak Building, Tritium Plant, Hot Cell, Radwaste Building, and Personnel Access Control Building are defined as radiological zones. The neutronics design analyses of these radiological zones include the shielding design and optimization analyses to evaluate the dose rates in the various rooms and satisfy the requirements for radiological protection. This subsection uses the shielding design for the Tokamak Building as an example to demonstrate the neutronics design for the building.

The main concern for the shielding design of the Tokamak Building is the radiation during the ITER plasma operation and maintenance. During plasma operation, the neutrons leaking from the tokamak machine and gamma rays emitted by the activated products will induce high-level radiation inside the building. The shielding analysis for the Tokamak Building evidenced that the electronics located in the rooms were sufficiently protected and the dose rate levels in each room satisfied the requirements for the specified radiological zone.

ITER defined the dose rate limits based on the radiological zoning and radiation risks. Figure 10.26 shows the radiological zoning for the Tokamak Building B1.

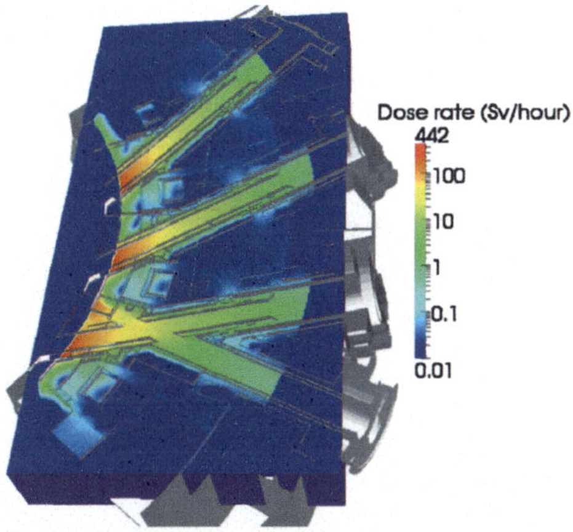

(a) Shutdown Dose Rate (after a decay period of 10^6 s)

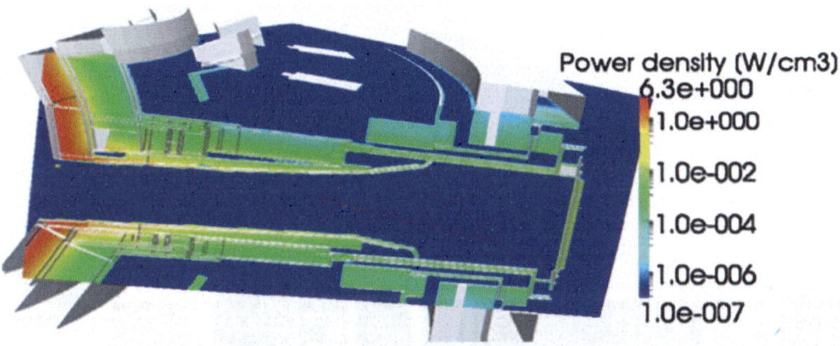

(b) Power Density (during operation)

Fig. 10.25 Dose rate distribution around the NBI ports after a decay period of 10^6 s (left) and the power density distribution during operation (right) [27]

Figure 10.27 shows the dose rate distribution when the activated divertor is placed in the port cell of B1, which is one of the fundamental data of the ITER radiation sources and is an important design input for other systems.

According to Fig. 10.27, when the activated divertor is placed in the port cell of B1, the maximum dose rate in the Tokamak Building was 7.65 Sv/h, and the maximum dose rates in the Tritium Building and Diagnostics Building were 2.5×10^4 and 2.4×10^3 μSv/h, respectively, thus exceeding the specified limits. Therefore, the inner shielding walls in the building needed to be optimized. For the dose rate outside the building, the maximum dose rate is 0.01 μSv/h, which satisfies the requirements [28].

3. Safety analysis and environmental impact assessment of ITER

The safety analysis and environment impact assessment in ITER include studies on the ORE, release of radiological nuclides and their impact on environmental safety,

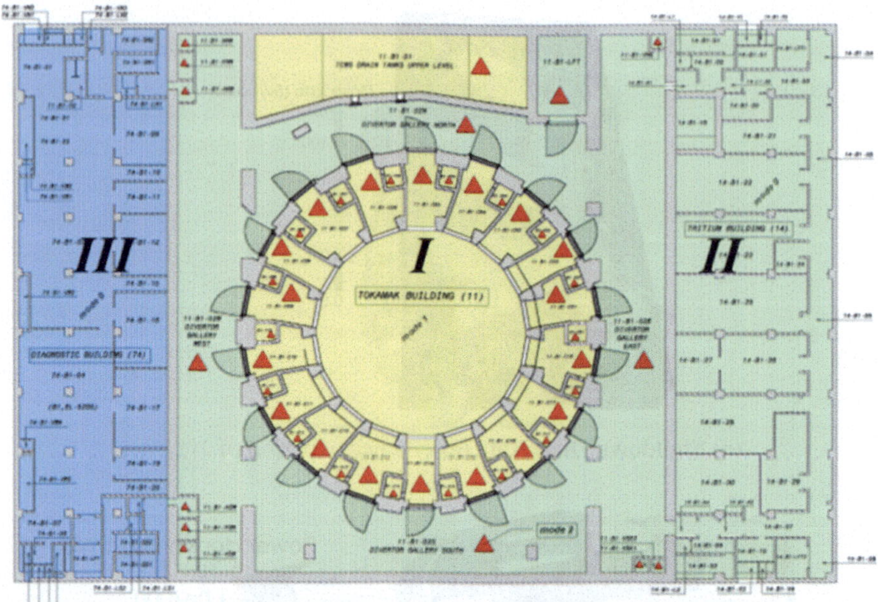

Fig. 10.26 Radiological zoning for Tokamak Building B1. (I Yellow zone effective dose rate <2 mSv/h; II Green zone effective dose rate <25 μSv/h; III Blue zone effective dose rate <7.5 μSv/h)

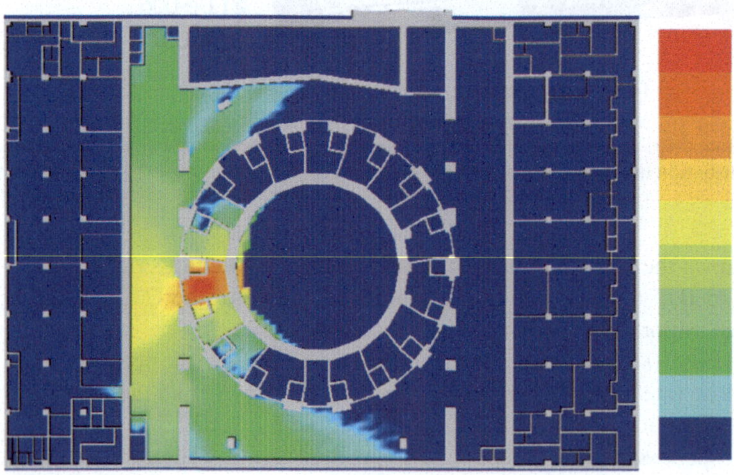

Fig. 10.27 The dose map in ITER Tokamak Building B1 during the transfer of the divertor (μSv/h) [28]

the management of radioactive wastes, the assessment of decay heat, which is an important procedure to demonstrate the safety and environment-friendly features of the design.

(1) Occupational radiation exposure

During the plasma operation of ITER, the structural and cooling materials in the tokamak machine will be activated by the fusion neutrons. The activation products, which can emit highly penetrating gamma radiation, are the major exposure sources for workers. In addition, another significant potential source of exposure for workers is the tritium, which exists across the ITER facility in the cooling water, plasma-facing materials, Hot Cell, and Tritium Plant Buildings.

ITER has estimated individual and collective doses for workers under each routine and maintenance operation based on the design and maintenance programs. The annual collective dose related to ITER operation (including maintenance) was estimated as 496 person-mSv [29]. Figure 10.28 illustrates the individual contribution of each system to the overall collective dose, showing that the collective dose (corresponding to the highest-dose operations) results mostly from the maintenance or repair operations on the in-vessel components (~41%), in terms of contribution per system, the highest-dose operations concern mainly the diagnostics systems (~17%) and electron cyclotron heating and current drive systems (~7%), and maintenance operations on TCWSs (~15%).

To apply the ALARA principle to protect workers from radiation exposure, a series of protection approaches are being adopted at the design stage such as defining access controls, radiological zoning, placing fixed or mobile radiation shields, and replacing hands-on operations with RH tools, to protect workers against external exposure to neutrons or photons. To protect against internal exposure caused by tritium and active dusts, confinement barriers between the radioactive substances and operating personnel should be set, and personal protective means such as self-contained masks or pressurized suits with air supplies should be provided.

(2) Release of radiological nuclides and their impact on the environmental safety

The radioactive effluents in ITER include tritium in the vapor form (mainly HTO), gaseous tritium release (mainly HT), and tritiated dust, which originate from the detritiation system, ventilation systems, and TCWS, respectively. The amount of radioactive effluents released is low (for example, normally the annual release of tritium is approximately 2.2×10^{14} Bq), and these radionuclides will be diluted by air and water, minimally impacting the environment. According to the latest estimation, the maximum annual effective dose for the public due to radioactive release during ITER operation would be as low as 2 μSv, lower than the specified limit [29].

(3) Radioactive waste management

The radioactive wastes in ITER include activated materials such as the blankets, divertors, and the vacuum vessel, and tritiated materials such as components of the tritium systems.

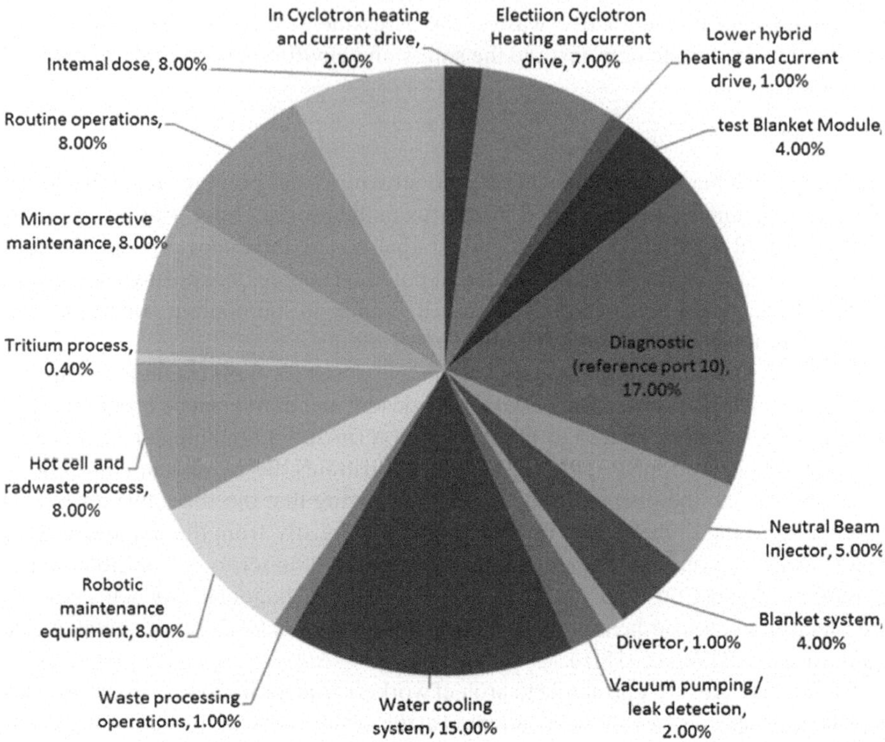

Fig. 10.28 Contribution of the various systems to the collective dose (person-mSv/y)

According to the waste classification system managed by the ANDRA (French National Radioactive Waste Management Agency), the radioactive waste can be classified into four categories based on its activity and radiotoxicity:

1) TFA waste (very low-level),
2) FMA-VC waste (low and intermediate-level short-lived) or type A waste,
3) MA-VL waste (intermediate-level long-lived) or type B waste, and
4) HAVL waste (high-level long-lived).

According to the estimation, ITER would produce approximately 38,000 tons of radioactive waste, most of which will be FMA-VC or TFA waste. The operational and decommissioning waste inventories are summarized in Table 10.11.

In addition, 500 tons of tritium-containing waste would be generated during the operation and decommission of ITER. Although such wastes are not neutron activation products, it will still be necessary to detritiate this waste.

(4) Shutdown decay heat

On the basis of ad hoc "New SA1" scenario defined on purpose by the safety group for decay heat analyses, the decay heat was found to be ~11 MW (the instant after

Table 10.11 ITER operational and decommissioning waste inventories before processing [30]

	TFA	FMA-VC	MA-VL	HA-VL
Operational waste				
Process and housekeeping	80 m^3a	200 m^3a	<1 m^3a	None
Component replacement	<10 t	20–100 t	1070 t	None
Liquids effluents	None	164 m^3a	None	None
Deactivation and decommissioning				
Components	18,300 t	8920 t	390 t	None
Buildings	2350 t	4300 t	None	None

shutdown) and 0.6 MW (cooling 1 day after shutdown), respectively [31]. Compared to a fission reactor, the density of decay heat for ITER is very small.

References

1. Wu YC (2017) Fusion neutronics. Springer Nature Singapore Pte, Ltd
2. Qiu LJ (2008) Fusion energy and its application. China Science Publishing and Media Ltd
3. Dolan TJ (ed) (2013) Magnetic fusion technology. Springer, New York
4. Bigot B (2016) Status of the ITER project in relation to TBM program activities. In: 16th meeting of the TBM program committee, Sain-Paul-lez-Durance, 19–21 Oct 2016
5. Ishizaka S (2010) ITER_D_2DY7NG Project specification. Private Communication
6. Maisonnier D, Campbell D, Cook I et al (2007) Power plant conceptual studies in Europe. Nucl Fusion 47:1524–1532
7. Federici G, Bachmann C, Barucca L, et al (2018) DEMO design activity in Europe: progress and updates. Fusion Eng Des 136(Part A):729–741
8. ARIES Program Public Information (2017) http://www-ferp.ucsd.edu/ARIES
9. Kikuchi M, Seki Y, Nakagawa K (2000) The advanced SSTR. Fusion Eng Des 48:265–270
10. Wu YC, Team FDS (2006) Conceptual design activities of FDS series fusion power plants in China. Fusion Eng Des 81(23–24):2713–2718
11. Wu YC, Wang HY, Ke Y et al (2006) Status of development and design of magnetic confinement fusion reactors and ITER test blanket modules. Nuclear Phys Rev 23(2):89–95 (in Chinese)
12. Wu YC, Team FDS (2008) Conceptual design of the China fusion power plant FDS-II. Fusion Eng Des 83(10–12):1683–1689
13. Wu YC, Team FDS (2009) Fusion-based hydrogen production reactor and its material selection. J Nucl Mater 81(386–388):122–126
14. Qiu LJ, Wu YC, Xiao BJ et al (2000) A low aspect ratio tokamak transmutation system. Nucl Fusion 40(3Y):629
15. Iida H (2004) ITER_D_22F2ST Nuclear analysis report. Private Communication
16. Baker D (2010) ITER_D_33AMDD General safety principles. Private Communication
17. Fischer U, Bachmann C, Palermo I et al (2015) Neutronics requirements for a DEMO fusion power plant. Fusion Eng Des 98–99:2134–2137
18. Zheng SL, Wu YC (2003) Neutronic comparison of tritium-breeding performance of candidate tritium-breeding materials. Plasma Sci Technol 5(5):1995–2000
19. Zucchetti M, Pace LD, Guebaly LE (2009) The back end of the fusion materials cycle. Fusion Sci Technol 55:109–139

20. Wang H, Wang W, Bai YQ et al (2005) Preliminary thermal-hydraulics design of the dual-cooled. Fusion Eng Des 75–79:841–845
21. Li JJ, Zeng Q, Chen ML et al (2008) Comparison analysis of 1D/2D/3D neutronics modeling for a fusion reactor. Fusion Eng Des 83:1678–1682
22. Li JJ, Zheng SL, Zeng Q et al (2005) Preliminary neutronics design of the dual-cooled lithium lead blanket for FDS-II. Fusion Eng Des 75–79:715–719
23. Swan ME, Abdou M (2006) Physics and technology conditions for attaining tritium self-sufficiency for the DT fuel cycle. Fusion Eng Des 81:1131–1144
24. Chen ML, Huang QY, Li JJ et al (2005) Activation analysis and waste management for dual-cooled lithium lead breeder (DLL) blanket of the fusion power reactor FDS-II. Chinese J Nuclear Sci Eng 25:178–183
25. Santoro RT (2000) Radiation shielding for fusion reactors. J Nucl Sci Technol 37:11–18
26. Leichtle D, Colling B, Fabbri M, et al (2018) The ITER tokamak neutronics reference model C-Model. Fusion Eng Des
27. Pereslavtsev P, Fischer U, Loughlin M et al (2015) Modelling and shielding analysis of the neutral beam injector ports in ITER. Fusion Eng Des 98–99:1585–1588
28. Ying DC, Zeng Q, Qiu YF et al (2011) Assessment of radiation maps during activated divertor moving in the ITER building. Fusion Eng Des 86(9–11):2087–2091
29. Taylor N (2011) ITER_D_3ZR2NC Preliminary safety report. Private Communication
30. Rosanvallon S (2016) Waste management plans for ITER. Fusion Eng Des 109–111:1442–1446
31. Perrault D (2016) Safety issues to be taken into account in designing future nuclear fusion facilities. Fusion Eng Des 109–111:1733–1738

Chapter 11
Neutronics Design of Hybrid Nuclear Systems

Hybrid nuclear systems, rather than a self-sustained chain reaction, are subcritical nuclear systems using additional neutrons from an external neutron source to sustain a chain reaction. Once the external neutron source supply stops, the fission reaction in the subcritical reactor/blanket soon stops. Hybrid nuclear systems include accelerator-driven subcritical system (ADSs) and fusion-driven subcritical system (FDSs). These systems can not only produce energy but also offer an effective approach to the settlement of accumulated of nuclear waste and the shortage of fissile fuel. In particular, FDS is considered as a potential path to the early application of fusion energy [1].

In this chapter, hybrid nuclear systems are introduced with descriptions of their principles, features, and several typical concepts. Additionally, in the subsequent section, the FDS-I is used as an example to describe the neutronics design of a hybrid nuclear system.

11.1 Introduction to Hybrid Nuclear Systems

11.1.1 Basic Principles

A hybrid nuclear system includes an external neutron source, a fission subcritical reactor/blanket, and related auxiliary systems. The fission subcritical reactor/blanket can achieve the goals of nuclear waste transmutation, nuclear fuel breeding, and energy production.

The nuclear wastes transmuted in hybrid nuclear systems include minor actinides (MA) and long-lived fission products (LLFP). There are two main methods of transmuting MA: one method involves fission reactions with fast neutrons, and another method is where the MA are transmuted into fissile nuclides through initial capture reactions with thermal neutrons and then fission reactions of the produced fissile nuclides with thermal neutrons. In hybrid nuclear systems, MA are mainly transmuted

Y. Wu, *Neutronics of Advanced Nuclear Systems*,
https://doi.org/10.1007/978-981-13-6520-1_11

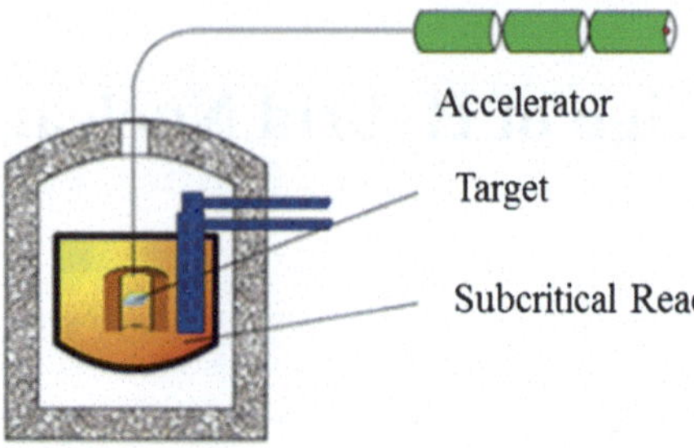

Fig. 11.1 Schematic diagram of ADS

through fission reactions with high-energy neutrons. The harder the neutron spectrum is and the higher the neutron energy is, the higher the transmutation efficiency can be, which is why MA are usually loaded in the areas close to the external neutron source. To reduce the requirements for the external neutron source and improve the transmutation efficiency of a hybrid nuclear system, plutonium can be used in the subcritical reactor/blanket for neutron multiplication. The transmutation of LLFP (such as ^{129}I, ^{135}Cs, and ^{99}Tc) is achieved through neutron capture reactions, which transmute LLFP into short-lived or stable nuclides.

Both tritium and fissile nuclides can be bred in hybrid nuclear systems. The breeding of fissile nuclides can be achieved through capture reactions by adding fertile nuclides (^{232}Th and ^{238}U) to the subcritical reactor/blanket. Tritium is bred by adding lithium-containing materials to the subcritical reactor/blanket. The lithium nuclides in these materials interact with neutrons to produce tritium.

The subcritical reactor/blanket of a hybrid nuclear system is an amplifier of the external neutron source. The energy production capacity can be represented by the energy multiplication factor M. Increasing the external neutron multiplication factor is an effective way to increase M. For the function of breeding or transmutation, a hybrid nuclear system needs to expend more neutrons, which results in the reduction of the energy production capacity of the system.

The main difference between ADSs and FDSs is the disparity in the external neutron source. The external neutron source of ADS is composed of the accelerator and the target, as shown in Fig. 11.1. A proton accelerator is usually the primary option. A proton accelerator provides a stable, high current proton beam to initiate the spallation reaction of heavy metals in the target and produces a large number of spallation neutrons to induce fission reactions in the subcritical reactor.

Unlike ADSs, FDSs use fusion driver as an external neutron source, and the schematic diagram of tokamak-based FDS is shown in Fig. 11.2. The tokamak fusion driver mainly consists of a vacuum vessel, magnet coil, and divertor. The blanket mainly consists of the fission zone, tritium breeding zone, and shielding zone. The tritium breeding zone and the shielding zone are similar to those of fusion systems.

The blanket is the key component in an FDS. The fusion core provides an independent external neutron source that allows the fission zones to be operated subcritically

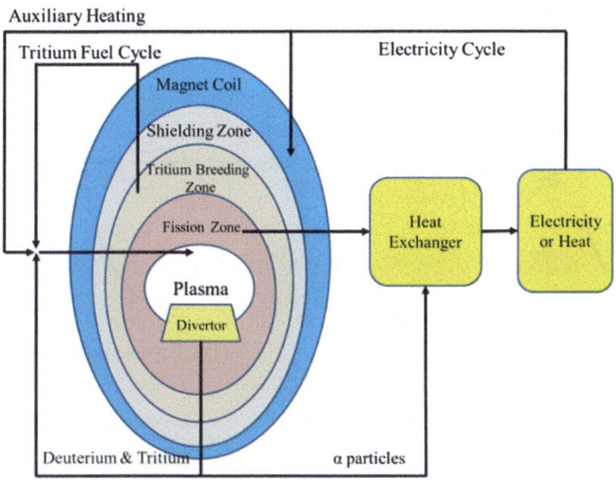

Fig. 11.2 Schematic diagram of a tokamak-based FDS

Table 11.1 Comparison of three kinds of blanket concepts

Blanket concepts	Neutron energy	Functions
Fission-suppressed blankets	Intermediate-energy neutrons	Transmutation/breeding
Fast-fission blankets	High-energy neutrons	Transmutation/breeding
Thermal-fission blankets	Thermal neutrons	Energy production

and achieves the goals of nuclear waste transmutation, nuclear fuel breeding, and energy production. From the perspective of fission reactions, the blanket concepts of FDSs can be categorized into fission-suppressed blankets, fast-fission blankets, and thermal-fission blankets, as shown in Table 11.1.

Fission-suppressed blankets use fissionable nuclides that have large capture reaction cross sections with intermediate-energy neutrons. Moderators and multipliers are added to the blankets to slow the fast neutrons and breed neutrons, respectively, eventually increasing the capture reaction rates. Fast-fission blankets use fissionable nuclides that have large fission reaction cross sections with high-energy neutrons. Two to four neutrons can be released by each fission reaction, which increases the neutron flux in a fast-fission blanket and thereby increase the capture reaction rate. Thermal-fission blankets use fissile nuclides that have large reaction cross sections with thermal neutrons to produce energy. Water exhibits very good performance for moderating neutrons and is usually used as a neutron moderator in thermal-fission blankets. Fission-suppressed blankets and fast-fission blankets are applied in FDSs for the transmutation of nuclear waste and the breeding of nuclear fuel, whereas

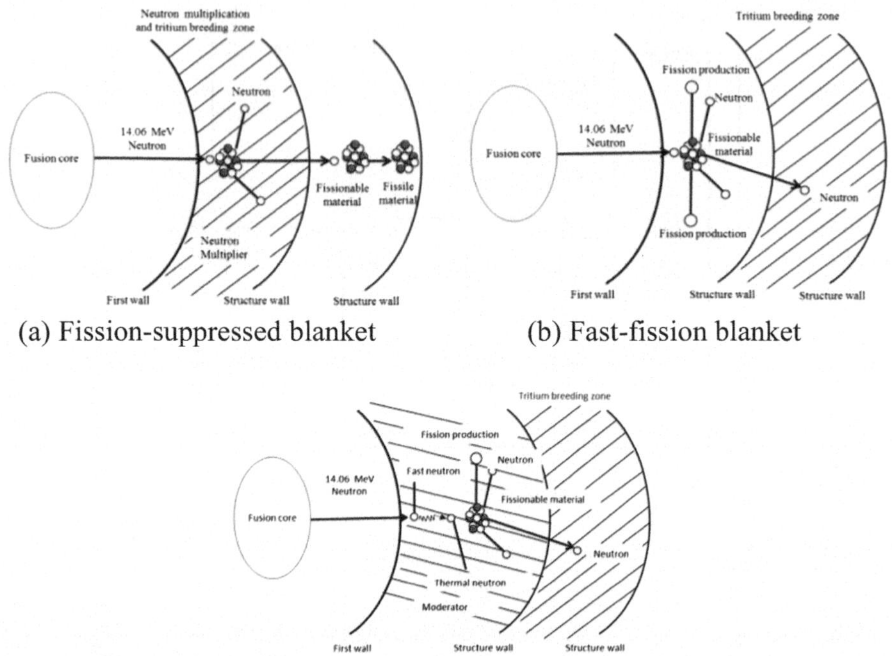

(a) Fission-suppressed blanket

(b) Fast-fission blanket

(c) Thermal-fission blankets

Fig. 11.3 Schematic diagrams of FDS blankets

thermal-fission blankets are often used in FDSs for energy production. Figure 11.3 shows schematic diagrams of fission-suppressed blankets, fast-fission blankets, and thermal-fission blankets.

11.1.2 Technical Features

Hybrid nuclear systems have the following advantages compared with fission systems.

1. Better fuel breeding and nuclear waste transmutation performances

Hybrid nuclear systems can achieve higher breeding ratios (BRs) and higher transmutation support ratios (TSRs) than current fission systems. The BRs of typical liquid metal fast breeder reactors (LMFBRs) and hybrid nuclear systems are listed in Table 11.2. The TSRs of various types of nuclear systems are listed in Table 11.3.

2. More flexible fuel loading

Fuel loading needs to be restricted in fission systems for safety reasons. For example, the fraction of plutonium in a thermal reactor should not exceed 13%. How-

Table 11.2 BRs of LMFBRs and hybrid nuclear systems

Reactor type	Fuel cycle	BR
LMFBR	U/Pu	1.2–1.4
LMFBR	Th/U	1.03–1.15
FDS (fast-fission blanket)	U/Pu	2.5–3
FDS (fast-fission blanket)	Th/U	2.8–3.5
FDS (fission-suppressed blanket)	Th/U	1.6–1.85
ADS	Th/U	1–2.3

Table 11.3 TSRs of various types of nuclear systems

Reactor type	TSR for MA	TSR for LLFP
Thermal reactor	1–3	>2
Fast reactor	2–20	–
FDS	15–55	>10
ADS	8–30	4–5

Table 11.4 Typical fuel options for hybrid nuclear systems and critical fission systems

Reactor type	Uranium fuel	Transuranium fuel	
	Uranium enrichment (wt%)	Pu (%HM[a])	MA (%HM)
Pressurized water reactor	>3	<13	–
Heavy water reactor	>0.7	–	–
Fast reactor	>7	–	<5
High-temperature gas-cooled reactor	>7	<13	–
Hybrid nuclear systems	0–100	100	100

[a]HM% denotes the mass fraction of heavy metals in the fuel

ever, hybrid nuclear systems operate subcritically and have a large safety margin. Therefore, hybrid nuclear systems can be loaded with various fuels, such as natural uranium, highly enriched uranium, and depleted uranium. Table 11.4 lists typical fuel options for hybrid nuclear systems and critical fission systems.

3. Deeper burnup and longer refueling period

In a fission system, excess reactivity is reserved when fresh fuel is loaded to compensate for the reactivity loss caused by burnup and the accumulation of fission products.

Table 11.5 Plasma parameters for various types of systems

Systems	Fusion power (MW)	Neutron wall loading (MW m^{-2})	Q
JET/JT-60U	16	0.2	~1
ITER	500	0.57	10
Fusion power plants	2000–3500	4–7	10–30
FDS	10–100	0.5–1	~1

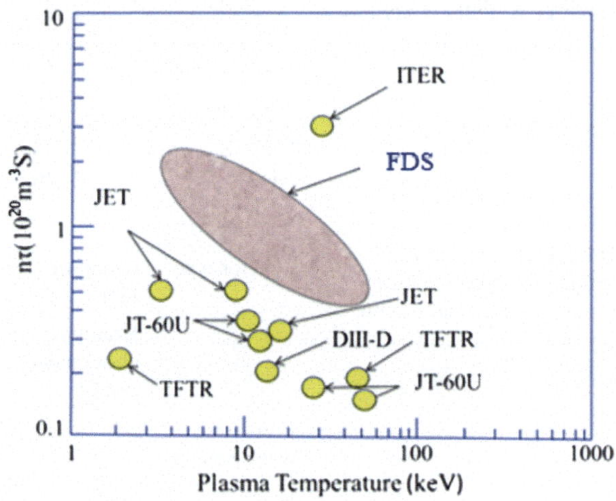

Fig. 11.4 Fusion triple products for a series of experimental facilities versus the required parameters for an FDS

Refueling is needed when this excess reactivity is exhausted. By contrast, in a hybrid nuclear system, the power can be controlled by adjusting external neutrons, and the system can achieve deeper burnup and longer refueling period.

In addition, FDSs have the following advantages compared with fusion systems.

1. Lower plasma requirements

The plasma parameter requirements of FDSs are much lower than those of fusion systems because of the energy multiplication achieved by using fission materials in the blankets, which means the requirements for fusion power and for neutron wall loading of FDSs are lower than those of fusion systems. The fusion energy gain (Q) of FDS is approximately 1, which is close to the Q achieved in existing fusion experiments. The plasma parameters for various types of systems are listed in Table 11.5. The fusion triple products for a series of experimental facilities and the required parameters for FDS are shown in Fig. 11.4.

2. Higher tritium breeding ratio

Tritium self-sustaining is the main challenge of the commercial operation of fusion systems. FDS usually has a higher tritium breeding ratio (TBR) than that of a fusion

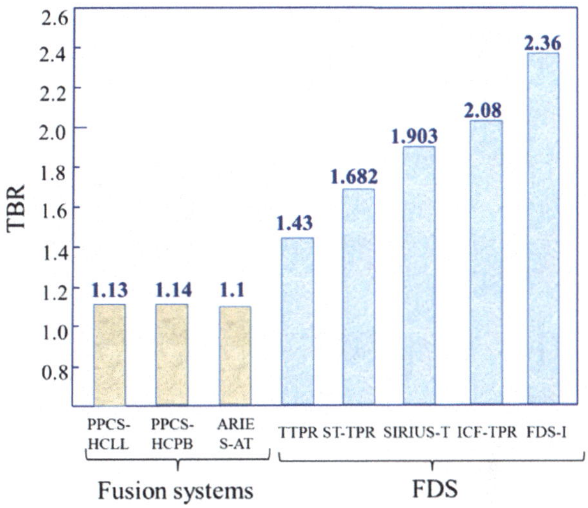

Fig. 11.5 TBR of a series of fusion systems and FDSs

system because of the higher neutron flux in the blanket. Figure 11.5 shows the TBR of a series of fusion systems and FDS. The TBR of a fusion system is approximately 1.1, whereas that of an FDS can be greater than 2.0.

Nevertheless, since hybrid nuclear systems combine the functions of external neutron source and subcritical reactors/blankets, the subsystems in hybrid nuclear systems are more complex than those in fusion systems and fission systems, which increases the difficulty of radiation containment and residual heat removal and introduces challenges in terms of safety and economic performance.

11.1.3 Typical Conceptual Designs

11.1.3.1 Typical ADS Concepts

The ADS concept combining an accelerator and subcritical reactor was proposed in the 1950s. It was not until the 1980s that the research on ADSs began to flourish because of the limitation of accelerator technology. In 1993, a group of scientists from European Organization for Nuclear Research (CERN) led by Nobel Laureate Carlo Rubbia presented the basic idea, called Energy Amplifier (EA). A subcritical nuclear system based on the U-Th cycle and fed by a high-intensity proton accelerator could be used for energy production and for the transmutation of MAs and LLFP. Many projects supporting the development of ADSs were proposed in many countries and regions, such as the MYRRHA Project (Multipurpose hYbrid Research Reactor for High-end Applications) in the EU, the ATW program (Accelerator-driven Transmutation of Waste) in the US, and the OMEGA program (Options Making Extra Gains from Actinides and fission products) in Japan. China has carried out ADS studies

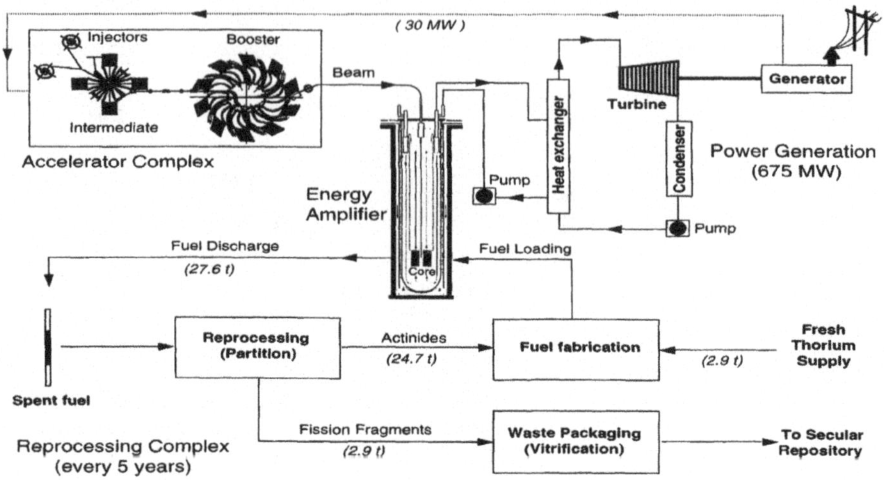

Fig. 11.6 General layout of the energy amplifier complex [2]

since the 1990s, and preliminary studies were performed with the support of the Ministry of Science and Technology (MOST), the Natural Science Foundation of China (NSFC), and the Chinese Academy of Sciences (CAS). At the beginning of 2011, CAS launched the strategic priority research program "The Future Advanced Nuclear Fission Energy—ADS transmutation system". The typical ADS programs of the different countries and regions are introduced below.

1. EA concept (EU)

Conceptual design of a fast-neutron-operated high-power energy amplifier was proposed by CERN in 1993, as shown in Fig. 11.6. An EA module consists of a 1500 MW unit with a dedicated 1.0 GeV proton accelerator of 12.5 mA. A compact, highly reliable, and modular cyclotron has been designed. A plant may be composed of several such modules. For instance, a cluster of three such modular units produces approximately 2000 MWe of primary electrical power. A relevant feature of the design is that it relies on natural convection to remove the heat generated inside the core. The EA is a large, passive device where a proton beam is dumped, and the heat generated by nuclear cascades is extracted, without other major elements of variability. The delivered power is controlled exclusively by the current of the accelerator. The fuel does not need to be accessed throughout the whole burnup and may be kept sealed as a nonproliferation safeguard.

2. MYRRHA project (EU)

The MYRRHA project was launched by the Belgian nuclear research center SCK•CEN in 1998 and intends to start construction in 2024. The EUROTRANS (EUROpean Research Program for the TRANSmutation of High-Level Nuclear Waste in an Accelerator-Driven System) partners have accepted the SCK•CEN offer to use

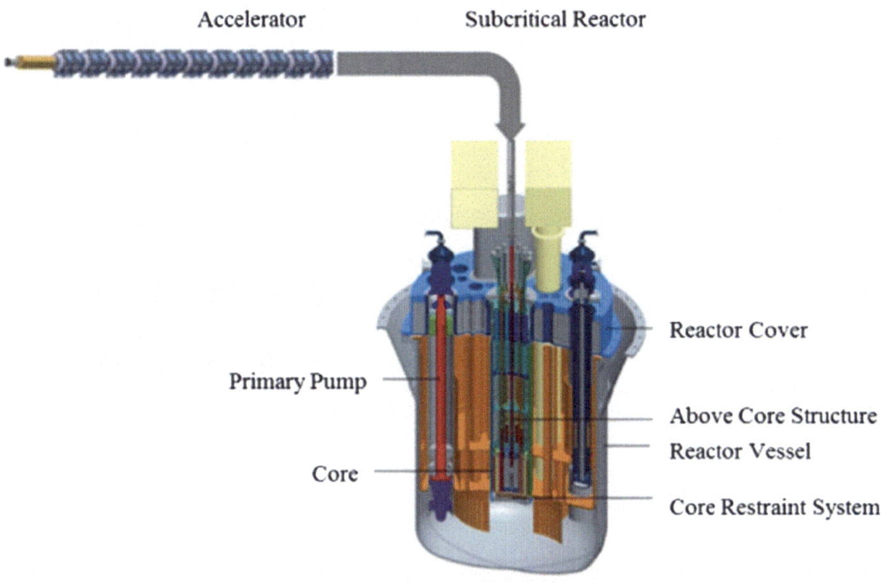

Fig. 11.7 Overall structure of MYRRHA [3]

MYRRHA as a starting basis for the XT-ADS (eXperimental ADS for Transmutation) design.

MYRRHA has been designed to be operated in two modes: critical and subcritical mode. The project consists of a 600 MeV/4 mA superconducting proton accelerator delivering a beam to a windowless LBE spallation target without a cooling loop that, in turn, couples to a 100 MW_{th} pool-type subcritical fast reactor cooled with LBE. k_{eff} is designed as 0.95. The overall structure of MYRRHA is shown in Fig. 11.7.

3. ATW program (US)

The ATW program was proposed by Los Alamos National Laboratory (LANL) in 1992. The roadmap of ATW was proposed by the US Department of Energy (DOE) after an assessment of the program. The reference ATW plant concept is shown in Fig. 11.8. Two 45 mA/1 GeV proton linear accelerators (linacs) can each provide beam for four subcritical fission assemblies to assure the continuity of the electric power exported, with the latter sized at 840 MW_{th}. At an assumed thermal-electricity efficiency of 0.38, these transmuters produce a total of 2175 MWe of electricity. Approximately, 380 MWe is used to operate the plant, and the balance of the electric power (approximately 2175 MWe) is exported to the grid. The reference accelerator design is derived from the high-power CW proton linac proposed for the Accelerator Production of Tritium (APT) project. The spallation target uses a lead or lead–bismuth eutectic (LBE) option. k_{eff} is designed as 0.97. Pu, MA, and LLFP fuels are burned in ATW, and the annual transmutation ability could reach 200 kg.

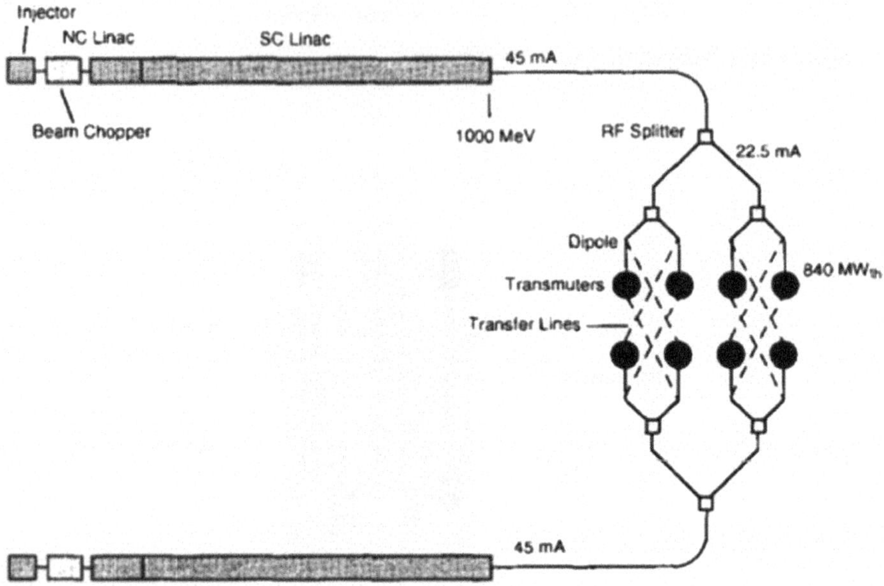

Fig. 11.8 Overall structure of ATW reference concept [4]

4. OMEGA program (Japan)

Research on partitioning and transmutation in Japan started in 1988. The Japan Atomic Energy Commission launched the OMEGA program, and the Japan Atomic Energy Agency (JAEA) proposed several concepts for ADSs such as a sodium-cooled ADS, molten-salt-cooled ADS, and LBE-cooled ADS. Considering the chemical properties of the coolant and the positive void coefficient of reactivity in the sodium-cooled ADS, the compatibility of fuel and material in the molten-salt-cooled ADS, and the weakened defence in the depth strategy, the LBE-cooled ADS was chosen as the basic reference design of the OMEGA program. The overall structure of OMEGA is shown in Fig. 11.9.

The components of the reference ADS are a 30 MW (1.5 GeV/20 mA) superconducting proton linac, an LBE spallation target, and an 800 MW_{th} subcritical core using nitride fuel and LBE coolant. The initial k_{eff} is designed as 0.95, and the highest k_{eff} is 0.97. The initial loading amount of MA is 2.5 tons, and the transmutation ability is 250 kg per 300 days.

5. CLEAR program (China)

Accelerator-driven subcritical system studies have been performed since the 1990s in China. From 1999, the project "The physical and technical foundation research on accelerator-driven clean nuclear power system" was launched with the support of the national basic research program of China ("973" Program), and a thermal-fast-neutron spectrum-coupling subcritical experiment platform China ADS subcritical experimental assembly (VENUS-1) was established.

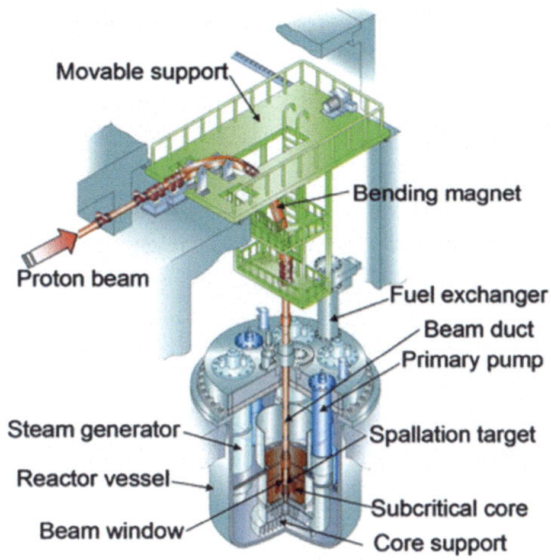

Fig. 11.9 Overall structure of OMEGA [5]

The China LEAd-based Reactor (CLEAR) was selected as the reference reactor for the strategic priority research program of "The Future Advanced Nuclear Fission Energy—ADS transmutation system" and lead-cooled fast reactor (LFR) technology. Currently, the FDS Team has completed the conceptual design of the first phase of the 10 MW Advanced External Neutron Source-Driven Nuclear Energy Experimental System CLEAR-A10, and the overall structure is shown in Fig. 11.10. CLEAR-A10 adopts a proton accelerator with 900 MeV/1 mA and lead as the window target and coolant. CLEAR-A10 has a negative reactivity coefficient, large thermal inertia, and passive residual heat removal capability, which provides good safety advantages.

11.1.3.2 Typical FDS Concepts

The study of FDSs dates back to the 1950s. With the development of fusion technology and the flourishing of fission nuclear power plants, research regarding FDSs has attracted considerable attention worldwide. In the late 1980s, research in FDS slowed under pressure from anti-nuclear movements, nuclear proliferation risks, and lower prices for oil and natural gas. However, in the late 1990s, because of the increase in nuclear waste accumulation with the development of fission energy, interest in research into FDSs rebounded, and many studies have since been conducted in the US, Japan, Turkey, and other countries.

The study of FDSs has a history of nearly 30 years in China. In the early stages, these studies were supported by the National High-tech R&D Program ("863" Program). The related research units include the Chinese Academy of Sciences (CAS), the China Academy of Engineering Physics (CAEP), the Southwestern Institute of Physics (SWIP), and several universities and colleges. China joined the ITER project in 2006, and since then, national support for fusion-related research has greatly strengthened and has played an important role in promoting research on

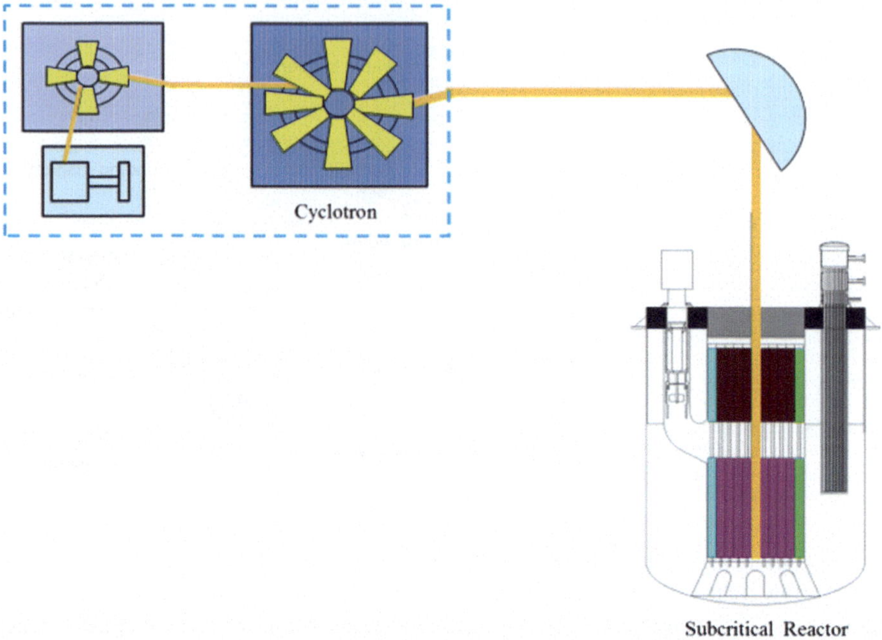

Fig. 11.10 Overall structure of CLEAR-A10 [6]

FDSs. Many FDS concepts have been proposed by the FDS Team with different functions based on different types of fusion drivers. In addition to those conceptual designs, a series of R&D studies in material and experimental loop technology have been performed. A conceptual design of FDS was presented by SWIP for fissile fuel breeding and nuclear waste transmutation based on conventional tokamaks. An FDS concept proposed by CAEP was driven by Z-pinch.

Several typical FDS concepts based on different types of fusion drivers are introduced below.

1. FDS-I (China)

The Fusion–fission Hybrid Reactor for Spent Fuel Burning (FDS-I) proposed by the FDS Team is a multifunctional hybrid nuclear system for the purposes of nuclear waste transmutation, nuclear fuel breeding, and energy production. This system is suitable for early applications of fusion energy with relatively low plasma parameter requirements. The level of the plasma parameters for FDS-I rests between the levels for fusion experimental facilities (such as JET and JT-60) and ITER. The overall and blanket configurations of FDS-I are shown in Fig. 11.11. The blanket of FDS-I is divided into two parts, the inboard blanket and the outboard blanket. Because of the limited space and the difficulty of the complex operations involving the inboard blanket, the system is designed with simple structures to realize the functions of tritium breeding and radiation shielding. By contrast, the outboard blanket consists of

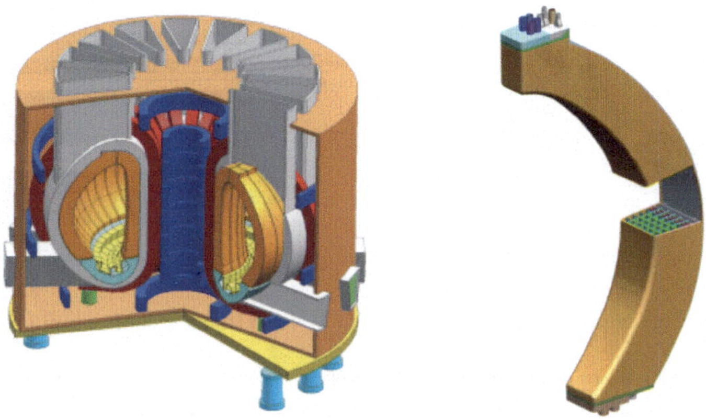

(a) Overall configuration (b) Blanket configuration

Fig. 11.11 Configurations of FDS-I and blanket [7]

MA zones, LLFP zones, and shielding zones with relatively complex configurations, which enables improved transmutation efficiency because of the distinct neutron energy spectra in different functional zones.

2. GDT-Hybrid (China)

The Gas Dynamic Trap (GDT)-based fusion–fission hybrid reactor (GDT-Hybrid) proposed by the FDS Team was intended to explore the potential applications of a fusion neutron source [1]. The fusion neutron source based on a GDT magnetic mirror features inherent steady-state operation, compact structure, easy construction, convenient maintenance, relatively low tritium consumption, and flexible operation. This reactor can be used to drive a hybrid nuclear system to produce energy and to transmute radioactive waste.

The overall and blanket configurations of GDT-Hybrid are shown in Fig. 11.12. The cylinder-shaped blanket is divided into a waste transmutation zone and a tritium breeding zone. The waste transmutation zone is used to transmute nuclear waste and produce energy. The length of the blanket exceeds the length of the neutron reaction zone to reduce neutron leakage at both ends of the blanket.

3. Z-FFR (China)

A Z-pinch-driven fusion–fission hybrid power reactor (Z-FFR) proposed by CAEP has been designed for energy production. This reactor is operated in pulsed mode. The available space in the blanket of the Z-FFR is greater than that in magnetic confinement facilities. The blanket is divided into four components: the FW, a fission zone, a tritium breeding zone, and a shielding zone. The FW material is steel. U-10Zr is used as the fuel. Water is used as the coolant and moderator, and Li_4SiO_4 as the tritium breeder. The overall and blanket configurations are shown in Fig. 11.13.

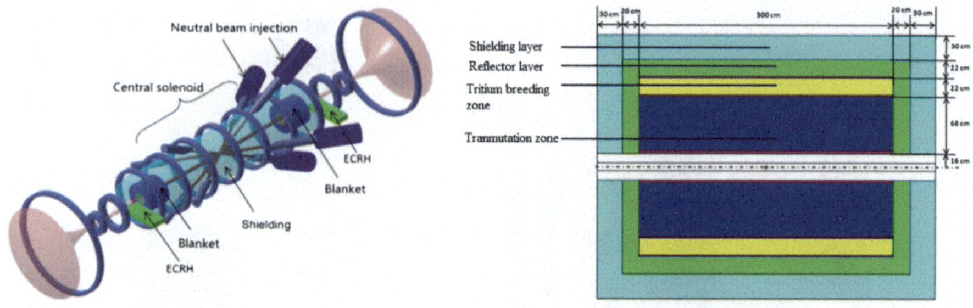

(a) Overall configuration (b) Blanket configuration

Fig. 11.12 Configurations of FDS-GDT and blanket

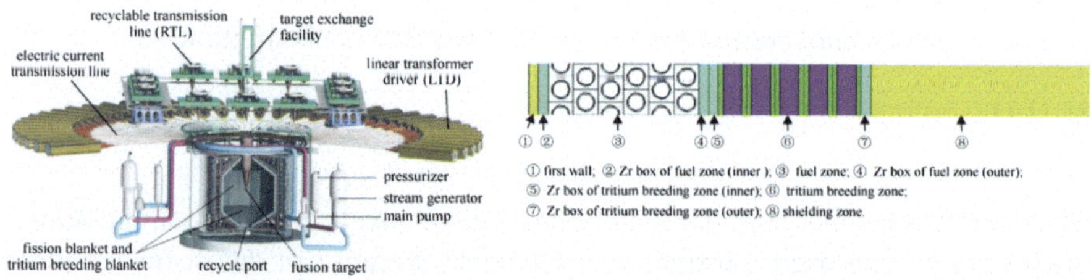

(a) Overall configuration (b) Blanket configuration

Fig. 11.13 Configurations of Z-FFR and blanket [8]

4. LIFE(US)

A hybrid nuclear system with a laser inertial confinement fusion core for producing energy and burning nuclear waste, called the Laser Inertial Fusion Engine (LIFE), was proposed by Lawrence Livermore National Laboratory (LLNL). The driver of LIFE, which is designed to use the National Ignition Facility (NIF), focuses on an ultraviolet laser of 1.4 MJ on a deuterium–tritium pellet, thereby inducing a nuclear fusion reaction inside the fusion target and consequently releasing a large amount of energy. The spherical fusion chamber is surrounded by the blanket. The possible fuels include depleted uranium, spent nuclear fuel, thorium, and weapons-grade plutonium. Tristructural isotropic (TRISO) nuclear fuel particles are suspended in liquid FLiBe, which is used as a coolant, neutron multiplier, and tritium breeder. The overall and blanket configurations of LIFE are shown in Fig. 11.14.

11.2 Principles and Requirements of Neutronics Design

The neutronics design of hybrid nuclear systems should consider not only the functions of the subcritical reactor/blanket but also the safety and economic performance of the whole system. The general design objectives for hybrid nuclear systems are illustrated as follows.

1. Subcriticality

Hybrid nuclear systems operate in subcritical states. The deeper the subcriticality is, the safer the hybrid nuclear system is. However, smaller k_{eff} values decrease the multiplication capacity. Therefore, it is necessary to determine the optimal k_{eff} value.

2. Power density

The subcritical reactor/blanket around the external neutron source has high power density, so the subcritical reactor/blanket design should analyze the heterogeneous power distribution and the cooling capacity of the coolant and define the maximum power density (MPD) limit to ensure safe operation. The nuclear heat deposited in the subcritical reactor/blanket needs to be transferred in time, and the temperature of the subcritical reactor/blanket materials cannot exceed the design limit.

Except for the above-stated parameters, the following parameters for different functions should be considered.

1. Energy multiplication factor

Hybrid nuclear systems for energy production should be designed with a highest energy multiplication factor (M). Each fission reaction induced by fusion neutrons releases 200 MeV of energy. A larger number of fission reactions induced by fusion neutrons increase the M of the subcritical reactor/blanket and eventually increase the energy production capacity.

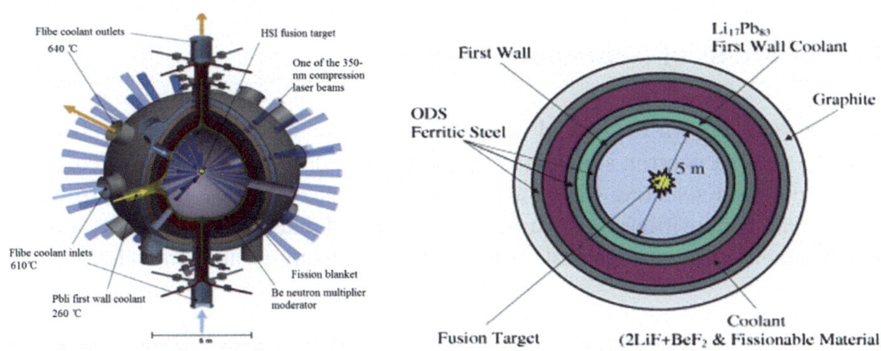

(a) Overall configuration (b) Blanket configuration

Fig. 11.14 Configurations of LIFE and blanket [9]

2. Waste transmutation ratio

The waste transmutation ratio (WTR) is defined as the number of long-lived isotopes converted to stable or short-lived isotopes by a single external neutron in the nuclear waste transmutation subcritical reactor/blanket. A higher WTR represents higher transmutation efficiency.

3. Fuel breeding ratio (Including TBR)

The fuel breeding ratio (FBR) is defined as the average number of fissile nuclides produced by a single external neutron in the fuel breeding subcritical reactor/blanket. A higher FBR represents higher breeding efficiency. Especially for FDSs, the important function of the blanket is to maintain tritium self-sufficiency for the nuclear system. Considering the tritium combustion, tritium decay, and tritium retention, TBR is generally required to be greater than 1.

4. Burnup and refueling period

The burnup and refueling period depend on the function of the hybrid nuclear system. For energy production, hybrid nuclear systems require a long refueling period and deep burnup to improve the economic performance. For fuel breeding or waste transmutation, hybrid nuclear systems require a reasonable refueling period considering the breeding or transmutation efficiency.

The neutron source of the hybrid nuclear system is designed to meet the different requirements of the subcritical reactor/blanket. According to the different kinds of drivers, neutron sources can be divided into two types: spallation neutron sources and fusion neutron sources. The appropriate spallation neutron source parameters include neutron intensity, energy spectrum, and spatial distribution, which can be obtained by the design of the beam flow and target of the accelerator. The appropriate fusion neutron source parameters including neutron intensity and spatial distribution can be obtained by adjusting the parameters of the fusion driver.

11.3 Methods of Neutronics Design

The neutronics design for hybrid nuclear systems consists of the selection of the parameters of the neutron source and the neutronics design of the subcritical reactor/blanket. After selecting the parameters of the neutron source and the structural and material compositions of the subcritical reactor/blanket, the neutronics model is built, and then, the neutronics calculations and optimizations are conducted.

11.3.1 Parameters Selections of Neutron Sources

11.3.1.1 ADS Spallation Neutron Source

In ADSs, the spallation neutrons are produced by the bombardment of protons accelerated to the middle-high energy to the heavy metal target. The performance of the spallation neutron source is mainly affected by two kinds of parameters: one kind includes the energy, intensity, and shape of proton beam; the other kind includes the structure, size, and material of the spallation target. Heavy metals are usually chosen as the target material, because the heavier the target nucleus is, the higher the probability of the spallation reaction is. The heavy metals and their alloys widely used include tungsten, tantalum, uranium, lead, and lead–bismuth alloy. With the same incident proton energy, the spallation neutron yield of the uranium target is the highest.

According to the results of a 10.2 cm * 60 cm cylindrical target measured by J S Fraser in 1965, the spallation neutron production yield can be expressed by Eq. (11.1).

$$Y(\mathrm{E}) = \begin{cases} 0.1(A+20)(E_{\mathrm{GeV}} - 0.120), & (A > 9, \mathrm{except}\ ^{238}\mathrm{U}) \\ 50(E_{\mathrm{GeV}} - 0.120), & (^{238}\mathrm{U}) \end{cases}$$
$$0.2\,\mathrm{GeV} \le E_{\mathrm{GeV}} \le 1.5\,\mathrm{GeV} \tag{11.1}$$

where A is the atomic weight, and E_{GeV} is the energy of the incident proton in units of GeV.

According to the current of the incident proton, the spallation neutron source intensity S_{ADS} can be calculated with Eq. (11.2).

$$S_{\mathrm{ADS}} = \frac{I_{\mathrm{p}}}{1.602 \times 10^{-19}} \cdot Y \tag{11.2}$$

where S_{ADS} is the spallation neutron source intensity in units of n/s, I_{p} is the proton current in units of A, and Y is the neutron yield in units of n.

The neutron spectrum in an ADS is complex, and the energies range from thermal neutrons to ultrahigh energy neutrons. The peak width is relatively wide, and the peak amplitude is relatively small. In general, the higher the incident proton energy is, the harder the spallation neutron spectrum is. Take the 10 mA proton current as an example, the energy spectrum characteristics of the spallation neutrons corresponding to the different energies of the incident protons were analyzed, as shown in Fig. 11.15.

To achieve the best ADS performance, the influence of the spallation neutron source parameters on the system fuel loading, discharge burnup, burnup reactivity loss, power peak factor, and material irradiation damage should be analyzed, and optimizing the target size parameters is carried out. And according to the optimal design of the target, the neutron characteristics of the different incident proton parameters are calculated, and the optimal design of the spallation neutron source is obtained.

11.3.1.2 FDS Fusion Neutron Sources

The first step of fusion neutron source design is to choose a suitable fusion reaction approach. Deuterium and tritium fusion reactions are generally accepted as the most feasible ways to achieve fusion currently. Then, the types of fusion reaction devices should be defined; if the priority is given to the maturity of the engineering technology, a tokamak is preferred as the driver. If the economy of the system is considered, a more compact fusion neutron source, GDT, may be chosen. According to the different design goals, the appropriate fusion device is selected as the neutron source. Lastly, the parameters of the fusion core are designed.

Because of the strong energy multiplication factor of the FDS blanket, the fusion neutron source is generally designed by using a small fusion core to achieve the application of fusion technology as soon as possible. The parameters of the fusion neutron source, such as the fusion power and first wall loading, need to be designed according to the output power, subcritical degree, and power density of the blanket. The fusion power can be changed to meet the demand of the blanket. The influence of the neutron source intensity and neutron energy spectrum on the design of blanket should be considered. The fusion neutron source intensity S_{FDS} can be used to calculate the intensity of the fusion neutron source according to the fusion power P_{fusion} as

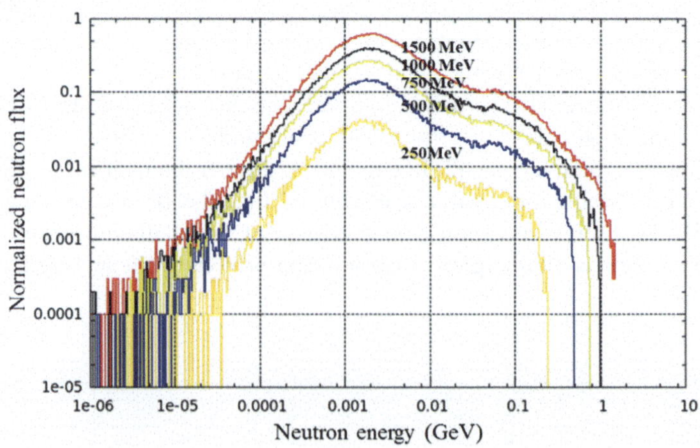

Fig. 11.15 Spallation neutron spectrum corresponding to different energies of the incident protons

shown in Eq. (11.3). The fusion neutron spectrum is relatively simple. The spectrum width is narrow, and the energy distribution ranges between 11 and 17 MeV. The peak value is 14.06 MeV. The distribution of typical deuterium and tritium fusion neutron sources is shown in Fig. 11.16.

$$S_{FDS} = \frac{P_{fusion}}{17.6 \times 1.602 \times 10^{-19}} \quad (11.3)$$

where S_{FDS} is the fusion neutron source intensity in units of n/s, and P_{fusion} is the fusion power in units of MW.

11.3.2 Design for Subcritical Reactors/Blankets

The basic process of the subcritical reactor/blanket neutronics design is as follows: the subcritical reactor/blanket structure and material selections are preliminarily determined according to the overall objectives and basic parameters, and then, the subcritical reactor/blanket neutronics design is determined by optimizing the target parameters and safety objectives.

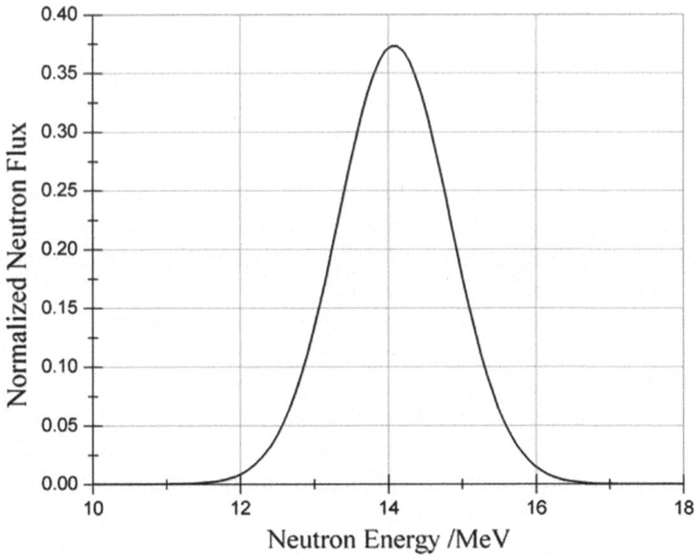

Fig. 11.16 Normalized neutron energy spectrum of the D-T fusion neutron source

11.3.2.1 Material Selections and Configuration

Compared with fusion systems, the structures of hybrid nuclear systems are more complex, and the layouts and options for the materials, such as the fuels, coolants, and moderators, are much more diverse. In a hybrid nuclear system, for nuclear waste transmutation, the MA are generally arranged to be near the external neutron source, and the LLFP are placed in a position where the thermal neutron flux is high for improving the transmutation efficiency. For nuclear fuel breeding, the fissionable nuclides are loaded in either the inboard or outboard subcritical reactor/blanket according to the structure of the breeding zone. The subcritical reactor/blanket can be designed as a fission-suppressed or fast-fission subcritical reactor/blanket because of the different neutron spectra in the subcritical reactor/blanket. For energy production, selecting different fuels and moderators also lead to different subcritical reactor/blanket structures.

ADS subcritical reactors are similar in material selection to fission systems in terms of structural materials, coolants, moderators, and shielding materials. The material selections of FDS blankets are similar to these of fusion systems in tritium breeders, neutron multipliers, coolants, and shielding materials.

The fuels for hybrid nuclear systems include heavy metals (U, Pu, MA, etc.) and LLFP (^{99}Tc, ^{129}I, ^{135}Cs, etc.). The fuel properties, including thermal properties, mechanical properties, chemical stability, material compatibility, irradiation resistance, irradiation creep, and irradiation swelling, must be considered. Meanwhile, the technical maturity of the fuel manufacturing and spent fuel reprocessing are also important factors that affect the fuel selection. The heavy metal fuels include oxide, nitride, carbide, and metal fuels. Generally, the metal fuels are selected to transmute ^{99}Tc, CaI_2, and NaI fuels are selected to transmute ^{129}I, and CsCl fuel is selected to transmute ^{135}Cs.

The subcritical reactor/blanket coolants for hybrid nuclear systems mainly include liquid metals, helium, water, and molten salt, and liquid metals are the most used followed by helium. With a high boiling point, a liquid-metal-cooled reactor can operate in condition of high temperature and low pressure.

Liquid metals can play multiple roles in subcritical reactor/blanket except for the coolant. For example, LBE can be used as the target in an ADS, and LiPb can be used as the tritium breeder in an FDS. The compatibility between the liquid metals and materials is more serious, and the problem of materials corrosion needs special consideration.

The advantages of helium are a small neutron absorption cross section and good neutron economy. A hard neutron energy spectrum can be obtained in the subcritical reactor/blanket. As an inert gas, helium is not easy to activate, and does not corrode equipments and pipelines at high temperature and has good compatibility with the materials and water . The disadvantage of helium is that the heat capacity is low and needs to be operated at high pressure to ensure that the heat can be exported smoothly. In a loss of pressure or coolant accident, it is very difficult to remove the decay heat of the subcritical reactor/blanket in time.

11.3.2.2 Neutronics Calculations and Optimization

As shown in Fig. 11.17, hybrid nuclear systems need perform more neutronics calculations and more optimization pathways than fusion systems. The main differences are shown as follows.

1. Neutron transport calculation

The main parameters of the particle transport calculation include k_{eff}, MPD, M, TBR, FBR, and WTR. The design objectives for these parameters can be achieved by adjusting the structure selection and material compositions. The optimization design is as follows.

The structures and material compositions of the hybrid nuclear system affect the k_{eff}, which can be optimized by adjusting the fuel enrichment, fraction of the fuel and moderator, and size of the subcritical reactor/blanket. The M is related to and can be derived from the k_{eff}.

During the optimization of the power density, the MPD of the subcritical reactor/blanket is set as high as possible with the precondition that the function requirements are satisfied. The higher the compaction of the subcritical reactor/blanket structure is, the better the economic performance of the hybrid nuclear system is.

The optimization process of the TBR for hybrid nuclear systems is similar to that for fusion systems. The FBR (or WTR) refers to the number of reactions induced by the fusion neutrons. The optimization methods for the FBR and WTR are similar to those for the TBR. Take an FBR as an example, harder neutron spectra and higher neutron fluxes are required for a higher FBR. Generally, measures are taken to increase the fraction of fissionable nuclides in the subcritical reactor/blanket to increase the FBR. However, an exorbitant fraction leads to rapid burning of the fissionable nuclides. Therefore, reasonable material compositions should be selected.

2. Burnup calculation

The composition and density of the nuclides in the subcritical reactor/blanket change with time. By calculating the burnup, the density of each nuclide in the subcritical reactor/blanket and the values of the FBR and WTR are obtained. After comparing these parameters with the design objectives, further optimization should be performed. For example, a hybrid nuclear system for the purpose of waste transmutation requires a higher power density and a lower burnup reactivity loss to improve the transmutation efficiency.

3. Other neutronics parameter calculations

In the design of the subcritical reactor/blanket, radiation shielding performance and transient parameters need to be evaluated. The radiation shielding performance

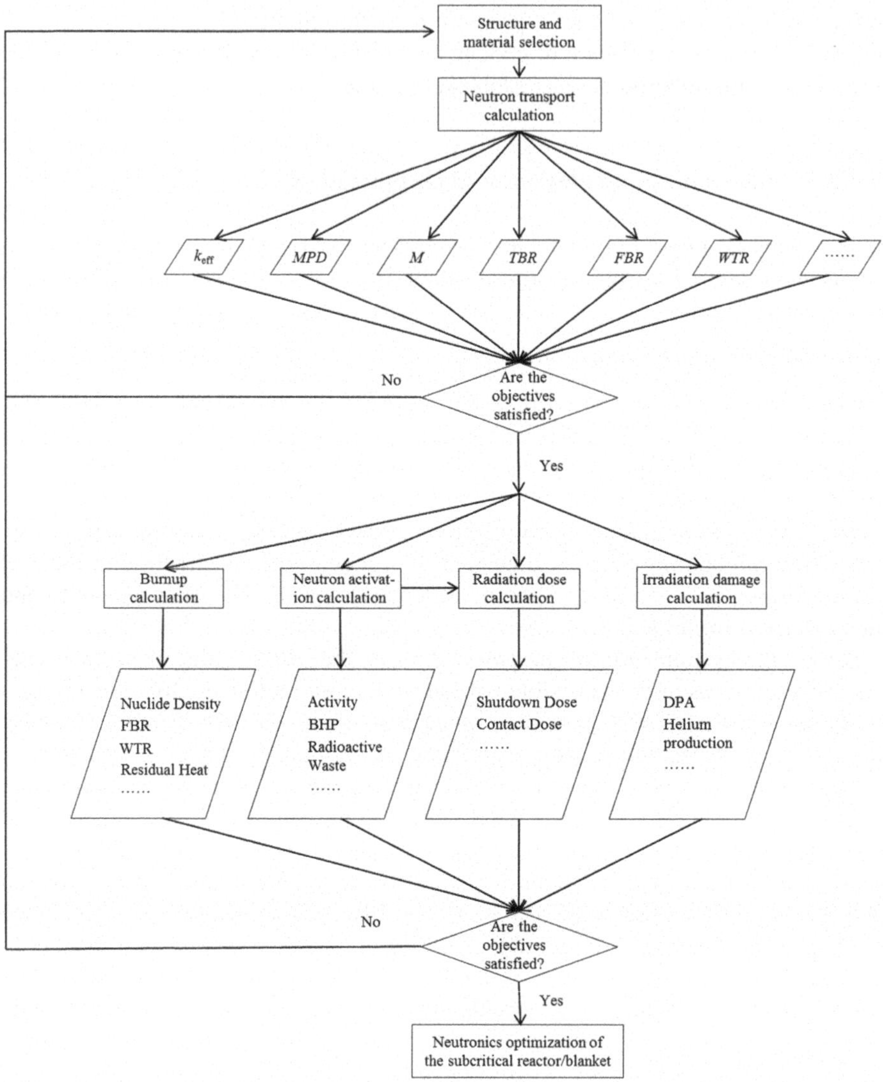

Fig. 11.17 Neutronics calculation and optimization procedure for hybrid nuclear systems

parameters include displacement per atom, helium production, total radioactive waste, and activity. The transient parameters include the reactivity coefficient, effective delayed neutron fraction, and neutron lifetime.

The physical and engineering parameters of subcritical reactor/blankets have a wide range of values, and their relationships are complex. It is difficult to optimize all these parameters through a simple qualitative analysis or single objective optimization. With the development of computer technology, inverse algorithms (which auto-

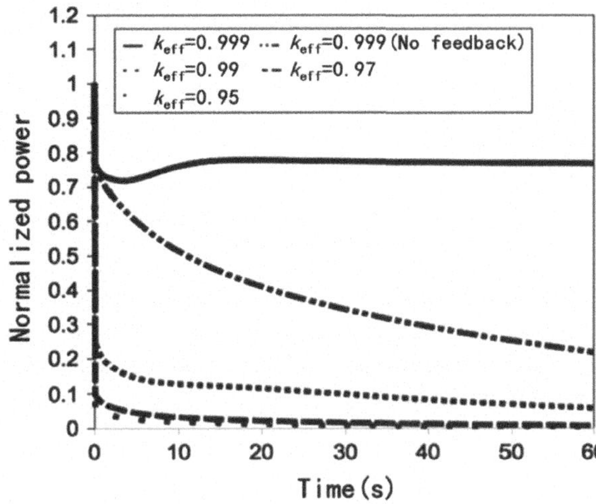

Fig. 11.18 Change in the subcritical reactor power without the external neutron source (XADS)

matically adjust the input variables according to the design objectives and then specify the design parameters) and multi-objective optimization methods have become effective approaches for optimizing the parameters.

11.3.2.3 System Shutdown and Power Control Method

The hybrid nuclear systems are subcritical, and the shutdown and power can be controlled by the regulation of the external neutron source and the subcritical reactor/blanket reactivity. The external neutron sources regulated through the parameters of the fusion plasma in an FDS and by adjusting the accelerator beam in an ADS. The subcritical reactor/blanket reactivity is usually regulated using control rods and burnable poison, such as in commercial fission systems. Due to the complex structure of FDS blankets, the design of control rods is difficult, while Li has been proposed as burnable poison. For a subcritical ADS reactor with a structure basically similar to a fission system, the reactivity control can be fully referred to those of the fission systems.

Taking EXperimental Accelerator-Driven System (XADS) as an example, without the consideration of the reactivity feedback, the shutdown of the subcritical reactor can be realized by stopping the external neutron source under different k_{eff}, as shown in Fig. 11.18. However, considering the reactivity feedback, the subcritical reactor cannot be shut down only by stopping the external neutron source with $k_{eff} = 0.999$. Therefore, when designing a hybrid nuclear system with a high k_{eff} value, it is necessary to introduce other reactivity control methods in the subcritical reactor.

In a hybrid nuclear system, the control ability of the control rod depends on the subcritical depth of the subcritical reactor/blanket [10]. When the external neutron source intensity is fixed, the power of system is

$$P \propto \frac{k_{\text{eff}}}{1 - k_{\text{eff}}} \tag{11.4}$$

When Δk is the change in k_{eff}, the relative power variation in the subcritical reactor/blanket is

$$\frac{P'}{P} = \frac{k_{\text{eff}} + \Delta k}{1 - k_{\text{eff}} - \Delta k} / \left(\frac{k_{\text{eff}}}{1 - k_{\text{eff}}}\right) = \frac{(k_{\text{eff}} + \Delta k)(1 - k_{\text{eff}})}{k_{\text{eff}}(1 - k_{\text{eff}} - \Delta k)} \tag{11.5}$$

The power output of an FDS is mainly related to the fusion power and the k_{eff} of the blanket, which can be expressed by Eq. (11.6). According to this equation, FDS power regulation can be achieved by adjusting the power of the fusion core and the reactivity of blanket.

$$P_{\text{FDS}} = P_{\text{fusion}} \frac{14.1\overline{\nu}(1 - k_{\text{eff}}) + k_{\text{eff}} E_{\text{fission}}}{17.6\overline{\nu}(1 - k_{\text{eff}})} \tag{11.6}$$

where E_{fission} is the average energy released per fission reaction in units of MeV and $\overline{\nu}$ is the average number of neutrons released per fission reaction.

For an ADS, the power mainly depends on the energy and the current of the incident protons and the k_{eff} of the subcritical reactor, which can be expressed by Eq. (11.7). According to this equation, ADS power regulation can be achieved by adjusting the energy and the current of the incident protons and the reactivity of the subcritical reactor.

$$P_{\text{ADS}} = \frac{k_{\text{eff}}}{1 - k_{\text{eff}}} \cdot \frac{E_{\text{fission}} Y \varphi^*_{\text{ADS}}}{\overline{\nu}} \cdot I_{\text{P}} \tag{11.7}$$

where $*_{\text{ADS}}$ is the ratio of the average spallation neutron source importance to the average fission importance.

11.3.3 Shielding Design

For the shielding design of FDSs, the design content and method are similar to those of fusion systems. For the shielding design of ADSs, compared with traditional fission systems, the radiation field of the system is more complex due to the coupling between the accelerator, the target and the subcritical reactor, which raises new problems and challenges for the shielding design [11].

In most ADSs, the beam tubes of the accelerator are inserted from the top of the subcritical reactor to the core, which results in the leakage of the neutrons and the photons in the system to the outside of the core, not only through the top of the system but also through the vacuum tube. In addition, when the accelerator operates, particles deviate from the original beam direction and react with the wall of the beam tube, resulting in secondary radiation (secondary neutrons). These leakages and secondary

Table 11.6 Main plasma parameters of FDS-I

Parameters	Value
Fusion power (MW)	150
Major radius (m)	4
Minor radius (m)	1
Aspect ratio	4
Plasma elongation	1.78
Triangularity	0.4
Plasma current (MA)	6.3
Toroidal magnetic field (T)	6.1
Safety factor	3.5
Auxiliary power (MW)	50
Energy multiplication	3
Average neutron wall load ($MW \cdot m^{-2}$)	0.5
Average surface heat flux ($MW \cdot m^{-2}$)	0.1

radiation not only make the activation of certain stainless steel parts on the top of the pile more serious but also cause certain important parts of the stack, such as the deflecting magnet and focusing magnet, to be damaged by neutron irradiation, thus endangering the safe operation of the accelerator. Therefore, secondary radiation is a key point that should be considered.

In the top shielding design of an ADS, the dominant factors in the radiation fields of different regions should be determined according to the design objectives and constraints and the characteristics of the radiation field at the top. The emphasis is placed on strengthening the local shielding for the more serious leakage areas (such as the outer area of the beam pipe above the top of the system) to achieve effective top shielding of the ADS.

11.4 Example of Neutronics Design

The neutronics design principles of ADSs and FDSs are similar, so FDS-I is taken as an example. The neutronics design of a hybrid nuclear system is elaborated in this section, and the shielding design content and method are similar to those of fusion systems, and will not be elaborated here.

FDS-I is a multifunctional hybrid nuclear system with the purposes of nuclear waste transmutation, nuclear fuel breeding, and energy production with relatively low plasma parameter requirements. Therefore, the main plasma parameters of FDS-I are between those of experimental facilities (JET, JT-60, etc.) and ITER, as shown in Table 11.6, and the design objectives of FDS-I are shown in Table 11.7.

Table 11.7 Neutronics design objectives of FDS-I [12, 13]

Items	Objective
k_{eff}	≤0.95
MPD (MW · m^{-3})	≤100
TBR	≥1.2
Initial MA/Pu/FP inventory (UPWR)[a]	≤300
Fuel consumption balance	To keep equivalence of consumed Pu, MA, FP
Transmuted waste per year	To maximize

[a]The unit UPWR is defined as the equivalent amount of identified isotopes from the spent fuel at the burnup of 33,000 MWD/MTU from a standard 3000 MW$_{\text{th}}$ PWR in a full-power year

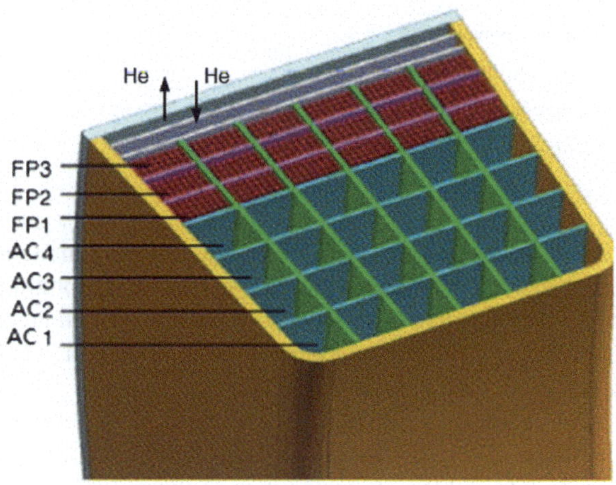

Fig. 11.19 Reference module of the DWTB

1. Material selections and configuration

A liquid blanket was selected as the blanket for FDS-I, as it has many advantages, such as an adaptable geometry, an easy manufacturability, a high engineering feasibility, and a high TBR, and enables the online extraction of tritium.

The helium gas and liquid lead–lithium (LiPb) eutectic dual-cooled waste transmutation blanket (DWTB) of FDS-I is presented based on previous studies [13], and the module structure of the outboard blanket is shown in Fig. 11.19. Helium gas is used to cool the first wall, structural walls, and long-lived fission product transmutation zones (FP zones). Liquid LiPb eutectic with tiny particle fuel is used in the actinide zones (AC zones). The LiPb in the AC zones serves as a coolant, tritium breeder, and fuel circulating carrier. The Pb presented in the LiPb is also used as a neutron multiplier.

The outboard blanket consists of four AC zones, three FP zones, and two helium manifolds in order of inside to out, and this blanket is used for transmutation, breeding, and energy production. The inboard blanket consists of two tritium breeding zones, a neutron reflector and a helium manifold. No heavy metal materials (MA

Table 11.8 Material compositions and radial sizes of the DWTB

Zones	Material component	Thickness (cm)
Inboard blanket		
FW	CLAM steel + He	3
Tritium breeding zone	LiPb	11 × 2
Structural walls	CLAM steel + He	1 × 2
Neutron reflector	Graphite	13
Helium manifold	CLAM steel + He	10
Shield layer	CLAM steel + H_2O	30
Outboard blanket		
FW	CLAM steel + He	3
AC1/AC2/AC3/AC4	MAC + PuC + LiPb	11/11/11/11
Structural walls	CLAM steel + He	1 × 6
FP (CsCl/NaI/Tc)	FP + graphite + He	7/7/7
Helium manifold	CLAM steel + He	11
Shield layer	CLAM steel + H_2O	60

or U) are arranged in the inboard blanket, which is used to breed tritium only. The basic material compositions and radial sizes of the DWTB in FDS-I are given in Table 11.8.

The blanket of FDS-I adopts a banana-shaped design, which is beneficial for avoiding the fuel loading and discharging problems of rod- or plate-type fuel designs. In view of its technical maturity and fuel design feasibility, a coated particle fuel that is similar to that in a high-temperature gas-cooled reactor (HTGR) has been adopted in FDS-I. This fuel circulates in the liquid LiPb coolant channel.

Actinides are designed in the form of carbide fuel with a SiC coating in FDS-I, considering the chemical and high-temperature compatibility between the fuel and LiPb.

For the LLFP, based on the available spent fuel separation technologies and material properties, metal technetium, NaI and CsCl are used in FDS-I. To increase the transmutation efficiency of the LLFP, a neutron moderator is added to the FP zones to form a thermal neutron spectrum. There are several types of neutron moderators, including hydrides ($ZrH_{1.7}$, CaH_2), graphite, and heavy water. Considering different aspects such as feasibility, economic competitiveness, and commercial scale, graphite is determined to be the most attractive choice for the LLFP transmutation in FDS-I.

The spent fuel in the hybrid nuclear systems is discharged from pressurized water reactors (PWR). To avoid complex separation technologies during nuclear fuel reprocessing, the heavy elements and fission products initially loaded in the blanket have the same fractions as the compositions of the spent fuel. The composition of spent fuel from a typical 3000 MW_{th} PWR is shown in Table 11.9.

China Low Activation Martensitic Steel (CLAM), vanadium alloys and SiC_f/SiC composites are potential candidate structural materials for the FDS-I blanket, since these materials have advantageous properties, such as irradiation swelling resistance, low-activation, high-temperature resistance, and irradiation resistance. CLAM steel was the first choice for the structural material of FDS-I, and Al_2O_3 was selected as the coating material to reduce tritium permeation and prevent LiPb corrosion.

2. Neutronics calculation and optimization

Calculations and optimizations related to neutron transport, burnup, and irradiation damage were conducted based on the FDS-I model, as shown in Fig. 11.20 [13].

(1) MA and Pu loading without separation

The fractions of MA and Pu in the AC zones of FDS-I were the same as those of the spent fuel from the PWR. The objective of the optimization was to achieve the maximum transmutation capacity with the minimum initial MA loading and maintain the transmutation amounts of MA and Pu at similar levels.

The neutronics performance analysis after 1-year operation was performed by changing the initial loading amounts of MA and Pu, and the results are shown in Fig. 11.21. With an initial loading amount of 200 UPWR of MA and Pu, the k_{eff}, *MPD*, and *TBR* were obtained to be approximately 0.946, 126 MW/m^3, and 4.3, respectively. In addition, the transmutation amounts of MA and Pu were 24.4 UPWR and 30.5 UPWR, respectively. Therefore, it was difficult to achieve similar transmutation amounts for MA and Pu loading without separation [13].

Table 11.9 Composition of spent fuel from a typical 3000 MW_{th} PWR (after 10 years of cooling)

Spent Fuel	Nuclide	UPWR (kg)		Notes
MA	^{237}Np	14.5	34.7	Collectively refers to as TRU
	^{241}Am	16.6		
	^{243}Am	3.0		
	^{244}Cm	0.6		
Pu	^{238}Pu	4.5	288.1	
	^{239}Pu	166.0		
	^{240}Pu	76.7		
	^{241}Pu	25.4		
	^{242}Pu	15.5		
LLFP	^{99}Tc	25.69	41.65	–
	^{129}I	5.96		
	^{135}Cs	10		

Fig. 11.20 Calculations and optimization model of FDS-I

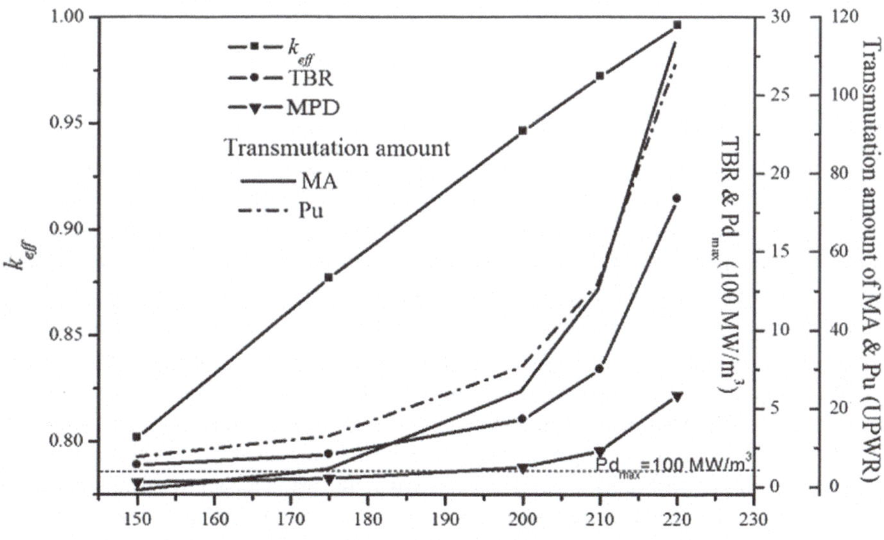

Fig. 11.21 Neutronics performance as a function of the initial loading of MA and Pu

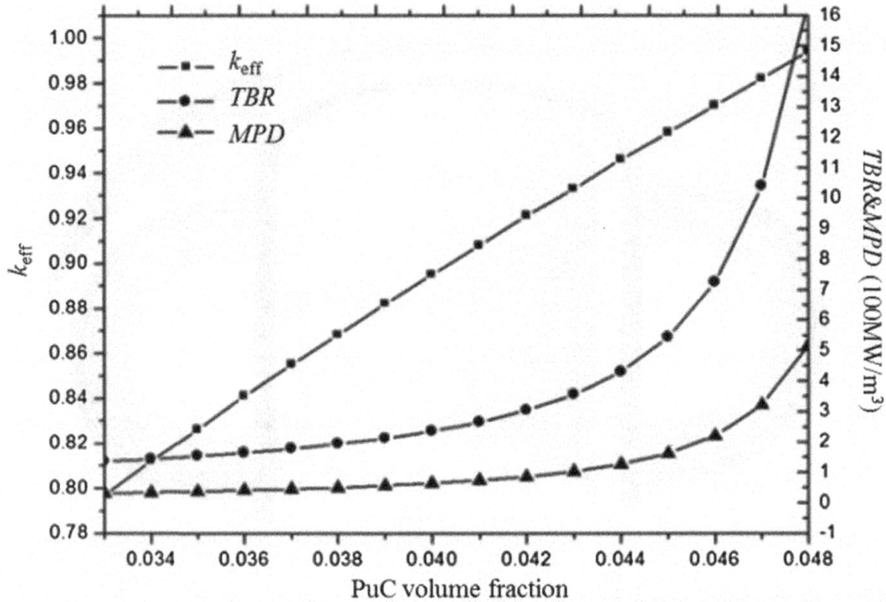

Fig. 11.22 Neutronics performance as a function of the PuC volume fraction (with a fixed MAC volume fraction)

(2) MA and Pu loading with separation

The neutronics performance is more sensitive to the loading of Pu than that of MA. The impact of the initial loading of Pu on the neutronics performance was analyzed by setting the initial loading of MA to 200 UPWR. Figure 11.22 shows the influence of the PuC volume fraction on the neutronics performances in the AC zones. *MPD* reached the maximum value with a PuC volume fraction of 4.3% (approximately 197 UPWR). k_{eff} was approximately 0.93, TBR was approximately 3.6 and the transmutation amounts of MA and Pu were 17 UPWR and 26 UPWR, respectively.

To maintain similar transmutation amounts of MA and Pu, the influence of the MAC volume fraction on the neutronics performance was assessed, as presented in Fig. 11.23. The initial loading volume fraction of PuC was 4.3%. When the initial loading amount of MA was 238 UPWR (point A in Fig. 11.23), the transmutation amount was 24 UPWR for both MA and Pu; moreover, the TBR was 3.4, and MPD was 100 MW/m^3. The main neutronics design objectives of FDS-I were achieved [13].

(3) Optimization of LLFP loading

The transmutation of the LLFP in FDS-I mainly benefits from the thermal neutrons through the AC zones. The neutron capture cross sections of ^{135}Cs, ^{129}I, and ^{99}Tc under fast and thermal neutron spectra are given in Table 11.10. To maintain similar

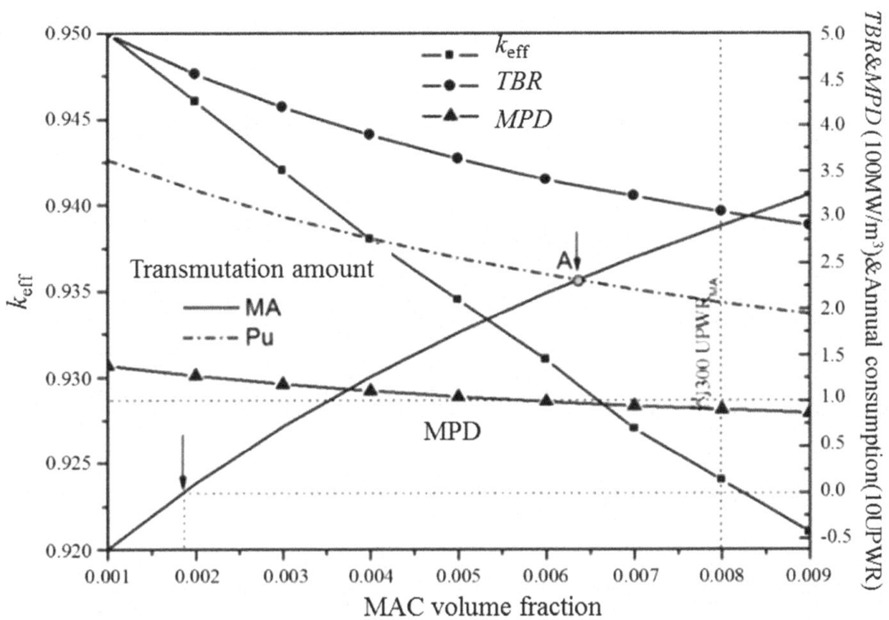

Fig. 11.23 Neutronics performance as a function of the MAC volume fraction (with a fixed PuC volume fraction)

Table 11.10 Neutron capture cross sections of the LLFP

Neutron capture cross section (b)	^{99}Tc	^{129}I	^{135}Cs
Fast neutron spectrum	0.2	0.14	0.07
Thermal neutron spectrum	4.3	4.3	1.3

transmutation amounts of the three isotopes, the three LLFP zones (CsCl, NaI, and metallic Tc) were successively arranged in the radial direction.

The influence of the initial loading of LLFP and graphite on the transmutation of the LLFP is presented in Fig. 11.24. The transmutation capacity increased with increased amounts of LLFP and graphite loading. However, the burnup levels of the three main LLFP isotopes could not reach the same magnitude. When the initial loadings of MA, Pu, and LLFP were approximately 200 UPWR each, the transmutation amounts of ^{135}Cs, ^{129}I, and ^{99}Tc were 7, 9, and 13 UPWR, respectively, after 1-year operation.

The neutronics performance of FDS-I after 3-year operation based on the optimization approach above is presented in Table 11.11.

The transmutation capacity significantly decreased after 1-year operation without refueling, as shown in Table 11.11. The TBR decreased below 1.0 after a 2-year operation period, which did not meet the requirements for tritium self-sufficiency. During the third year, ample quantities of MA were produced that eventually sur-

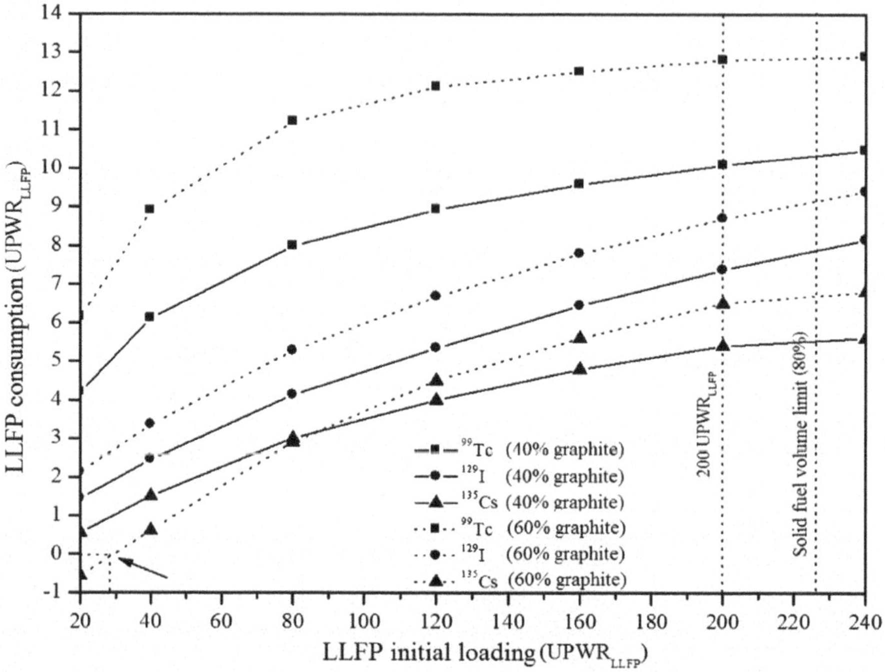

Fig. 11.24 Transmutation of the LLFP as a function of the initial loading

Table 11.11 Main neutronics performance of FDS-I after 3-year burnup

Unit: year	k_{eff}	*MPD* (MW · m^{-3})	*TBR*	Initial loading/(UPWR)		Transmutation capacity (UPWR)	
				MA	Pu	MA	Pu
0	0.93	100	3.4	238	197	–	–
1	0.80	31	1.17	214.3	173.2	23.7	23.8
2	0.76	25	0.98	214.2	167.7	0.1	5.5
3	0.73	22	0.88	215.2	164.2	−1	3.5

passed the burned quantity of MA. Thus, the scenario did not satisfy the design objectives. Therefore, the recommended refueling period was 1 year [13].

(4) Multi-objective optimization

A multi-objective optimization of the neutron parameters of FDS-I was carried out based on a genetic algorithm. Considering the flow characteristics of the LiPb coolant in the AC zones, the total volume fractions of minor actinide carbide (MAC) and plutonium carbide (PuC) would not exceed 10%. The waste transmutation ratio of MA (WTR_{MA}) and the annual burnup depth (ABUD) were classified as the optimization parameters.

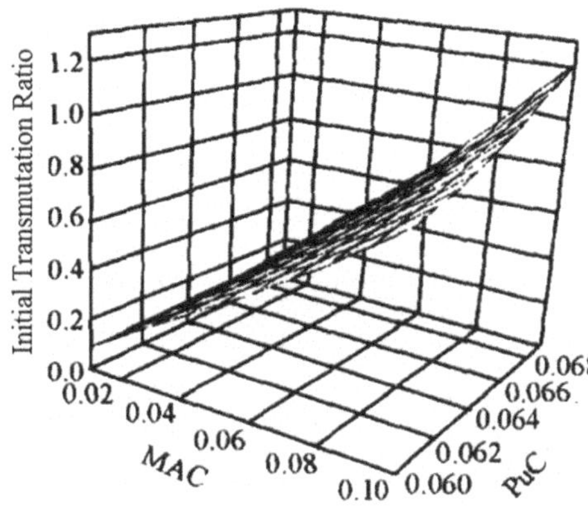

Fig. 11.25 Influence of the fraction of MAC and PuC on the WTR

1) WTR

After increasing the initial loading amounts of MAC and PuC in the FDS-I blanket, the WTR was improved, and the neutron design objectives were satisfied. There was a positive correlation between the WTR and the loading of MAC and PuC, as shown in Fig. 11.25 [14].

2) ABUD

Because of its strong radioactivity and long half-life, it is necessary to minimize the MA loading to improve the safety of hybrid nuclear systems. At the same time, it is necessary to minimize the initial TRU loading and improve the waste transmutation efficiency. Here, we define the ABUD of MA as the ratio of the annual average burnup of MA to the loading amount of transuranic nuclides.

The optimized neutronics parameters and material compositions of the AC zones are shown in Table 11.12 for refueling periods of 1 year, 3 years, and 30 years. When the refueling period was 30 years, the TBR was lower than 1.2, which would not achieve the design objectives.

The dependence of the ABUD on the MAC and PuC loadings for a refueling period of 1 year is shown in Fig. 11.26. As shown in the figure, the ABUD is more sensitive to the MAC fraction than to the PuC fraction. However, the heat released during the transmutation of MA exceeds the maximum power density if the MAC fraction is very high [14].

(5) Activation and dose analysis

One of the design objectives of FDS-I is that the production of additional long-lived radionuclides should be as low as possible. No extra MA is generated with a proper blanket design. The use of low-activation steel as the structural material also reduces the production of long-lived activation products. The irradiation time was set to 30 years for the activation calculations, and the exposure dose rates of the first wall are shown in Fig. 11.27. The residual heat distribution in the main zones of the blanket is shown in Fig. 11.28.

Table 11.12 Neutronics parameters and material compositions of the AC zones

Refueling period (year)	Initial loading fraction (%)			Neutronics performance			Actinide initial loading (kg)	Pu initial loading (kg)	ABUD (%/a)
	MAC	PuC	LiPb	k_{eff}	*TBR*	*MPD* ($MW \cdot m^{-3}$)			
1	3.0	6.6	90.4	0.85	1.50	98	17169	37053	2.5
3	3.4	6.6	90.0	0.84	1.39	95	19387	37321	2.3
30	4.3	5.6	90.1	0.76	0.98	63	24571	31635	0.7

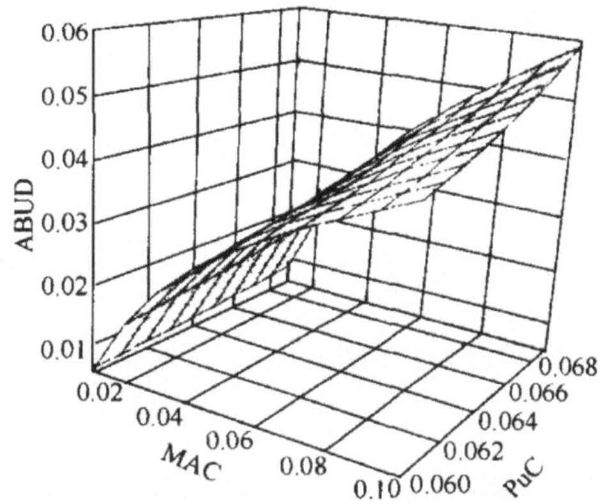

Fig. 11.26 Dependence of BUDA on the MAC and PuC loadings

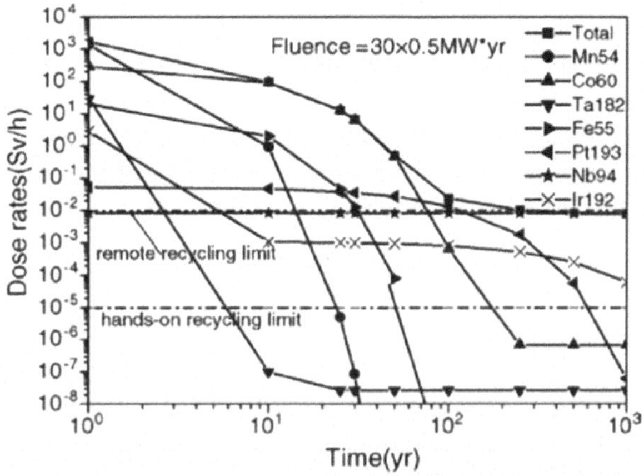

Fig. 11.27 Exposure dose rates of the first wall

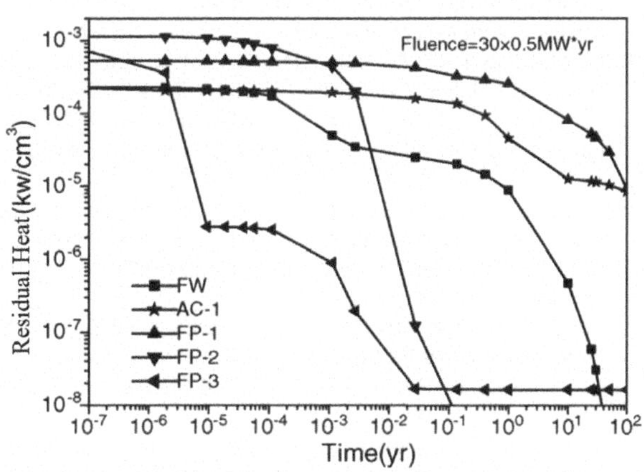

Fig. 11.28 Residual heat in the main zones of the DWTB [12]

Table 11.13 Irradiation damage to materials in FDS-I

	Shielding zone of the inboard blanket	First wall of the outboard blanket
Structure material	CLAM	CLAM
Atom displacement damage/(dpa/year)	4.84×10^{-2}	28.76
Helium production rate/(appm/year)	4.34×10^{-2}	40.94
He/DPA ratio/(appm/dpa)	0.90	1.42

(6) Irradiation damage to materials

The DPA and helium production rates in the structural material of FDS-I after 1-year full-power operation are presented in Table 11.13. As indicated in the table, the irradiation damage to the shielding zones of the inboard and outboard blankets differs considerably because of the different neutron spectra in those areas. The shielding zone of the inboard blanket is far from the plasma, resulting in slight irradiation damage, whereas the first wall of the outboard blanket suffers the most severe irradiation damage.

References

1. Wu YC (2017) Fusion neutronics. Springer Nature Singapore Pte. Ltd
2. Rubbia C et al (1995) Conceptual design of a fast neutron operated high power energy amplifier. CERN/AT/95-44(ET)
3. Abderrahim HA, Kupschus P, Malambu E et al (2001) MYRRHA, a multipurpose accelerator driven system for research and development. Nucl Instrum Meth A 463(3):487–494
4. Hill D, Van Tuyle G, Beller D (1999) A roadmap for developing atw technology: systems scenarios & integration. Office of Scientific &Technical Information Technical Reports
5. Takizuka T, Nishida T, Mizumoto M et al (1995) JAERI R&D on accelerator-based transmutation under OMEGA program. In: AIP conference American Institute of Physics, vol 346, pp 64–73
6. Wu YC, Bai YQ, Song Y et al (2016) Development strategy and conceptual design of China lead-based research reactor. Ann Nucl Energy 87:511–516
7. Wu YC (2006) Conceptual design activities of FDS series fusion power plants in China. Fusion Eng Des 81(23–24):2713–2718
8. Peng XJ, Wang Z (2014) Conceptual research on Z-pinch driven fusion-fission hybrid reactor. High Power Laser Part Beams 26(09):1–6 (in Chinese)
9. Sahin S, Sahin HM, Acir A (2012) LIFE hybrid reactor as reactor grade plutonium burner. Energy Convers Manag 63:44–50
10. Wang MH (2005) Neutronics design and study on subcritical spent fuel burner. Dissertation, University of Science and Technology of China (in Chinese)
11. Li B, Yang Q, Chang B et al (2016) Preliminary analysis of radiation characteristic at upper section of accelerator driven subcritical system. Ann Nucl Energy 90:410–416
12. Wu YC, Zheng SL, Zhu XX et al (2006) Conceptual design of the fusion-driven subcritical system FDS-I. Fusion Eng Des 81(8):1305–1311

13. Zheng SL (2005) Study on neutronics safety characteristic of the fusion-driven subcritical system with the dual-cooled transmutation blanket. Dissertation, Institute of Plasma Physics Chinese Academy of Sciences (in Chinese)
14. Li JJ, Wu YC (2004) Neutronics optimization based on genetic algorithm for the initial loading long-lived minor actinides in the blanket for the fusion-driven subcritical system. Chin J Nucl Sci Eng 24(2):171–177 (in Chinese)

Part III
Neutronics Experimental Technologies of Advanced Nuclear Systems

Chapter 12
Neutronics Experimental Facilities and Technologies

Neutronics experiments are important parts of neutronics research for nuclear systems. Neutronics experimental facilities contain accelerator-based neutron sources and nuclear systems. Neutronics experiments include nuclear data measurement and validation, validation of methods and software, validation of reactor core physics design of nuclear systems, etc. To carry out various types of neutronics experiments, it is necessary to build neutronics experimental facilities and develop neutron control and measurement technologies.

This chapter is divided into three parts: the first part is about facilities for neutronics experiments, the second part illustrates neutron control technologies, and the third part introduces neutron measurement technologies. The experimental technologies for nuclear systems will be introduced in the following chapters.

12.1 Facilities for Neutronics Experiments

The facilities for neutronics experiments can be divided into isotope neutron sources, accelerator-based neutron sources, and nuclear systems. Isotope neutron sources, seldom used in neutronics research for advanced nuclear systems due to low neutron intensity, will not be introduced in this chapter.

12.1.1 Accelerator-Based Neutron Sources

Accelerator-based neutron sources use accelerated ions to bombard specific targets for neutron production. Neutrons with different energies can be obtained by changing the type and energy of ion beam and the type of target material. According to the characteristics of the neutron spectrum, accelerator-based neutron sources can be divided into monoenergetic neutron sources and white neutron sources. Monoenergetic neutron sources produce monoenergetic neutrons. For example, an accelerator-based

Y. Wu, *Neutronics of Advanced Nuclear Systems*,
https://doi.org/10.1007/978-981-13-6520-1_12

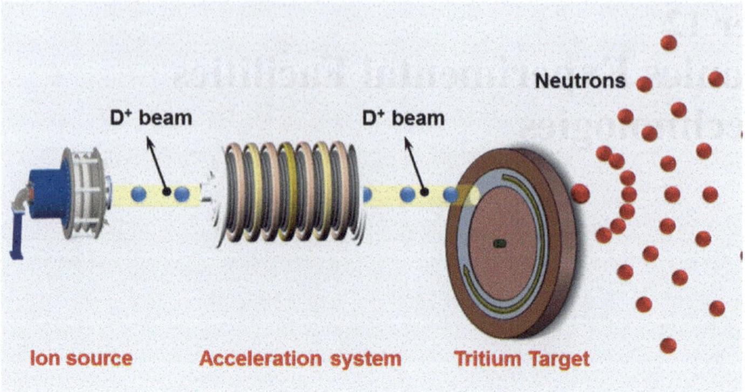

Fig. 12.1 Accelerator-based fusion neutron source

D-T or D-D fusion neutron source is a monoenergetic neutron source, which produces monoenergetic fusion neutrons of approximately 14 MeV/2.5 MeV. White neutron sources can produce neutrons with a wide range of energies, including spallation neutron sources (from the ultracold energy region to the GeV energy region) and electron linac-driven neutron sources (10^{-3} eV–10 MeV). The advantages of an accelerator-based neutron source include high-intensity neutron yield, low gamma-ray background, and steady or pulsed mode operation.

1. Accelerator-based fusion neutron sources

Accelerator-based fusion neutron sources usually refer to accelerator-based D-T or D-D fusion neutron sources, in which deuterium ions accelerated to several hundred keV bombard a tritium or deuterium target to produce monoenergetic neutrons of approximately 14 MeV/2.5 MeV (Fig. 12.1). The main components of an accelerator-based fusion neutron source include the ion source system, the acceleration system, the target system, etc. The accelerator-based fusion neutron source can be designed to operate in steady or pulsed mode.

Accelerator-based fusion neutron sources that produce the same neutron energy are a suitable platform for fusion neutronics experiments. The typical running accelerator-based D-D and D-T fusion neutron sources in the world are listed in Table 12.1.

In the following section, HINEG will be introduced briefly. In order to carry out the research of the neutron physics and radiation safety, key technologies of advanced nuclear systems, and nuclear technology application, the FDS Team has launched HINEG (High Intensity D-T Fusion Neutron Generator) project [3], as shown in Fig. 12.2, which produced a D-T fusion neutron yield of up to 6.4×10^{12} s^{-1}. HINEG has the highest level neutron yield in the word among accelerator-based fusion neutron sources under operation, while the NG-430 with 5.0×10^{12} s^{-1} is the second highest under operation. The control technology for neutron yield based on the nuclide component stratification has been developed. The target was developed with enhanced heat dissipation technology based on the array jet coupled strong shear

Table 12.1 Worldwide list of typical running accelerator-based D-D or D-T fusion neutron sources [1, 2]

Institute	Facility	Neutron yield (s^{-1})	Beam energy on target (keV)	Beam current on target (mA)
China • The FDS team	HINEG	6.4×10^{12}	300–400	35
Russia • BINP	NG-430	5×10^{12}	430	20
Germany • TUD	TUD	1×10^{12}	320	10
China • CIAE	CPNG-6	3×10^{11}	200–600	~1
China • CAEP	PD-300	2×10^{11}	300	2
Italy • ENEA	FNG	1×10^{11}	230	1

Fig. 12.2 HINEG facility

flow field. In addition, a high-power high-voltage electrostatic ion acceleration tube with double-electron-clouds space charge lens was adopted. The operational stability of HINEG facility is enhanced. At the same time, HINEG has been operated to drive the Lead-based Zero Power Critical/Subcritical Reactor CLEAR-0. Thanks to the neutron control technologies, the complex neutron spectra of multi-type advanced nuclear systems can be reproduced. Neutronics experiments of fusion nuclear systems, lead-based fission nuclear systems, hybrid nuclear systems, and performance test of nuclear instruments, neutron radiation resistance test of electronic components and new materials, shielding properties of new composite materials, neutron breeding, neutron radiography nondestructive testing, and biological effects of radiation have been carried out in this facility.

Table 12.2 Worldwide list of running spallation neutron sources [4–8]

Country	Facility	Proton beam power (MW)	Proton beam energy (GeV)	Repetition rate (Hz)
USA	SNS	1.4	1	60
Japan	JSNS	1	3	25
Switzerland	SINQ	1	0.59	CW
UK	ISIS	0.16	0.8	50
China	CSNS	0.1	1.6	25
USA	LANSCE	0.056	0.8	20

Furthermore, a newtype multilayer composite target with different functional regions was developed, which has excellent tritium storage capacity and higher tritium utilization. At the same beam condition, the neutron yield of this target is more than 30% higher than the traditional target. To conduct research on high neutron flux material irradiation experiments and component nuclear performance tests, the FDS Team has been working on the second phase of HINEG, which aims to achieve a neutron yield of 10^{15}–10^{18} s^{-1}.

2. Spallation neutron sources

For spallation neutron sources, light ions (protons or deuterons) with an energy range from hundreds of MeV to several GeV are used to bombard heavy nuclei to produce neutrons with white neutron spectra through the spallation reaction. Spallation neutron sources, which produce neutrons with a wide range of energies, can be propitious to materials and life science research by neutron scattering methodology, nuclear data measurement study, etc. Spallation neutron sources have been developed by the United States, Japan, Switzerland, China, and other countries. Table 12.2 lists the main parameters of the running spallation neutron sources around the world.

In the following section, the China Spallation Neutron Source (CSNS) will be introduced.

CSNS was designed and constructed by the Institute of High Energy Physics, Chinese Academy of Sciences (Fig. 12.3). In the CSNS, a proton beam with energy of 1.6 GeV bombards a tungsten target to produce neutrons. This facility operates in a pulsed mode at a pulse frequency of 25 Hz. The CSNS is a typical facility providing the platform for research on many frontier disciplines related to physical properties, microstructures, and dynamics of materials, such as high-temperature superconductivity materials and mechanisms, rare earth permanent magnet materials, nanoscience and nanotechnology, gene and protein engineering, etc. In addition, the CSNS can also be used as a white neutron source (1 eV–20 MeV) to perform nuclear data measurement and other nuclear research.

3. Electron linac-driven neutron sources

In an electron linac-driven neutron source, a linac is used to generate a high-energy electron beam that bombards a thick metal target with a high atomic number to produce strong bremsstrahlung radiation. Neutrons are produced by the (γ, n) reactions.

Fig. 12.3 CSNS facility [9]

Table 12.3 Worldwide list of electron linac-driven neutron sources

Country	Facility	Neutron yield (peak) (s^{-1})	Target	Beam energy (MeV)	Beam current (peak) (A)	Pulse width (ns)	Repetition rate (Hz)
Belgium	GELINA	6×10^{19}	U	120	~100	<1	900
USA	ORELA	4.0×10^{18}	Ta	140	15	3	1000
USA	LIVERMORE	2.3×10^{18}	Ta	115	10	5	1440
USA	HELIOS	2.2×10^{18}	U	94	6	5	2000
Japan	JAERI	1.4×10^{18}	Ta	120	6	10	600
USA	N. B. S	1.0×10^{18}	Ta	120	4	5	720
Russia	KURCHATOV	2.9×10^{17}	U	60		50	900
USA	R. P. 1	1.6×10^{17}	Ta	80	1	10	720
Ukraine	KIPT	3.3×10^{14} [a]	U/W	100	0.6	2700	625
China	TPNS1	1.2×10^{11} [a]	W	15		3	266

Note [a] The average neutron yield

This kind of neutron source can provide white neutrons and reach a peak neutron yield of 10^{19} s^{-1}. The neutron pulse width can reach nanoseconds, which is suitable for high-resolution time-of-flight measurement. In the past 20 years, a large amount of high-resolution neutron cross section data has been obtained on electron linac-driven neutron sources. Table 12.3 lists the main parameters of the electron linac-driven neutron sources across the world.

In the following section, the Geel Linear Accelerator (GELINA) in Belgium will be presented briefly [10]. To conduct radionuclide measurements, the GELINA has

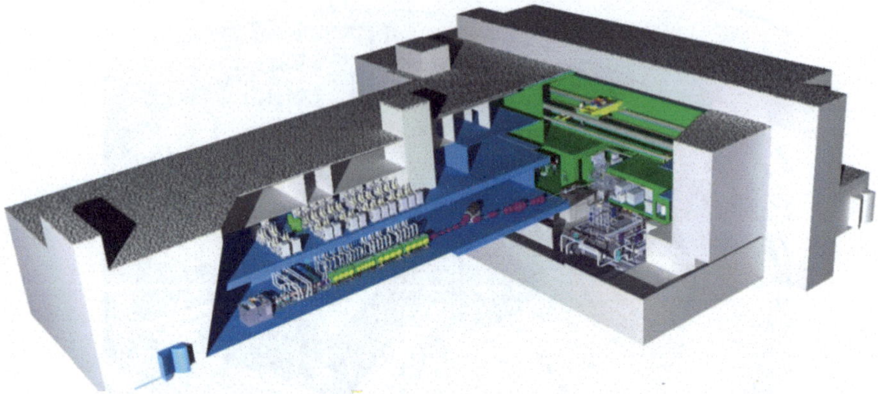

Fig. 12.4 The design diagram of IFMIF-DONES facility [11, 12]

been developed at the IRMM (Institute of Reference Materials and Measurement). For the GELINA, the intense pulsed electron beams are accelerated up to 120 MeV in an electron linac. After compression, these high-energy electrons bombard a rotating mercury-cooled uranium target. Colliding with the target, the electrons produce bremsstrahlung radiation in the target and subsequently neutrons by the (γ, n) and (γ, f) reactions. The distribution of neutrons has a peak at 1–2 MeV. The GELINA facility is used mainly for high-resolution nuclear data measurement and code validation experiments.

4. Other accelerator-based neutron sources

In addition to the neutron sources described above, some other accelerator-based neutron sources have been built or are now being built. Based on the preliminary engineering design of IFMIF, the EU has proposed a development plan for the IFMIF-DONES (DEMO Oriented Neutron Source) facility (Fig. 12.4) for studying the irradiation of fusion materials. In the DONES, neutrons will be generated through the Li(d, xn) reaction by bombarding liquid lithium with accelerated deuterium ions. The average energy of the neutrons produced by the IFMIF-DONES is approximately 14 MeV. The energy of the beam (40 MeV) and the current of the accelerator (125 mA) have been tuned to maximize the neutron flux (10^{14} $cm^{-2} \cdot s^{-1}$) and obtain the irradiation conditions. The DONES will be used to simulate the equivalent irradiation conditions of materials in commercial fusion systems and test the long-term service behavior of materials under extreme conditions.

12.1.2 Nuclear Systems for Neutronics Experiments

Nuclear systems for neutronics experiments are an important platform for the research in advanced nuclear systems and can also promote the development of materials, information, life sciences, and other subjects. Nuclear systems for neutronics experiments can be divided into fission nuclear systems, fusion nuclear systems, and hybrid nuclear systems. In fission systems for neutronics experiments, the neutron radiation field is formed by fission reactions. The neutron radiation field is formed by D-T or D-D fusion reactions in fusion systems for neutronics experiments of fusion nuclear technologies and experiments. Hybrid nuclear systems for neutronics experiments usually use fusion facilities or accelerator neutron source to drive a subcritical reactor core.

1. Fission systems for neutronics experiments

Fission systems for neutronics experiments utilize mainly the high-flux neutron field generated by the nuclear systems to carry out isotope production, material irradiation, testing of nuclear components, and nuclear data measurement. The typical fission systems for neutronics experiments in the world include the HFIR (High-Flux Isotope Reactor) of America, the HFETR (High Flux Engineering Test Reactor) of China, the CARR (China Advanced Research Reactor) and the CMRR (China Mianyang Research Reactor), the BOR-60 of Russia, etc. In the following section, HFIR will be introduced briefly.

The HFIR was a multipurpose 85 MW isotope production and test reactor built by Oak Ridge National Laboratory in the 1960s (Fig. 12.5). HFIR is a beryllium-reflected reactor using highly enriched uranium as fuel. The reactor core assembly is found in a pressure vessel 2.44 m in diameter in a pool. The reactor core consists of a series of concentric rings, 380 mm in diameter, and 0.61 m in height. The fast neutron flux of the HFIR irradiation channel is $1.1 \times 10^{15}\ \mathrm{cm}^{-2} \cdot \mathrm{s}^{-1}$, and the thermal neutron flux is $2 \times 10^{15}\ \mathrm{cm}^{-2} \cdot \mathrm{s}^{-1}$. HFIR was used mainly for isotope production, material irradiation, experiments on neutron activation, and neutron scattering, and it can be used for basic research and applied research in the fields of materials, bioscience, etc.

2. Fusion systems for neutronics experiments

Fusion systems for neutronics experiments are important platforms for the research on materials testing for future fusion systems. Although some accelerator-based white neutron sources can provide high-energy neutrons with high neutron flux, the shape of the neutron source and the neutron spectrum are quite different from the fusion system environment, and the fusion systems of higher neutron flux are needed for the research and development of the nuclear technology and materials for fusion systems. The typical fusion systems for neutronics experiments in the world include the ITER, the FVNS (Fusion Volumetric Neutron Source) of China, the FNS-ST (a compact tokamak Fusion Neutron Source based on a small Spherical Tokamak) and the DTNS (Dynamic Trap Neutron Source) of Russia, Fusion Nuclear Science Facility (FNSF) of America, etc. At present, these fusion systems for neutronics experiments are in the design stage. The ITER facility has been introduced in the previous chapter. In the following section, FVNS will be introduced briefly.

Fig. 12.5 Aerial photo of the HFIR facility [13]

To develop the fusion nuclear technology and materials, the FDS Team proposed the development plan of the Fusion Volumetric Neutron Source (FVNS). The designed D-T neutron yield could reach above 10^{18} s^{-1}. The FVNS facility (Fig. 12.6) has a total length of 20 m, fusion power of approximately 3 MW, a neutron wall load of more than 2 MW · m^{-2} in the high-flux test area, a test volume of more than 10 L for a neutron wall load greater than 2 MW · m^{-2}, and a test volume of 100 L for the neutron wall load range of 1–2 MW · m^{-2}. The FVNS design adopts the design proposals of high magnetic field neutral beam injection and large mirror ratio (R ~ 100) to improve the fusion power density and energy gain. The FVNS is used mainly to test fusion materials and blanket in the complex and extreme environment of fusion nuclear systems and to solve the key scientific and technological problems in the engineering application of fusion energy.

3. Hybrid nuclear systems for neutronics experiments

To carry out experimental research on hybrid nuclear systems, hybrid nuclear systems for neutronics experiments usually use neutrons generated by a fusion driver or an accelerator neutron source to drive a subcritical reactor. The typical hybrid nuclear systems for neutronics experiments in the world include the KUCA (Kyoto University Critical Assembly) of Japan, the YALINA of Belarus, the VENUS-F of Belgium, the MASURCA (Mock-up facility for fast breeder reactor studies at CADARACHE) of France, the CLEAR-0 (China Lead-based Zero Power Critical/Subcritical Reactor) and the "VENUS" facility of China, etc. The purpose of KUCA and MASURCA was to carry out fission systems experiments in the initial stages of construction.

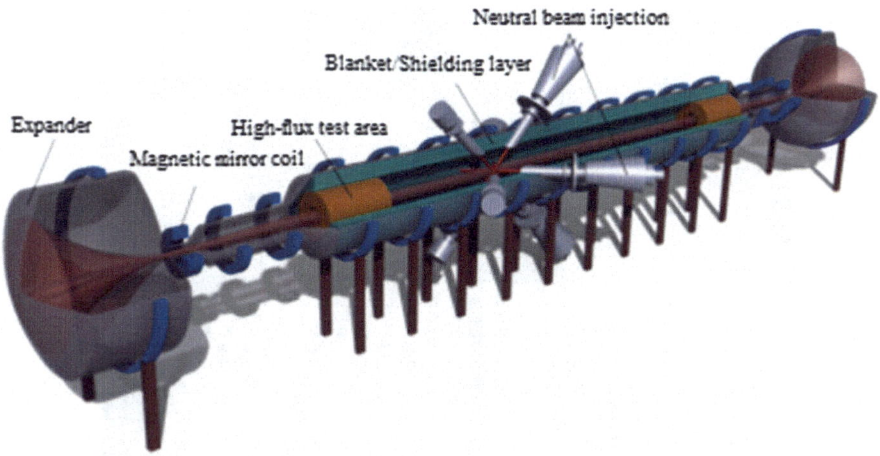

Fig. 12.6 The design diagram of the FVNS facility

In the later stages, hybrid nuclear systems experimental studies were carried out by upgrading facilities. In the following section, KUCA will be introduced briefly.

The KUCA facility [14] (Fig. 12.7) at Kyoto University, Japan, is a multicore-type critical device with two cores with solid moderator and one core with light water moderator. The KUCA facility is a zero-power experimental facility. The core of KUCA can be coupled with an accelerator to study the accelerator-driven subcritical system ADS. Currently, there are two types of accelerators in the KUCA facility: a 150 MeV proton accelerator and a D-T accelerator that produces neutrons of approximately 14 MeV. The core height of the KUCA facility is 40 cm, and there are 50-cm-thick polyethylene reflectors above and below the fuel zone. The fuel assembly is a cuboid, and the fuel is enriched uranium dioxide. To shield high-energy neutrons, some shielding materials, including polyethylene and ^{10}B, were put into the target and active regions. The KUCA facility can carry out spatial distribution of neutron flux and neutron spectrum measurement experiments in the reactor core, neutron spectrum evolution measurement experiments with time at different positions in the reactor, etc.

12.2 Neutron Measurement Technologies

There is no essential difference between the neutron measurement methods of the advanced nuclear energy systems and the traditional nuclear energy systems. The neutron radiation field of advanced nuclear energy systems is more complex, and the average neutron energy is higher, which increases the difficulty of neutron measurement and the complexity of measurement systems. It is necessary to improve the performance of neutron measurement instruments of traditional nuclear energy

Fig. 12.7 KUCA core [15]

systems. This section focuses on the neutron measurement methods, measurements of neutron flux, and measurements of the neutron spectrum.

12.2.1 Principles

Different from charged particles, neutrons can neither directly ionize nor excite atoms because they are electrically neutral. The electrical neutrality of neutrons makes it easy to approach closely and interact with nuclei. In some cases, this interaction produces charged particles, and in other cases, the nucleus recoils or emits radiation. All these occurrences make the detection of neutrons possible. In the following sections, we present the most common measurement methods and distinguish the following four different classes of methods based on the physical processes, namely, secondary particle emission method, the nuclear recoil method, the fission method, and the activation method.

1. The secondary particle emission method

The most commonly used method of detecting neutrons is the measurement of the secondary particles (e.g., charged particles or γ rays) emitted in neutron-induced reactions. The most frequently used nuclear reactions include $^{3}He(n, p)^{3}H$, $^{6}Li(n, t)^{4}He$, and $^{10}B(n, \alpha)^{7}Li$. The cross sections of these reactions are presented in Fig. 12.8, which shows that the cross sections in the low and thermal neutron energy regions are proportional to $1/v$. Detectors that follow the $1/v$ pattern are usually called $1/v$ detectors, such as ^{3}He proportional counters, Li-glass scintillator detectors, and scintillators doped with gadolinium. Alternatively, threshold nuclear reactions are used,

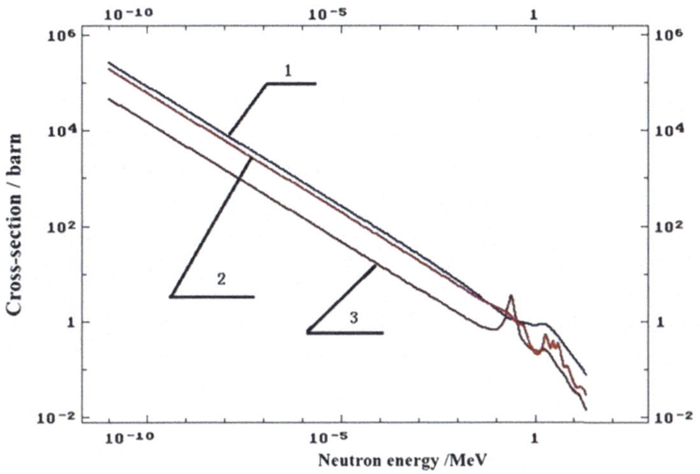

Fig. 12.8 Cross sections for the neutron-induced production of ^{3}He, ^{6}Li, and ^{10}B, 1 ^{3}He(n, p), 2 ^{10}B(n, α), 3 ^{6}Li(n, t)

such as ^{12}C(n, α)^{9}Be in a diamond detector. For the detection of fast neutrons (with energies of >1 MeV), the neutrons are usually moderated into thermal neutrons to improve the detection efficiency.

$$n + {}^3\mathrm{He} \rightarrow p + {}^3\mathrm{H} + 0.765\,\mathrm{MeV} \quad \sigma_0 = 5333 \pm 7\mathrm{b} \tag{12.1}$$

$$n + {}^6\mathrm{Li} \rightarrow \alpha + {}^3\mathrm{H} + 4.786\,\mathrm{MeV} \quad \sigma_0 = 940 \pm 4\mathrm{b} \tag{12.2}$$

$$n + {}^{10}\mathrm{B} \rightarrow \alpha + {}^7\mathrm{Li} + 2.792\,\mathrm{MeV} \quad \sigma_0 = 3837 \pm 9\mathrm{b} \tag{12.3}$$

2. The nuclear recoil method

The fundamental principle of the nuclear recoil method is that when a neutron interacts with a nucleus via elastic scattering, the nucleus acquires part of the energy of the neutron. Then, the matter around it will be ionized or excited if the nucleus acquires sufficiently high energy, producing a detectable signal enabling the detection of the neutron. Since the atomic mass of the neutron is $A = 1$, the conservation of kinetic energy in the elastic scattering process dictates that the atomic mass of the recoil nucleus should be as small as possible to maximize the energy that the recoiling nucleus acquires. Therefore, hydrogen is the most frequently used medium for detecting fast neutrons. Existing detectors that employ the nuclear recoil method include proportional counters filled with hydrogen, proton recoil telescopes, organic scintillators nuclear emulsions, etc.

3. The nuclear fission method

The principle of the nuclear fission method is that a neutron interacts with a heavy nucleus and induces a fission reaction, and the fission fragment can be detected to identify the neutron. Because an average energy of approximately 200 MeV is released in the fission of a heavy nucleus such as ^{235}U, these fission fragments will have a very high energy and are thus capable of inducing ionization in the surrounding materials. The classic detector based on the nuclear fission method is called fission chamber. The fission chamber has characteristics of high-amplitude pulse output, anti-γ-ray background, and low electric noise. The heavy nuclei that are sensitive to thermal neutrons include ^{235}U and ^{239}Pu, whereas ^{238}U and ^{237}Np are sensitive to fast neutrons.

4. The activation method

The fundamental principle of the activation method, which is passive, is that some of the nuclei in a material interact with neutrons and are transformed into a new radioactive element. Information about the incident neutrons can be acquired by measuring the activity of the radioactive element that is produced. This method can be used to detect neutrons with various energies via the selection of appropriate materials based on the magnitudes of their cross sections. The most frequently used activation materials include indium, gold, copper, aluminum, and niobium. Activation materials can be fabricated in various forms, depending on the experimental requirements.

12.2.2 Measurements of Neutron Flux

The neutron flux is a basic parameter that describes a neutron field or a neutron beam. The accuracy of the neutron flux measurements directly influences the accuracy of other neutronics parameters that are associated with neutron flux such as absorbed dose, cross section, and angular distribution. There are many technologies for measuring neutron flux, such as the long counter based on the secondary particle emission method, proton recoil telescopes based on the nuclear recoil method, fission chambers based on the nuclear fission method, and activation foils based on the activation method.

The neutron long counter is used mainly to transfer the value of the neutron reference radiation field. The recoil proton telescope is suitable mainly for fast neutron measurement above 1 MeV. The fission chamber has a large output pulse amplitude, strong anti-gamma background, and electronic noise ability, and can be used for neutron flux measurement of fission systems, fusion systems, accelerator-based neutron sources, etc. The activation method is an offline measurement method for neutron field measurement and has a small volume and strong anti-jamming ability with wide applications on various occasions.

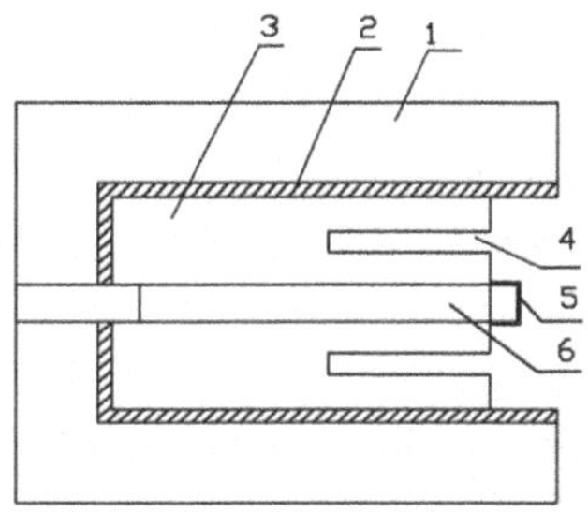

Fig. 12.9 Schematic diagram of a classical long counter, 1 inner moderator, 2 boron-containing layer, 3 outer moderator, 4 air column, 5 layer of cadmium, 6 proportional counters

1. Long counters

The design of long counters is based on the secondary particle emission method. A long counter is simple to construct and highly reliable, and it offers a detection efficiency that is insensitive to the incident neutron energy and a high n/c discrimination capability. Long counters are commonly found in neutron metrology laboratories and are often used as secondary standard instruments for measuring the neutron flux.

A long counter consists mainly of a centrally located ^{3}He or BF_3 proportional counter, an inner moderator, a boron-containing layer, and an outer moderator. There is usually also a layer of material such as cadmium to absorb thermal neutrons and an empty space in front of the inner moderator to improve the detection efficiency for low-energy neutrons. A schematic diagram of a classical long counter is shown in Fig. 12.9.

The neutron flux measured by a long counter can be expressed as

$$\phi = \frac{N}{\varepsilon \cdot S} \tag{12.4}$$

where N is the net count rate of the detector after background subtraction, ε is the detection efficiency, and S is the effective cross-sectional area.

The detection efficiency of a classical long counter is high and insensitive to the incident neutron energy ranging from 1 keV to 6 MeV. However, the detection efficiency decreases significantly above 6 MeV on account of the sharp drop of cross section of nuclear reactions of interest in neutron detection. Take ^{10}B(n, α) reaction, for example, the cross section difference between 6 MeV neutron and thermal neutron is more than 4 orders of magnitude. The low-efficiency response in the high-energy regions prevents long counters from being used in applications involving high-energy neutrons such as fusion systems and spallation neutron sources. One method of improving the efficiency response in the high-energy region is to add compensating materials that are capable of neutron multiplication.

The FDS Team has developed a compensatory neutron long counter [16] based on multilayer heavy metal structure that can be used to measure neutron flux in high-energy and wide energy-domain neutron radiation fields. Based on the (n, xn) reaction with heavy metal elements, the detection efficiency in the high-neutron-energy region can be improved by adding a compensating structure consisting of tungsten into the

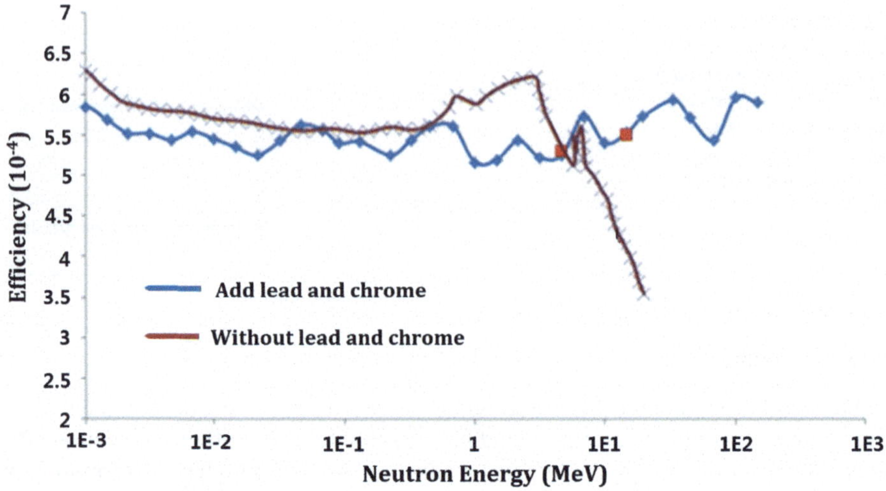

Fig. 12.10 The efficiency curves with and without tungsten as a function of neutron energy of a compensatory long counter

modulation material. The detection efficiency of the neutrons with different energies is shown in Fig. 12.10. This plot shows that the detection efficiency discrepancy in the energy region between 1 keV and 20 MeV is less than 7.8%, and the detection efficiency of 14 Mev neutrons is increased by 50%, and the detection efficiency of 20MeV neutrons is increased up to 86%. The efficiency response function can be further improved by adding lead and chrome into the modulator, which results in an efficiency differential of less than 11% in the energy region between 1 keV and 150 MeV. The compensatory neutron long counter has been applied for neutron flux measurement in HIENG and small neutron tubes.

2. Proton recoil telescopes

Proton recoil telescopes are designed based on the nuclear recoil method. The flux of neutrons with energies of 10 keV–200 MeV can be measured based on the ^{1}H(n, n′)^{1}H cross section, since the scattering total and differential cross sections for neutrons interacting with hydrogen nuclei can be determined precisely. There are three types of detectors for measuring the neutron flux based on the ^{1}H(n, n′)^{1}H cross section, namely, proportional counters filled with hydrogen gas, semiconductor telescopes, and scintillator telescopes. The applicable energy regions are 40 keV–3 MeV, 1–5 MeV, and 5–200 MeV, respectively.

The flux of fusion neutrons can be measured by using a scintillator proton recoil telescope, which consists mainly of a radiator, a scintillator, and a beam-defining slit. The radiator is used to generate recoil protons, and the scintillator is used to detect protons. Such a detector commonly uses the coincidence method to discriminate between scattered neutrons and the strong γ-ray background. Specifically, a recoil proton first passes through a thin detector called the ΔE detector and loses a small

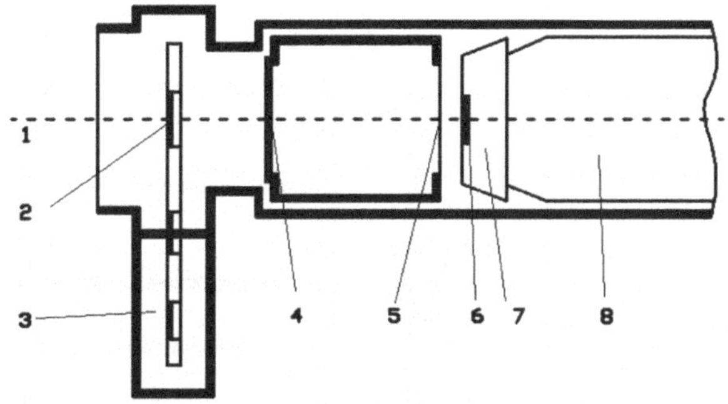

Fig. 12.11 Schematic diagram of a scintillator telescope, 1 neutron source, 2 radiator, 3 rotating disk, 4 $\Delta E-$Si detector, 5 tantalum diaphragm stop, 6 CsI crystal, 7 light guide, 8 photomultiplier tube

fraction of its energy there. The recoil proton then enters a thick detector called the E detector and dissipates the remainder of its energy there. A coincidence measurement is performed using the pulse signals generated from both the ΔE and E detectors. The ΔE detector may consist of either one or two thin-window gas proportional counters or a semiconductor detector. The E detector is usually a CsI crystal or a depletion-type surface barrier Au–Si detector. Depending on the number of ΔE detectors, either twofold or threefold $\Delta E-E$ coincidence is applied for background subtraction.

A schematic diagram of a classical scintillator proton recoil telescope is shown in Fig. 12.11. The telescope consists of a polyethylene film, a ΔE detector, a CsI crystal, and a photomultiplier, where the polyethylene film has a mass thickness of 40 ~ 60 mg/cm^2, the ΔE silicon semiconductor detector has a diameter of 25 mm and a thickness of 50–300 μm, and the CsI crystal has a diameter of 25 mm and a thickness of 1–3 mm. The distance between the radiator and the CsI crystal is 80 mm. Radiators are located at four positions on an automatically rotating disk. Three of the radiators are polyethylene films of different thicknesses, and the fourth radiator is a dummy radiator that is used to measure the background.

If the number of recoil protons is N_p, then the neutron flux can be expressed as $\phi = N_p/\varepsilon$, with $\varepsilon = 10^{-27}\ P \cdot m$, where P is the number of hydrogen atoms per unit area of the transfer target, and m is a parameter related to the geometry of the telescope.

The total uncertainty of this telescope originates mainly from the calculation of the m value, the weight of the sample, the distance measurement, neutron scattering, background, and statistics.

The proton recoil telescope is the most widely used instrument for measuring the neutron flux because of its high precision, low cost, and ease of construction.

3. Fission chambers

Fission chambers are designed based on the nuclear fission method, usually using ^{235}U and ^{238}U as fissionable materials. ^{235}U is used for thermal neutron detection, whereas ^{238}U is used for detecting neutrons with energies higher than 6 MeV since the reaction threshold for the ^{238}U fission is approximately 1 MeV.

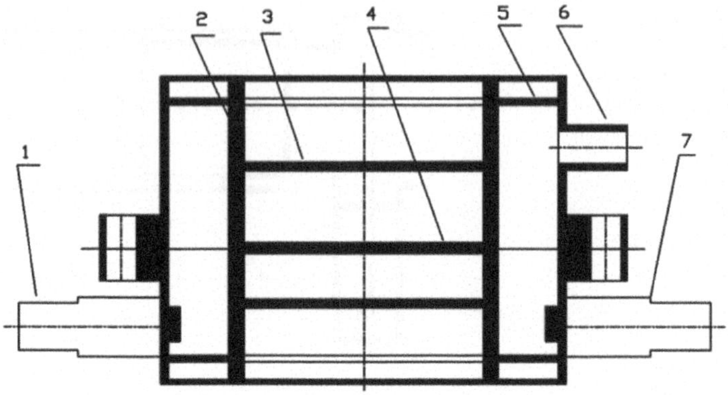

Fig. 12.12 Schematic diagram of a planar ^{238}U fission chamber, 1 power connector, 2 insulating column, 3 collector, 4 ^{238}U pole, 5 protection ring, 6 inflatable mouth, 7 signal interface

To determine the neutron flux using a fission chamber, it is necessary to know certain features of the neutron spectrum. The neutron flux can be expressed as

$$\phi_n = \frac{N_f}{n \cdot \sigma_f} \cdot K \tag{12.5}$$

where N_f is the net count rate of fission fragments, n is the number of nuclei of fissionable material inside the detector, σ_f is the mean fission cross section, and K is a correction factor.

The uncertainty arises mainly from the purity and content of the fissionable materials, the uncertainty of the mean fission cross section, the scattering neutron background, and the decay correction.

The basic structure of a fission chamber consists of a central pole with fission material cladding, a collector, a power connector, a signal interface, and inlet–outlet ports. A fission chamber can be fabricated with either a planar or cylindrical form, depending on the characteristics of the neutron field to be measured. Planar-type fission chambers (Fig. 12.12) are used mainly to measure the fluxes of parallel neutron beams such as those produced by accelerator-based D-T fusion neutron sources, whereas cylinder-type fission chambers are used mainly for measuring neutrons for which no specific direction is favored such as the neutron flux of fission systems.

Fission chambers are commonly used in neutronics experiments for measuring the neutron flux, because of their characteristics of high-amplitude pulsed output, anti-γ-ray background, and resistance to electrical noise. A mini-type fission chamber has been developed by the FDS Team for monitoring the power of a fusion system, since the environment inside a fusion apparatus has a high neutron flux and limited space for monitor installation.

4. Activation foils

Activation foils are used to measure neutron flux based on the activation method. The neutron flux is determined by measuring the particles emitted by radioactive nuclides generated during an experiment. An activation foil cannot perform the online monitoring of the neutron flux because it is a passive detector. The neutron flux as measured using the activation foil technology can be expressed as

$$\phi = \frac{\lambda AC}{MN_A\eta S\sigma D\varepsilon I_\gamma} \tag{12.6}$$

where $S = 1 - e^{-\lambda T_0}$ is the saturation factor, $D = e^{-\lambda(t_1 - T_0)} - e^{-\lambda(t_2 - T_0)}$ is the collection factor, T_0 is the radiation time, t_1 is the start time of the measurement, t_2 is the end time of the measurement, M is the sample mass, η is the abundance of the target nuclei, A is the atomic weight, N_A is the Avogadro constant, σ is the cross section of the target nuclei, λ is the decay constant, C is the net count of emitted particles to be detected, I_γ is the branching ratio, and ε is the detection efficiency.

If the neutron flux changes during the irradiation period, the integral method is used to correct the saturation factor S. For a monoenergetic neutron field, σ is the cross section of the target nuclei, whereas for a continuous energy field (or for neutrons with a broad energy distribution), the spectral information is required to calculate the mean cross section.

The activation foil technology is usually used for integral measurements of neutron flux in fission systems and is also used to determine the yield of fusion neutrons. The neutron field inside a fusion system is a mixed field of neutrons from the D-D and D-T reactions, and it is difficult to distinguish them from each other. One way to discriminate neutrons from the D-T reaction is to select an activation foil material with a threshold higher than 2.45 MeV, to utilize reactions such as $^{197}Au(n, 2n)^{196}Au$, $^{93}Nb(n, 2n)^{92}Nb$, $^{59}Co(n, 2n)^{58}Co$, $^{46}Ti(n, 2n)^{45}Ti$, or $^{45}Sc(n, 2n)^{44}Sc$.

5. Other measuring devices

In addition to the detection technologies mentioned above, one can also use a solid-state nuclear track detector such as a CR-39 detector or a nuclear emulsion detector to detect neutrons with energies ranging from approximately 10^{-2} eV to 20 MeV. The fission materials are usually clad on the detector surface to interact with the incident neutrons, and the neutron flux is determined by measuring the number of tracks left by heavy charged particles, which are fission fragments produced from the fission materials.

With the development of semiconductor technology, diamond detectors have gained attention because they are based on a newly developed detection technology that offers good radiation resistance, a compact detector size, and anti-γ-ray background. Diamond detectors have the potential to play an essential role in the detection of high-intensity fusion neutrons. The threshold of the $^{12}C(n, \alpha)^9Be$ reaction for neutrons interacting with the carbon in diamond is approximately 5.7 MeV. Therefore, diamond detectors are suitable for measuring neutron flux of D-T fusion facilities by the pulse amplitude discrimination method. Diamond detectors have been used to measure the neutron flux of the JET (Joint European Torus), the TFTR in America, and the JT-60U in Japan.

For an isotropic point neutron source, the neutron flux can be obtained by the counts of the neutron detectors located near the neutron source, and the intensity of the point neutron source can be obtained by combining the distance from the measuring point to the neutron source, as shown below:

$$S = 4\pi R^2 \phi \tag{12.7}$$

where S is the yield of the neutron source, R is the distance from the measuring point to the neutron source, and ϕ is the neutron flux at the measuring point.

12.2.3 Measurements of Neutron Spectrum

The neutron spectrum is the distribution of the number of neutrons versus energy, which provides important information for studies of nuclear structure and nuclear reaction mechanisms. It is very important for fusion neutronics experiments.

There are several methods of measuring the neutron spectrum, including the time-of-flight method, the proton recoil method, the activation foil method, the multi-sphere method, and the nuclear reaction method. The important parameters in the measurement of the neutron spectrum are energy resolution, measurement range of the energy spectrum, and detection efficiency.

1. The time-of-flight method

The time-of-flight method [17] is one of the most important measurement technologies in neutronics experiments and plays an essential role in measuring the neutron spectrum, angular distribution, and total cross section as well as the evaluation of neutron nuclear databases and the validation of engineering designs.

The principle of the time-of-flight technology is that the energy of a neutron can be obtained by measuring its speed. It is necessary to measure the neutron flight distance l over a flight time t. When the neutron energy is less than 20 MeV, the relationship between the flight time t and the neutron energy E is

$$t = 72.3l \cdot E^{-1/2} \tag{12.8}$$

where t is the time in units of ns, l is the distance in units of m, and E is the neutron energy in units of MeV.

The energy resolution of a time-of-flight spectrometer is given by the following equation:

$$\frac{\Delta E}{E} = \frac{2\Delta t}{t} + \frac{2\Delta L}{l} = \left[\frac{2.8\sqrt{E}}{l}\Delta t + \frac{2\Delta l}{l}\right]\% \approx \frac{2.8\sqrt{E} \times \Delta t}{l}\% \tag{12.9}$$

where Δt is the time resolution of the spectrometer in units of ns, which can be calculated as the full width at half maximum of the signal peak in the time-of-flight spectrum generated by the reaction of the accelerated particle with the target nucleus, and Δl is the measurement error on the flight distance l.

The energy resolution $\Delta E/E$ is proportional to Δt and the square root of the neutron energy and is inversely proportional to the flight distance l. Δt usually depends on the neutron source, the detector, and the electronics of the spectrometer instrument and other laboratory equipment. l is adjusted depending on the energy resolution requirements of different experiments.

Fast neutron time-of-flight spectrometers are used to measure the spectra of fusion neutrons and can be classified as either steady or pulsed-beam fast neutron time-of-flight spectrometers, depending on the neutron source. Pulsed-beam fast neutron time-of-flight spectrometers, which can achieve high energy resolution, have important applications in the collection of high-precision nuclear data and the measurement of neutron energy spectra. Therefore, in this section, this type of spectrometer will be discussed in detail.

To obtain the flight time when using a pulsed-beam time-of-flight spectrometer to measure the neutron spectrum, it is necessary to identify the start and termination signals. These signals are usually sent to a multichannel analyser to determine the number of neutrons. The start and termination signals are commonly provided by the beam pick-up system and the neutron detector at the end of the flight, respectively. The detectors used in fast neutron time-of-flight spectrometers need to exhibit a fast time response (on the scale of at least ns). Applicable detectors include liquid scintillator detectors and plastic scintillator detectors.

When the neutron spectrum of an accelerator-based D-T fusion neutron source is measured using a time-of-flight spectrometer, the following requirements must be met:

(1) High energy resolution. It is necessary to accurately reflect the changes in the neutron energy structure from each reaction channel of interest. For example, to achieve a neutron energy resolution of 2% for the D-T reaction, the neutron flight distance needs to be greater than 8 m, when the detector thickness is 5 cm and the pulse beam width is 1 ns.
(2) Wide energy range. It is necessary for the electronic instruments and detectors to cover a wide dynamic range in energy. In particular, when pulse shape discrimination is used to eliminate the γ-ray background, the dynamic range of the detector determines the minimum neutron energy that can be measured.
(3) Stable long-term operation of the facility. This requirement arises from the need for a long measuring period because of the low count rate due to the long flight distance of the neutrons and the broad neutron energy distribution.

Each fusion system requires a specialized time-of-flight spectrometer designed to be compatible with its own characteristics. The time-of-flight spectrometers used in HINEG and JET will be introduced below.

The layout of the time-of-flight spectrometer at HINEG is shown in Fig. 12.13. The available flight distance at HINEG is 3–15 m, and the scattering angle range is

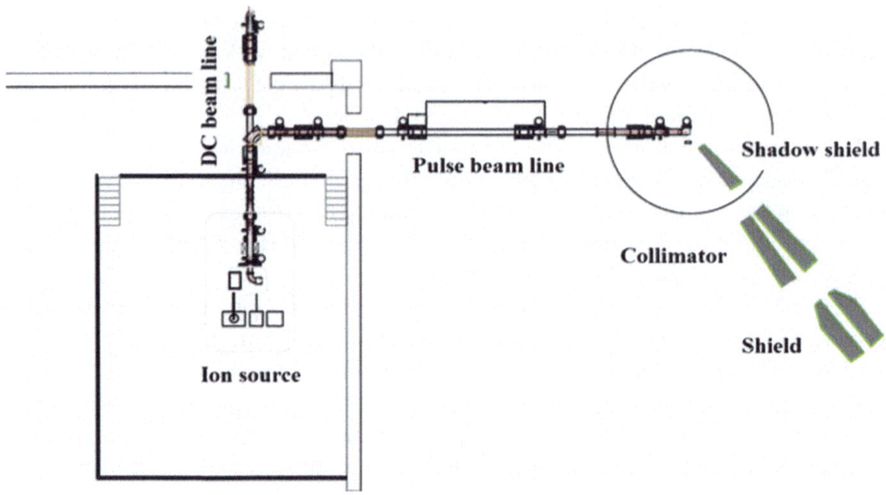

Fig. 12.13 The time-of-flight spectrometer at HINEG

0°–170°. The frame structure of the time-of-flight spectrometer includes a pulse beam pick-up system, shadow shielding, a collimator, a shielding body, a detector, and an electronics system. The performance of the HINEG time-of-flight spectrometer is as follows:

(1) energy resolution better than 3%,
(2) dynamic range of 0.8–15.7 MeV, and
(3) signal-to-background ratio better than 10:1 throughout the entire energy range.

A time-of-flight spectrometer for the measurement of D-D neutrons called TOFOR has been developed for JET, as shown in Fig. 12.14. It was optimized for operation at high rates to perform neutron emission spectroscopy for diagnosing deuterium plasmas. The initial signal is produced when a neutron enters the first detector layer, and the termination signal is produced when it enters the second layer. The flight time of the neutron can be obtained from the time delay between the two signals. The count rate and energy resolution of TOFOR can reach up to 10^5 s^{-1} and approximately 6%, respectively.

2. The proton recoil method

The energy of incident neutrons can be measured by detecting the energy of recoiling protons using a proton recoil energy spectrometer. The probability of the production of protons with different energies E_p can be taken to be constant when the incident neutron energy is less than 10 MeV, since the neutron–proton scattering is approximately isotropic in this case.

The ideal proton recoil spectrum should be of a rectangular shape in the case that the recoil protons are recorded at all angles for monoenergetic neutrons. For the case of non-monoenergetic neutrons, the proton recoil spectrum can be treated

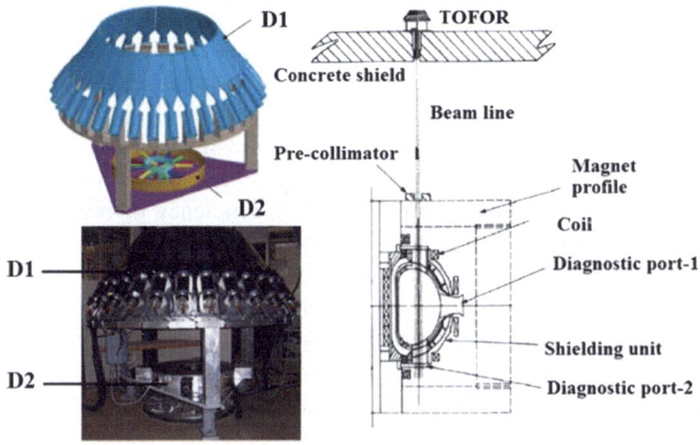

Fig. 12.14 TOFOR neutron spectrometer used at JET, D1, and D2 are the detector sets of TOFOR [18]

as a superposition of the rectangular spectra generated by a series of monoenergetic neutrons, which can be decomposed to obtain the incident neutron spectrum. If the experimental proton recoil spectrum $J(E_p)$ and the incident neutron spectrum $\phi(E)$ are divided into n intervals, then the number $J_j(E_p)$ of recoiling protons in the jth interval can be expressed as

$$J_j(E_p) = \sum_{i=j}^{n} f_{j,i}\phi(E_i) \quad j = 1, 2, \ldots, n \; (i = 1, 2, 3, \ldots, n) \tag{12.10}$$

where the $f_{j,i}$ are matrix elements that can be calculated from the number of hydrogen nuclei in the detector and the (n, p) elastic scattering cross section. The incident neutron energy $\phi(E)$ can be obtained by taking the inverse of Eq. 12.10.

There are many types of proton recoil spectrometers, including hydrogen proportional tube proton recoil spectrometers, organic scintillator proton recoil spectrometers, magnetic proton recoil spectrometers, coincidence spectrometers, and nuclear emulsions. As representative facilities, organic scintillator proton recoil spectrometers and magnetic proton recoil spectrometers will be described in detail in this section.

An organic scintillator proton recoil spectrometer has high detection efficiency for protons and thus can be used to measure high-energy neutrons. A typical organic scintillator proton recoil spectrometer consists mainly of a scintillator, a photomultiplier, and a voltage divider circuit. When the charged particles enter the scintillator, they lose either some or all of their energy through the ionization process, and the consequently excited atoms and molecules in the scintillator are able to decay and emit light of a certain wavelength, called fluorescence photons. Most of these fluorescence photons will reach the photocathode of the photomultiplier tube via a light guide and reflector, where they produce photoelectrons via the photoelectric effect. These photoelectrons are then multiplied in the photomultiplier tube and are finally collected by the anode to create a voltage pulse signal that is recorded by the subsequent electronics. The neutron spectra of Am–Be, Cf, D-T, and D-D sources can be measured by using organic scintillator proton recoil spectrometers through the decomposition of the detected proton recoil spectra with the help of the calculated detector response matrix. Although the detection efficiency of organic scintillator proton recoil spectrometers is high, their energy resolution is not ideal. For example, the energy resolution is 15% for a 14 MeV neutron source.

Magnetic proton recoil spectrometers are another commonly used type of energy spectrometer. The performance of magnetic spectrometers is much better than the performance of traditional neutron energy spectrometers, especially in terms of the energy resolution, count rate, and detection efficiency. In 1996, the first magnetic proton recoil spectrometer [19] (Fig. 12.15) was built at JET. The collimator and the shield are of the same size, 1.9 m × 2.6 m × 0.8 m, and CH_2 conversion targets with mass thickness of 1.6, 8, 18, and 50 mg/cm^2 are used. The intrinsic detection efficiency of this system is 10^{-7}–10^{-6}. The magnetic analysis system consists of two separate dipole magnets with a deflection angle of 135° for a proton beam at the reference energy. An array detector consisting of 37 plastic scintillators on the focal plane, with widths of 8 mm for scintillators in central positions and 20 mm for scintillators at the edge, is used to measure protons at positions covering approximately ±20% of the central energy region (16 MeV, adjustable by means of a magnetic field). The spectrometer can achieve an experimental energy resolution of 2.5%.

3. The activation foil method

Some nuclear reactions induced by neutrons have threshold energy. Therefore, the neutron spectrum can be obtained based on the active products detected in a series of activation foils with different thresholds together with the dependence of the cross section on the neutron energy.

The principles listed below are usually applied when selecting the materials to be used as activation foils.

(1) Complete, high-precision data regarding the nuclear reaction cross sections should be available. The cross sections should have relatively large values. There should be good linear independence among the cross sections of different materials to be used simultaneously.
(2) For a reasonable energy response range, the sensitivity regions of the nuclear reactions should be properly distributed throughout the spectrum to be measured.
(3) The half-lives of the reaction products should be suitable, and the abundance of the interested isotope should be as high as possible to enable separation from competing activation processes in the case of a low neutron flux.
(4) The materials should have stable physical and chemical properties.

In the case that different types of activation foils are placed in the neutron radiation field to be measured for some period of time, the activation foils are then removed to

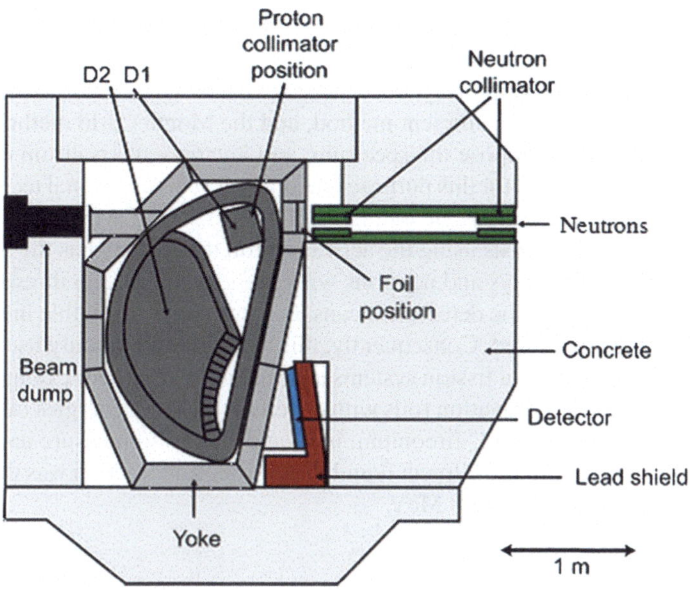

Fig. 12.15 Schematic diagram of a magnetic proton recoil spectrometer

measure their activity. The activity A_i of the ith irradiated activation foil, the reaction cross section σ_i of the ith activation foil, and the neutron spectrum $\phi(E)$ have the following relationship:

$$A_i = \int_0^{\infty} \sigma_i(E)\phi(E)dE \quad (i = 1, 2, 3, \ldots, n) \tag{12.11}$$

The neutron spectrum $\phi(E)$ can be obtained by solving Eq. 12.11 since A_i and $\sigma_i(E)$ can be obtained by measuring the activities of the activation foils and from the corresponding evaluation database, respectively. In practice, the number of activation foils that can be used is limited; in other words, the number n of available nuclear reaction channels is limited. Therefore, it is necessary to discretize the measured neutron spectrum and the corresponding nuclear reaction cross sections and then to transform Eq. 12.11 into the form of a sum, as shown in Eq. 12.12.

$$A_i = \sum_{j=0}^{m} \phi_j(E)\sigma_{ij}(E)\Delta E_j \quad (i = 1, 2, 3, \ldots, n) \tag{12.12}$$

where $\phi_j(E)$ is the average neutron flux in the jth interval of the neutron spectrum to be measured, $\sigma_{ij}(E)$ is the average nuclear reaction cross section of the ith activation foil in the jth neutron energy interval, ΔE_j is the width of the jth neutron energy interval, and m is the number of energy intervals into which the neutron spectrum is divided.

Based on the above principles, special mathematical methods (such as the iterative method, the undetermined coefficient method, and the Monte Carlo method) can be used to analyze and decompose the spectrum, and a variety of spectrum extraction codes have been developed for this purpose that integrate computational technologies with spectrum analysis methods, including SAND-II, MSIT, RDMM, and SANDBP.

Neutron spectrum analysis using the activation foil technology has the advantage of being insensitive to γ-rays and neutrons with energies lower than threshold energies. The small volume of the detectors means that they cause negligible interference with the field to be measured. Consequently, this method is commonly used to measure the neutron spectrum of fission systems and fusion systems. For example, Jonah used a set of high-purity activation foils with different threshold energies (aluminum, indium, nickel, niobium, gold, zirconium, iron, and cobalt) to measure and analyze the D-T neutron spectrum [20]. Jonah found that the neutron fraction was the highest in the energy range of 13.2–15.8 MeV.

4. The multi-sphere method

A multi-sphere neutron spectrometer, also known as a Bonner sphere spectrometer, consists of a series of thermal neutron sensors located at the center of several spheres with different diameters made of polyethylene. The thermal neutron sensor is usually a ^{3}He proportional counter or ^{6}Li detector, sometimes a BF_3 proportional tube or a thermal-neutron-sensitive activation foil. The number of spheres is usually between 5 and 18. The diameters are in the range of 0–35 cm, depending on the maximum energy of the neutrons to be measured: the larger the sphere diameter, the higher the neutron energy that can be measured.

The energy response function of the probe is fixed for a given probe structure and material. The energy spectrum $\phi(E)$ of a particular neutron radiation field can be obtained from the number of detector counts A and the response function $R(E)$ according to the relationship $\int R(E)\phi(E) = A$.

For a multi-sphere system, the count for each probe can be represented by the following equation:

$$A_i = \int_{E_{min}}^{E_{max}} R_i(E)\phi(E)dE \quad (\mathrm{i} = 1, 2, 3, \ldots \mathrm{n}) \tag{12.15}$$

where n is the number of spherical shells of different sizes, A_i is the neutron count for the ith spherical shell, $R_i(E)$ is the neutron response function of the ith shell for neutrons of energy E, $\phi(E)$ is the neutron spectrum, and $E_{\max}(E_{\min})$ is the highest (lowest) energy of the neutrons to be measured. The spectral analysis principle will not be explored here since it is like the spectral analysis principle of the activation method, as introduced in the previous section.

A Bonner sphere spectrometer, whose dynamic measurement range spans from the thermal neutron energy to 200 MeV, is used mainly for the protective monitoring of neutron radiation. The energy resolution is affected by the number of Bonner spheres and the accuracy of the efficiency curve of each sphere. The energy resolution of Bonner sphere spectrometers for high-energy neutrons is poor, since the energy response is low, and the response curves of the different spheres are similar for neutron energies above 20 MeV. The energy response can be optimized by adding heavy metals such as lead, iron, or copper to the polyethylene. Such an improved Bonner sphere spectrometer can be used to measure the neutron spectrum over a wide energy range.

5. The nuclear reaction method

The principle of a nuclear reaction spectrometer is to obtain the neutron energy by measuring the total kinetic energy of the reaction products induced by the incident neutrons. It is necessary that the nuclear reactions should produce charged particles, and the energy of these charged particles could be fully deposited in the detector. Several types of spectrometers have been developed based on this method, including ^{3}He ionization chamber spectrometers, ^{3}He sandwich spectrometers, and ^{6}Li spectrometers.

(1) ^{3}He ionization chamber spectrometer

The basic principle of a ^{3}He ionization chamber spectrometer relies on the ^{3}He(n, p) T reaction. There is a one-to-one correspondence between the output pulse amplitude and the neutron energy for neutron energies below 1 MeV. In other words, neutrons with a single energy produce only a linear spectrum. The maximum energy of the neutron measured by a ^{3}He ionization chamber spectrometer is usually less than 3 MeV.

(2) ^{3}He sandwich spectrometer

The basic structure of a ^{3}He sandwich spectrometer consists of two facing semiconductor detectors with a ^{3}He ionization chamber (or proportional tube) in the middle. The two charged particles generated in the ^{3}He(n, p)T reaction lose a small fraction of their energy in the ^{3}He ionization chamber and then deposit their remaining energy in the two semiconductor detectors. The pulses from the two semiconductor detectors and the ^{3}He ionization chamber are summed to obtain the total energy of the reaction products after normalization. At present, high-energy-resolution ^{3}He sandwich spectrometers can measure neutrons with energies in the range of 0.1–15 MeV. Commercially available ^{3}He ionization chambers have achieved an energy resolution of 85 keV (^{3}He under 5 atmospheres of pressure) and an intrinsic detection efficiency of approximately 10^{-2} for thermal neutrons. For fast neutrons, the pulse amplitude resolution is slightly degraded with the increased neutron energy, and the intrinsic detection efficiency is approximately 3×10^{-6} (for 2.5 MeV).

(3) ^{6}Li spectrometer

In a ^{6}Li spectrometer, the neutron energy is obtained by measuring the energies of alpha particle and tritium in the ^{6}Li(n, α)T reaction. ^{6}LiI(Eu) scintillator detectors and ^{6}LiF sandwich-type semiconductor detectors are currently the most commonly used. A ^{6}LiI(Eu) scintillator can only measure neutrons in the MeV energy range because of its relatively poor energy resolution. The reason is that the Q-value of the ^{6}Li(n, α)T reaction is relatively large (4.78 MeV). A ^{6}LiF sandwich neutron spectrometer is suitable for the case of high neutron flux. The lower energy detection limit for a ^{6}LiF sandwich neutron spectrometer is approximately 10 keV. The energy resolution of a ^{6}LiF sandwich neutron spectrometer is also good. For example, it is 14.3 keV (3.75 keV) for neutrons with an energy of 100 keV (6 keV).

6. Multi-threshold neutron spectrometer

To obtain the neutron spectrum of large energy regions of the radiation field, a multi-threshold neutron spectrometer was developed by the FDS Team. The spectrometer uses a set of detectors with different responses to neutron threshold energy. The technique of spectral feature recognition is developed to construct a sufficiently accurate preset spectrum based on the idea of machine learning to mine the function relationship between counting rate ratio of the multi-threshold neutron spectrometer and the key peaks of the energy spectrum in the range of 0–14 MeV. The preset spectrum is critical information that can enormously promote the accuracy and efficiency of an online spectrum unfolding. According to the counting of multiple detectors, the neutron energy spectrum of the 0–14 MeV wide energy range can be quickly solved with the relative deviation reduced by more than 34%, compared with the traditional preset spectrum in the TRIGA reactor validation.

12.3 Neutron Control

The neutron transport is a key technology affecting the safety and economy of the nuclear systems. The regulation and control of neutrons is an important part of the development of nuclear systems. To verify the service performance of key equipment in nuclear systems, it is often necessary to construct an experimental environment in which the neutron spectrum distribution meets specific requirements. Furthermore, in the aspects of nuclear technology, some explicit requirements are placed on the energy spectrum of the neutron beam. For example, in boron neutron capture cancer treatment (BNCT), the epithermal neutron intensity and the ratio of different energy segments are specified. It is very hard to meet the target with manual experience.

Based on systematically summarizing the work of the predecessors, this book proposed the concept of "neutron control", that developed the technology of regulating and controlling the transmission of neutrons in the system through nuclear reaction. In this book, it is called neutron control technology. Neutron control is an important way to achieve neutron angle, energy and its time, and spatial distribution characteristics. The neutron regulation process involves three main elements: source neutron, regulation module, and target neutron spectrum.

This section introduces the principles and methods of traditional neutron regulation technology and neutron control technology.

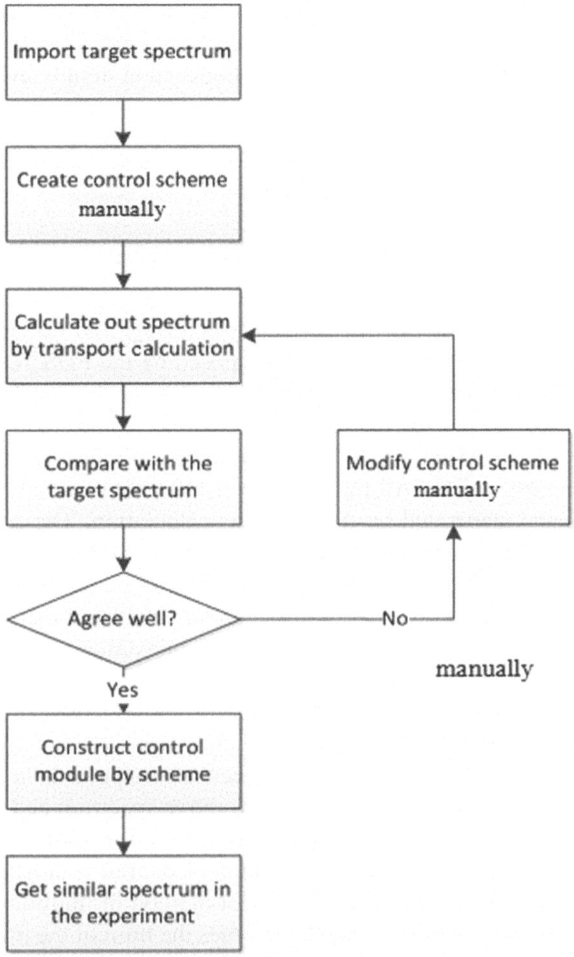

Fig. 12.16 The schematic diagram of traditional technology of neutron regulation

12.3.1 Traditional Neutron Regulation Technology

Traditional technology of neutron regulation is based on the source neutron spectrum. The basic regulation module design scheme is determined according to the manual experience. The output energy spectrum is obtained by neutronics calculation. Compared to the target neutron spectrum, the material selection and configuration arrangement are continuously adjusted in the regulation module until the output energy spectrum is close to the requirements. The process of traditional technology is shown in Fig. 12.16.

In practical applications, traditional technology relies on the personnel experience and repeatedly iteration, making it difficult to achieve precise regulation of the neutron spectrum. According to the requirements of different target energy spectra, the design of the regulation modules will vary greatly. For the same target energy spectrum requirements, the regulation schemes by different personnels will also be different. This method of continuously and iteratively selecting materials and modifying the regulation module arrangement is time-consuming and difficult with low precision of regulation.

12.3.2 Neutron Control Technology

To control the neutron spectrum precisely and efficiently, the idea of the neutron control technology has been proposed by the FDS Team. This technology employs reverse optimization method for determining the arrangement of the basic energy response modules which is defined as control module with different materials can support the moderation, multiplication, and absorption of neutron in multiple energy regions. All the control modules are to be combined in the energy response matrix generated by the transport calculation. The energy response matrix should be imported into the database and can be used to represent the relationship between input spectrum and output spectrum after neutrons pass through the corresponding module. The output spectrum can be calculated extremely rapid by combining the response matrixes of all control modules after each of them decouple its position influenced by the correction methods.

In the process of neutron control, the divergence between source spectrum and target spectrum is analyzed, and the proper element component from the database is selected. The control scheme selects the shape with the permutation and combination method. The difference between source spectrum and target spectrum is eliminated, and the procedure continues until the output spectrum is similar to the target spectrum. A final control scheme is accomplished, after it satisfies the given requirement. The method is the quantitative control method of multidimensional response matrix by step backtracking, which overcomes the limit of the traditional technologies that can get the more accurate output spectrum automatically.

For the users, they only need to import the target spectrum, and the system that employed the control technology can generate the effective arrangement of control modules immediately. The real control modular can be built by the arrangement, and neutrons that match the target spectrum can be generated for the experiment. The work process of the system is shown in Fig. 12.17. For the application in nuclear technology, the neutron control can meet more targets. Not only outputting the closed neutron spectrum, the neutron intensity of specified energy segments can be promoted. Therefore, the neutron spectrum reverse control technologies can support the development of the BNCT, neutrongraphy, etc.

The FDS Team has built a wide energy-domain reverse neutron control system (Fig. 12.18) based on neutron control technologies. The system is composed of several control modules including moderation, multiplication, and adsorption. The control modules can be combined arbitrarily in the neutron spectrum control terminal obtaining the target spectra for verification experiments of different advanced nuclear

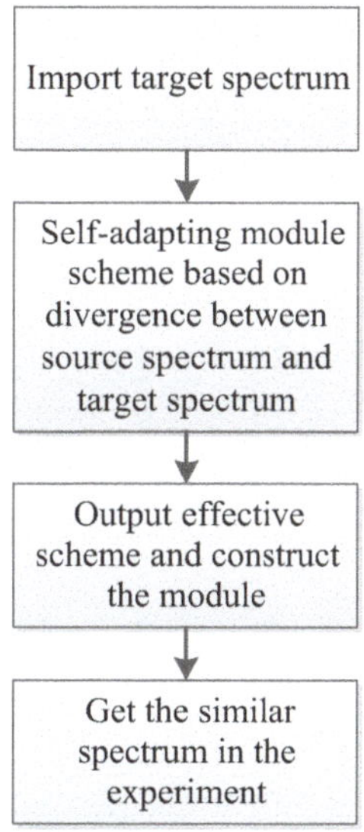

Fig. 12.17 The schematic diagram of neutron control process

Fig. 12.18 Wide energy-domain reverse neutron control system at the FDS Team

systems. The system has been applied in the design of advanced nuclear systems such as the lead-based reactors and fusion systems. The pioneering research enables to produce different neutron spectrums of advanced nuclear systems including advanced fission systems, fusion systems, and hybrid nuclear systems by using the same facility.

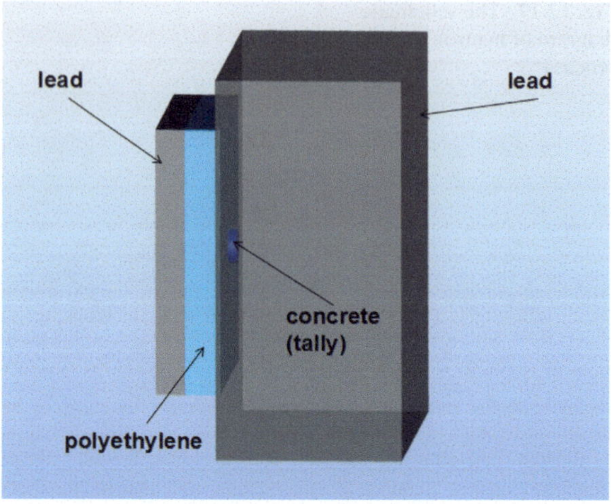

Fig. 12.19 The control modular design by traditional method

12.3.3 Example of Neutron Control

The neutron control technology has been applied in series neutronics experiments, such as the ITER TBM testing and ITER concrete activation analysis. The maximum deviation of the neutron spectrum can be reduced by more than 5 times in contrast with the traditional technology. The case of the ITER concrete activation analysis is shown briefly as follows.

Concrete is a common outer structural material used in ITER devices. During the operation of the reactor, the fast neutrons inside the core have been sufficiently moderated to reach this area. Therefore, the real spectrum is greatly different from the 14 MeV neutron source used in the experiment, which increases the uncertainty of the results.

To reconstruct the real energy spectrum of the region under the experimental conditions, the experts may need several days to obtain a reasonable control module by their experiences and iterative computations, which also bases on the traditional method (Fig. 12.19). In this case, the control module consists of 5-cm lead for multiple neutrons and 6-cm polyethylene to moderate them down. Behind the experimental module, a lead reflective module is set to promote the neutron flux of the experimental region.

The traditional method can only apply to the simple cases. Otherwise, the multiple materials and requirements for the complicated cases will cause the difficulty to exceed the limits of human trials.

Under the same conditions, the neutron control system just needs to import the target spectrum and source spectrum directly, and then an effective control module is generated, which is composed of many more materials. As shown in Fig. 12.20, front to back, the materials are 1.6 cm boron carbide, 5 cm lead, 1.6 cm beryllium,

Fig. 12.20 The control module generated by the neutron control method

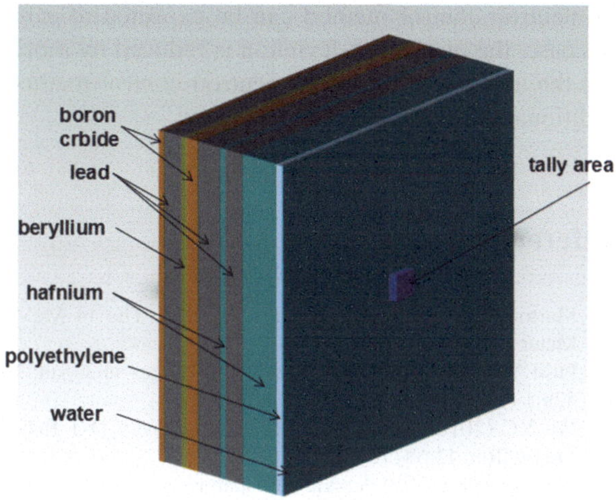

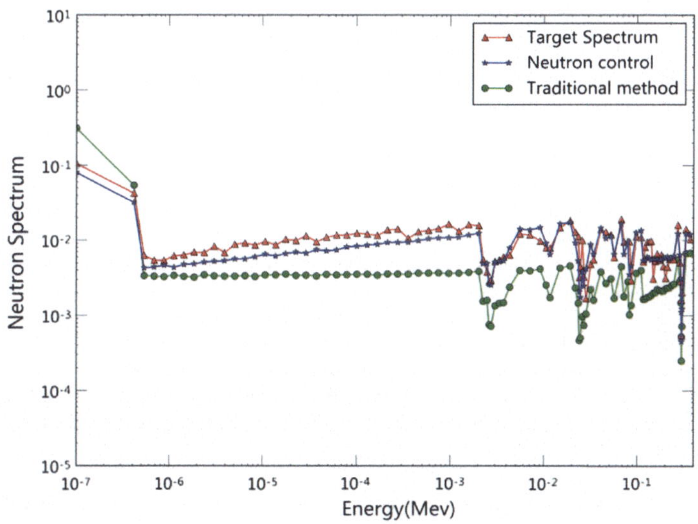

Fig. 12.21 The comparison of the results of the neutron control method and the traditional method

3.2 cm boron carbide, 1.6 cm lead, 1.6 cm hafnium, 5 cm lead, 10 cm hafnium, 1.6 cm polyethylene, and 0.3 cm water.

Figure 12.21 shows the comparison of the results of the traditional method and the neutron control method. The energy spectrum constructed by the neutron control method is in good agreement with the trend and fluctuation, except that the neutron of the hottest point is slightly higher than the target. However, the spectrum generated by the traditional method has a much higher thermal share than the target, and the other parts are seriously low. Therefore, compared with the neutron control method, the neutron control method can be expected to achieve much better accuracy. In this case, the maximum deviation is reduced by more than one order of magnitude, and the average deviation of neutron control method is two times lower than the traditional method.

References

1. Martone M, Angelone M, Pillon M (1994) The 14 MeV Frascati neutron generator. J Nucl Mater 212–215:1661–1664
2. Fujii Y (2004) Overview on neutron sources in Japan. Nucl Instrum Methods Phys Res A 529:1–4
3. Wu YC (2018) Development of high intensity D-T fusion neutron generator HINEG. Int J Energy Res 42:68–72
4. Gardner ISK (1998) A review of spallation neutron source accelerators. EPAC 98:98–102
5. Ding DZ, Ye CT, Zhao ZX et al (2005) Neutron physics, 1st edn. Atomic Energy Press, Beijing (in Chinese)
6. Naito S, Tomyo A (1999) KENS keV neutron flux for nuclear astrophysics. Nucl Instrum Methods Phys Res A 423:404–413
7. Atchison F, Fischer WE, Pepin M et al (1986) The spallation neutron source project SINQ. In: Proceeding of the international conference on neutron scattering, vol 136, pp 97–99
8. Wu YC (2017) Fusion Neutronics. Springer Nature Singapore Pte. Ltd., Singapore
9. http://pic.ihep.cas.cn/tpk/dkxzz_tpk/pic_CSNS/201808/t20180824_5059011.html
10. Ene D, Borcea C, Kopecky S et al (2010) Global characterization of the GELINA facility for high-resolution neutron time-of-flight measurements by Monte Carlo simulations. Nucl Instrum Methods Phys Res A 618(1):54–68
11. Moeslang A, Heinzel V, Matsui H et al (2006) The IFMIF test facilities design. Fusion Eng Des 81:863–871
12. Ibarra A, Dzitko H (2016) Overview of IFMIF-DONES and testing of materials for DEMO. In: Town meeting on IFMIF/ELAMAT complementary scientific program, 14–15 April 2016
13. http://lssf.cas.cn/lssf/kpyd/zsk/kyjd/201105/t20110520_4513354.html
14. Pyeon CH, Hirano Y, Misawa T et al (2007) Preliminary experiments on accelerator-driven subcritical reactor with pulsed neutron generator in Kyoto University critical assembly. J Nucl Sci Technol 44(11):1368–1378
15. https://www.rri.kyoto-u.ac.jp/en/facilities/ca
16. Li Y, Li T, Song G (2015) Response improved for neutron long counter. Radiat Prot Dosim 164:93–96
17. Shen G (2007) The neutron time-of-flight method and applications. Atomic Energy Press (in Chinese)
18. Johnson M, Giacomelli L, Hjalmarsson A et al (2008) The 2.5-MeV neutron time-of-flight spectrometer TOFOR for experiments at JET. Nucl Instrum Methods Phys Res A 591:417–430
19. Kallne J, Ballabio L, Conroy S et al (1999) New neutron diagnostics with the magnetic proton recoil spectrometer. Rev Sci Instrum 70(1):1181–1184
20. Jonah S, Ibikunle K (2003) Neutron spectral distribution from a sealed-tube neutron generator by multiple-foil activation unfolding method. Nucl Instrum Methods Phys Res A 501:514–516

Chapter 13
Neutronics Experiments of Advanced Fission Systems

Neutronics experiments can be classified into three main categories: measurement and validation experiments for nuclear data, validation experiments for neutronics methods and codes, and validation experiments for nuclear systems and component designs. Measurement and validation experiments for nuclear data are performed to obtain nuclear data or validate the applicability of a nuclear data library for specific conditions. Validation experiments for neutronics methods and codes are employed mainly to validate the applicability of neutronics methods and codes in specific conditions. Validation experiments for nuclear systems and component designs are utilized to confirm whether the systems, component designs, equipment, and nuclear materials satisfy the design requirements [1, 2].

The following three chapters will discuss the neutronics experiments of advanced fission systems, fusion systems, and hybrid systems. Experimental studies focus on different aspects for different development stages of nuclear systems. After more than 70 years of development, fission systems have been applied to engineering. The corresponding fundamental research of neutronics experiment has matured. The experimental methods of nuclear data measurement and radiation protection are extensively employed. The neutronics experiments of advanced fission systems focus on the key points of the characteristics of core physics. Fusion systems and hybrid systems are relatively new, and they remain in the research and development stage. According to the development stage of each system and the characteristics, important neutronics experiments will be introduced in this book. This chapter introduces typical core physics experiments of advanced fission systems, which includes criticality experiments, power measurement experiments, kinetic parameter measurement experiments, and reactivity measurement experiments. Chapter 14 introduces neutronics experiments of fusion systems, which includes the material activation experiments, tritium breeding experiments, material irradiation damage experiments, and shielding experiments. Chapter 15 introduces neutronics experiments of hybrid systems, which includes the subcriticality measurement experiments and transmutation nuclear data measurement experiments.

Y. Wu, *Neutronics of Advanced Nuclear Systems*,
https://doi.org/10.1007/978-981-13-6520-1_13

13.1 An Overview

By conducting neutronics experiments in fission systems, the neutronics characteristics of fission systems can be understood, and the core physical design, calculation software, and nuclear data in fission systems can be validated. With respect to the methods and principles, the existing neutron experimental methods and techniques have matured in the application of traditional nuclear systems. The components and materials of advanced fission systems are complex, and neutron energy is higher than traditional fission systems with a more complex spectrum. These new characteristics require new neutrons experiments. To satisfy the needs of an advanced fission system, the principle and method of a neutronics experiment of an advanced fission system should be investigated, new experimental platforms should be developed, and the corresponding experiments should be conducted to verify neutronics theory, calculation program, and material properties. For example, in the experiment of measuring the prompt neutron decay constant, the prompt neutron decay constant of a fast reactor is large; thus, a detection device that can quickly identify a pulse is needed. For example, a detector should not be placed in the core of nuclear systems, such as a high-temperature gas-cooled reactor, and an experimental method for out-of-reactor detection needs to be developed.

With the vigorous development of fission energy applications, research on nuclear data, systems, and components has remarkably developed, which shows similarities between advanced fission systems and traditional nuclear system. The existing data or similar research methods can be employed in advanced nuclear systems. The main difference is the physical characteristics of the core. Therefore, criticality experiments, power measurement experiments, kinetic parameter measurement experiments, and reactivity measurement experiments are the main topics introduced in this chapter.

1. Criticality experiments

The main goal of criticality experiments is to determine the actual critical load of a nuclear system and provide an experimental platform for the critical operation of subsequent experiments. Criticality experiments enable the nuclear system to gradually approach the critical state by increasing the fuel loading and other means.

2. Power measurement experiments

Power measurement experiments determine the total power or power distribution in a reactor and determine whether the key parameters, such as core power, radial power peak factor, and axial power offset, are within the established limits. The reaction rate is measured by in-core and ex-core detectors in the power measurement experiment, and the total power or power distribution in the reactor is derived.

3. Kinetics parameters measurement experiments

The measurement experiments of kinetics parameters include the measurement of the prompt neutron decay constant and effective delayed neutron fraction, which directly affects the prediction of the kinetics response characteristics of fission systems.

4. Reactivity measurement experiments

Reactivity measurement experiments are performed to determine how far the system is from criticality or to derive parameters such as control rod worth and reactivity coefficients. These experiments are directly related to the safe operation and control of fission systems and can be conducted to verify the appropriateness of core design and the accuracy of physical calculations.

13.2 Criticality Experiments

One of the goals of the design, test, and operation of a fission system is to enable the system to attain the critical state. The criticality of a fission system implies that the system can maintain a self-sustaining chain fission reaction. The loading at the critical state is usually determined by criticality experiments, in which the critical mass, critical volume, critical water level, and critical rod position are determined.

13.2.1 Methods

In the first stage of the criticality experiment, the reactor is in the subcritical state and can approach the critical state via increments of the reactivity. Assuming that external neutron source of strength $S(\mathrm{s}^{-1})$ is placed in the subcritical system with the effective multiplication $k_{\mathrm{eff}} < 1$ and neutron lifetime $l(\mathrm{s})$, the total number of neutrons in the system will approach a stable value after a sufficient amount of time

$$N = \frac{S_0 l}{1 - k_{\mathrm{eff}}} \tag{13.1}$$

i.e., the total neutron number will be $1/(1 - k_{\mathrm{eff}})$ times $S_0 l$ after a sufficient number of neutron generations. As k_{eff} approaches 1 in the criticality experiments, the total number of neutrons will approach infinity.

The general method for approaching criticality is described as follows: first, determining the subcritical state with safe initial system parameters (fuel loading, volume of core, position of control rod, and moderator level); second, increasing the reactivity of the system and determining the system parameters for the next step by plotting the extrapolated curve according to the neutron flux in the two states; third, repeating these steps and gradually approaching the criticality. In addition, reactivity interpolation is also applied to approach criticality. The system is slightly supercritical, and approaches criticality by adjusting the system parameters. This section introduces the first method.

In the criticality experiments, the traditional method for increasing reactivity is to increase the fuel in the system or gradually increase the moderator. An alternate

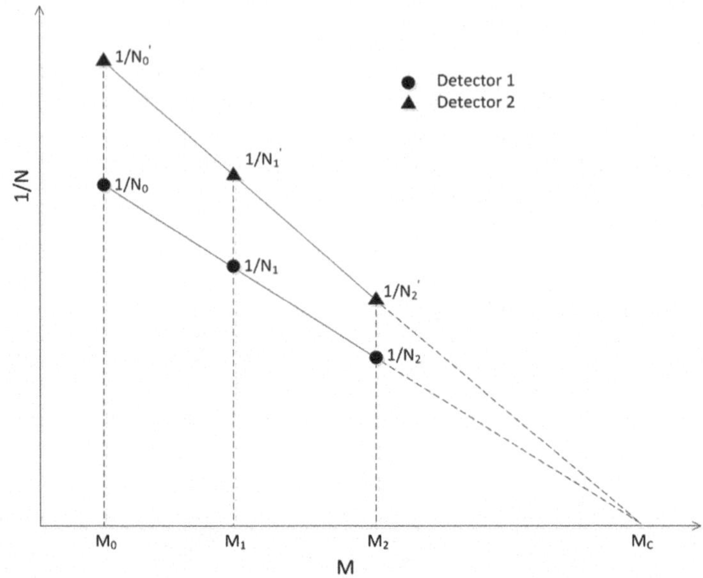

Fig. 13.1 Extrapolation curve of criticality approach

method is to reduce the neutron absorption in the system by withdrawing the control rods or reducing chemical poisons such as boric acid. The method of adding fuel will be introduced as follows.

In the criticality experiments, the amount of fuel is gradually increased and the neutron counting rate is measured to construct an extrapolation curve, as shown in Fig. 13.1. The x-axis represents the amount of fuel and the y-axis represents the reciprocal of the neutron counting rate. Theoretically, as the criticality is approached, the neutron counting rate N tends to infinity, and its reciprocal $1/N$ tends to zero. Ideally, the extrapolation curve is a straight line. Two sets of counting devices are employed. When the fuel loading capacity is M_1, the counting rate is N_1, N_1'. When the loading capacity is M_2, the counting rate is N_2, N_2'. Theoretically, the extrapolation curve intersects the x-axis at point M_c. Taking the result of the first counting device as an example, a similar triangle relation can be deduced

$$\frac{1}{N_1} : \frac{1}{N_2} = (M_c - M_1) : (M_c - M_2) \tag{13.2}$$

The typical arrangement of the criticality experiments includes the start-up neutron source, the core, the detectors, the signal processing system, and the data acquisition systems. The neutron sources that are commonly employed in the experiments consist of isotopic neutron sources and neutron generators. The isotopic neutron sources are small with simple production and they are often utilized during start-up. Several types of isotopic neutron sources exist. One neutron source is based on the

(α, n) reaction. The α particles from ^{210}Po, ^{241}Am, and ^{238}Pu will hit targets such as ^{7}Li or ^{9}Be. For example, the Am–Be source is usually employed in a laboratory, and the Po–Be source can serve as the primary neutron source for start-up of the system. Another type of isotopic source is based on the (γ, n) reaction. The γ particles from ^{24}Na, ^{88}Y, or ^{124}Sb will hit the targets. As an example, the Sb–Be source can serve as a secondary neutron source when the fission system is restarted. The spontaneous fission neutron sources are isotopic neutron sources that undergo spontaneous fission of heavy nuclides. For example, ^{252}Cf can produce fission neutrons near the fission spectrum by spontaneous fission. Neutron generators such as the D–T neutron generator are commonly employed in neutronics experiments, especially for measurements of the kinetics behavior of the system.

Criticality experiments require at least two independent detection devices. The devices should be stable, reliable, and have relatively high neutron resolution ability and are resistant to γ-ray interference. The BF_3 counter tube and fission chamber are commonly employed as neutron detection devices. Considering the method of processing the detector signal, pulse mode detectors are usually applied when the neutron count is low. The current mode detector is employed when the neutron count is high and single counting cannot be distinguished. The output signal of the current mode detector is the average effect of the interaction between the detectors and a large number of neutrons. The different neutron detector signals, such as the neutron pulse signal and the current signal, can be independently collected by a data acquisition card, and multichannel digital signals can be obtained by real-time automatic processing.

The shape of the extrapolation curve will be affected by the relative position of the external neutron source and the nuclear detection device. The neutron flux of the nuclear system can be characterized by the neutron counting rate N recorded by the detector:

$$N = AS_0 + B\frac{S_0 l}{1 - k_{\text{eff}}} \tag{13.3}$$

where the first term is the neutron counting rate contributed by the source neutron and the second term is contributed by the fission neutron in the nuclear system. The coefficients A and B are determined by the relative position of the nuclear system, the neutron source, the detector, and the detection efficiency of the neutron detector. When A is considerably larger than B, the neutron counting rate is dominated by the external neutron source, which cannot reflect the real situation in the reactor. When B is substantially larger than A, the neutron counting rate is dominated by the neutrons produced in neutron multiplication of the system, which is beneficial to the experimental measurement. In designing the experiment, the positions of the detector and neutron source (i.e., the start-up source) should be carefully selected to ensure that A is as small as possible and B is as large as possible to increase the credibility of the extrapolation curve.

In the criticality experiments with fuel loading, the following phenomena may occur:

(1) The neutron counting rate may be less than 1 s^{-1} after loading the neutron source.
(2) The counting rate may undergo slight changes before and after the first fuel loading.
(3) The counting rates for two adjacent fuel loadings undergo slight changes (minimal difference in counting rates).
(4) The extrapolated critical mass may be substantially greater than the calculated value and may even be infinite.

These phenomena are referred to as blind zone. Regarding the safety concerns, the blind zone indicates that the neutron counting rate is too low to perform extrapolation, and the loading process is unable to perform monitoring, which may cause potential safety hazards. The blind zone may be attributed to one or all of these following causes: the source is too low, the detector is not sufficiently sensitive, or the detector is located too far from the neutron source. As the blind zone occurs in the criticality experiment, the principle of 1/3 loading should be adopted, i.e., 1/3 of the difference between the last loading and the estimated value of the critical mass (minimum value of critical masses obtained by theoretical calculation and extrapolation from several counting devices). The principle of 1/2 loading is adopted when no blind zone exists. When the system approaches criticality, the rate of reactivity introduced in the core should not be limited.

13.2.2 Examples

The criticality experiments were performed at the KUCA experimental facility of the Institute of Atomic Energy at Kyoto University in Japan [3]. The actual approach to criticality experiments of the reactor will be introduced. KUCA began operation in 1974 with three different cores. Core A and Core B employed solid moderators with different materials such as polyethylene and graphite. Core C employed light water moderators. The control systems enabled only one core to be functional at one time. Taking Core A as an example, every fuel assembly contained a certain number of fuel plates with U–Al alloy. The number of fuel plates was counted when estimating the criticality loading.

In the actual measurement, the shape of the extrapolation curve will differ due to the influence of spatial harmonics, which may be concave or convex. As shown in Fig. 13.2 [3, 4], the extrapolation curve for detector 1 is concave; the curve for detector 2 is straight, and the curve for detector 3 is convex. For the concave extrapolation curve, the extrapolated mass is less than the practical mass, which is more conservative. Conversely, the convex extrapolation curve introduces a certain amount of error due to the larger extrapolated mass. The extrapolation curve of the straight line is the best curve.

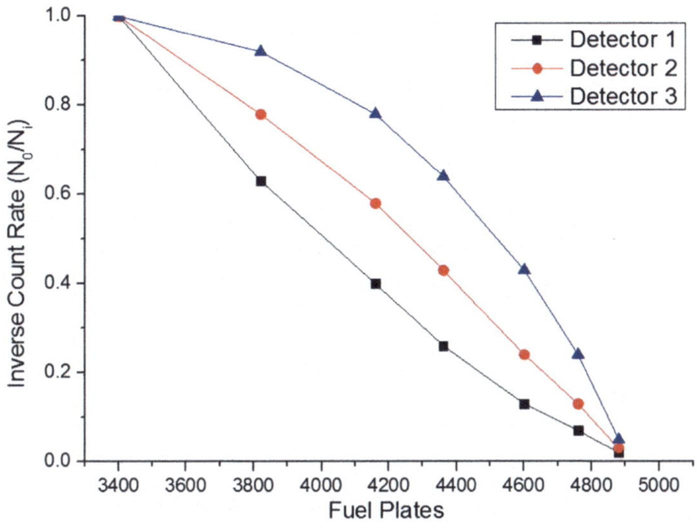

Fig. 13.2 Criticality approach example of KUCA experiments

To study the effect on the core physics caused by lead, lead material was added into the fuel plates in the KUCA experiments. The criticality experimental process is introduced as follows.

(1) A total of 3000 fuel plates existed in the initial fuel loading with the initial neutron flux N_0. The y-axis in Fig. 13.2 is N_0/N_i, which indicates that the inverse count rate is normalized by $1/N_0$. The inverse count rate of the initial fuel loading becomes 1.0.
(2) According to the loading principle, the second fuel loading was 3600 fuel plates, and the neutron flux N_1 was measured.
(3) According to the ratio of N_0 to N_1 and the fuel plate's number of the first two steps, the critical loading was extrapolated, and the estimated number is 4300. According to the 1/2 principle of fuel loading, 350 fuel plates should be loaded, which indicates that the number of fuel plates after the next loading is 3950. After fuel loading, the neutron flux was measured as N_2.
(4) Repeat these three steps. The extrapolated criticality fuel loading will approach the real criticality. The final critical loading was 4800 fuel plates. The estimation was 5% from the experimental result.

13.3 Power Measurement Experiments

Power measurement is essential to the control, regulation, and protection of a fission system. Accurate measurement of the power and power distribution is essential for the safety of a fission system. Power measurement methods can be generally classified into thermodynamic measurement methods and physical methods. The thermodynamic measurement method is suitable for large fission systems. The system power is deduced by measuring the temperature difference between the inlet and outlet of the system coolant and the flow rate. The physical measurement method is employed to measure the fission rate of nuclear fuel in a fission system. The power of a system is obtained according to the energy released from each fission reaction or can be estimated by the neutron flux distribution detected by the detector outside the reactor. This method is suitable for both large fission systems and small fission systems. This section introduces the physical measurement methods.

13.3.1 Methods

The power measurement of a fission system includes two physical measurement methods: in-core measurement and ex-core measurement. In-core measurement can be classified into online measurement and offline measurement, which deduce the total power/power distribution by direct measurement of the fission rate or the reaction rate with detectors in the core. Ex-core measurement is an online measurement method that derives the total power or power distribution according to the neutron flux distribution measured by detectors outside the core. The basic principles of the in-core measurement method and ex-core measurement method are described in the following section.

1. In-core measurement

For online measurement, detectors are set up in the core to measure the power distribution. After the neutron flux or power of the measured positions is obtained by the detectors, the power of other unmeasured points can be determined by an interpolation method, and the power distribution of the whole reactor can be obtained. Typical in-core detectors include micro traveling fission chambers or fixed self-powered neutron detectors. These instruments are small with a fast response speed and tolerance in the strong radiation environment in nuclear systems. For example, rhodium self-powered neutron detectors are used in the China Advanced Research Reactor (CARR). Rhodium metals can release electrons when they interact with neutrons. Neutron flux can be obtained by detecting the electron currents [5].

Offline measurements include the activation foil method and solid-state nuclear track detectors. The activation foil method has the characteristics of flexible arrangement. Activation foils with different threshold energies are placed in several areas for irradiation. The radioactivity of the activated foils is measured by a detector in the laboratory, and the reaction rates are obtained. As the counting is directly proportional to the radioactivity and the radioactivity is related to the density and volume of the foil, the method is limited by the manufacture of materials. Due to the influence of size error, material heterogeneity, and other factors, the shape of foils differ. Thus, an error is produced in the obtained power spatial distribution, which needs correction. The neutron field is also disturbed by the detectors. The main reasons for the disturbance are the neutron absorption of foils and the neutron scattering of media. The disturbance effect has not been well researched, and research on the disturbance of anisotropic field is relatively insufficient.

A solid-state nuclear track detector is commonly employed to measure the fission rate. The advantage of this method is that measurements are not restricted by the neutron flux. A solid-state nuclear track detector is an integral detector for recording charged particle tracks. If the solid-state nuclear track detector is placed at the position r_0 in the fuel element, then:

$$N_T = F_0 T \varepsilon \mu \tag{13.4}$$

where N_T is the number of the tracks in the detector; F_0 is the fission rate at the position r_0 in units of s^{-1}; T represents the irradiation time in units of s; ε is the efficiency of detectors for detecting fission fragments; and μ is the self-absorption correction factor for fission fragments that pass through the fission targets.

ε is known and can be obtained by a table lookup. For thin targets, μ can be deduced from an empirical formula; for thick targets, calibration is required in standard neutron fields. If the parameters T, ε, and μ are known, the fission rate at the position r_0 can be obtained by measuring the number of tracks.

2. Ex-core measurement

The majority of the detectors in a power reactor cannot be placed in-core for a long time due to the high temperature and radiation in the core. Thus, online measurement, which drives the development of an ex-core measurement method, is difficult. Limited by the detector numbers, obtaining the power distribution by relying on the ex-core detectors is also difficult. However, ex-core detectors can be used to measure the discrepancy of power distribution in the axial and radial directions, which can be helpful to assess the homogeneity. The in-core detectors, which can combine ex-core detectors to perform measurements, can be used to calibrate the ex-core detectors and to calculate the weight of detector counts and transforming coefficients of total power.

The traditional methods use ex-core detectors to obtain the power distribution inside the core. Recent studies focus on establishing the linear mapping relationship between the power distribution of the core and the detector counts outside the core by the response function of the detector. New methods such as detector response function methods and harmonic synthesis methods are commonly employed.

In the detector response function method, the variation law of power distribution in the reactor is transformed into the variation law of the detector counts, and information about the power distribution in the reactor is obtained. The spatial response function and core power distribution can be defined as $w(\mathrm{r})$ and $P(\mathrm{r})$, respectively, and the corresponding response of the detector R can be expressed as follows:

$$R = \int_V P(r) w(r) \mathrm{d}r \tag{13.5}$$

Obtaining the complete reactor power distribution $P(\mathrm{r})$ using the response R of the limited external detectors is difficult. With the development of computer technology, a conjugate transport method has been developed to solve the spatial response function. If this method is applied to an advanced fission system, additional neutronics experiments are needed to verify it.

If R and $w(\mathrm{r})$ are known, $P(\mathrm{r})$ can be obtained. For example, the harmonic synthesis method, which is based on the detector response function method, assumes that the combination of core power distributions for different operating conditions is limited. After the combination of characteristics distributions is determined in advance, a small number of ex-core reactor detectors can be relied on to calculate the possible power distribution using the response function. However, this method is only suitable for normal conditions and cannot be applied to predict the power distribution for special conditions after an accident.

In addition, the FDS Team had proposed a neutron field inversion method, which retrieves the neutron field layer by layer from the source response of discrete measurement points. The method calculates the penetration coefficients of neutrons with different energy in different modules, coolants, and structural materials, and constructs the response matrix in energy groups. An inversion method of relay amplification of low energy neutron counting response is proposed. According to the neutron penetration probability, the relay layers between the detector and the neutron field to be measured are innovatively designed. The response of neutron source in the neutron field is amplified layer by layer, and the contribution of the space distribution of neutron field to the detector is divided into energy groups. The neutron field is biased to each counting area according to the nonuniform distribution characteristics, and the fast calculation of whole space neutron field is realized. The neutron field distribution is derived based on ex-situ measurement data, which avoids the disadvantage of traditional in situ measurement method which disturbs the neutron field to be measured. The online radiation safety evaluation of the wide-energy neutron field is realized. The application of the method for the simplified reactor models shows that the inversion results agree well with the Monte Carlo simulation results with 11% average relative deviation. This method is of great significance to improve the safety and economy of nuclear power system.

13.3.2 Examples

1. In-core measurement

Reactor core power has been measured by the activated foil method in CEFR. The foil included metals of a specific thickness (Al, Ni, Au, and other high-purity metals), 90% high-enriched uranium sheets, 20% uranium sheets, natural uranium sheets, depleted uranium discs, high-purity uranium sheets, uranium spheres, and highly depleted uranium spheres. The main detector in the activation measurement is a high-purity germanium γ spectrometer. The detectors were placed in their respective irradiation

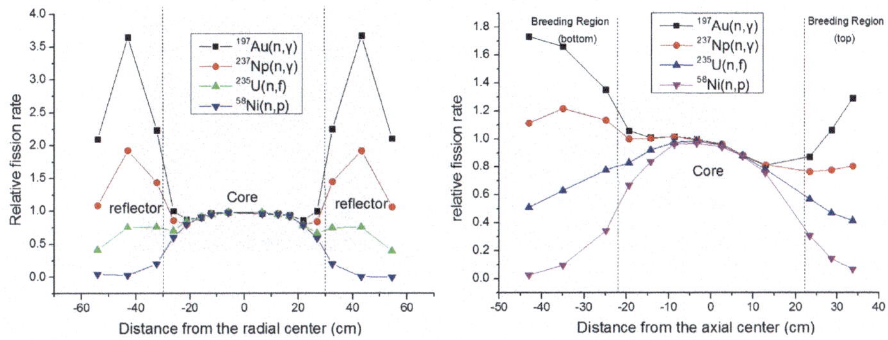

Fig. 13.3 Relative fission rate distribution in radial (left) and axial (right)

devices and installed at the core measuring points for irradiation. After they were irradiated for a specified period of time, the irradiation devices were removed and prepared for counting; the foil packages were separated and placed in small lead tanks. The small lead tanks were transported to an activation measurement laboratory, and the γ radiation count for each activation plate sample was determined using the high-purity germanium γ spectrometers. The reaction rate was estimated. The measurement encompassed several different positions of the reactor with different activation samples to obtain the relative value of the reactivity, and then the power distribution was obtained from the relationship between the reaction rate and power. The radial and axial rate distribution curves obtained after data processing are shown in Fig. 13.3. The relative deviations between the measured values and the calculated values show agreement in the core, with deviations less than 4%. The deviations gradually increase near the core edge and exceed 30% in the reflector zone.

2. Ex-core measurements

Taking the TRIGA reactor as an example, the ex-core power measurement method is introduced. Based on the slit imaging principle, the sensitive region of the detector was wrapped inside shielding materials to ensure that neutrons only enter and count from the thin band created by the slit collimator. By changing the scanning direction of the detector, the neutron counting in different directions were obtained. The two-dimensional power distribution of the reactor core was obtained by an iterative optimization algorithm. Figure 13.4 provides an overlook diagram of the TRIGA core and the measured neutron flux outside the reactor at 40, 100 and 250 kW power [6]. This figure is from the paper "High-intensity power-resolved radiation imaging of an operational nuclear reactor" by Beaumont, J., used under CC BY/cut from original.

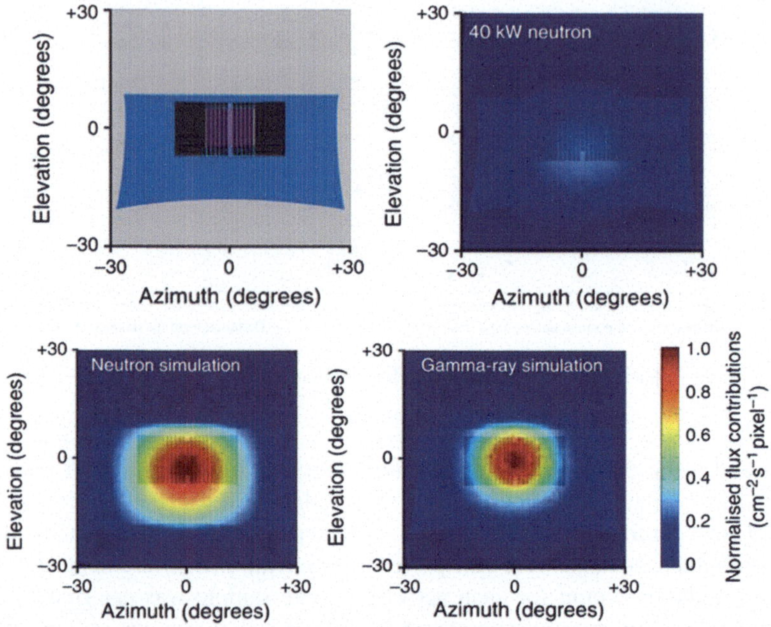

Fig. 13.4 Overlook of the TRIGA core and the flux distribution at different power levels

13.4 Kinetic Parameters Measurement Experiments

Kinetic parameters, including neutron lifetime, prompt neutron decay constant, and effective delayed neutron fraction, reflect the dynamic characteristics of a nuclear system. For a fission system, the accurate measurement of kinetic parameters will directly affect the prediction of the dynamic response characteristics of the system and the credibility of the reactivity measurement. Therefore, the kinetic parameter measurement is an important part of a neutron experiment. This section introduces the kinetic parameter measurement experiment by considering the important kinetic parameter prompt neutron decay constant and effective delayed neutron fraction.

13.4.1 Methods

13.4.1.1 Measurement of Prompt Neutron Decay Constant α

The prompt neutron decay constant (α) represents the number of prompt neutrons in the core along with time and also provides the baseline data for operating and controlling. Numerous methods for measuring the prompt neutron decay constant, such as the Rossi-α method, pulsed neutron source method, variance to mean method, interval distribution method, and ^{252}Cf method, have been developed. The Rossi-α method is a classical method with a simple procedure and equipment. This subsection introduces the principle for the measurement of the prompt neutron decay constant by taking the Rossi-α method as an example.

In the fission process, which is driven by the external source in a fission system, the neutrons in the same fission chain are time-correlated. The prompt neutron decay constant can be obtained by analyzing this correlation. The Rossi-α method measures the correlation with the neutron pulse statistical method, which is suitable for the delayed critical state or slightly subcritical state. In the Rossi-α method, two detectors are deployed inside the core or outside the core. When one counter exists in detector I at $t = 0$, the possibility that one counter is detected within dt interval at t time in detector II can be expressed as

$$P(t)\mathrm{d}t = C\mathrm{d}t + Ae^{-\alpha t}\mathrm{d}t \tag{13.6}$$

where C is the uncorrelated counter rate in units of s^{-1}, and A is the correlated count rate in units of s^{-1}. Based on the measured function of P(t) with time, α, C, and A can be obtained with the least squares fitting method.

Figure 13.5 shows an experimental schematic of Rossi-α [7]. If only one detector is employed, its counter signal is transmitted to both input ports, but it needs to be postponed before entering the input ports. When two detectors are used in the experiment, their signals will separately enter the two input ports.

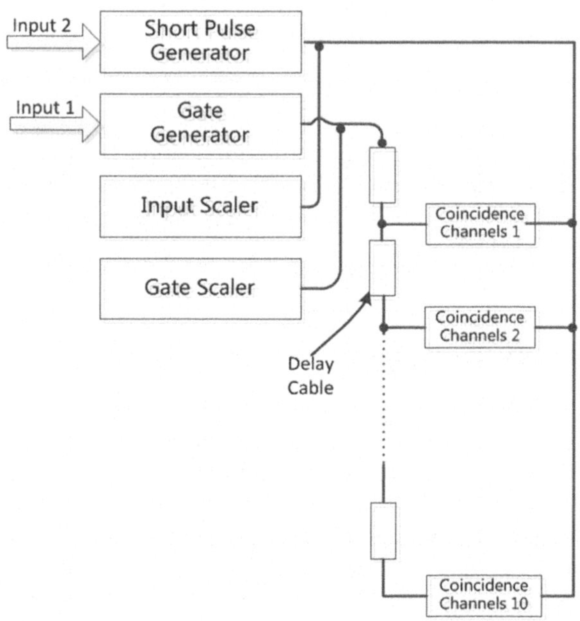

Fig. 13.5 Diagram of the analyzer of Rossi-α method

13.4.1.2 Measurement of Effective Delayed Neutron Fraction β_{eff}

The effective fraction of decayed neutrons β_{eff} is the characteristic parameter representing the contribution of decayed neutrons in the dynamic process of fission system. The β_{eff} is dependent on the specific system due to the influences of neutron leakage and spectra. The measurement methods of β_{eff} include Nelsonm method, Feyman method, Cf source method, Bennet method, covariance-to-mean method, noise analysis method, etc. Recently, some new methods such as "Determine β_{eff} by measuring reactivity and fixing k_{eff}" have been proposed. This new-proposed method with high accuracy and easy operation will be introduced.

Based on the prompt neutron decay constant α, which has been measured, the reactivity ρ of this subcritical system in unit of β_{eff} can be obtained as

$$-\frac{\rho}{\beta_{\text{eff}}} = \frac{\alpha_c - \alpha}{\alpha_c} \tag{13.7}$$

where α_c is the prompt neutron decay constant for the decayed critical system in units of s^{-1}. For systems with different subcriticality, the relationship between α_i and the detector's average counter rate $\overline{N_i}$ is

$$\alpha_{\text{i}} = \alpha_c + \frac{A}{\overline{N_i}} \tag{13.8}$$

where A is a constant in units of s^{-1}. With the measured α_i and $\overline{N_i}$, α_c can be calculated using the least squares fitting method. Thus, the β_{eff} can be expressed as

$$\beta_{\text{eff}} = -\frac{\rho_i}{(\alpha_c - \alpha_i)/\alpha_c} = \frac{(k_{\text{eff},i} - 1)/k_{\text{eff},i}}{(\alpha_c - \alpha_i)/\alpha_c} \tag{13.9}$$

where $k_{\text{eff,i}}$ is the effective multiplication factor for a subcritical system and can be obtained by calculation. The uncertainty of the calculated $k_{\text{eff,i}}$ is small, and uncertainties from the calculation methods, models and nuclear data can be partially canceled.

13.4.2 Examples

Rossi-α experiments have been performed in the MUSE-4 Reference core in France [8, 9]. Eight runs of 2800 s were performed with time-marking acquisition systems. The Rossi-α experimental data series were analyzed using a least squares fitting method. To evaluate the goodness of fit, graphical and numerical indicators were applied. The scatter plot should not display any pattern or trend. Based on Eq. (13.6), the Rossi-α fitting model was $p(\text{t}) = P/C$. The estimate of α was $8017 \pm 673\ s^{-1}$.

Table 13.1 Measured results of β_{eff} in FCA (unit: pcm)

Organization (Meas. Tech)	XIX-1 (U core)	XIX-2 (Pu/NU core)	XIX-3 (Pu core)
CEA (Noise)	743 ± 19	–	250 ± 6
IPPE (Rossi-α)	771 ± 25	–	–
IPPE (Cf)	706 ± 30	351 ± 10	244 ± 7
JAERI-KAERI (Cf)	735 ± 20	358 ± 10	249 ± 7
JEERI (Cov. to Mean)	724 ± 13	–	252 ± 5
Lanl (Nelson)	737 ± 20	–	–
Nagoya (Bennett)	782 ± 16	368 ± 6	256 ± 4
Calculation	745	366	244

From 1995 to 1998, to improve the accuracy of predicting β_{eff} and decrease its uncertainty, several laboratories in France, the United States, Russia, Japan, and South Korea performed the benchmark experiment of β_{eff} on the French zero-power-facility MASURCA and Japan's Fast Critical Assembly (FCA) [10]. The employed methods include the Nelson method, Feyman method, Cf source method, Bennet method, covariance-to-mean method, and noise analysis method. The measurement results are listed in Table 13.1. The experimental uncertainty was less than 3%. The calculated results and experimental results showed good consistency.

13.5 Reactivity Measurement Experiments

Reactivity is important to the safe operation and control of a fission system. A reactivity measurement is one of the core contents in the physics experiment of fission systems. Accurate measurement of reactivity is very important to the safety of a fission system. Based on the measurement of the reactivity, the control rod worth can be calibrated, and the reactivity coefficient can be calculated. This section will introduce the principles of several commonly employed reactivity measurement methods and illustrate how to conduct rod worth and reactivity coefficient measurement [11, 12].

13.5.1 Methods

The reactivity cannot be measured directly, but can be indirectly obtained based on reactor dynamic theory by using some measurable physical quantities such as neutron flux and nuclear power. In general, the reactivity measurement methods can be classified into three categories: static measurement, dynamic measurement, and noise measurement, which are only intended for the steady state around the reactivity

Table 13.2 Methods of reactivity measurement

Methods	Characteristics
Source Multiplication Method	Static measurement with criticality extrapolation
Period Method	Dynamic measurement with period above certain limit. Online method
Rod Drop Method	Dynamic measurement that can calibrate the rod worth. Offline method
Source Jerk Method	Dynamic measurement that can be used in high power reactor
Pulsed Neutron Source Method	Dynamic measurement that can be used in the case of low power or reactor start-up
Inverse Kinetics Method	Dynamic measurement that can consider delayed neutrons and be performed online
Noise Method	Noise measurement with low power level and long measuring time

transient, the transient process and steady state at a very low power level, respectively. Some common measurement methods are listed in Table 13.2 [13]. In this subsection, some typical reactivity measurement methods are discussed.

13.5.1.1 Period Method

The period method determines the reactivity by the measuring reactor period, which is easily performed. The theoretical basis of this method is the reactivity equation (also referred to as the inhour equation), which relates to the reactor period T and the reactivity. For example, the reactivity equation, which is based on the space-independent point kinetics equation, can be written as

$$\rho = \frac{\Lambda}{T} + \sum_{i=1}^{6} \frac{\beta_{\text{eff},i}}{1 + \lambda_i T} \tag{13.10}$$

where β_{eff} is the delayed neutrons fraction ($\beta_{\text{eff,i}}$ for group i), λ_i is the decay constant of a delayed neuron for group i in units of s^{-1}, Λ is the neutron generation time in units of s, and ρ is the reactivity.

Since β_{eff}, λ_i, and Λ are known, the reactivity can be calculated after measuring the reactor period T. The doubling time is defined as the time in which the neutron flux varies 100%. The limit of the system stable period T or doubling time is an important operating limit for a critical system, which must exceed a certain lower limit. For example, the doubling time should exceed 20 s for a power reactor.

Without additional measurement facilities, the reactor period can be measured using existing facilities to monitor power, which is necessary for safe operation and is a distinct advantage for the period method. Two approaches are available to measure the reactor period: offline approach and online approach.

(1) Offline approach: use a second chronograph and meters monitoring power to perform the measurement. Note that, the γ-radiation-resistant detector should be used to measure the rising period due to high-intensity γ-radiation. The measurement of period should not be affected by the change in power. Zero-point correction should be performed for these meters, and zero shift could not exceed the indication error within the range of measurement.
(2) Online approach: The measurement meters are similar to those mentioned in Sect. 13.2.1. The requirements for the detectors are identical to those for offline measurement.

13.5.1.2 Pulsed Neutron Source Method

After injecting a bundle of pulsed neutrons into the fission systems to be measured, the decay characteristics of the prompt neutron and delayed neutrons generated by fission are related to the reactivity of the system. Thus, the reactivity can be reconstructed according to the decay characteristics of the neutron flux with time. This method is the pulsed neutron source method. Based on the point kinetics equations, several methods can be applied to study the decay characteristics of the neutron flux, such as the prompt neutron decay method and the area method.

The prompt neutron decay method determines the reactivity based on the decay characteristics of the prompt neutron in fundamental mode. After the pulse neutron is injected into the fission system, the decay constant of neutrons in fundamental mode can be obtained according to the characteristics of the decay curve. The evolution of the prompt neutron flux $\phi_{\mathrm{p}}(r,t)$ over time t can be written as

$$\phi_{\mathrm{p}}(r,t)=\sum_{n=0}^{\infty}A_{n\mathrm{p}}\phi_{n\mathrm{p}}(r)e^{-\alpha_{n\mathrm{p}}t}\approx A_{0\mathrm{p}}\phi_{0\mathrm{p}}(r)e^{-\alpha_{0\mathrm{p}}t}=A_{0\mathrm{p}}\phi_{0\mathrm{p}}(r)e^{\frac{\rho-\beta}{\Lambda}t} \tag{13.11}$$

where A_p is the amplitude of the prompt neutrons (A_{ip} is the amplitude for $n=i$). If β/Λ is known, the reactivity in units of dollar ($) (the value of the reactivity of 1 $ is equivalent to the effective delayed neutron fraction) can be obtained. If β/Λ is unknown, the $\alpha_{0\mathrm{p}}$ value can be obtained by measuring the value of β/Λ at different subcriticalities and extrapolating to the critical condition. The prompt neutron decay method is simple and has a minimal effect on the spatial effects, but cannot be applied in deep subcritical cases.

The area method determines the reactivity by calculating the ratio of the areas under the count curves of the prompt neutrons and the delayed neutrons. The prompt (delayed) neutron flux is integrated over time.

$$I_{\mathrm{p(or\ d)}} = \int_0^{1/R} \phi_{\mathrm{p(or\ d)}} \mathrm{d}t \tag{13.12}$$

Divide the prompt portion by the delayed portion to obtain the reactivity in units of $

$$\rho_{\$} = \frac{\rho}{\beta} = \frac{I_{\mathrm{p}}}{I_{\mathrm{d}}} = \frac{\int_0^{1/R} \phi_{\mathrm{p}} \mathrm{d}t}{\int_0^{1/R} \phi_{\mathrm{d}} \mathrm{d}t} \tag{13.13}$$

As the area method does not require knowledge of the kinetic parameters, it can be used in cases of deeper subcritical levels. When the repetition frequency R of the periodic pulse is sufficiently high, the delayed neutron flux almost remains constant for the subcritical system in the quasi-steady state; thus, it can be distinguished from the prompt neutron flux.

After selecting the experimental parameters, some preparations for the experiment are required, including [14, 15]:

(1) Instrument pre-operation. Before the formal measurements, the experimental instruments should not work for a period of time until the stable working state is attained.
(2) Background measurement. After the pre-operation of the experimental instrument, the measurement of the background of the fission system is performed. The accuracy of the background measurement directly affects the accuracy of the final results. In the measurement, the multichannel analyzer (such as 100-channel), whose channel width can be set to 10 s, is used to obtain background counters, which are summed to obtain the background count per second.
(3) Parameter adjustment. The pulsed neutron source and the multichannel analyzer are not operated until sufficient data in each channel is accumulated to verify whether the parameter selection is appropriate. If appropriate, the formal experiment is launched. Otherwise, readjustment of the experimental parameters is needed.

After launching the formal experiment, the number of neutrons in the fission system is required to attain a dynamic equilibrium state, that is, in a neutron pulse period, the newly generated neutrons and the disappeared neutrons are equal. This neutron balance is the delayed neutron balance. Since the sufficient neutron pulses are injected into the reactor in the experiment preparations, the number of neutrons in the system has attained a state of dynamic equilibrium. After the preparations, therefore, the experiment can be directly launched without adjusting the experimental parameters. After launching the pulsed neutron source, the neutron flux is measured, and the measurement steps are the same as those in the background measurement. For a certain subcriticality, measurement should be repeated three to five times to minimize the statistical errors of measurement.

During the formal measurement, a small change in the background neutron will occur due to the slight change in the state of the fission system, which is caused by temperature and other reasons. After the neutron pulse injection is stopped, measurement of the background neutron is not conducted again until the precursors of the delayed neutron are decayed to approximately 15–30 min. During the data processing, the measured background neutrons before and after formal measurements are averaged by weight as the level of reactor background neutrons.

13.5.1.3 Rod Drop Method

The rod drop method is an extensively employed approach to measuring reactivity. In the rod drop method, the reactivity is measured by the decay characteristics of the neutron flux with time after the control rod is instantaneously inserted into the core. The response function can be derived from two analytical models based on point kinetics equations: integral model and differential model. In the integral model, the reactivity is calculated by the steady-state neutron count n' before dropping the rod and the neutron counts integrated over the infinite time after dropping the rod. In the differential model, the reactivity is obtained with the two steady-state neutron counts n_0 and n_1 before and after dropping the rod.

$$\rho_{\$} = -\frac{n' \sum_i^N \frac{\beta_{i\mathrm{eff}}}{\beta_{\mathrm{eff}} \lambda_i}}{\int_0^\infty n \mathrm{d}t} \tag{13.14}$$

$$\rho_{\$} = \frac{n_1 - n_0}{n_1} \tag{13.15}$$

Since the reactivity measured in the previously mentioned period method is positive, the reactivity in each measurement should not be too high. However, as the negative reactivity is inserted, the dropping rod will not cause critical safety problems. To obtain accurate measurement results, the control rod should be completely dropped into the core within a short time (order of ~0.1 s).

13.5.2 Examples

The method of measuring reactivity, as previously introduced, can be employed in the measurement of control rod worth (CRW) and reactivity coefficient, which will be introduced in the following section.

13.5.2.1 Measurement of Control Rod Worth

As a strong absorber, the control rod is an important device for adjusting the reactivity of the fission system. When the control rod is inserted from an initial reference position to a certain height, the reactivity introduced in the process equals the control rod integral worth at the insertion height. The reactivity change is caused by moving the control rod by a unit distance at a certain height. It is defined as the differential worth of the control rod near this height.

In the design and operation of fission systems, it is necessary to know not only integral CRW but also the differential CRW at different positions. The CRW needs to be measured before the nuclear systems are officially put into operation as the CRW is used to validate the nuclear design and provide the necessary data for the safe operation of the nuclear system.

1. Period method for rod worth measurement

The period method can be extensively employed to measure the CRW and has the advantage that additional equipment is not required. The whole rod can be divided into several sections for measurement, and each measured reactivity is generally less than the reactivity that corresponds to the doubling time of 20 s. Since the period method is applied to measure the reactivity corresponding to the stable period, a certain waiting time is needed in the actual measurement to enable the nuclear system period to approach its progressive value. Note that, the waiting time should be limited to avoid an excessive increase in the power of the nuclear power system, which is not conducive for measurement. In general, the waiting time is the key to the error of the period method. If the waiting time is taken as several periods, the measurement accuracy can be achieved by a level of 1%. The choice of the period should not be too short and must be greater than the specified value (e.g., doubling time should be larger than 15 s for research reactors) to comply with safety regulations. In general, the doubling time can be larger than 20 s.

For example, the calibration of the VENUS-F control rods was performed according to the period method. Figure 13.6 shows the reference critical core F01/27 [16]. Special experimental lead assemblies were installed at the four corners and in position (−2, −6). Two control rods (designated CR1 and CR2) were placed in positions (−6, −2) and (6, 2), respectively.

From a critical state, the reactor became supercritical by extracting CR1. After measurement of the period, the reactivity insertion worth at one position of CR1 was determined, in pcm/mm. The reactor became critical again after the insertion of CR2. This action provided a measurement point for the control rod calibration curves of both CR1 and CR2. This action was repeated by extracting CR2 step by step, which provided measurement points for both calibration curves. As the C2 compensated for the reactivity of the C1, the worth of C2 was simultaneously measured while measurement for C1 was conducted. The method of measuring C2 is also referred to as the "compensation method". To avoid interference effects, the distance between the C2 rod and the C1 rod should be limited, and the distance of the control rod should be larger than the thermal neutron diffusion length. Afterward, an extrapolation was made for the remaining curve. By integrating the CR1 and CR2 control rod calibration curves over the length of the control rod, the total worth of the control rods can be determined. In Fig. 13.7, the control rod calibration in the most reactive VENUS-F core is shown. When extrapolated, the worth of 443 pcm and 448 pcm are obtained for CR1 and CR2, respectively.

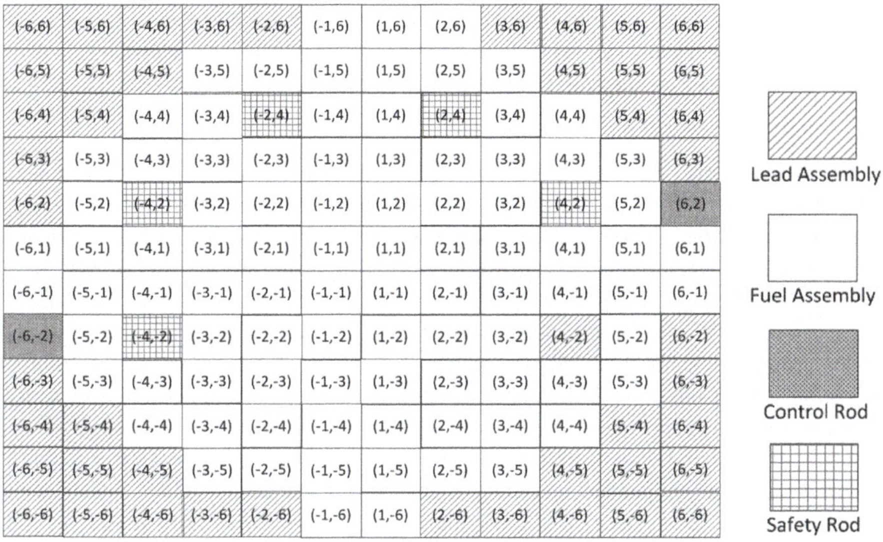

Fig. 13.6 VENUS-F reference critical core F01/27

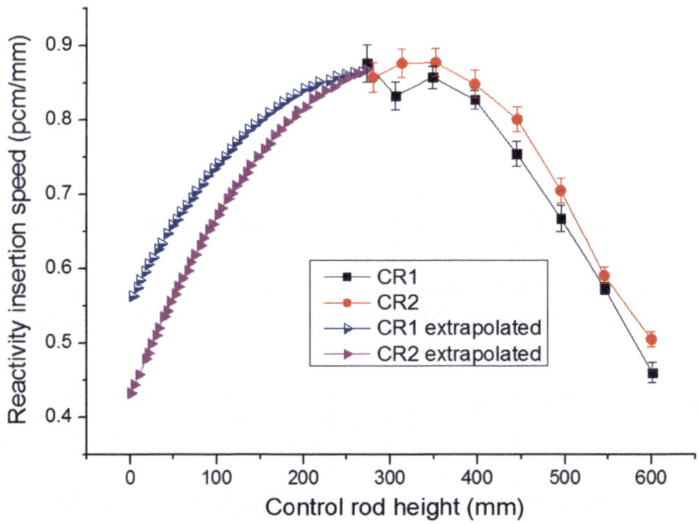

Fig. 13.7 Control rod calibration curve of VENUS-F

The period method is commonly applied to a control rod with small worth since the doubling time should not be shorter than 20 s. For control rods with large worth, only segmentation measurements can be employed.

2. Dynamic rod worth measurement

In the early 1990s, Westinghouse developed a dynamic rod worth measurement method and successfully applied it in countries such as the United States, the United Kingdom, and Japan. The Daya Bay Nuclear Power Plant and the Ling Ao Nuclear Power Plant in China introduced this technology in 2009. Unlike traditional methods, in which the system had to attain a certain steady state or a relatively "static" balance, the CRW can be measured in dynamic conditions with the dynamic rod worth measurement method. At the preparation stage of measurement, all control rods were lifted to the top of the core, which must have a certain positive reactivity. The control rods to be measured were continuously inserted at the bottom of the core from the top and were withdrawn to the core top. During this process, the CRW was measured using specific reactivity measuring instruments. After withdrawing the control rods, the system was maintained until positive reactivity was achieved, and the neutron flux was recovered. The other bundles were measured by repeating these steps until the value measurement of all control rods was completed. The CRW was obtained based on the calculated correction factor.

13.5.2.2 Measurement of Reactivity Coefficients

When the nuclear system is in a normal operation state, the reactivity of the system will change with the operating parameters, such as power, pressure, temperature, and void. The reactivity coefficient is defined to describe these phenomena, the safety state of the system, and the neutronics characteristics via the measurement of the reactivity coefficients. This section discusses the experimental measurements of the reactivity coefficients, including temperature coefficient, power coefficient, and void coefficient.

1. Temperature coefficient

The temperature coefficient represents a reactivity change caused by a temperature variation of 1 degree at a certain temperature; it is an important parameter that reflects the inherent safety of nuclear systems. The moderator temperature coefficient is generally negative. In large nuclear power systems, the experiments of temperature coefficient measurement are performed before the system power is increased to a higher level, usually after a criticality experiment, to eliminate the disturbance of poisons. The change in reactivity caused by the temperature variation of the moderator can be counteracted by moving the regulating rod. By measuring the moving distance of the regulating rod, the value of the temperature-induced reactivity can be obtained. The experimental procedures are described as follows.

(1) Start-up the system, and lift power to the level at which the regulating rod can be automatically adjusted. The position of the regulating rod is adjusted to the

lower part of the position where the rod worth is linear with the rod length, and the rod position is recorded.

(2) Record the inlet and outlet temperature of the coolant as T1.
(3) Increase the coolant temperature to ensure that the moderator temperature in the core increases to T2 and stabilizes for a period at this temperature. The position of the regulating rod is recorded.
(4) Continue to increase the moderator temperature to T3 and stabilize at this temperature for a period of time. Record the position of the regulating rod.
(5) Repeat step (4) until the regulation rod moves to the upper end of its linear segment.
(6) Adjust the position of the control rod back to the lower end of its linear segment and repeat steps (4) and (5).

During the experiments, it is important to ensure that the positions of the regulating rod that correspond to all measuring points are in its linear segment. If some of the experimental points are not in the linear segment of the regulating rod, they should be abandoned. One rod position should have at least three experimental points. Otherwise, the least square method cannot be employed.

For example, the temperature coefficients were measured with a power of 30 kW and 9 MW at the beginning and half-life of the HTTR in Japan. In the experiment, the power was maintained at 30 kW or 9 MW and the core temperature was increased. The positions of the control rods were adjusted according to ensure that the control rod R3 was pulled out and the position difference of C, R1 and R2 rods was less than 2 mm. According to the positions of the control rod before and after the change in temperature, the variation in reactivity was obtained from the reactivity worth curve of the control rods, and the temperature reactivity coefficient can be obtained by dividing the variation in reactivity by the change in temperature.

2. Power coefficient

The reactivity change caused by the change in unit power is referred to as the power reactivity coefficient or the power coefficient. The variation in power can cause changes in the temperature of moderators and the fuel in nuclear power systems, which cause changes in the reactivity. Therefore, using the power coefficient to represent the change in reactivity with power is more intuitive than directly using moderator temperature and fuel temperature coefficient.

To measure the power reactivity coefficient, the system is adjusted to a given power level, and the position of the control rod is recorded after the temperature of the medium is stable in the reactor. The power of the system is adjusted to another power level. After the temperature of the medium is stable in the reactor, the position of the rod is recorded. By repeating these steps, the relationship between different power levels and rod positions can be obtained, and the power coefficient can be obtained.

For example, when increasing the power level of CARR to the maximum, the power coefficient measurement experiments were repeated three times. The steps of each experiment are described as follows: first, the system operates at low power and record parameters such as power and position of the control rod, as a reference for measurement; second, the system power is lifted by steps, and after the power is stabilized, the relevant parameters are rerecorded; last, the power reactivity coefficient is obtained by calculating the change in reactivity before and after the movement of the control rod and using the reactivity worth curve of the control rod. The power coefficient of the CARR core when fully loaded is -3×10^{-8} $\Delta k/k \cdot kW^{-1}$.

3. Void coefficient

The void coefficient is defined as the reactivity variation when the volume of void varies at a given position. The local boiling of liquid coolant will generate bubbles, and its density will be smaller than the density of the coolant. The decrease in the liquid density will reduce the number of atoms in the liquid per unit volume, which reduces the slowdown power of the coolant. Generally, the influence of the void on the reactivity can be positive or negative. When the core runs from the beginning of the life cycle to the end of the life cycle, the void coefficient may change.

For example, the experimental method of the CEFR is to replace the fuel assembly at the measurement point with an experimental assembly that simulates the sodium void and measure the reactivity changes caused by replacement of the assembly. The structure of the experimental assembly and fuel assembly are similar. The influence of the flux caused by the control rod should be avoided at these measuring points. The positions of the measuring points were selected from the positions of the fuel assemblies of the four types of pins. The principle was that the ^{235}U content of the fuel assemblies at the measuring points was the least different from that of the experimental assemblies.

The experiment is performed after conducting the criticality experiment. Before the beginning of the experiment, the sodium void experimental assembly needs to be preheated. The preheated sodium void experimental assembly is transported into the core and placed in the storage location. After preparation, the experimental steps are summarized as follows:

(1) Pull out one fuel assembly, place it in another storage location, and place the sodium void assembly in the position of the fuel assembly for the criticality experiment.
(2) Pull out the sodium bubble assembly and place it in the storage position. Return the fuel assembly that was pulled out in the previous step back to the original position and restore the core to the original state.

The same method is applied for other measuring points. Both the period method and the inverse kinetics method are used to measure the reactivity before and after the assembly substitution. Considering one measuring channel with the period method as an example, the reactivity variation for five measuring points are −40.3, −47.4, −40.8, −43.4, and −34.5 pcm. The results are consistent with the calculated results from the CITATION software, which are −38.5, −39.4, −36.5, −35.1, and −32.3 pcm. The reactivity variation for the sodium void effect from the fuel assemblies of the whole core is −3159 pcm, which has 15% discrepancy from the calculated value of −3644 pcm.

13.6 Prospects

Regarding the development of a fission system, a large number of neutron experiments for fission systems have been performed. Important results have been achieved, which provides an important foundation for the commercial application of fission systems. The development of advanced fission systems has introduced new requirements for neutronics experimental research. The prospect of neutronics experiments on advanced fission systems is discussed in this section.

1. Criticality experiments

The criticality experiments of an advanced fission system remain based on extrapolation but should be adjusted according to the characteristics of nuclear systems. For example, a liquid molten salt reactor contains mobile fuel; thus, the fuel loading process differs from the traditional method, and the criticality extrapolation method needs further studies.

2. Power measurement experiment

Advanced fission systems require additional study of the corresponding reactor power distribution characteristics. Since the deployment of detectors inside the core is difficult in advanced nuclear systems, a new trend exists for the development of an advanced fission system to retrieve the core power distribution by the measured results of ex-core detectors. This method needs an appropriate response matrix and limited number of detectors. Regarding the detector techniques, it is hard to satisfy the requirement of efficiency and resolution for the detectors used in the fast reactors, which means the more research is needed for detector techniques.

3. Kinetic parameters measurement experiment

Advanced fission systems pose higher requirements on the measurement instruments and measurement methods for kinetic parameters. For example, in a pulsed neutron source experiment that employs a neutron tube, the neutron pulse width is generally expressed in microseconds. The waveform is not ideal, and the front and rear edges are not sufficiently steep. If the pulsed neutron source method is applied to measure the decay constant α for prompt neutrons in thermal neutron systems, the neutron tube has a minimal effect on the measurement as α is small in thermal neutron systems. For fast reactor systems, however, α is large and the width of a neutron pulse waveform and front and rear edges are required on the order of nanoseconds, which poses high demand for experimental facilities. In this case, the pulsed neutron source is not always suitable for these systems. The development of an accelerator neutron source with suitable pulse width is necessary to satisfy the experimental requirements, and a detector with a fast response and short shaping time is needed.

4. Reactivity measurement experiment

Experiments for measuring the reactivity of advanced fission systems are also needed. For example, a liquid molten salt reactor contains liquid fuel, and a delayed neutron precursor can flow in and out of the reactor core. The traditional point kinetics equations cannot be described, and the corresponding experimental methods need to be developed. The complex distribution of the fast neutron spectrum and the more complex spatial effect will affect the reactivity measurement, which deserves further studies [17].

Advanced nuclear systems impose new requirements on the nuclear experimental data. For example, the cross sections of actinide and minor actinide in resonance regions are very important in the performance assessment of transmutation and proliferation for fast reactors. The inelastic scattering cross sections of Pb and Bi in lead-based reactors and the elastic and inelastic scattering cross sections of ^{23}Na in sodium fast reactors are important to the critical safety calculations. However, these experimental data remain scarce and the uncertainty is large. Additional measurements are required in the future.

References

1. Wu YC (2017) Fusion neutronics. Springer Nature Singapore Pte. Ltd.
2. Wu YC (2018) Multi-functional neutronics calculation methodology and program for nuclear design and radiation safety evaluation. Fusion Sci Technol 74(4):321–329
3. Pyeon C, Hervault M, Misawa T et al (2008) Static and kinetic experiments on accelerator-driven system with 14 MeV neutrons in Kyoto University critical assembly. J Nucl Sci Technol 45(11):1171–1182
4. Tsuyoshi M, Hironobu U, Pyeon C (2010) Nuclear reactor physics experiments. Kyoto University Press
5. Jammes C, Chapoutier N, Filliatre P et al (2014) Neutron flux monitoring system of the French GEN-IV SFR: assessment of diverse solutions for in-vessel detector installation. Nucl Eng Des 270:273–282
6. Beaumont JS, Mellor MP, Villa M et al (2015) High-intensity power-resolved radiation imaging of an operational nuclear reactor. Nat Commun 6:8592
7. Orndoff JD (1957) Prompt neutron periods of metal critical assemblies. Nucl Sci Eng 2:450–460
8. Villamarin D, Perret G, Gonzalez E et al (2011) First measurements of the kinetic response of the MUSE4 fast ADS mock-up to fast neutron pulse. In: Proceedings of 7th exchange meeting actinide and fission product partitioning and transmutation
9. Soule R, Assal W, Chaussonnet P et al (2004) Neutronic studies in support of accelerator-driven systems: the MUSE experiments in the MASURCA facility. Nucl Sci Eng 148(1):124–152
10. Okajima S, Sakurai T, Lebrat JF et al (2002) Summary on international benchmark experiments for effective delayed neutron fraction (β_{eff}). Prog Nucl Energy 41(1):285–301
11. Ott KO, Neuhold RJ (1985) Introductory nuclear reactor dynamics. American Nuclear Society
12. Rineiski A, Maschek W et al (2005) Kinetics models for safety studies of ADS. Ann Nucl Energy 32:1348–1365
13. Zhang XX (2017) Research on subcriticality reconstruction based on inverse kinetics method for subcritical systems. University of Science and Technology of China (in Chinese)
14. Baeten P, Abderrahim HA (2003) Reactivity monitoring in ADS, application to the MYRRHA ADS project. Prog Nucl Energy 43(1–4):413–419
15. Shi Y, Zhang W, Cao J et al (2007) China ADS sub-critical experimental assembly-Venus-1 and preliminary experiment. Front Energy Power Eng Chin 1(2):150–157

16. Uyttenhove W, Baeten P, Ban G et al (2012) Experimental results from the VENUS-F critical reference state for the GUINEVERE Accelerator Driven System project. In: International conference on advancements in nuclear instrumentation measurement methods and their applications. IEEE
17. Alan EW, Donal RT, Pavel VT (2012) Fast spectrum reactors. Springer Science

Chapter 14
Neutronics Experiments of Fusion Systems

Since the 1950 s, with the development of fusion systems, many fusion neutron source facilities have been built around the world. A series of experiments have been performed to support the development of fusion systems with these facilities that aim to validate the reliability of neutronics design, the precision of nuclear data and the correctness of computational methods and tools [1].

In this chapter, a general introduction of fusion neutronics experiments is first presented. Then, certain typical experiments such as activation experiments, tritium breeding experiments, irradiation experiments, and shielding experiments are introduced. Last, the prospects for the development of fusion neutronics experiments are discussed.

14.1 Contents

In a fusion energy system, the neutrons released by the fusion reaction interact with the materials of the components, such as the first wall, divertors, blankets, vacuum vessels, superconducting magnets and bio-shielding, resulting in the activation of materials, breeding the tritium in the blanket and inducing irradiation damage in components such as superconducting magnets and vacuum vessels. In addition, a fraction of neutrons can escape from the fusion system, causing potential radiation hazards for personnel and the environment.

The early neutronics experiments focused on the fission system, obtaining numerous experimental data related to the fission neutron spectrum. With the development of fusion energy systems, new requirements have been put forward for experimental data. The D−T reaction in a fusion system produces neutrons with an average energy of 14.06 MeV, while the average neutron energy in the fission system with ^{235}U as primary fuel is approximately 2 MeV, and the fraction of neutrons with energy higher than 10 MeV is very low. Compared with low energy neutrons, high-energy neutrons have stronger penetrating capability, and the physical processes of the interactions with media are also different. Therefore, fusion neutronics experiments that investi-

Y. Wu, *Neutronics of Advanced Nuclear Systems*,
https://doi.org/10.1007/978-981-13-6520-1_14

gate and validate the distribution of high-energy neutrons over various subsystems and their interactions with matter are crucial.

Compared with the experiments for fission systems, the neutronics experiments for fusion systems currently focus on basic research. Therefore, experiments focusing on material activation, tritium breeding, material irradiation damage, and shielding are necessary to validate the accuracy of computational methods/codes, the precision of nuclear data libraries and the accuracy and reliability of neutronics design schemes.

1. Material activation experiments

The neutrons in fusion systems have high flux and energy, which easily results in the activation of materials. These materials include the steel of structural materials used as blankets, tungsten, or tungsten alloys used as first walls and divertors, and copper alloys used as divertors. For example, for a 1000 MW fusion demonstration reactor, there may be approximately 3500 t of low activation steel for blankets. To ensure material disposal or recycling, it is necessary for activation experiments to validate whether the materials used in the design satisfy the requirements of low activation.

2. Tritium breeding experiments

Tritium self-sufficiency is a pre-condition of the steady state operation of fusion systems. The scientific design of blanket to breed the tritium with neutrons together with the tritium extraction technique is required to achieve a self-sustaining fusion reaction. It is necessary to conduct tritium breeding experiments to validate whether the blanket neutronics design satisfies the requirements of tritium self-sufficiency.

3. Material irradiation damage experiments

In fusion systems, the components, such as the first wall, divertors, blankets and superconducting magnets, are subjected to the severe irradiation of a mixed field of neutrons and photons (e.g., ITER produces a neutron yield of 1.8×10^{20} s^{-1} at an operation of 500 MW power). The superconducting magnets that confine the high-temperature plasma are sensitive to neutron irradiation. There are issues with the effects of neutron irradiation effects on these parts, and so material irradiation damage experiments are one of the most important activities in fusion neutronics research.

4. Shielding experiments

In fusion systems, the complex geometry together with large-scale space and thick shielding raises challenge for shielding design and simulation, resulting in calculated results that may not always agree with the actual ones. To ensure the radiation safety of both the components and the personnel, it is necessary to perform shielding experiments to validate the accuracy of the calculation methods/codes and the reasonability of the design schemes.

14.2 Material Activation Experiments

The material of each component in a fusion system can be activated by high-energy neutrons. These activated materials have key influences on personal radiation safety and waste disposal; in particular these materials are significant radiation sources for external exposure of workers during maintenance tasks after shutdown. The activation characteristics (such as activity, decay heat and contact dose rate) of these activated materials are key factors that decide whether these materials are suitable for fusion systems.

Based on the design and construction plan for ITER, the materials used in the systems and components are mainly plasma facing materials, structural materials, heat deposition materials, neutron multiplicative materials, tritium breeding materials, and magnetic materials (including superconductor and insulation materials). The common candidates for the above materials are the following:

(1) Plasma facing materials: W and W alloys, Be, graphite, SiC, and carbon fiber are used for the first wall and divertor.
(2) Structural materials: austenitic steels, reduced activation ferritic/martensitic steel (such as CLAM, Eurofer, and F82H), vanadium alloys, and SiC are used for the blankets, support of the ports and port plugs.
(3) Copper alloys: ODS-Cu and CuCrZr are used for heat sink material of divertor.
(4) Neutron multiplicative materials: Be and Pb are used for the neutron multiplication in the blankets.
(5) Tritium breeding materials: LiO_2, LiPb, Li_2TiO_4, and Li_2CO_3 are used for the tritium breeder in the blankets.
(6) Superconductor materials: NbTi and Nb_3Sn are used for the superconductor in the superconducting magnet system.
(7) Insulation materials: epoxy resin is used for coil insulation layer of the superconducting magnet.
(8) Coolants: Li, LiPb, He, H_2O, and LiFBe are used for removal of the heat from the blankets/divertors.

To control the activation characteristics of these candidate materials, it is necessary to acquire certain parameters, such as the material compositions, neutron reaction channels, reaction cross sections and half-lives of the nuclides. The activation experiment methods and typical experiments are presented next.

14.2.1 Methods

The material activation experiments are conducted mainly to measure the activation products of the components and coolant during operation. The basic method of the material activation experiment is to determine the material characteristics by measuring the activity of the radionuclides that are produced via the interaction of neutrons with the nuclei in materials.

In general, the material activation experiments mainly include the preparation of the material samples, irradiation of the samples in a specific neutron radiation field, and activity measurements of the irradiated samples. The radioactivity of an irradiated sample can be measured by an HPGe spectrometer and can be described as follows:

$$A_i = \frac{C\lambda_i B}{\gamma_d \varepsilon \left(1 - e^{-\lambda_i t_0}\right) e^{-\lambda_i (t_1 - t_0)} \left(1 - e^{-\lambda_i (t_2 - t_1)}\right)} \tag{14.1}$$

where A_i is the ith total activity of the activated sample; λ is the decay time constant of the activated nuclei; C is the net count rate measured for the full-energy peak during the time interval t_2-t_1; γ_d is the probability of γ-ray emission; ε is the detection efficiency of the HPGe spectrometer; B is the self-absorption factor of the sample, $B = 1$ if the sample is thin; t_0 is the irradiation time of the sample; t_1 is the cooling time and t_2 is the measurement time.

By comparing the measured results with the calculated ones, both the computational methods/codes and nuclear data libraries can be validated. If good agreement is found, the activation calculations for a full-scale model of a fusion system based on the design scheme can be performed with these validated methods/codes and nuclear data libraries, to evaluate the applicability of these materials by extrapolating the total inventories of the radioactive activation products within the lifecycle of the system for the specific operation scenario.

14.2.2 Examples

In this subsection, the ITER materials, which were performed on JET (Joint European Torus), are taken as a typical example to introduce the activation experiments [2]. Figure 14.1 shows photographs of certain sample materials irradiated including Nb_3Sn (used for the poloidal and toroidal field coils), stainless steels (SS316L, SS304B4, etc.), Eurofer and Al alloys (660 Al alloy, AlCu alloy, etc.). These samples were placed on the inner side of the VV in JET for irradiation. The measurements for certain key nuclides were performed with an HPGe spectrometer. The measured specific activity results are shown in Table 14.1 and were compared with the results calculated by FISPACT. As shown in Table 14.1, except for the larger deviations (>15%) for the ^{90}Zr(n, 2n)^{89}Zr, ^{56}Fe(n, p)^{56}Mn and ^{24}Mg(n, p)^{24}Na reactions, good consistency was found for the other reactions with deviations of less than 10%, indicating that there may be certain problems with the neutron activation cross section data of ^{90}Zr, ^{56}Fe and ^{24}Mg and that further investigations are still needed.

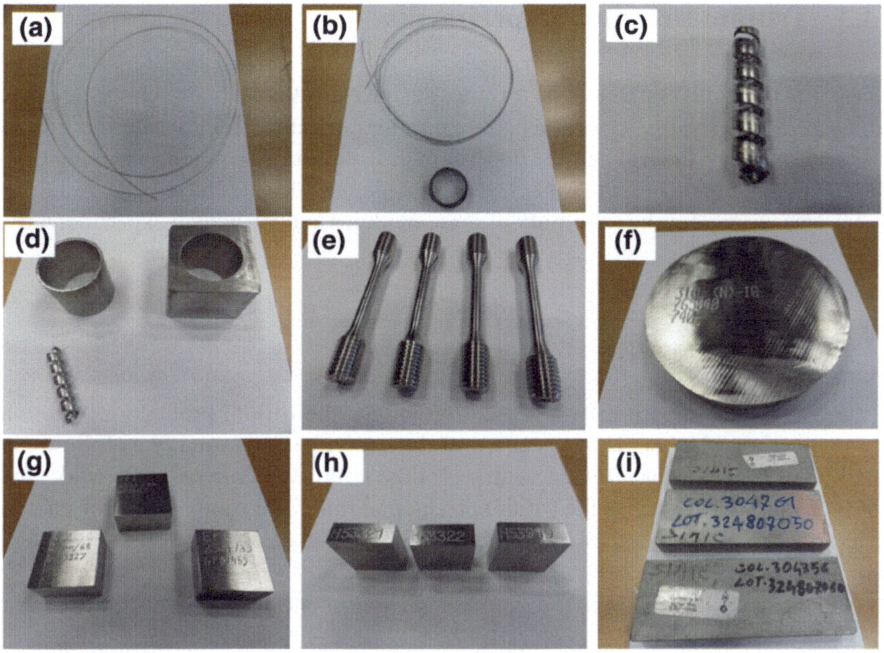

Fig. 14.1 Pictures of materials samples used for ITER **a**, **b** PF and TF Nb3Sn strands; **c** TF cooling spiral (316L steel); **d**: TF jacket (316L(N)), top left, PF jacket (316L), top right, TF cooling spiral (316L), bottom; **e**–**i** various 316L(N) steel specimens [2]

Table 14.1 Comparison of experimental results with calculated results [2]

Experimental condition: JET pulse: 87217, neutron yield: 2.77×10^{16} n/s			
Nuclear reaction	Measured specific activity (Bq/g)	Calculated specific activity(Bq/g)	Relative deviation (%) [a]
^{27}Al(n, p)^{27}Mg	472 ± 20	471 ± 47	0.2
^{27}Al(n, α)^{24}Na	6.0 ± 0.3	6.1 ± 2.8	−1.7
^{56}Fe(n, p)^{56}Mn	62.5 ± 2.2	14.7 ± 0.44	76.5
^{24}Mg(n, p)^{24}Na	7.52 ± 1.03	8.70 ± 4.1	−15.7
^{58}Ni(n, 2n)^{57}Ni	0.127 ± 0.019	0.139 ± 0.007	−9.4
^{90}Zr(n, 2n)^{89}Zr	1.02 ± 0.06	0.63 ± 0.19	38.2
^{93}Nb(n, 2n)^{92m}Nb	0.425 ± 0.027	0.42 ± 0.04	1.1

Note [a]The relative deviation was calculated as (measured value-calculated value)/measured value $\times$ 100%

14.3 Tritium Breeding Experiments

One of the major tasks of fusion neutronics design is to attain tritium self-sufficiency, which generally requires a tritium breeding ratio of tritium ≥ 1.1 considering tritium loss such as decay and leakage and uncertainties from data and model approximation. Based on numerical computations, several test blanket modules have been designed, and it is necessary for the feasibility of these modules to be validated experimentally.

14.3.1 Methods

Tritium breeding experiments are used to validate the tritium production function of the breeding blanket. The procedure of the experiments is usually as follows. A mock-up is placed near a fusion neutron source with an energy of 14.06 MeV, and physical quantities, such as the tritium production rate (TPR) and neutron or photon spectrum, are measured after irradiation to validate the calculated results or design schemes.

Generally, the process of tritium breeding experiments is as follows:

(1) Based on the specific blanket design scheme, the blanket experiment module is first made in the proportions of 1:1 or the scaling model. There are several equal spacing measurement channels that are oriented with the incident direction of the neutron beam.
(2) The samples are placed into the measurement channels and irradiated with fusion neutron sources to measure the TPR and neutron spectrum. Since the TPR is related to the neutron spectrum (TPR can also be calculated based on the neutron spectrum), the measurements can be mutually validated.

The TPR is defined as the number of tritium atoms generated per source neutron [3], P_T, and can be calculated as

$$P_T = \frac{N_T}{N_n} \tag{14.2}$$

where N_n is the intensity of the source neutrons, and N_T is the number of tritium atoms produced for the given source intensity.

There are two methods of measuring the tritium production rate for macroscopic samples. One method is to directly measure the number of tritium atoms produced in the sample, and the other is to detect the TPR based on the tritium production reaction in the tritium-producing material.

For the direct method, there are usually two ways to directly measure tritium production, the internal gas proportional counting method, and the liquid scintillation method. In the former method, tritium is extracted by heating the sample, and then is pumped into the proportional counter through the vacuum system. To meet the previous requirements of the measurement, there should be sufficient tritium produced in the sample and the proportional counter with high precision and the complex vacuum system, which means that a neutron generator with high intensity or the longer irradiation time is required to reduce statistical errors. In the latter method, the tritium in the irradiated sample is first treated with a complex conversion process and then is injected into the scintillation liquid to be measured. The direct method has the advantages of high sensitivity and high efficiency and can be used in the offline measurements of TRP.

In the tritium production reaction method, the TPR can be deduced from the number of pulses in the detector produced by the energetic charged particles (α and T) released by the ^{6}Li (n, α) T reaction. The effect of γ-rays should be deducted by distinguishing the neutron and gamma such as by the frequency analysis method, when a liquid scintillation detector is used due to its sensitivity to γ-rays. This method can be applied in the online measurements of TRP.

The neutron flux spectrum is usually measured by a series of activation foils with different nuclear reaction thresholds. By detecting the amount of activated products, the incident neutron energy spectrum is calculated according to the corresponding reaction cross sections.

With the neutron spectrum obtained by the activation foil, the TPR P_T in the sample can be calculated as

$$P_T = \emptyset^{-1} \sum_{i=0}^{m} \emptyset_i(E)\sigma_i(E)\Delta E_i \tag{14.3}$$

where Φ is the total neutron flux; $\Phi_i(E)$ is the ith grouped neutron flux; $\sigma_i(E)$ is the ith grouped cross section for lithium; ΔE_i is the ith group energy interval and m is the number of energy groups.

(3) The measured TPR and neutron spectrum are compared with the simulation results. If the measured and calculated results are in good agreement, then a full-scale TBM model is established to calculate, and the tritium production capacities of various types of TBM are analyzed, which provides a theoretical basis for the analysis of tritium production in the fusion systems.

In addition, the blanket tritium production capacity is related not only to the TPR but also to the recovery of tritium. To measure the tritium recovery capability of the system, a TBM device needs to be established to evaluate the online tritium recovery capability.

14.3.2 Examples

Since the end of the twentieth century, several tritium breeding experiments have been performed at the fusion neutron source for test blanket modules, such as the water-cooled ceramic breeding (WCCB) blanket in the FNS (Fusion Neutron Source) in Japan, helium cooled lithium lead (HCLL) blanket in the FNG (Frascati Neutron Generator) in Italy [4] and dual-functional lithium-lead (DFLL) blanket in China's HINEG [5]. In this subsection, two typical tritium breeding experiments are introduced.

1. HCLL Experiment

The TPR was measured in the mock-up of HCLL at the FNG with a long irradiation time. The mock-up (Fig. 14.2) consisted of Li–Pb eutectic alloy ingots and Eurofer steel plates. Several pellets were used at the same time to measure the TPR as a function of depth along the mock-up axis. The lithium carbonate (Li_2CO_3) pellets consisted of natural lithium (Li-nat with 7.5 w.t. % ^{6}Li) and lithium enriched to 90% in ^{6}Li, and the LiF thermoluminescent detectors (TLDs) contained natural lithium (TLD-100) and lithium enriched to 99.93% in ^{7}Li (TLD-700). The mock-up was irradiated by neutrons with a total yield of 5.03×10^{15}. The activities of the tritium measurements in each group of pellets were measured, and then, the tritium breeding quantity with distance from the neutron source and the contribution portion of TPR for ^{6}Li and ^{7}Li were acquired. During the experiment, the uncertainty of the cumulative irradiation of the neutron source was 3%, and the uncertainty of the ^{6}Li measurement was 3.7%.

To validate the calculation results, the calculation was performed by using MCNP5 with the JEFF-3.1.1 nuclear data library. The uncertainty of the nuclear data cross-section was approximately 3%, and the relative statistical uncertainty of the calculations was generally lower than 1%. The ratio of the calculated (C) over measured (E) reaction rate was analyzed, as shown in Fig. 14.3. For the pellets with highly enriched ^{6}Li, the deviation between the calculated and measured TPRs from ENEA, Technical University of Dresden (TUD) and JAEA was lower than 10%, while a maximum deviation of ~20% was found for the pellets with natural enrichment of ^{6}Li.

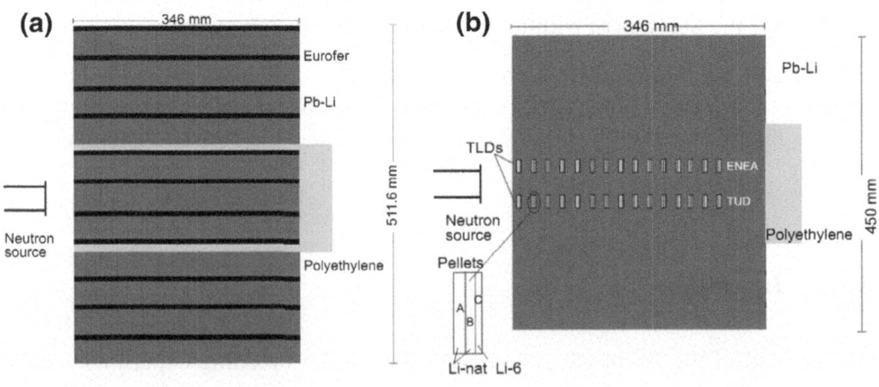

Fig. 14.2 **a** Vertical cross section of the HCLL mock-up, **b** horizontal cross section of the HCLL mock-up [3]

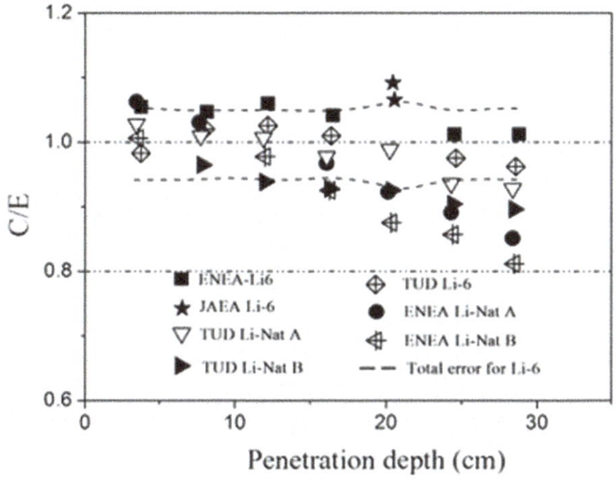

Fig. 14.3 C/E comparison of tritium measurements with increasing depth [3]

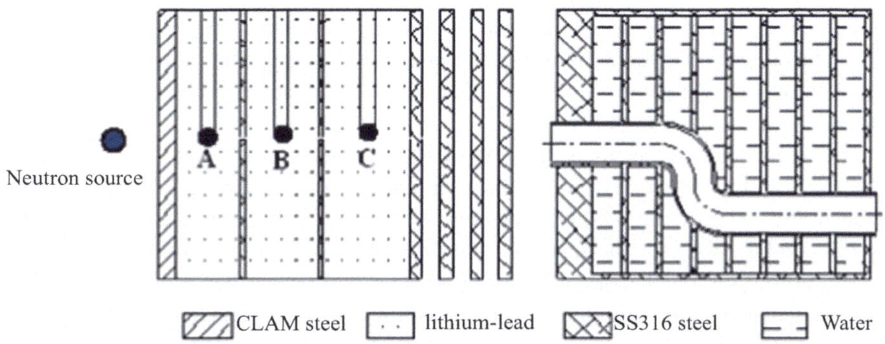

Fig. 14.4 Schematic diagram of the mock-up for the DFLL experiment [5]

2. DFLL Experiment

The mock-up of the DFLL blanket was designed by the FDS Team. The neutronics experiment was carried out on the full-scale mock-up of the DFLL. The activities of the activation pellets at a given position was measured, and the reaction rate of neutron activation and TPR were estimated. The C/E values were obtained.

The DFLL mock-up, as shown in Fig. 14.4, mainly consisted of TBM and shielding cladding. The TBM consisted of CLAM steel plates and Li-Pbeutectic alloy ingots, and the shielding cladding consisted of S316 steel plates, water boxes, water, and curved tube. The inner mock-up had certain holders for the placed detectors such as activation pellets and Li-glasses.

Two independent techniques, i.e., offline and online measurements, were used at the same time to measure the TPR and neutron activation reactive rate as a function of depth along the mock-up axis. Each group of detectors was equipped with In, Nb

Table 14.2 Comparison of calculations and measurements for TPR [5]

Position	Li glass detector/*N*	TPR		C/E
		Experimental	Calculated	
A	184300 ± 0.7%	3.51×10^{-28} ± 7.1%	3.28×10^{-28}	0.93
B	82852 ± 1.8%	1.66×10^{-28} ± 7.3%	1.60×10^{-28}	0.96
C	39661 ± 2.2%	6.47×10^{-29} ± 7.4%	6.10×10^{-29}	0.94

and Al activation foils for offline measurements, and groups were placed at the A, B and C positions. The activity of each activation foil was measured using an HPGe spectrometer after irradiation. The calculation reaction rate of the activation foil was based on the reaction channel of ^{115}In (n, γ) ^{116}In,^{93}Nb (n, 2n)^{92m}Nb and ^{27}Al (n, α) ^{24}Na. A ^{6}Li glass detector for online measurement was placed at the A, B, and C positions to measure TPR. The counts of the ^{6}Li glass detector corresponded to the nuclear reaction event of ^{6}Li, i.e., the total counts of the ^{6}Li glass detector was equal to the tritium production quantities.

The total yield of the neutronics experiment performed was approximately 4.0×10^{15}. The uncertainty mainly included the neutron intensity, detection efficiency and statistics, which were approximately 3, 5, and 5%, respectively. The total combined uncertainty was lower than 8%.

The neutronics experiment for the DFLL blanket was simulated by using SuperMC, and the Fusion Evaluated Nuclear Data Library (FENDL3.1) was adopted because the experiment was a fusion reaction. The International Reactor Dosimetry File (IRDF2002) for the reaction rate of activation foils was used. The angular distribution of the D–T neutron source was accurately described in the simulation.

A comparison of the calculations and measurements for TPR is listed in Table 14.2, a showing that the C/E values varied from 0.93 to 0.96. The values were almost 1, and the deviations of the experiment and calculations were lower than 10%, which validates the DFLL design.

This experimental verification showed that the design of the TBM was feasible, and the deviations of the design and calculated results were provided for further reference for the engineering design of tritium breeding.

14.4 Material Irradiation Damage Experiments

One of the most serious challenges for materials in fusion systems is the irradiation damage induced by 14.06 MeV fusion neutrons. For example, the material irradiation damage in the first wall of fusion systems is dozens of times higher than that in the cladding of fission systems. Additionally, the damage doses of the important components can reach to 150–200 dpa, accompanied by accumulations of helium, hydrogen, and other impurities transmuted by fusion neutrons. However, there are still no facilities simulating actual irradiation fields, and then materials irradiation experiments are performed in fission systems or spallation neutron sources.

14.4.1 Methods

During material irradiation damage experiments, the material is first exposed to a certain radiation source, and the then damage properties of the material are analyzed. Key components are irradiated under specific environmental conditions. After reaching a certain dose of irradiation, the mechanical and physical properties are tested in a hot chamber. The irradiation resistance, low activation properties, and compatibility with liquid metals can be evaluated for judging the applicability of materials by the comparisons between the previous and the late neutron irradiations.

The complete process of the material irradiation damage experiments includes preparing special shaped samples, irradiating the samples under an environment with a certain temperature and neutron flux, putting for a long time to reduce activities of samples, and analyzing the microstructure and mechanical properties.

The changes in the microstructure and properties of the irradiated materials are obtained by comparison those to the pre-irradiated materials. However, neither fission systems nor spallation neutron sources can simulate the real irradiation environment of fusion systems, since the neutron energy spectrum and flux are different. The applicability of existing materials for fusion systems can be evaluated through extrapolating the changes in the microstructure and mechanical properties of materials under irradiation at equivalent facilities, which is called an equivalence study between fission or spallation neutrons and fusion neutrons. For example, the damage behavior of samples under the irradiation environment of fusion systems can be predicted by combining the damage behaviors and transmutation gas effects caused by fission neutrons and spallation neutrons.

14.4.2 Test Blanket Modules Experiments

There are three main candidates for structural materials in fusion systems: reduced activation ferrite/martensitic (RAFM) steels, vanadium alloys, and SiC_f/SiC composites. RAFM steel has low-activation properties, anti-radiation swelling, and compatibility with liquid metals, and uses fundamental research and metallurgy techniques. RAFM steel is generally considered to be a leading candidate material for the construction of future fusion systems or the first fusion power plants. RAFM steel mainly includes CLAM (China), F82H (Japan), Eurofer (European), 9Cr-2WVTa (United States), Rusfer EK-181 (Russia), ARAA (Korea) and IN-RAFM (India). The main reasons to perform structural material irradiation damage experiments are to study the microstructure and property changes of RAFM steels under irradiation of fusion

Fig. 14.5 High-flux engineering experiment reactor (left) assembly of the samples (right) [3]

neutrons and thermal neutrons and establish data libraries including physical properties, microstructure, mechanical properties and irradiation properties of materials.

CLAM steel was developed by the FDS Team under close collaboration with many institutes [6]. This material has been chosen as the candidate structural material in the design of the China test blanket module for ITER and the China fusion engineering test reactor blanket. A series of low dose neutron irradiation experiments for CLAM steel have been performed. The irradiation experiments in fission systems and spallation neutron sources are introduced as follows:

1. CLAM steel irradiation experiments in fission system

Specimens of CLAM steel in the fission system were irradiated at the High-Flux Engineering and Test Reactor (HFETR), as shown in Fig. 14.5. The irradiation tank was located in the middle section of the reactor bore, where 18.7% of the fast neutrons had energies greater than 1 MeV.

There are neutron irradiation experiments such as NPIC-0, NPIC-I, NPIC-II, and NPIC-III in HFETR [7, 8].

In the NPIC-II irradiation experiment, the irradiation dose was 2.6 dpa and the irradiation temperature was 430 °C. The impact tests of the mechanical properties after irradiation are shown in Fig. 14.6. The experimental results of the CLAM steel before and after irradiation showed that, after irradiation to 1.5 dpa, the impact absorption on the work platform reduced (~50 J), while the ductile to brittle transition temperature increased after the irradiation and was less than 10 °C. Compared with other RAMF steels under similar irradiation parameters, the irradiation embrittlement of the CLAM steel was lower. After irradiation to 2.6 dpa, the impact absorption energy

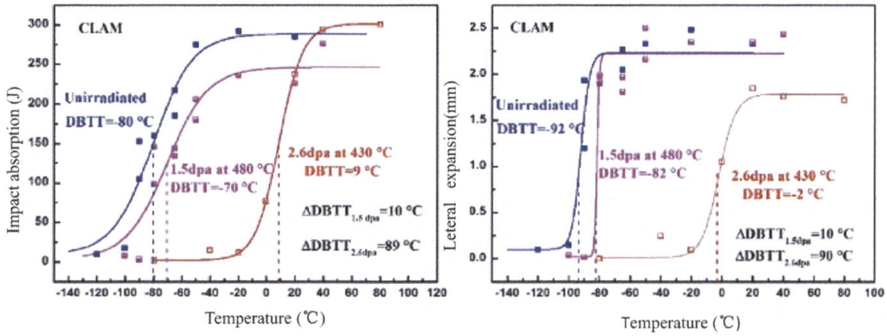

(a) 1.5 dpa (b) 2.6 dpa

Fig. 14.6 Impact curves of CLAM steel before and after neutron irradiation at 1.5 and 2.6 dpa [7]

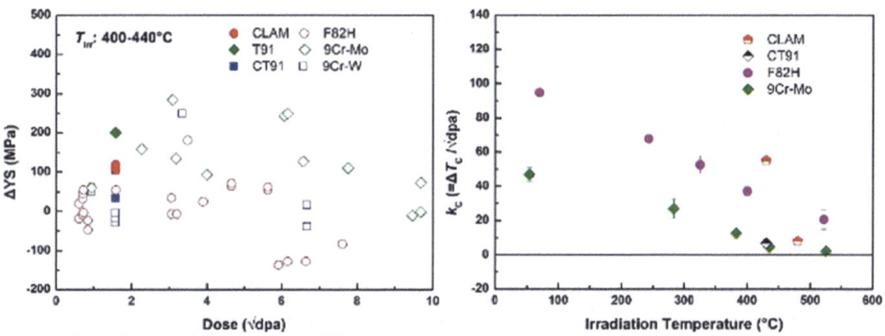

(a) Yield strength increment at different dose (b) Embrittlement degree at different temperatures

Fig. 14.7 Mechanical properties of CLAM steel before and after neutron irradiation compared with the RAMF steel [7]

platform of the CLAM steel was basically consistent with that before irradiation, but the ductile to brittle transition temperature increased significantly (~90 °C).

Under similar irradiation conditions, the mechanical properties of CLAM steel were equivalent to those of other RAFM steels. As shown in Fig. 14.7a, under irradiation conditions of 2.6 dpa and 430 °C, the yield strength increment caused by CLAM steel hardening was equal to that of F82H steel and lower than that of T91 steel. Figure 14.7b shows the embrittlement degree of materials at different temperatures after dose-removing. Under the irradiation temperature of 430 °C, the embrittlement of the CLAM steel was slightly higher than that of F82H, while the embrittlement of the CLAM steel was significantly lower than that of F82H under an irradiation temperature of 480 °C. The experimental results indicated that the strength of the CLAM steel changed slightly (<20 MPa) after irradiation to 1.5 dpa at high temperature (200 and 300 °C) but changed greatly (>100 MPa) at room temperature. The change in the ductility was small (<4%) after irradiation and was small with the test temperature. After irradiation to 2.6 dpa, the strength and ductility with the tested temperature changed similar to that of the irradiation to 1.5 dpa, while the strength rose significantly. Among them, the yield strength rose more significantly than the strength of extension. Under three different temperatures, the yield strength was similar to the strength of extension. The change in ductility was small and was similar to that at 1.5 dpa.

2. CLAM steel irradiation experiments in spallation source

To study the irradiation hardening behavior of CLAM steel under high-dose neutron irradiation, a series of neutron irradiation experiments were carried out in the spallation neutron source (SINQ) in Switzerland, including the STIP-V, STIP-VI, and STIP-VII spallation-target irradiation experiments [9]. In the STIP-V irradiation experiment, the HEAT 040SB batch material was used to carry out the neutron irradiation experiment of the CLAM steel spallation, and the samples of the HEAT 040SB batch were subjected to tensile tests, impact tests and microscopic analysis. The samples were irradiated under condition of from 56 to 471 °C and from 5.4 to 20.9 dpa. The effects of temperature and dose on the tensile strength and microstructure are described below.

(1) The strength analysis results after the tensile tests showed that the yield strength growth rate and dose of the CLAM steel induced by hardening were similar to the *E* exponential relationship. The results were similar to the irradiation experiment results of RAM steel. In addition, when the irradiation dose reached the highest level in the experiment (20.1 dpa), the intensity continued to increase, but the embrittlement phenomenon arose. The tensile curves of the CLAM steel under room temperature testing in the STIP-V irradiation experiment are shown in Fig. 14.8.

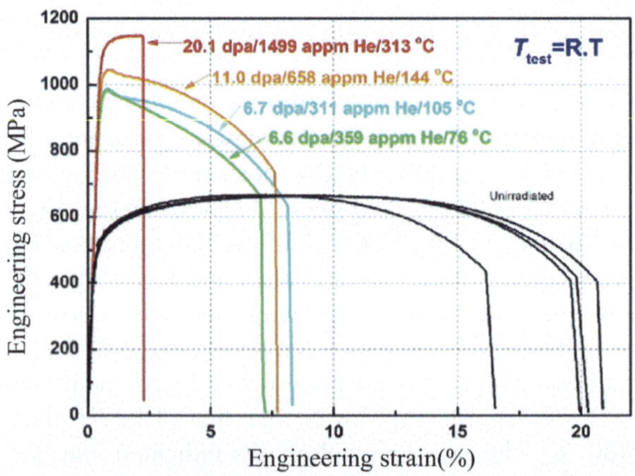

Fig. 14.8 Tensile curves of the CLAM steel under room temperature testing in the STIP-V irradiation experiment [9]

(2) The microhardness analysis results after irradiation showed that the microhardness values and the dose were similar to the *E* exponential relationship. The increase in the microhardness slightly reduced at doses of from 12 to 20.8 dpa. Compared with the RAFM steel, the hardness increments at the same temperature and dose were also similar to the *E* exponential relationship, which showed that the temperature for hardening was not very obvious when the temperature was lower than 313 °C, and the influence of the dose was the main factor influencing the hardening.
(3) Based on TEM microstructure observation, the radiation defects and helium bubbles in the CLAM steel were quantitatively analyzed. Under the irradiation condition of from 5.4 to 20.8 dpa and from 56 to 328 °C, the dislocation blocking factor of the radiation defects was approximately 0.7, and the dislocation blocking factor of helium bubbles was approximately 0.15. Irradiation hardening was mainly caused by irradiation defects, but a high density of small helium bubbles also had certain effects on the hardening.

14.4.3 First Wall and Divertor Experiments

The plasma facing material of the first wall and divertor mainly comprises W and W alloys. Pure tungsten samples have been irradiated by neutrons in HFIR (High Flux Isotope Reactor, Oak Ridge National Laboratory, USA) at from 90 to 850 °C and from 0.03 to 2.2 dpa [10]. The results indicated that voids and dislocation loops contributed to hardness increases in at low dose (<0.3 dpa), while the formation of intermetallic second phases from transmutation, dominated the radiation-induced strengthening beginning with a relatively modest dose (>0.6 dpa). Fukada et al. investigated the effect of neutron irradiation on the functional properties of pure tungsten (W) and advanced tungsten alloys (e.g., lanthanum (La)-doped W and potassium (K)-doped W) tested in Japan's experimental fast reactor JOYO [11]. The irradiation temperature and damage were in the range from 804 to 1073 K and from 0.1 to 0.47 dpa, respectively. TEM images of all the samples after 0.42 dpa irradiation at 1023 K showed voids, black dots, and dislocation loops, indicating that similar damage structures formed in pure W, La-doped W and K-doped W. The electrical resistivity of all the specimens increased following neutron irradiation. Nearly identical electrical resistivity and irradiation hardening were observed in pure W, La-doped W and K-doped W.

The heat deposition material of the divertor includes CuCrZr alloys [12, 13]. The alloy specimens were irradiated in the IVV-2 reactor core at 200 °C with irradiation damage of 0.15 and 0.27 dpa and were also irradiated in HFIR at 80 °C with irradiation damage of 0.15 and 1.5 dpa. Post-irradiation tests were carried out to assess the tensile properties and fracture toughness of the materials. The tests results showed that irradiation hardening was obvious at low dose and the fracture toughness decreased slightly.

14.4.4 Superconducting Magnet Experiments

Superconducting magnets are one of the key components of fusion systems. Neutron and photon irradiation on superconducting magnets may change their superconducting properties and result in quench that may make the plasma rupture.

The neutrons produced in a fusion system scatter elastically with the first wall and shielding blanket. At the location of the superconducting magnet, the neutron spectrum is extended by seven orders of magnitude and the flux is greatly reduced. The neutron energy spectrum is similar to those of fission systems (as shown in Fig. 14.9), so the superconducting magnets can be tested experimentally with the neutron sources of fission reactors [14]. In this case, the structural irradiation effects of superconducting magnets comes mainly from fast neutron and photon activation, which affects the physical properties of the superconducting magnets and the mechanical properties of the insulators, such as changing the critical current of the superconducting coils or the critical temperature of superconducting coils and decreasing the tensile strength of the coil insulation materials. Examples of superconducting coil and coil insulation materials are given to introduce superconducting magnet irradiation experiments.

The irradiation experiments related to the Nb_3Sn and NbTi superconductors are introduced as follows:

Nb_3Sn superconductor was irradiated with 14 MeV neutrons in the FNS with a cumulative neutron flux of 1.78×10^{21} m^{-2}, and the superconductor current with the change of the magnetic field was measured [15]. Compared with the variation of the critical current under non-irradiation, as shown in Fig. 14.10, when the magnetic field was less than 20 T and was decreased, a higher critical current was required to reach the same magnetic field. When the magnetic field was greater than 20 T and

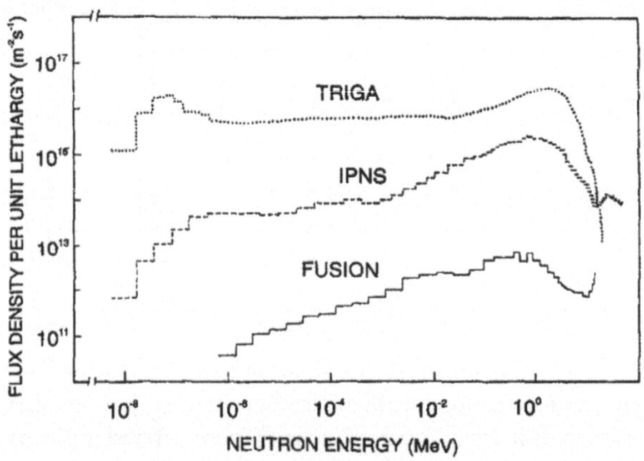

Fig. 14.9 Neutron energy spectra for three different neutron sources, TRIGA: fission reactor channel, IPNS: Intense Pulsed spallation source and FUSION: the magnet location in fusion system [15]

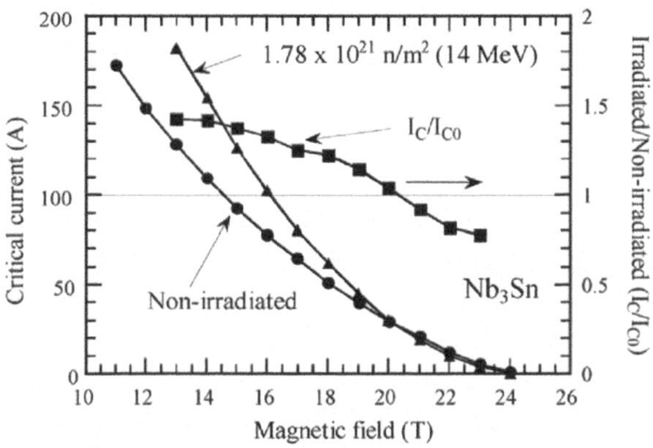

Fig. 14.10 Change in the critical current of Nb_3Sn wire against magnetic field after 14 MeV neutron irradiation [15]

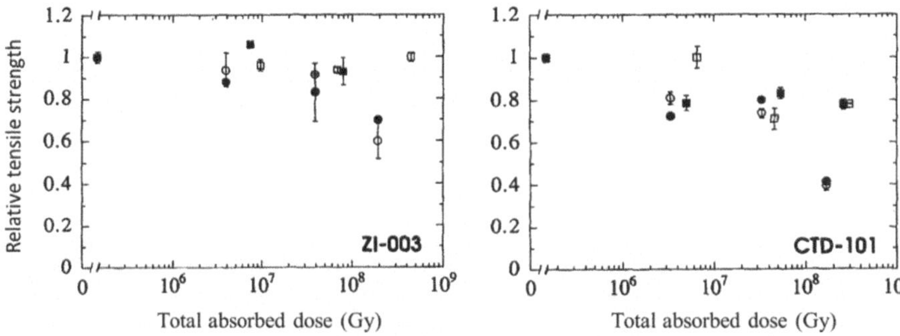

Fig. 14.11 Relative tensile strengths of two insulating materials (Z-003 and CTD-101) as a function of total absorbed dose [16], ○ OTRIGA source, ■IPNS source, □ 60Co source, ●2 meV electron accelerator

was increased, a low critical current was required to meet the same magnetic field. The results showed that both the superconducting critical current I_C and the superconducting critical temperature Tc changed after being irradiated by fast neutrons. The high flux has a significant inhibitory effect on the superconducting properties of low magnetic field superconductors.

The irradiation effects on the insulation material of the superconducting magnets in fusion system are another concerning issue. Figure 14.11 show the variation of the relative tensile strength (RTS) of two epoxy resin insulating materials (ZL-003 and CTD-101) commonly used in coils at room temperature with irradiation dose, where the RTS was the ratio of the tensile strength after irradiation to that without irradiation. The results indicated that the RTS of the Z-003 insulating material decreased with increasing irradiation dose no matter the type of irradiation, and the radiation resistance of Z-003 was better than that of CTD-101. In addition, the scaled-down resin components were placed in the reactor to measure the mechanical properties of the components after the cumulative dose of irradiation, such as tensile strength experiments, interlaminar stress experiments, electrical breakdown experiments, and expansion experiments. However, these scaled-down experiments still need to consider engineering validation.

14.5 Shielding Experiments

In fusion systems, the flux of high-energy neutrons is greatly attenuated through the blanket that is the most important component of the thermal shield and nuclear shield, which plays an important role in the protection of the vacuum vessel and other components. The concrete shielding is the outermost shielding of the facility and primarily shields neutrons and photons leaking from the fusion system. In addition, the surrounding environment may be affected by the neutrons and photons leaking from the top of the fusion system through the sky. Therefore, it is necessary to perform shielding experiments to validate the shielding designs of the blankets and concrete shielding and to assess the impact of the skyshine.

14.5.1 Methods

Shielding experiments are used to validate the shielding design of the fusion system. The method of the shielding experiment is as follows. A combined shielding module based on the specific design scheme as well as the composition and type of shielding material is built up and irradiated by a fusion neutron source, and then the relevant physical quantities (such as the flux/dose of neutron/photon flux and nuclear heating) in the region of interest are measured. The measured results and calculated are compared to validate the shielding performance for the specific design scheme and to evaluate whether the radiation level at the specific region during the operation of the fusion system satisfies the limit requirements of the relevant physical quantities.

Fusion shielding experiments mainly focus on blanket shielding experiments, concrete shielding experiments and skyshine scattering experiments. The detailed procedure for these experiments is introduced as follows:

1. Blanket shielding experiment

(1) Design and fabrication of the experiment module: based on the specific blanket design scheme, the equivalent experimental module is designed and fabricated at the size of 1:1 scaling or scaled-down. Several measurement channels are set along the incident direction of the neutron beam in the blanket. The position of these channels may be set up at steps equal to the increase of the neutron penetration depth.

(2) Preliminary simulation: Before performing the irradiation of the experiment module, a preliminary simulation should be carried out to provide the basis

(such as the optimization of the experiment configuration and the determination of the irradiation, cooling and measurement time) for the experiment proposal/plan by considering the parameters of given neutron source and measurement instrument.

(3) Irradiation and measurement: According to the experiment proposal/plan, the experiment module with irradiation samples (such as activation foils) or detectors is irradiated in the fusion neutron source for a period of time. The specific physical parameters at the position of interest are measured.

For the measurement of the nuclear heat deposition, small detectors such as thermo-luminescent detectors (TLDs) are usually applied to avoid distortion of the radiation field in the module. If a TLD-300 (CaF_2:Tm) is selected, the relationship between the dose D_{TLD} in the irradiation sample and dose D_{mat} in the surrounding material (such as stainless steel or copper) can be expressed by the following:

$$D_{mat} = D_{TLD} + D_\gamma\left[(1 - C_n K_n)/D_n/D_\gamma + (1 - C_\gamma)\right] \tag{14.4}$$

where C_γ is the photon Kema factor of the TLD; C_n is the neutron Kema factor of the TLD; D_γ and D_n are the photon absorption dose and neutron absorption dose, respectively, (both are measured by two high temperature peaks in the TLD300 luminescence curve) in the surrounding materials, and K_n is the neutron sensitivity of the TLD-300.

(4) Result analysis and comparison: A detailed calculation based on the actual experiment may be performed. The newly calculated results are compared with the measured results to validate the calculation methods/codes and nuclear data libraries. If good agreement is reached, the numerical simulation with the validated tools and nuclear data libraries for the specific blankets can be performed to assess whether the radiation level at a specific region satisfies the design requirements.

2. Concrete shielding experiment

(1) Determining the experiment proposal: Based on the experimental hall and entrance constructed for the fusion experimental device, the experiment proposal can be determined by considering the properties of existing facilities in the hall (such as the position, structural parameters, and materials composition). The measurement positions are usually set at the access corridor where the shielding is weak.

(2) Irradiation and measurement: According to the neutron radiation measurement scheme, the detectors are arranged in the specific measurement points. After irradiation by the fusion experimental device, the dose or fluence distribution in the experimental hall and the entrance can be obtained. The small detectors, such as TLDs with high enriched ^{6}Li or natural Li, are usually used to measure the dose or fluence of the neutrons and photons in the mixing field. TLDs with high enriched ^{7}Li can be used to measure the dose or fluence of photons. Thus, the dose or fluence of neutrons can be obtained by the difference.

(3) Result analysis and comparison: The shielding performance of the concrete shield is evaluated by comparing the measured results with the calculated ones to validate whether the dose level satisfies the limits specified in the radiological area, which provides the basis for the design of other fusion systems.

3. Skyshine scattering experiment

(1) Experiment proposal: Based on the experiment purpose, the measurement positions and the detectors are specified first. The detectors are usually deployed outside the facility building in the radial direction with different distances from the center, which comprise the components producing the neutrons in the experiment system. A preliminary simulation may be required to provide guidance for the experiment proposal.
(2) Irradiation and measurement: According to the specified experiment proposal, detectors are deployed at specific positions. The distributions of the parameters of interest (such as dose and flux) are measured for the case of the specific operating parameters for the experiment facility. Portable instruments, such as multi-sphere neutron spectrometers and dosimeters, are usually used to measure the neutron spectrum and dose, respectively.
(3) Results analysis and comparison: Comparing the measured results with the calculated ones, the shielding design due to skyshine scattering is assessed. The impact of skyshine scattering on the environment and the public is analyzed to provide a basis for the design of fusion systems.

14.5.2 Shielding Blanket Experiments

In the 1980 s, according to the research needs of ITER, a series of experiments were conducted at the FNG, such as ITER blanket bulk shielding and stainless steel shielding experiments. In 1999, an integral bulk shielding experiment was performed at the FNG to validate the ITER shield system. Neutron and γ-ray spectra and various responses were measured inside this mock-up when irradiated by 14 MeV neutrons. The measured quantities were predicted within an uncertainty margin of approximately 15% in the shielding blanket and vacuum vessel and 30% up to the region of the toroidal field coil, which did not meet the requirements of ITER.

In recent years, a new validation experiment has been carried out in the FNG [17]. As shown in Fig. 14.12, the experimental module consisted of the shielding blanket module and TF coil module. The shielding blanket module consisted of 316 stainless steel plate and plexiglass plate. The TF coil module consisted of 316 stainless steel, copper and four Nb_3Sn cylinders. To measure the nuclear heat deposition and neutron energy density at various parts of the module, a high sensitivity thermoluminescence meter and an activation sheet group composed of niobium, aluminum, nickel, gold, and tungsten were arranged in the central axis of the module and in the small room of the TF coil module. The experimental results are shown in Fig. 14.13.

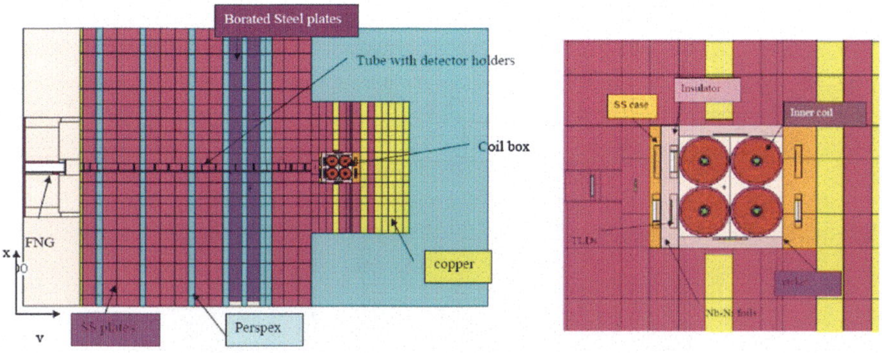

(a) Detector layout in the ITER shielding blanket module (b) TF coil module area enlargement

Fig. 14.12 Section of the shielding module inside ITER [17]

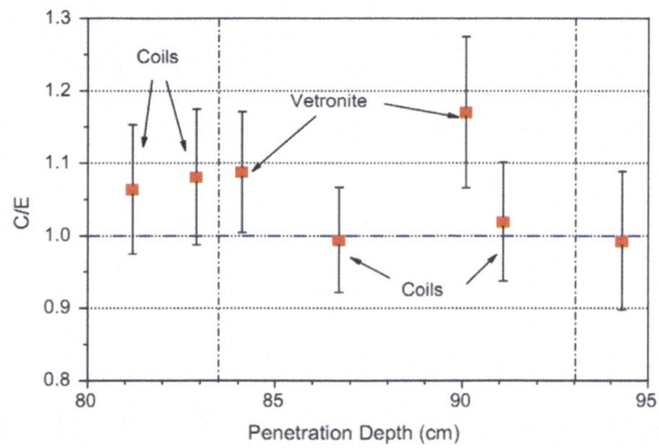

Fig. 14.13 Variation of the ratio of the simulated and experimental measurements of the nuclear heat deposition (C/E) with penetration depth in the TF coils [17]

The measurement error of the TF coil nuclear heat deposition was lower than 10%, which met the requirements of ITER.

The equivalent models were adopted in the above experiments. There will be many gaps in the blanket due to component installation and maintenance needs. The existence of these gaps will weaken the shielding effect of the blanket and make neutron, photons and so on produce penetrating radiation through these gaps, resulting in increased local radiation from the shielding body and increased material activation around the gaps, which increase the radiation risk to the staff. Therefore, it is necessary to evaluate the effect of gaps on engineering design and safety and carry out corresponding verification experiments.

In view of the demands of ITER, the FNG was used to carry out experiments on the effect of gaps on shielding [18]. As shown in Fig. 14.14, the experimental module was composed of shielding blanket modules and superconducting magnet modules. The superconducting magnet module consisted of 316 stainless steel plate and copper plate, and the shielding blanket module consisted of 316 stainless steel

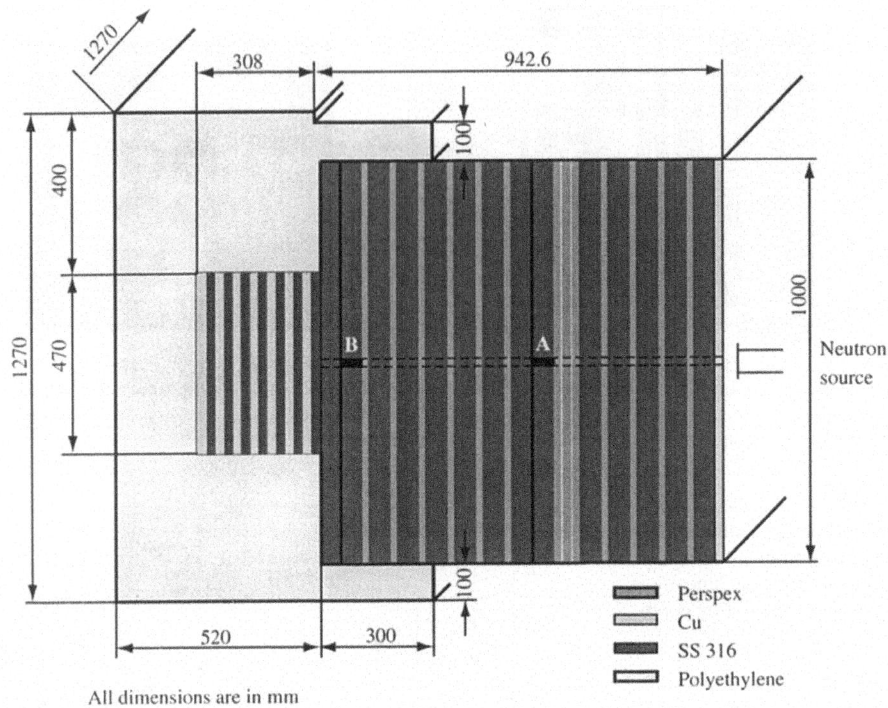

Fig. 14.14 Section of the experimental module for the experiments on the effect of gaps on the shielding [18, 19]

plate and plexiglass plate with a diameter of 28 mm, a length of 389 mm and a cavity of 52 mm × 148 mm × 48 mm. The neutron energy densities and energy spectra in the pipes, cavities, shielding modules and superconducting magnet modules were measured by using niobium, aluminium, nickel, gold and tungsten activation plates, and the nuclear heat deposition in shielding modules and superconducting magnet modules was measured by thermoluminescence detector. The results are shown in Fig. 14.15, the deviation between the measured and calculated neutron energy densities in the cavity was lower than 10%, and the deviation between the measured and calculated nuclear heats in the superconducting magnet was lower than 30%.

14.5.3 Concrete Shielding Experiments

The neutron fluxes in fusion systems greatly attenuate through blankets, VV and other components, resulting in a decrease in the specific activity and dose from the activated products. The concrete shielding is the last barrier for environmental and personal protection. Therefore, the study of concrete shielding performance for neutrons and

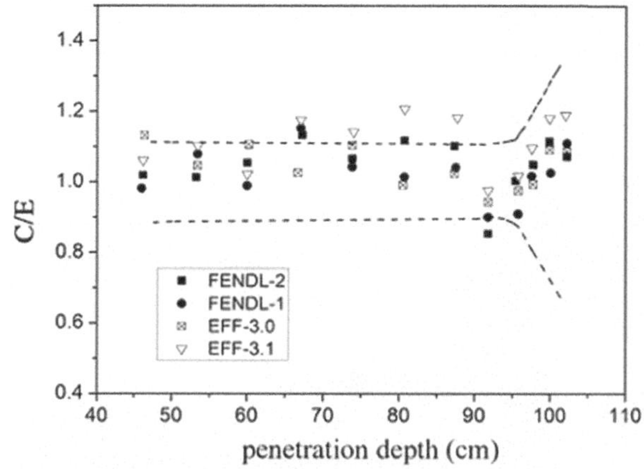

Fig. 14.15 Variation of C/E with penetration depth in nuclear heat deposition in superconducting magnet modules [3]

photons is not only an important method to ensure the safety of personnel but also an important embodiment for the three principles of radiation protection.

The concrete shielding experiments are focused on the measurement of the neutron and photon dose in specific environments. The attenuation of neutrons and photons in the concrete of different thickness is usually studied by changing the composition of the concrete. Since the radiation field is a mixture of neutrons and photons, the distinctions between two types of radiation must be considered. Current monitoring tools include mature instruments and dosimeters for either active detection or passive detection. An example of a concrete shielding experiment in JET is introduced as follows [20].

Figure 14.16 shows the entrance labyrinth located at the southwest (SW) corner of the JET hall. The entrance labyrinth was a four right-angle turn labyrinth with a total length of 11.80 m (sums of the M1–M2, M2–M3, M3–M4, M4–M5, and M5–M6 sections as indicated in Fig. 14.16). The labyrinth height was 2.60 m. The labyrinth width varied between 0.90 and 1.10 m. The thickness of the concrete wall was 2.50 m, and the internal surface of the concrete was covered by a layer of borated concrete that was 0.30 m thick. The densities of the concrete and borated concrete were 2.43 and 2.20 $g \cdot cm^{-3}$, respectively.

To calculate the equivalent dose rate around JET in the labyrinth lane under the conditions of D–D operation, the Monte Carlo code MCNPX (v.2.5.0) and the JEFF 3.1.2 library cross section data were used for the calculations of the surface neutron source. The neutron incidence surface was a quasi-spherical source that was a quarter-sphere with the center at the SW Hall corner (1.0 m above the floor level) with a radius of 5.0 m and a cylindrical surface source below 1.0 m (see Fig. 14.17).

In D–D operations JET produces neutron emission rates on the order of $2.0 \times 10^{16}\ s^{-1}$, and the predicted neutron ambient dose equivalent H*(10) is shown in Fig. 14.18 (as a function of parameter $L/A^{0.5}$) for the labyrinth sections M1–M6.

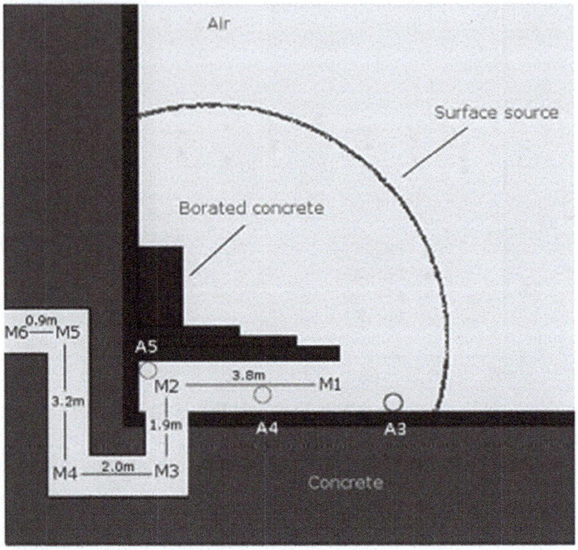

Fig. 14.16 Cross-sectional view of the SW entrance labyrinth [20]

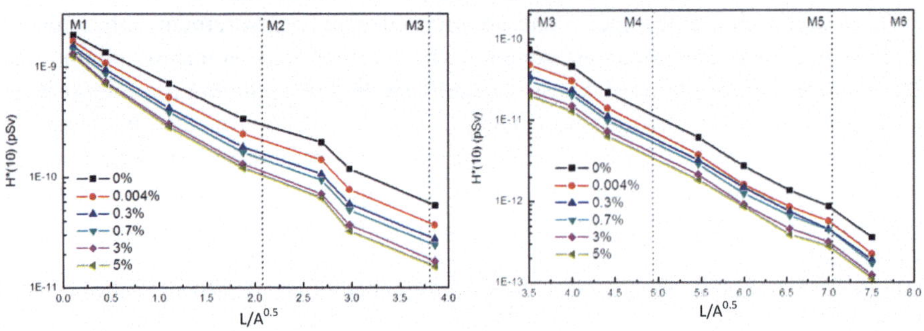

Fig. 14.17 Function between H*(10) and $L/A^{0.5}$ [20]

Different hydrogen contents in the concrete (0, 0.004, 0.3, 0.7, 3, and 5%) were found when H*(10) was plotted as a function of parameter $L/A^{0.5}$. Figure 14.18 shows that when the boron contents of the concrete were 0.3, 0.7 and 3%, the dosage decreased by 28, 32, and 42%, respectively, compared with that of the concrete without borax.

The boron content of concrete was 3% according to the "as-build" experimental conditions. MCP-N (natural lithium LiF:Mg,Cu,P) and MCP-7 (highly enriched lithium ^{7}LiF:Mg,Cu,P) type TLDs were selected for monitoring fluence at three points A3, A4, and A5 (see Fig. 14.17). The experiment was carried out when JET was in D–D operation mode, and the cumulative neutron yield was 2.76×10^{18}. Table 14.3 shows the calculated and the measured neutron fluence, as well as the

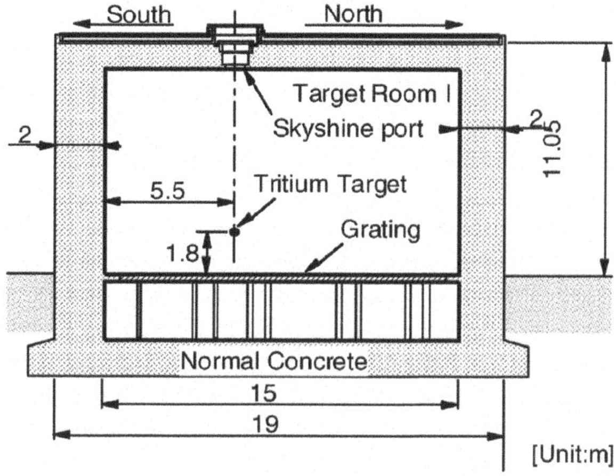

Fig. 14.18 Side view of the FNS target room [21]

Table 14.3 Comparison of calculated and measured neutron fluences at three labyrinth positions [21]

Detector	Fluence (cm^{-2} n^{-1})		C/E ratio
	Calculated	Experimental	
A3	1.94E+09(4.8)	1.03E+09(12.7)	1.88(13.6)
A4	1.12E+09(4.1)	3.86E+08(18.0)	2.89(18.4)
A5	6.54E+07(8.8)	2.01E+07(17.7)	3.25(19.8)

calculated result to experimental (C/E) fluence ratio, along with their relative errors, at three detector positions in the labyrinth region.

Table 14.3 shows that the calculations systematically overestimated the measurements at all the detector positions(approximately 1.88–3.25 times). The observed discrepancy in the C/E ratios was attributed to three factors.

(1) The system equipment of the JET hall was simplified in the calculation simulations and was quite different from the equipment in the JET hall at the time of the measurements, resulting in a large deviation in the contribution of neutron scattering.
(2) The assumed composition of the surrounding concrete materials (including the water content and boron content of the concrete) led to a large deviation between the calculated results and the experimental values. For example, a slight increase in the amount of boron from 0 to 1% (w/w) resulted in a reduction in the neutron fluence at the labyrinth exit (M6) by a factor of 2.0.
(3) For the TLD calibration, the average energy of the radiation field was much higher than that of the calibration field. When estimating an average high-energy radiation field using the natural lithium of MCP-N, the neutron fluences were lower than the actual values.

14.5.4 Skyshine Experiments

The buildings of fusion systems are constructed with thick walls for radiation shielding but generally with relatively thin roofs due to design limitations. The skyshine of radiation, leakage from the roof to the sky returning to the ground by several scattering events with air-constituent nuclei, is known as one of the environmental evaluation. Aspects of the skyshineis have been discussed for different types of facilities, and simple analytical formulas have been proposed to express the characteristics of the propagation of this radiation. It is important that the neutron radiation leaking from the tops of these facilities is accurately evaluated, and the transportation of neutrons in the sky has been researched for shielding design and radiation protection.

Since skyshine scattering experiments have large measurement fields and low neutron fluxes, environmental neutron dosimeters or spectrometers can generally meet the experimental requirements.

Skyshine experiments were performed at the FNS (Fig. 14.18) [21]. During the experiments, the port in the roof of the target room was opened. The fusion neutrons leaking out of the port collided with the air and flew towards the ground. The accuracies of the relevant neutron transport and dose calculation codes were assessed by measuring the neutron dose and spectrum on the ground from 20 to 550 m, as shown in Fig. 14.19.

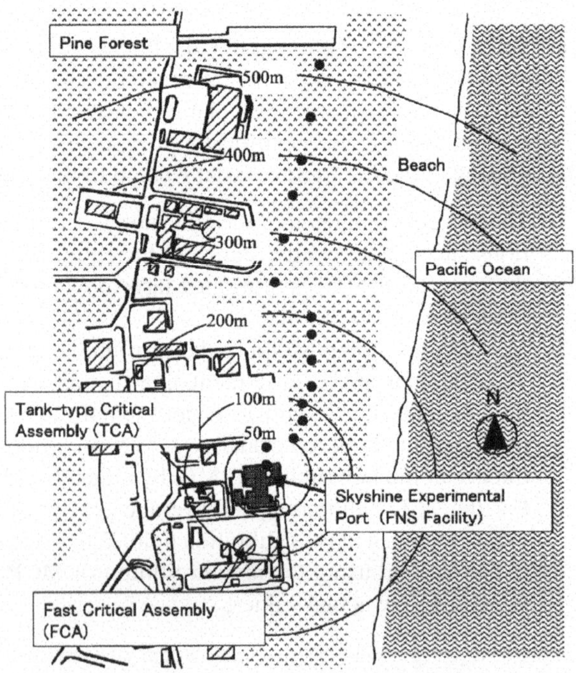

Fig. 14.19 Measurement points around the FNS building for the skyshine experiments [21]

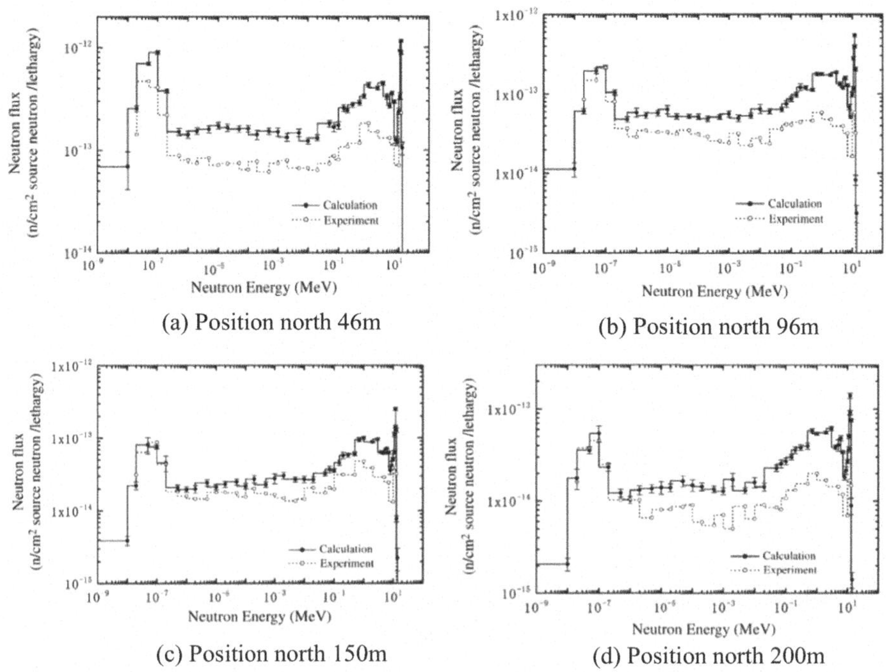

Fig. 14.20 Dose distribution calculated and unfolded spectra at several measurement points [21]

The neutron energy spectrum, measured by a multispeed neutron spectrometer with seven different Bonner spheres, is compared with the calculations in Fig. 14.20. This figure shows that there is a minor discrepancy when the neutron energy is lower than 13 MeV, but the total fluences of the experimental and calculated results are within 10% of each other. The discrepancy in the result may be attributed to the overestimation of the calculated results because the forests and certain buildings in the measured orientation cannot be accurately modeled.

The spatial distributions (north and south) of the neutron ambient dose measured by the ^{3}He counter with an 8.6-cm-diameter and are shown in Fig. 14.21 accompanied by the calculated results. Compared with the results measured by dosimeter, the average (E/C) was 0.8 to the north and 0.95 to the south. The (E/C) to the south was better than that to the north, due to the different situations of the measurement points where the pine forest and the building were not considered in the calculations. Meanwhile, the dose in the experiments was estimated based on the unfolded spectrum. The average (E/C) of the dose was lower by 20%, which may be attributed to the low evaluation of energy spectrum for the energy group (<13.5 MeV) of source neutrons (Fig. 14.20).

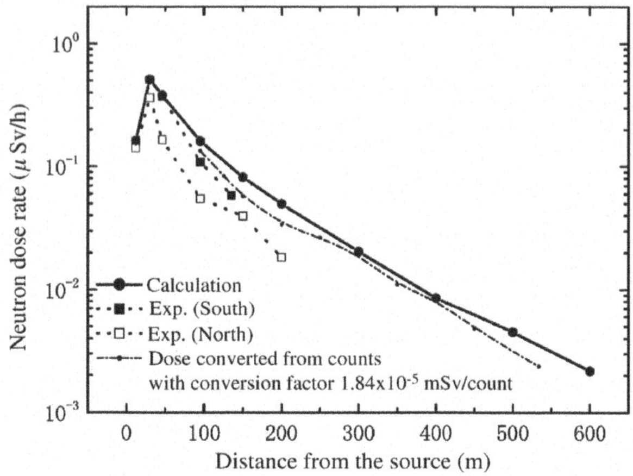

Fig. 14.21 Spatial distributions of the calculated ambient dose versus that measured by the ^{3}He counter with an 8.6-cm-diameter [21]

14.6 Prospects

In recent decades, numerous fusion neutronics experiments have been performed regarding material activation, radiation damage induced by neutrons, fuel production for fusion systems, and shielding of fusion neutrons. Several important achievements have been made and provided good foundations for the realization of ITER and commercial fusion systems in the future. However, the current research cannot fully satisfy the demands of future fusion systems in the following aspects:

(1) Material Activation Experiments: According to the design of ITER, the accumulated radioactive activation products will reach 38,000 tons after 20 years of operation, and the activity of radioactive materials will increase by a factor of 20–50 in future commercial fusion systems. Additionally, when performing decommission of fusion systems, certain nuclides with small neutron activation cross sections and long half-lives will make large contributions to the dose. In the limited experimental conditions, low-intensity neutrons resulted in low measurement accuracy for the cross sections of certain nuclides, such as (n, p), (n, α), (n, t), and (n, ^{3}He) cross sections of copper, iron, tungsten and nickel and (n, γ) cross sections of iron and lead, so the estimated radioactive source terms are very different from the practical situations. Thus, high-dose material activation experiments must be conducted to improve the accuracy of the relevant nuclear data and supplement the deficiencies in nuclear data of important nuclides.

(2) Tritium Breeding Experiments: Current tritium breeding experiments are still based on "point" neutron sources with low intensities, and only offline measurements of TPR can be performed. Compared with the engineering application of fusion systems, a large gap exists, and the verification of online tritium extraction technology is still needed. For example, according to the design of the tritium breeding modules of ITER, the neutron flux on the surface of the module should achieve a level of 3.5×10^{13} cm$^{-2} \cdot$ s^{-1}. Therefore, it is necessary to develop

"volumetric" neutron source experimental facilities with high yield level of 10^{13} $cm^{-2} \cdot s^{-1}$ for carrying out online tritium breeding and extraction experiments to verify the effectiveness of the design of the tritium breeding modules.

(3) Material Irradiation Damage Experiments: The neutron flux at the first wall of ITER can reach 10^{14} $cm^{-2} \cdot s^{-1}$ with a specific energy spectrum, and the first wall is operated under a complex environment with extreme temperatures, strong magnetic fields, and multiple irradiation particles. However, the radiation fields provided by the current fusion facilities cannot meet the requirements of such material irradiation damage experiments, and the required results can be obtained only through extrapolating from the measurements in similar neutron irradiation fields. In addition, the materials are usually faced with temperatures of 570–1270 K in fusion systems; during the start-up or shutdown of fusion systems, the material temperature changes dramatically. All these factors influence the microstructures of the materials. In the future, material irradiation damage experiments need to be performed under multiphysics synergistic effects based on high-intensity D–T neutron sources.

(4) Shielding Experiments: According to the design of ITER, the tokamak is sized dozens of meters in diameter and comprises thousands of components that can be described by complex surfaces such as sweeping surfaces and B-spline surfaces. Moreover, many slits and holes are reserved for installation and maintenance, which makes the geometry increasingly complex and increases the difficulty of shielding experiments. Additionally, certain components are usually operated under high neutron flux, and it is necessary to consider the dose exposure of workers due to the material activation induced by neutron irradiation. Most of the current experiments were performed with a single component in a single environment, so engineering validation experiments under a complex environment are an important challenge.

In summary, to meet the requirements of fusion neutronics research, nuclear data still need to be improved and supplemented with neutronics experiments. At the same time, fusion neutronics experiments should be performed in two aspects in the future to promote the engineering application of fusion systems. First, a high-intensity "point" neutron source experimental facility should be developed to obtain accurate neutron radiation fields with the help of neutron control technology, and material activation and irradiation damage experiments should be performed for the verification of activation performance, microstructure performance, calculation methods, codes, etc. Second, high-flux volumetric fusion sources must be built to provide the multiphysics coupling environments that are required to perform integration testing experiments.

References

1. Konnoa C, Maekawaa F, Kasugai Y et al (2001) Neutronics experiments for ITER at JAERI/FNS. Nucl Fusion 41(3):333–337
2. Packer LW, Batistoni P, Colling B et al (2017) Status of ITER material activation experiments at JET. Fusion Eng Des 124:1150–1155
3. Wu YC (2017) Fusion neutronics. Springer nature Singapore Pte Ltd, Singapore
4. Batistoni P, Angelone M, Fischer U et al (2012) Neutronics experiments for uncertainty assessment of tritium breeding in HCPB and HCLL blanket mock-ups irradiated with 14 MeV neutrons. Nucl Fusion 52(8):083014
5. Xiong HH, Chen SZ, Wang YF et al (2018) Neutronics experiment on DFLL-TBM mock-up irradiated with D-T fusion neutrons. Nucl Sci Eng 38(1):75–80 (in Chinese)
6. Huang QY, Li C, Li Y et al (2007) Progress in development of China low activation martensitic steel for fusion application. J Nucl Mater 367–370:142–146
7. Ge HE (2015) Hardening behaviour of CLAM steel induced by neutron irradiation. University of Science and Technology of China, Hefei (in Chinese)
8. Xin JP, Liu SJ, Li CJ et al (2016) Tensile and impact properties of CLAM steel after neutron irradiation to 2.98dpa at 430 °C. Nucl Sci Eng **36**(4):487–491
9. Ge HE, Peng L, Dai Y et al (2016) Tensile properties of CLAM steel irradiated up to 20.1 dpa in STIP-V. J Nucl Mater 468:240–245
10. Hu XX, Koyanagi T, Fukuda M et al (2016) Irradiation hardening of pure tungsten exposed to neutron irradiation. J Nucl Mater 480:235–243
11. Fukuda M, Hasegawa A, Tanno T et al (2013) Property change of advanced tungsten alloys due to neutron irradiation. J Nucl Mater 442:S273–S276
12. Kalinin GM, Artyugin AS, Yvseev MV et al (2011) The effect of irradiation on tensile properties and fracture toughness of CuCrZr and CuCrNiSi alloys. J Nucl Mater 417:908–911
13. Li MM, Sokolov MA, Zinkle SJ (2009) Tensile and fracture toughness properties of neutron-irradiated CuCrZr. J Nucl Mater 393:36–46
14. Weber HW (2011) Radiation effects on superconducting fusion magnet components. Int J Mod Phys E 20(06):1325–1378
15. Arata N, Takao T, Shigehiro N et al (2009) Neutron irradiation effects on superconducting wires and insulating materials. Fusion Eng Des 84:1425–1428
16. Humer K, Weber HW, Tschegg EK (1995) Radiation effects on insulators for superconducting fusion magnets. Cryogenics 35(12):871–882
17. Angelonea M, Batistonia P, Moroa F (2012) Neutronics analysis and nuclear heating measurement up to the TFC in a mock-up of the ITER in board shield. Fusion Eng Des 87:910–915
18. Batistoni P, Angelone M, Fischer U et al (2000) Neutron streaming experiment at FNG: results and analysis. Fusion Eng Des 51–52(1):653–661
19. Batistoni P, Angelone M, Fischer U et al (1999) Neutronics experiment on a mock-up of the ITER shielding blanket at the Frascati neutron generator. Fusion Eng Des 47(1):25–60
20. Vasilopoulou T, Stamatelatos IE, Batistoni P et al (2015) Neutron streaming along ducts and labyrinths at the JET biological shielding: Effect of concrete composition. Radiat Phys Chem 116:359–364
21. Wakisaka M, Kaneko J, Fujita F et al (2005) Analysis of neutron propagation from the skyshine port of a fusion neutron source facility. Nucl Instrum Methods Phys Res 554:347–355

Chapter 15
Neutronics Experiments of Hybrid Nuclear Systems

Since the concepts of hybrid nuclear systems were proposed in the 1950s, many designs and simulations of hybrid nuclear systems have been performed. Various neutronics experiments of hybrid nuclear systems have been carried out to verify the effectiveness of experimental methods, the validity of simulation methods, and the reliability of evaluated nuclear data. This chapter is organized as an overview, experimental methods for the measurement of subcriticality and transmutation nuclear data, typical experiments and prospects for future experiments.

15.1 An Overview

There are features of hybrid nuclear systems that differentiate them from fission and fusion systems [1, 2]. Hybrid nuclear systems run in a subcritical state can breed nuclear fuel and transmute nuclear waste driven by external neutron sources with high energy. Such features imply that hybrid nuclear systems require techniques of dealing with subcriticality and transmutation nuclear data measurement in related neutronics experiments [3–8]. The experiments described in this chapter mainly focus on the distinctive features of hybrid nuclear systems.

1. Experiments of subcriticality measurement

Subcriticality is a key reference parameter for designing, monitoring, and operationally controlling hybrid nuclear systems. To guarantee safe operation of such systems, subcriticality should be accurately monitored at all times during operation. Since a hybrid nuclear system has external neutron sources and runs at a subcritical level, the prompt neutron decay constant is large and the neutron flux level constantly deviates from that of the static state, which introduces the kinetic distortion effect. The classical subcriticality measurement method is based on the point kinetics model, which assumes that the system's neutron flux distribution is independent of time. However, this same assumption cannot be applied to hybrid nuclear systems, where neutron flux distribution is changing. Thus, a series of subcriticality measurement experiments have been carried out recently to test the applicability of classical methods and to verify the effectiveness of newly developed methods.

Y. Wu, *Neutronics of Advanced Nuclear Systems*,
https://doi.org/10.1007/978-981-13-6520-1_15

2. Experiments of transmutation nuclear data measurement

The transmutation data mainly includes transmutation reaction nuclear data, decay data and fission yields data. The accurately measuring of transmutation data is essential for the research on nuclear waste transmutation and nuclear fuel breeding of hybrid nuclear systems. Transmutation nuclear data measurement of long-lived nuclear wastes is one of the main objectives of performing experiments of hybrid nuclear systems. Some special requirements need to be met to carry out transmutation nuclear data measurement, including the preparation of target materials, neutron beams, and measurement platforms. The long-lived radioactive nuclides in nuclear waste mainly include minor actinides and fission products with long half-life. The transmutations of these nuclides are mainly executed by neutron capture reactions or fission reactions. The external neutron source of hybrid nuclear systems can be categorized into two types according to the neutron energy, i.e., spallation neutron sources with neutron energy reaching several hundreds of MeV, and fusion neutron sources with energy from several to tens of MeV. The transmutation nuclear data of fission products with long half-life is usually measured for some parts of the reaction channels and the energy range is limited to the region of thermal neutron energy and near the 14 MeV. Additionally, the data for some other long-lived fission products such as ^{121}Sn is even not measured at all.

The transmutation nuclear data of minor actinides is usually measured near the region of thermal neutron energy and 14 MeV for some parts of the reaction channels, and only a few nuclides such as ^{237}Np are fully covered through thermal energy to hundreds MeV. Additionally, the overall transmutation nuclear data does not have high accuracy. The reasons include the difficulty of preparing the radioactive nuclide target, and the insufficient measurement platforms and neutron beams with high intensity, high accuracy, and high energy. To resolve these problems, a series of transmutation measurement experiments have been carried out to improve accuracy, cover more types of nuclides, extend the energy coverage up to the keV region for fission products with long half-life, and extend the energy coverage up to the GeV region for minor actinides.

15.2 Subcriticality Measurement Experiments

As mentioned in the previous section, there are some problems concerning the applicability of the classical subcriticality measurement method in hybrid nuclear systems. Thus, the Rossi-Alpha Prompt Jump Analysis (RAPJA) method and the current-to-flux ratio method have been developed. In addition, the classical inverse kinetics method has been modified for hybrid nuclear systems. This section introduces the methods of the subcriticality measurement, and the typical experiments to illustrate the applicability of these methods.

15.2.1 Methods

1. Inverse kinetics methods

The inverse kinetics method is one of the most widely used methods for reactivity reconstruction of critical nuclear energy systems [9]. It is based on the point kinetics model and can also be theoretically applied to a hybrid nuclear system. The neutron flux $\phi(\boldsymbol{r}, t)$ can be expressed by using the multiplicity of the spatial related shape function and the time-related amplitude function

$$\phi(r,t) = n(t) \times \varphi(\boldsymbol{r}) \tag{15.1}$$

For a hybrid nuclear system with external neutron sources, the reactivity can be expressed as

$$\begin{aligned}\frac{\rho(t)}{\beta} &= 1 + \frac{\Lambda}{\beta}\frac{\mathrm{d}n(t)}{\mathrm{d}t}\frac{1}{n(t)} - \frac{\Lambda}{\beta n(t)}\sum_i \lambda_i c_i(t_0)\mathrm{e}^{-\lambda_i(t-t_0)} \\ &- \frac{1}{n(t)}\sum_i \lambda_i \frac{\beta_i}{\beta}\int_{t_0}^{t} n(\tau)\mathrm{e}^{-\lambda_i(t-\tau)}\mathrm{d}\tau - \frac{\Lambda S(t)}{n(t)\beta}\end{aligned} \tag{15.2}$$

where t_0 is the initial time in units of s; $\rho(t)$ is reactivity or subcriticality in units of pcm; β *is* the delayed neutron fraction with β_i as the sub-group fraction; Λ is the neutron generation time in units of s; $n(t)$ is the neutron density in units of cm^{-3}; λ_i is the decay constant of the precursors in units of s^{-1}; $c_i(t_0)$ is concentration of the delayed neutron precursor in units of cm^{-3}; and $s(t)$ is the intensity of the external neutron source in units of s^{-1}.

The value of $c_i(t_0)$ is determined by the initial condition. The system was running steadily when $t \le t_0$, and the system goes into a dynamic mode because of the change of reactivity when $t > t_0$, where $c_i(t_0)$ can be obtained as

$$c_i(t_0) = \frac{\beta_i}{\Lambda\lambda_i} n(t_0) \tag{15.3}$$

Then, Eq. (15.2) can be rewritten as,

$$\begin{aligned}\frac{\rho(t)}{\beta} &= 1 + \frac{\Lambda}{\beta}\frac{\mathrm{d}n(t)}{\mathrm{d}t}\frac{1}{n(t)} - \frac{n(t_0)}{n(t)}\sum_i \frac{\beta_i}{\beta}\mathrm{e}^{-\lambda_i(t-t_0)} \\ &- \frac{1}{n(t)}\sum_i \lambda_i \frac{\beta_i}{\beta}\int_{t_0}^{t} n(\tau)\mathrm{e}^{-\lambda_i(t-\tau)}\mathrm{d}\tau - \frac{\Lambda S(t)}{n(t)\beta}\end{aligned} \tag{15.4}$$

Since the relation $\lambda_i(t - t_0) \gg 1$ holds

$$c_i(t_0)e^{-\lambda_i(t-t_0)} = 0 \tag{15.5}$$

One can obtain

$$\frac{\rho(t)}{\beta} = 1 + \frac{\Lambda}{\beta}\frac{\mathrm{d}n(t)}{\mathrm{d}t}\frac{1}{n(t)} - \frac{1}{n(t)}\sum_i \lambda_i \frac{\beta_i}{\beta}\int_{t_0}^{t} n(\tau)\mathrm{e}^{-\lambda_i(t-\tau)}\mathrm{d}\tau - \frac{\Lambda\bar{s}(t)}{n(t)\beta} \tag{15.6}$$

The reactivity can be obtained by combining the measurement of $n(t)$ and the calculation of β_i and Λ.

The neutron flux in a hybrid nuclear system is a time-dependent quantity, because of reactivity perturbation or external source changing; thus, Eq. (15.1) no longer holds. The classical inverse kinetics method cannot be directly applied to the subcriticality measurement of hybrid nuclear systems. The FDS Team developed an inverse spatial kinetics method based on the most suitable detector response position for the subcriticality measurement of a hybrid nuclear system [10]. Via theoretical investigations, a specific region was found where the shape function of the neutron flux function has weak time-dependence on the time evaluation; in other words, Eq. (15.1) still can be applied in the region, which is illustrated in Fig. 15.1. The subcriticality can be measured by arranging detectors in the found region of a hybrid nuclear system.

This method was used to reconstruct the subcriticality of Primarily Design Studies of Experimental ADS. Figure 15.2 shows the subcriticality evolution by inducing 0.5$, 1.0$ reactivity when the initial k_{eff} is 0.97. One can find that the reconstructed subcriticality is consistent with the reference value, with a relative variance level of 1%. This study verified that the proposed inverse kinetics method can be applied for the online monitoring of subcriticality for hybrid nuclear systems .

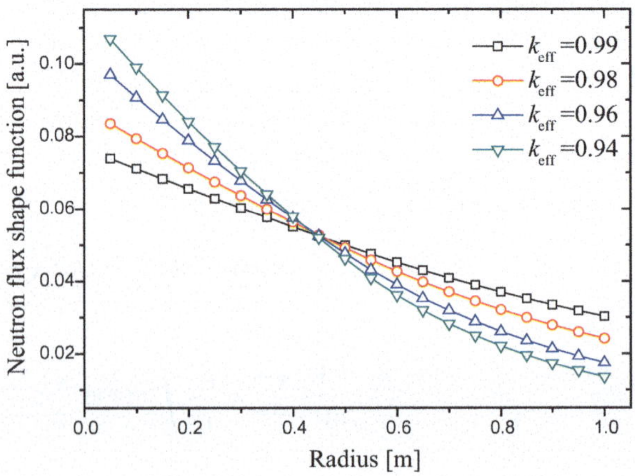

Fig. 15.1 The shape function of neutron flux functions [10]

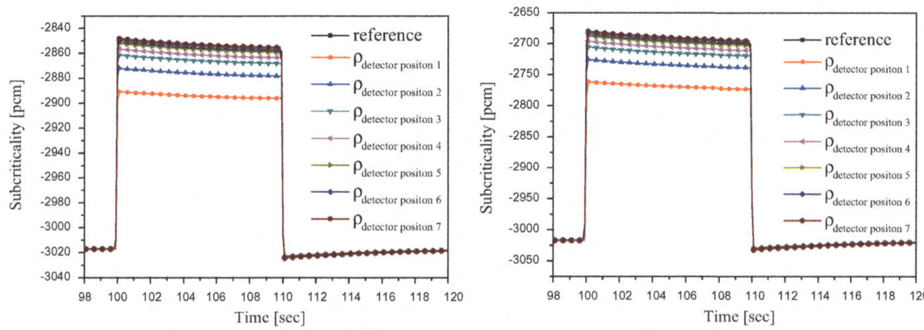

(a) Introduction of 0.5 $ reactivity perturbation

(b) Introduction of 1 $ reactivity perturbation

Fig. 15.2 Reconstruction of subcriticality with detectors in various positions [10]. **a** introduction of 0.5$ reactivity perturbation, **b** introduction of 1$ reactivity perturbation

2. RAPJA methods

The RAPJA method is a modified noisy measurement method that combines two well-known techniques, i.e., the measurement of a Rossi-alpha distribution and a prompt jump analysis [11, 12]. In the RAPJA method, the system runs in a steady critical state at first, and then the negative criticality is introduced, followed by the prompt jump analysis to obtain the subcriticality (in units of $). Finally, the β_{eff} is obtained by analyzing the Rossi-alpha distribution, and then the subcriticality is derived. The RAPJA method assumes the same subcritical level during the prompt jump and the measurement of the Rossi-alpha distribution.

Since the insertion of anti-reactivity during the prompt jump has an approximate linear relation with the time, the evolution of neutron flux during the prompt jump can be described by

$$n(t) = \begin{cases} n_0 \frac{\beta_{\mathrm{eff}}}{\beta_{\mathrm{eff}} - \gamma t}, & t \le t_{\mathrm{p}} \\ n_0 \frac{\beta_{\mathrm{eff}}}{\beta_{\mathrm{eff}} + |\rho|} e^{-\lambda(t - t_{\mathrm{p}})}, & t > t_{\mathrm{p}} \end{cases} \tag{15.7}$$

where t_p is the end of anti-reactivity insertion in units of s; n_0 is the initial count of the neutron detector at the initial time; β_{eff} is the effective delayed neutron fraction; γ is the rate of anti-reactivity insertion in units of s^{-1}; ρ is the total anti-reactivity insertion in units of pcm; and λ is the mean decay constant of the precursor in units of s^{-1}.

By fitting the measured response to this model, the anti-reactivity (in units of $) $\rho_{\$}$ can be directly obtained.

There are various ways to introduce anti-criticality, and here the rod drop method is used as an illustration. The rod drop method is based on the transient response of the system to a rapid insertion of high negative reactivity. Two analytical models were developed based on the point kinetics equation, i.e., the integral method and the differential method. The integral method needs to detect neutron counts of stable status before the rod drop (n'), and the total neutron counts after the rod drop. The differential method needs to obtain neutron counts of stable status before and after the rod drop (n_0, n_1). The corresponding calculations are listed as

$$\rho_\$ = -\frac{n' \sum_{i=1}^{6} \frac{\beta_{\mathrm{eff},i}}{\beta_{\mathrm{eff}}\lambda_i}}{\int_0^\infty n dt} \tag{15.8}$$

$$\rho_\$ = \frac{n_0 - n_1}{n_1} \tag{15.9}$$

As mentioned in Sect. 13.4, by using Rossi-alpha distribution method, the quantities of prompt neutron decay constant α, counts of unrelated signals C, and counts of related signals A can be obtained. These three quantities have relationships with the effective delayed neutron fraction β_{eff} and neutron life l as

$$C = \frac{\varepsilon_n g^* S(\rho_\$ \beta_{\mathrm{eff}} - 1)}{\rho_\$ \beta_{\mathrm{eff}}} \tag{15.10}$$

$$A = \frac{g\overline{\nu_{\mathrm{p}}}\mathrm{D}_\nu(1-\beta_{\mathrm{eff}})\rho_\$ \alpha C}{2g^*\beta_{\mathrm{eff}}^2(1-\rho_\$)^2 S} \tag{15.11}$$

$$\alpha = \frac{\beta_{\mathrm{eff}}(\rho_\$ - 1)}{l(1-\rho_\$ \beta_{\mathrm{eff}})} \tag{15.12}$$

where S is the intensity of neutron source in units of s^{-1}; $\mathrm{D}_\nu = \overline{\nu_{\mathrm{p}}(1-\nu_{\mathrm{p}})/\nu_{\mathrm{p}}^2}$ is defined with $\overline{\nu_{\mathrm{p}}}$ as the mean value of the prompt neutron numbers of each fission process; ε_n is the neutron detection efficiency, g is a factor that corrects for the detector efficiency; and g^* is a factor that corrects for the values of the source neutrons.

From Eqs. (15.10), (15.11), and (15.12), together with the condition of $\beta_{\mathrm{eff}} \ll 1$, one can obtain

$$\beta_{\mathrm{eff}} = \left(\frac{g\overline{\nu_{\mathrm{p}}}\mathrm{D}_\nu}{2g^*}\right)\frac{\alpha C \rho_\$}{AS(1-\rho_\$)^2} \tag{15.13}$$

$$l = \frac{\beta_{\mathrm{eff}}}{-a}\left(\frac{1-\rho_\$}{1-\rho_\$ \beta_{\mathrm{eff}}}\right) \tag{15.14}$$

Except for the correction factors g^* and g, all other parameters are either measured quantities ($A, R, \alpha, \rho_\$$) or from the nuclear database. By applying the RAPJA method, the quantities of prompt neutron decay constant α, neutron life l, effective delayed neutron fraction β_{eff}, and reactivity $\rho_\$$ can be obtained. Thus the subcriticality of the system is obtained.

RAPJA needs to carry out the prompt jump analysis, which cannot be applied to a hybrid nuclear system in steady operation mode. Moreover, the measurement of the Rossi-alpha distribution belongs to the noise analysis technique, which is based on the detection of a single neutron. This feature indicates that the system cannot run in a high output power; otherwise, detection will be saturated, as the decay chains would overlap.

3. Current-to-flux ratio methods

The subcriticality quantity should be accurately monitored to guarantee the safe operation of a hybrid nuclear system. Since, in a hybrid nuclear system, the neutron flux level of external sources is proportional to the beam current and the reactivity level, the current-to-flux ratio can be used to monitor the reactivity level. The current-to-flux ration method is based on the relationship between the power level (or the neutron population) in the system and the intensity of the external neutron source [13, 14].

Let N be the total number of neutrons produced by multiplication in the system after the introduction of source neutrons S from an external source, then the neutron source multiplication factor k_s can be defined as

$$k_s = 1 - \frac{S}{N} \tag{15.15}$$

In general, k_s depends on the characteristics and positions within the assembly of the neutron source. By analogy with the multiplicity source method, which is routinely used to evaluate rod worth in critical reactors, Eq. (15.15) is normally rewritten in terms of k_{eff} by introducing the concept of source efficiency, which is defined as

$$k_{eff} = 1 - \varphi^* \frac{S}{N} \tag{15.16}$$

where φ^* is the source efficiency. φ^* is defined as $\varphi^* = \frac{1-k_{eff}}{1-k_s}$.

From Eq. (15.16), it is possible to evaluate the k_{eff} of the system by measuring N and S or the ratio between them. However, this measurement requires that φ^* is calibrated at specific configurations. Furthermore, neither N nor S will be experimentally measured, but some related measurable magnitudes M and R (detector counting rates or current intensities) will be measured. Parameters M and R are respectively proportional to N and S, and related through corresponding efficiencies as $M = \varepsilon_D N$ and $R = \varepsilon_S S$. Hence, in terms of M and R, Eq. (15.16) becomes

$$k_{eff} = 1 - \varphi^* \frac{\varepsilon_D}{\varepsilon_S} \frac{R}{M} \tag{15.17}$$

In any case, these factors usually evolve slowly and they can be considered as constant during a period of time. Therefore, the current-to-flux ratio method is useful for the online monitoring of reactivity. The difference of reactivity between two configurations of a system is given by

$$\Delta k_{eff} = k_{eff,2} - k_{eff,1} = (1 - k_{eff,1})\left(1 - \frac{\varphi_2^* \frac{\varepsilon_{D2}}{\varepsilon_{S2}} \frac{M_1}{R_1}}{\varphi_1^* \frac{\varepsilon_{D1}}{\varepsilon_{S1}} \frac{M_2}{R_2}}\right) \tag{15.18}$$

If both configurations are similar enough, then Eq. (15.18) can be approximated as

$$\Delta k_{eff} \cong (1 - k_{eff,1})(1 - \frac{M_1}{R_1}\frac{R_2}{M_2}) \quad (15.19)$$

The deviation of online reactivity from the reference value can be monitored by measuring the ratio between the beam intensity of external sources and the neutron detection counts in a hybrid nuclear system. Since the methods of detecting beam intensity and neutron counts are relatively mature, the current-to-flux ratio method can monitor reactivity accurately. Although the current-to-flux ratio method measures the relative subcriticality and needs to obtain the absolute reactivity value with the help of other method, it has the advantages of being simple and practical.

15.2.2 Examples

Measurements were performed at the Yalina hybrid nuclear system to verify the applicability of the current-to-flux ratio method [13, 14].

A schematic view of the Yalina-Booster subcritical assembly is shown as Fig. 15.3. The subcriticality values were measured before and after the control rod insertion in or extraction out of the system. The detection channels were EC1B and EC2B at the booster zone and the EC5T at the thermal zone. The neutron detector was a ^{235}U fission chamber, and the intensity of the external neutron source was monitored using a BC501A scintillator. The current-to-flux ratio method has been applied using the approximate Eq. (15.18) and the main results are shown in Fig. 15.4.

The measured k_{eff} change between the two control rods positions was found to be approximately 200 pcm, which was lower than the numerical simulation result (300 pcm) and the changes measured with the source-jerk and the prompt decay constant technique. It must be taken into account that, in addition to statistical errors, systematic errors due to local effects are also present. These local effects had been neglected by assuming that the factor $\varepsilon_D\varphi^*/\varepsilon_S$ remained constant during the control rod movement. Actually, it was possible to estimate the variation of the factor $\varepsilon_D\varphi^*/\varepsilon_S$ with the control rod position using numerical simulation. By considering that ε_S did not depend on the control rod position (which is normally true), Eq. (15.18) became

$$\Delta k_{eff} = (1 - k_{eff,1})(1 - \frac{\varepsilon_{D2}\varphi_2^*}{\varepsilon_{D1}\varphi_1^*}\frac{M_1/R_1}{M_2/R_2}) \quad (15.20)$$

The value of $\frac{\epsilon_{D2}\varphi_2^*}{\epsilon_{D1}\varphi_1^*}$ could be determined with numerical simulation. Using Eq. (15.20), it had been observed that the variation in Δk_{eff} measured by the current-to-flux method increased to 260 pcm, which was closer to the simulation reference value. The current experiment preliminarily verified the applicability of the current-to-flux ratio method for the online monitoring of the subcriticality variation in hybrid nuclear systems.

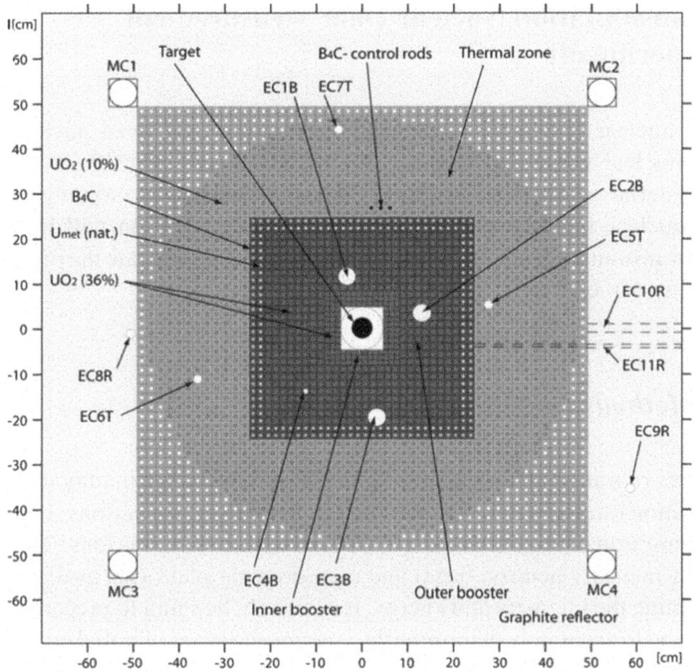

Fig. 15.3 The core configuration of the Yalina hybrid nuclear system [13]

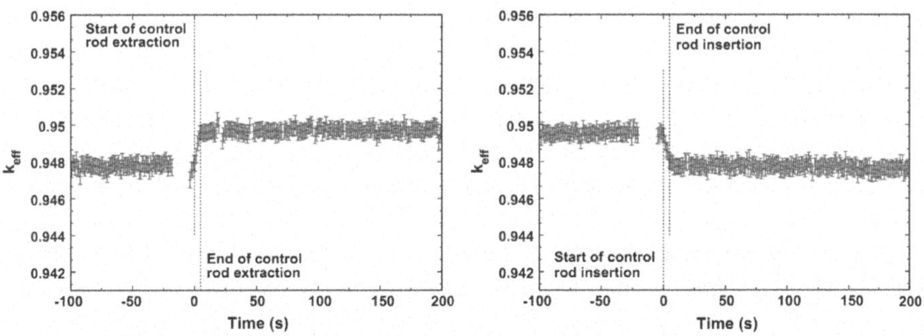

(a) Current-to-flux, control rod extraction

(b) Current-to-flux, control rod insertion

Fig. 15.4 Reactivity measurement of the Yalina hybrid nuclear system [13]. **a** current-to-flux, control rod extraction, **b** current-to-flux, control rod insertion

15.3 Transmutation Nuclear Data Measurement Experiments

The hybrid nuclear system is capable of transmuting long-lived nuclear waste to short-lived nuclear waste and breeding the nuclear fuel $^{233}U/^{239}Pu$ from the more abundant material $^{232}Th/^{238}U$, which is the key technique for solving the nuclear waste and nuclear fuel shortage problem. The present section will introduce the methods of transmutation nuclear data measurement, and illustrate the progress using the typical experiments.

15.3.1 Methods

The quantities of transmutation nuclear data measurements are mainly integrals such as transmutation rates and differentials such as neutron cross sections. It is challenging to measure transmutation nuclear data for the following reasons. There are not enough high-intensity neutron beam and measurement platforms available to accurately determine the intensity and energy. It is also challenging to prepare transmutation targets and to accurately determine the amount and types of initial nuclides before and after transmutation. The high-intensity neutron beams are available at high-flux experimental reactors and spallation neutron sources. The intensity and energy of neutron beams are determined using an offline activation method or an online time-of-flight method at spallation neutron sources. The transmutation samples can be obtained by partition and purification of spent fuel, or by irradiation at experimental reactors. The amount and types of nuclides can be accurately determined by mass spectrometer, and those quantities after transmutation can be determined by the measurement of characteristic gamma rays or by detecting reaction products such neutrons, protons, alpha particles, and gamma rays online at spallation neutron sources.

Neutron capture reaction is one of the best choices for transmuting LLFPs. However, the neutron capture data of LLFPs are rarely measured or have low precision because of the difficulty of preparing experimental samples of LLFPs, and the complexity of neutron capture reactions. For example, there are more than 100 resonance structures for the neutron capture cross section of ^{107}Pd in the energy region from thermal energy to 3.5 keV. Thus the high-quality neutron beam, and the measurement platform with high energy resolution and long-term stability to measure data of neutron capture reaction are both necessary. The pulse-height weight technique (PHWT) is usually used to measure the cross section of neutron capture reactions [15]. It is a method based on the conservation of energy

$$B_n + E_n = \sum E_\gamma \tag{15.21}$$

where B_n is the neutron binding energy of the target nuclide; E_n is energy of the incident neutron in the center of mass frame; E_γ is energy of gamma rays released from the neutron capture reaction; all these quantities are in units of MeV.

The yield of neutron capture reaction events Y can be expressed as

$$Y = \frac{\sum W(I)S(I)}{B_n + E_n} \tag{15.22}$$

where I is the signal pulse height of detected γ-rays; $S(I)$ is the pulse height spectrum, corresponding with the gamma ray energy in units of MeV; and $W(I)$ is the weight function, which is related with the detector response function and can be obtained by numerical simulation. The yield of neutron capture reaction events with the relation to the neutron's time of flight can be obtained using the PHWT technique. To obtain the neutron capture reaction data, the effects of background, target container, associated isotopes, and reference nuclide also need to be measured, as shown in Fig. 15.5. The absolute cross section data of neutron capture reactions are commonly acquired by comparing one simple resonance structure of the interested nuclei to some reference data from other experiments or evaluated data files [16].

15.3.2 Examples

1. Differential quantities of transmutation nuclear data measurements

^{99}Tc is one of the LLFPs and has the highest priority to be transmuted due to its high fission yield and high radio-toxicity. To design a nuclear transmutation system for ^{99}Tc, reliable neutron cross sections of ^{99}Tc in a wide range of neutron energies are

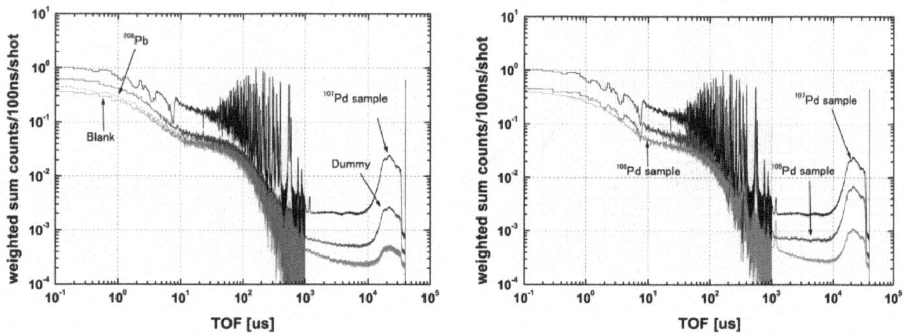

(a) TOF spectra of weighted sum for ^{107}Pd, the dummy container, ^{208}Pb and no sample

(b) TOF spectra of weighted sum for ^{107}Pd, ^{105}Pd,and ^{108}Pd

Fig. 15.5 The relation of neutron capture yield with time of flight [16]. **a** TOF spectra of weighted sum for ^{107}Pd, the dummy container, ^{208}Pb and no sample, **b** TOF spectra of weighted sum for ^{107}Pd, ^{105}Pd, and ^{108}Pd essential. Measurement of the ^{99}Tc neutron capture cross section has been carried out in Japan Proton Accelerator Research Complex (J-PARC) with time-of-flight (TOF) experiments using Accurate Neutron-Nucleus Reaction Measurement Instrument (ANNRI) [17, 18]. The neutron capture cross section of ^{99}Tc was measured from thermal energy to keV.

ANNRI has been built for nuclear data measurement by utilizing a high-intensity pulsed neutron beam from a spallation neutron source in J-PARC, as shown in Fig. 15.6. NaI(Tl) detectors of ANNRI were used for measurements. Two NaI(Tl) detectors were placed at detection angles of 90° and 125° with respect to the beam axis. The sizes of the NaI(Tl) crystals were 330 mm and 203 mm in diameter, respectively. A fast data acquisition system, based on the pulse-width analysis method was implemented to achieve measurement in the high energy region.

A ^{99}Tc metal sample with radioactivity of 52 MBq was used in the measurement. The sample was sealed in an aluminum container. The size of the sample was 6.3 mm in diameter and 0.28 mm in thickness. The sample was placed at a flight distance of 27.9 m from the neutron source. In addition to the ^{99}Tc sample, runs for a blank, dummy aluminum container, and a carbon sample were carried out for background measurements. The incident neutron spectrum was determined from detected counts of 478 keV γ-rays from the $^{10}B(n, \alpha)^{7}Li$ reaction. ^{197}Au was also measured as the reference measurement. The TOF method was employed to determine the incident neutron energy. The repetition rate of the pulsed neutron beam was 25 Hz. The incident proton beam power was approximately 200 kW. The pulsed proton beam was operated in the single-bunch mode. The single-bunch mode operation significantly benefitted the present work by allowing the analysis of resolved resonances with a fine time resolution of 100 ns in a simple way. Background of scattering neutrons was estimated from measurements with a graphite sample. The net TOF spectrum of ^{99}Tc was converted into the cross section by dividing the incident neutron spectrum determined from the boron measurement. The absolute cross section was determined from a saturated resonance of ^{99}Tc at 5.6 eV.

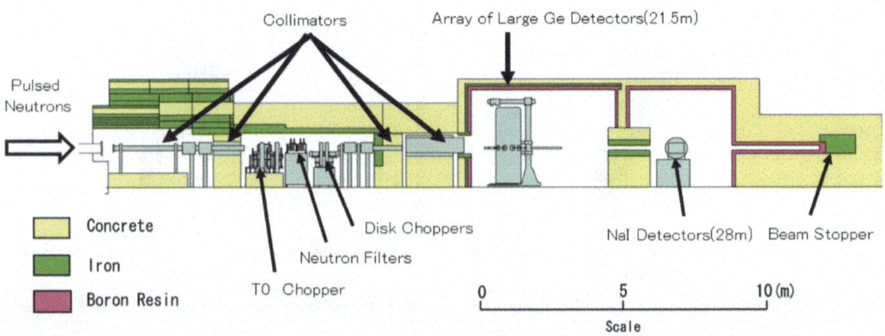

Fig. 15.6 Vertical cross-section of Accurate Neutron-Nucleus Reaction Measurement Instrument [17]

Table 15.1 Current status and future plans for neutron capture cross-section measurements at ANNRI [19]

	Analysis finished	Partly reported	Already measured	Future plan
MA	^{237}Np, ^{241}Am, ^{244}Cm, ^{246}Cm		^{243}Am	^{241}Am, ^{243}Am (new sample)
LLFP		^{93}Zr, ^{99}Tc, ^{107}Pd	^{129}I	^{135}Cs, ^{137}Cs
Stable Isotopes		^{74}Se, ^{77}Se, ^{90}Zr, ^{91}Zr, ^{112}Sn, ^{118}Sn, ^{105}Pd, ^{108}Pd, ^{106}Pd	^{54}Fe, ^{61}Ni, ^{76}Se, ^{78}Se, ^{117}Sn, ^{119}Sn, ^{120}Sn, ^{122}Sn, ^{127}I	^{58}Ni, ^{60}Ni, ^{62}Ni, ^{80}Se, ^{115}Sn, ^{116}Sn, ^{133}Cs, ^{133}Cs, ^{155}Gd, ^{157}Gd

The measurement results were compared with the evaluated cross section values from JENDL-4.0, as shown in Fig. 15.7. The self-shielding factors calculated from the resonance parameters were multiplied to the evaluated cross section for comparison. With a single-bunch mode time resolution (~100 ns), the count statistics were enough for resolved resonance analysis up to approximately 500 eV. Figure 15.8 shows a close-up of the high energy continuum region. The results were compared with those of previous measurements and evaluations of JENDL-4.0 and ENDF/B-VII.1. The ENDF/B-VII.1 evaluation was closer to the data than the JENDL-4.0 evaluation.

The measurements for MAs and LLFPs have been performed at ANNRI, including ^{244}Cm, ^{246}Cm, ^{241}Am, ^{237}Np, ^{129}I, ^{107}Pd, ^{99}Tc and ^{93}Zr, as shown in Table 15.1 [19]. These results significantly contribute to the development of hybrid nuclear systems.

2. Integral quantities of transmutation nuclear data measurements

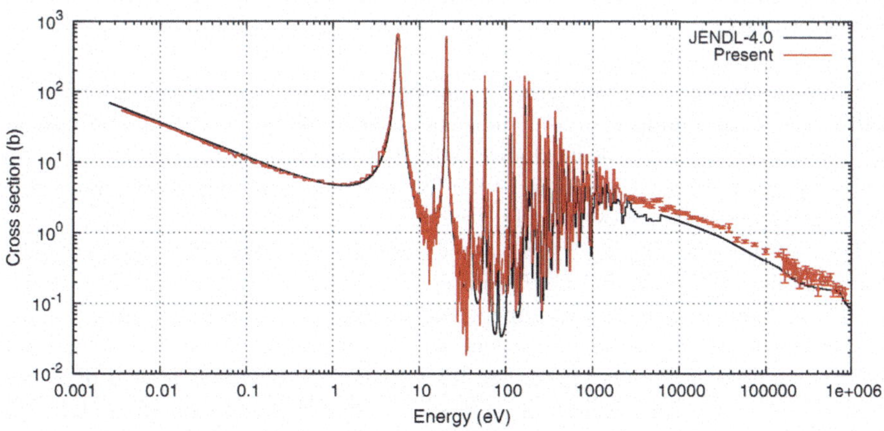

Fig. 15.7 Neutron capture cross section of ^{99}Tc [18]

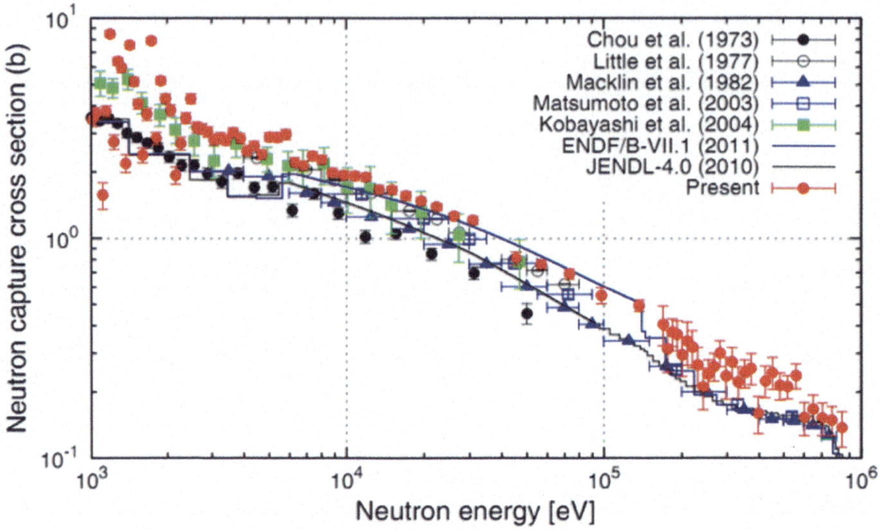

Fig. 15.8 Neutron captures cross section of ^{99}Tc in the high energy region [18]. The current experimental data points (red) are compared with previous measurements and evaluations. Only statistical errors of the current experimental data are shown

The elimination of MAs is one of the dominating aspects for the utilization of nuclear power as an energy resource extending into the future. The lack of integral experiment data on MAs makes it difficult to evaluate the uncertainty of the nuclear data, which directly affects the calculated properties of MA transmutation systems. Therefore, it is desirable to accumulate experimental data, including not only the differential data but also the integral data. On this basis, a series of integral experiments or the measurements of MA reaction rates in the thermal neutron system have been initiated at KUCA [20]. The results of ^{241}Am transmutation nuclear data measurement experiments are illustrated here.

The experiments were performed using polyethylene moderated and reflected, highly enriched uranium cores constructed at a solid moderated core in the KUCA. To vary the neutron spectra at the irradiation field, five cores having different H/^{235}U ratio were employed. The specifications of the five cores are summarized in Table 15.2, and neutron spectra are shown in Fig. 15.9.

The transmutation rate of ^{241}Am was measured relative to the ^{235}U reference. The atoms number of ^{241}Am and ^{235}U were $(1.73 \pm 0.02) \times 10^{16}$ and $(1.49 \pm 0.02) \times 10^{16}$, respectively. In the experiment, the samples were put into a back-to-back (BTB) type double fission chamber, with the ^{241}Am sample at one side of the BTB and ^{235}U sample at the other side. The BTB fission chamber was loaded in a void element and settled at the height of the core mid-plane by using aluminum tubes in the void element. Based on the counts of BTB, and Eq. (15.23), the relative fission rate ^{241}Am/^{235}U can be obtained as

Table 15.2 Specifications of the cores [20]

Core	Unit cell structure[a]	$H/^{235}U$	Spectrum index[b]
EEE1	EU + EU + EU + 1/8P	34	0.065
EE1	EU + EU + 1/8P	52	0.088
E1	EU + 1/8P	103	0.141
E2	1/8P + EU + 1/8P	207	0.220
E3	1/4P + EU + 1/8P	316	0.287

[a]EU is 93% Enriched Uranium (0.15875 cm), 1/8P is Polyethylene (0.3086 cm), 1/4P is Polyethylene (0.63 cm)
[b]Calculated fraction of the neutron flux in the energy region below 1 eV relative to the total neutron flux at fission chamber position

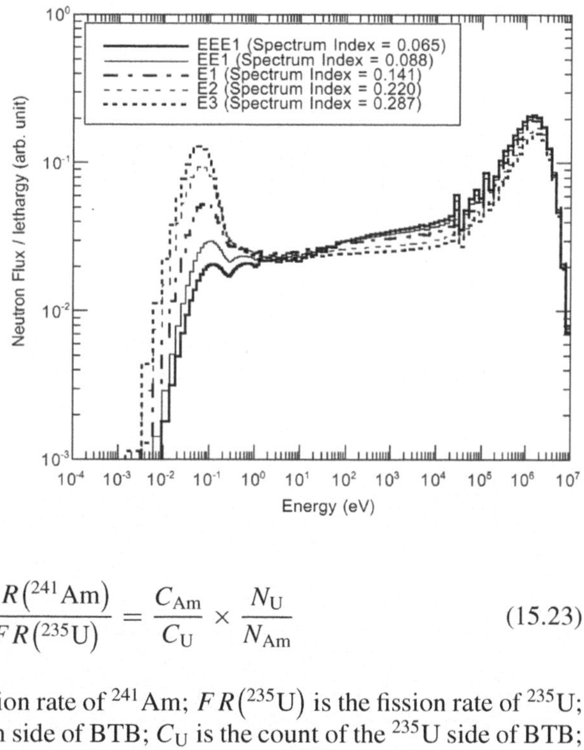

Fig. 15.9 Calculated neutron spectra for the experimental cores [20]

$$\frac{FR(^{241}\mathrm{Am})}{FR(^{235}\mathrm{U})} = \frac{C_{\mathrm{Am}}}{C_{\mathrm{U}}} \times \frac{N_{\mathrm{U}}}{N_{\mathrm{Am}}} \tag{15.23}$$

where $FR(^{241}\mathrm{Am})$ is the fission rate of ^{241}Am; $FR(^{235}\mathrm{U})$ is the fission rate of ^{235}U; C_{Am} is the count of the ^{241}Am side of BTB; C_{U} is the count of the ^{235}U side of BTB; N_{U} is atom number of ^{235}U; and N_{Am} is atom number of ^{241}Am.

Relative fission rates of ^{241}Am/^{235}U were obtained for the five different core configurations, and compared with the calculation with cross sections from JENDL-3.2, JEF2.2, and ENDF/B-VI files, as shown in Fig. 15.10. The C/E values obtained by JENDL-3.2 ranged from 0.86 to 0.90, with an average value of 0.88 and they were the smallest ones among the three libraries examined. The C/E values using ^{241}Am fission cross sections from ENDF/B-VI and JEF-2.2 were generally larger than those of the JENDL-3.2 results by approximately 2% and 7%–9%, respectively. Additionally,

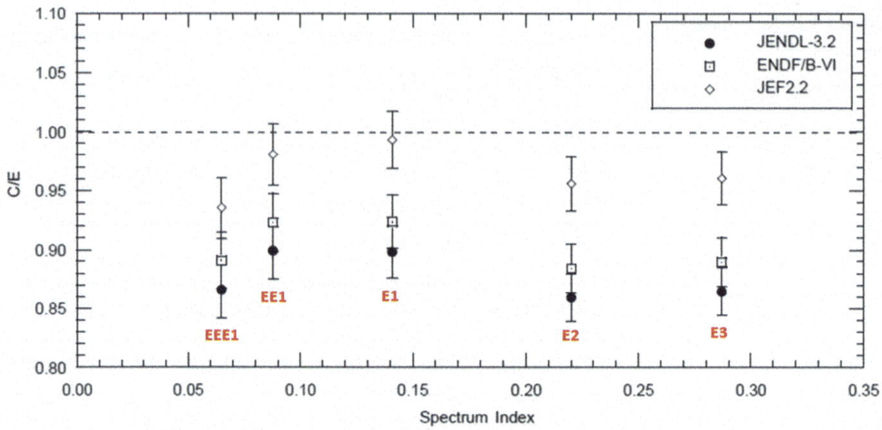

Fig. 15.10 C/E values of the ^{241}Am/^{235}U fission rate ratio [20]

C/E value in the condition of EEE1 (thermal neutron ratio is the smallest) and E3 (thermal neutron ratio is the biggest) were the furthest away from the unit value most for the five different core configurations.

The experiments of transmutation nuclear data measurements at J-PARC shows that it is necessary to accurately measure the fission cross section of ^{241}Am in the resonance energy 0.576 eV and thermal energy region (0.01–0.2 eV) and to evaluate these measured data, to meet the requirement for the high precision nuclear data in the neutronics analysis of hybrid nuclear systems.

15.4 Prospects

The hybrid nuclear system has the potential solutions to the two challenges of nuclear energy development, i.e., limited fuel resources and nuclear waste accumulation. Thus, significant manpower and financial support are continuously invested by many countries to research hybrid nuclear systems [1–3]. This section will forecast two of the most characteristic contents of the neutronics experimental research of hybrid nuclear systems, i.e., measurements of subcriticality and transmutation nuclear data.

1. Subcriticality measurements

The characteristics of a hybrid nuclear system are much different from those of a classical fission reactor, because the hybrid nuclear system has external neutron sources and runs at a subcritical level. The two most distinctive features indicate that the spatial effect of the neutron flux distribution cannot be ignored, and the neutron energy is relatively higher in a hybrid nuclear system. Thus the traditional subcriticality measurement cannot be directly applied, since the point kinetics model is approximate and invalid for a hybrid nuclear system. Newly developed methods such as the RAPJA method and the flux-to-current ratio method have limited scope of application and need to be further verified. There are urgent demands in subcriticality measurement for developing new methods, especially for the independent online method, and studying the sources of the spatial effect and the modification approach for fulfilling the requirements of hybrid nuclear systems under various conditions to guarantee safety.

2. Transmutation nuclear data measurements

With the development of advanced neutronics experiment facilities, a series of experiments need to be carried out to verify the newly proposed mechanisms and developing measurement methods related to nuclear waste transmutation. These experiments need to complement the transmutation nuclear data of LLFPs and MAs to cover the energy region from thermal energy to hundred MeV and even higher. Sufficient data will further promote the research of studying the dynamic changes of a hybrid nuclear system, such as the effect of burnup level for transmutation blanket to the subcriticality measurement. Additionally, it is important to obtain the precise transmutation nuclear data based on newly developed neutron field control technology.

In summary, the development and verification of subcriticality measurement methods and the measurement of transmutation nuclear data are still two of the most crucial components of neutronics experiment in hybrid nuclear systems. It is also necessary to promote further study on the effects of the kinematic parameters of hybrid nuclear systems, such as the condition change of the external source, the burnup level of transmutation blankets and the accumulation of newly produced nuclei. Additionally, it is necessary to perform engineering integration verification experiments under multi-physics phenomena circumstances to develop demo hybrid nuclear systems in the future.

References

1. Giulia B, Rob E, Maurizio V (2017) ADS conceptual design assessment. In: Enhanced European coordination for accelerator research and development-deliverable report (EuCARD2-Del-D4.4-Final)
2. Strategic Priority Research Programs (Category A) of the Chinese Academy of Sciences (2016) Future advanced fission energy-ADS transmutation system. Bull Chin Acad Sci 31:32–36 (in Chinese)
3. Cheol HP, Masao Y, Song HK (2017) Benchmarks of subcriticality in accelerator-driven system at Kyoto University Critical Assembly. Nucl Eng Technol 49:1234–1239
4. Dulla S, Nervo M, Ravetto P (2016) A method for the continuous monitoring of reactivity in subcritical source-driven systems. Ann Nucl Energy 87:1–11
5. Igashira M, Kiyanagi Y, Oshima M (2009) Nuclear data study at J-PARC BL04. Nucl Instrum Methods Phys Res, Sect A 600:332–334
6. Krivopustov MI, Pavliouk AV, Kovalenko AD (2009) First results studying the transmutation of ^{129}I, ^{237}Np, ^{238}Pu, and ^{239}Pu in the irradiation of an extended ^{nat}U/Pb-assembly with 2.52 GeV deuterons. J Radioanal Nucl Chem 279:567–584
7. Aliberti G (2004) Impact of nuclear data uncertainties on transmutation of actinides in accelerator-driven assemblies. Nucl Sci Eng 146:13

8. Chen C, Zou J, Xu D et al (2012) Development and benchmark of high energy continuous-energy neutron cross section library HENDL-ADS. Nucl Sci Eng 4:360–365 (in Chinese)
9. Silva SC, Palma D, Congcalves A (2010) Inverse kinetics for subcritical system with variating external sources. In: The 31st annual conference of the Canadian Nuclear Society & 34th CNS/CAN student conference, Quebec, Canada
10. Zhang XX (2017) Research on subcriticality reconstruction based on inverse kinetics method for subcritical systems. PhD thesis, University of Science and Technology of China. (in Chinese)
11. He RF, Deng MC (2012) The physics experiments of fast critical assemblies and pulsed reactors. National Defense Industry Press. (in Chinese)
12. Baeten P, Paepen J, Meer KV (2001) Absolute measurement of β_{eff} and l on weapon-grade MOX fuel at the VENUS critical facility by means of the RAPJA technique. Ann Nucl Energy 28:287–295
13. Bécares V, Villamarín D, Fernández-Ordóñez M (2013) Validation of ADS reactivity monitoring techniques in the YALINA-booster subcritical assembly. Ann Nucl Energy 53:331–341
14. Cao Y, Gohar Y (2010) YALINA-booster subcritical assembly pulsed-neutron experiments: data processing and spatial corrections. ANL-10/22
15. Russo PA, Vo DT (2005) Gamma-ray detectors for nondestructive analysis. Unclassified Los Alamos National Lab Report, LA-UR-05-3813
16. Kazushi T, Tatsuya K, Motoharu M (2015) Measurement of neutron capture cross sections of Pd-107 at J-PARC/MLF/ANNRI. Prog Nucl Energy 82:118–121
17. Maekawa F, Harada M, Oikawa K (2010) First neutron production utilizing J-PARC pulsed spallation neutron source JSNS and neutronics performance demonstrated. Nucl Instrum Methods Phys Res, Sect A 620:159–165
18. Tatsuya K, Motoharu M, Masayuki I (2017) Measurement of the neutron capture cross section of ^{99}Tc using ANNRI at J-PARC. In: EPJ Web of Conferences, vol 146, pp 11050
19. Atsushi K, Hideo H, Shoji N (2015) Current activities and future plans for nuclear data measurements at J-PARC. Eur Phys J A 51:180
20. Unesaki H, Iwasaki T, Kitada T (2001) Measurement and analysis of Am fission rate ratio relative to U fission rate in thermal neutron systems using Kyoto University Critical Assembly. J Nucl Sci Technol 38:600–606

The manufacturer's authorised representative in the EU is Springer Nature Customer Service Centre GmbH, Europaplatz 3, 69115 Heidelberg, Germany. If you have any concerns regarding our products, please contact ProductSafety@springernature.com

Printed and bound by CPI Group (UK) Ltd, Croydon, CR0 4YY
25/09/2026
02222886-0008